GEOLOGICAL EVOLUTION OF NORTHWESTERN INDIA

(A volume in honour of Professor Devendra Singh Chauhan)

Editor
B.S. PALIWAL
Professor and Head,
Department of Geology, Faculty of Science,
Jai Narain Vyas University,
JODHPUR (Raj.)

Published by
SCIENTIFIC PUBLISHERS (INDIA)
5 A, New Pali Road, P.O. Box 91
JODHPUR 342 001 (India)
E-mail: info@scientificpub.com
Website: www.scientificpubonline.com

Print, 2026

ISBN: 978-81-72332-10-5
e-ISBN: 978-93-87893-30-6

Printed in India.

PREFACE

Northwestern part of the Indian Peninsular Shield, encompassing the great Thar Desert has a lot more to narrate about the mother planet - its geological history engraved on a variety of rocks of the region, ranging from Precambrian to Recent. Journey from "Rodinia" to "Gondwana" possibly has its pivot some where in Rajasthan. The Pan-African Orogeny has stretched its tentacles in this region too. Several geological mysteries of global interest like Precambrian - Cambrian and Cretaceous - Tertiary boundary problems probably have the clue in this part of the earthcrust. In view of the above facts a feeling developed in the geoscience community at home and abroad that the data generated by a large number of academicians, research organizations and exploratory agencies engaged in scanning this region for a pretty long time, should be compiled.

Present endeavour imbibes a medley collection of papers presented at GEOWEST-R- National Seminar on Geological Evolution of Western Rajasthan, organized by the Department of Geology, Jai Narain Vyas University Jodhpur on October 7-8, 1996 in collaboration with University Grants Commission, Department of Science and Technology-Government of India, Oil and Natural Gas Corporation, Shell India Production and Development B.V. and Oil India Limited.

To collate a large number of papers on a wide spectrum of subjects related to the earth science, could have not been possible without active cooperation of all the authors and reviewers. I wish to extend my heartiest thanks to all of them.

With an apology for the some what jumped nature of the volume, I hope the keen readers will find updated information about this still unexplored and virgin land. A few contributions on fringe areas have also been included in the volume as a gesture of inter-disciplinary approach.

A liberal financial assistance received from University Grants Commission, Department of Science and Technology-Government of India, Oil and Natural Gas Corporation, Shell India Production and Development B.V. and Oil India Limited in organizing the National Seminar and in bringing out the volume, deserves sincere thanks.

I am thankful to my colleagues at the Department of Geology, Jai Narain Vyas University, Jodhpur and my students who put in every effort to make this mission successful. The commendable job done by Messrs Scientific Publishers (India), Jodhpur with a team of competent personnels, deserves a word of appreciation. In this yeoman task of editing such a large volume on diversely divers subjects my worries and agonies were shared patiently by my near and dear ones whose inspiration and encouragement is a matter of pride for me. I thank them all from the inner core of my heart.

JODHPUR
May 19th 1999

B.S. PALIWAL

GEOLOGICAL EVOLUTION OF NORTHWESTERN INDIA

Department of Geology, Faculty of Science
Jai Narain Vyas University,
Jodhpur (Rajasthan)

ʾrofessor D.S. Chauhan -- *a dedicated geoscientist*

We wish to dedicate this volume on "Geological Evolution of North Western India" to Professor Devendra Singh Chauhan who devoted his prime time in understanding the geological processes that took place in this part of the Indian subcontinent since Proterozoic period.

Born on 17th July, 1939 in a beautiful city of lakes surrounded by Aravalli hills, Prof. D.S. Chauhan completed his education in Udaipur city itself. Inspired by Prof. K.P. Roday and Prof. M.K Pandya, Prof. Chauhan did his doctoral degree in geology from University of Rajasthan in 1976.

Prof. Chauhan initiated his research with an attempt to understand the origin of Lead-Zinc deposits of Zawar. It was Prof. Chauhan who could successfully establish the synsedimentary origin of these base metal deposits. He was credited with the reporting of framboidal pyrite - an unequivocal product of diagenesis for the first time from the Lead-Zinc deposits of Zawar.

Late sixties was the time when the Precambrian algal structures popularly known as "Stromatolites" were in lime light in the field of geoscience. These algal structures were used in biostratigraphy and basin analysis of the Early Proterozoic rocks that were otherwise free from any evidence of early life. At that time Prof. Chauhan got inspiration from Prof. K.S. Valdiya — "a great stromatophile" and took keen interest in the study of phosphate bearing unusual stromatolites of the Aravalli province. He proposed a progressive replacement of accreting biogenic carbonate structures by phosphate bearing water under the persistant influence of microbial life. He also tried to establish tidalites of biogenic origin in the Aravalli sediments. Prof. Chauhan contributed significantly in this field of Earth Science.

In the year 1979, Prof. Chauhan joined the faculty of Science in the University of Jodhpur (now Jai Narain Vyas University) where, he took the challenge of building-up of a new Department of Geology. By his persistant efforts the department was upgraded to post-graduate level in the academic year 1988-89.

Splendid beauty of Ummaid Palace and Mehrangarh Fort attracted Prof. Chauhan to study the sandstone of Jodhpur, particularly its sedimentation and depositional environment. Vast terrain of the trans - Aravalli region provided him an opportunity to do so. His long affiliation with stromatolites and phosphorite, Prof. Chauhan prompted to under take investigation of far fatched phosphorite occurrence of the Thar Desert around Birmania in Jaisalmer area. To create afresh awakening amongst the scientific community towards the Vedic Saraswati and its relation with the ancient civilization, Prof. Chauhan is keen to trace the burried course of this holy river in the sands of the great Thar Desert. He has not only appreciated the burried course of Saraswati River but also tried to search its description in Vedas, Epics and Puranas. It was his rigorous effort that he could provide a scientific approach to the religious narrations about the lost river and our ancient Vedic Civilization that flourished along its banks.

During his stay at Jodhpur, Prof. Chauhan, as member of the Indian delegation to the United States of America, participated in the International Symposium on Neogene Phosphorite in May, 1985.

As a visiting scientist, Prof. Chauhan went to Federal Republic of Germany during Sep.-Nov., 1987 under the Cultural Exchange Programme between the two countries. Prof. Chauhan delivered a series of well attended lectures on "Stratabond Basemetal Sulphide Deposits and Phosphorites of Rajasthan at the Institution of Minerology and Petrology, Heidelberg University.

Prof. Chauhan as a devoted social worker, is associated with "Vidya Bharti" -- a prime organization dedicated to school education. As a dedicated researcher and sincere teacher, Prof. Chauhan inspired several scholars. I am proud of, being one of his students.

JODHPUR
19th May, 1999 **Bhawani Shanker Paliwal**

Some Important Publications of Professor D.S. Chauhan

Antahsalila (Hindi) Editor Vedic Saraswati Nadi Sodha Sansthan, Jodhpur (Jointly with Shishindra Vishvabadhu 1995).

Some Evidences about synsedimentary nature of Lead-Zinc-Pyrite deposits of Zawar. *Jour. Min. Metal and Fuel,* 18 : 81 - 86, 1970.

The Pyrophyllite deposit of Village Barors, Distt. Udaipur, *Jour. Ind. Asso. Geol.* 4 : 63 - 69.

Stromatolites from the Precambrian phosphorite bearing stratum of the Aravalli region of Udaipur, Rajasthan, (India) *Doklady Akad Nauk, USSR (MOSCOW)* 208 : 1429 - 1431, 1973.

Diagenetic pyrite from the Lead-Zinc deposits of Zawar, India. *Mineralium Deposita (Heidelberg)* 9 : 69 - 73, 1974.

The Dariba Main Lode of Rajpura-Dariba Zinc-Lead-Copper belt, Udaipur District, Rajasthan *Jour. Geol. Soc. India,* 18 : 611 - 616, 1977.

Phosphate bearing stromatolites of the Precambrian Aravalli phosphorite deposits of Udaipur region, their environmental significance and genesis of phosphorite. *Precambrian Research (Amsterdam),* 8 : 95 - 126, 1979.

Stromatolites of Nimch Mata and Bargaon phosphorite deposit and genesis of rock phosphate. *Geol. Surv. India, Misc. Publication* 44 : 314 - 329, 1980.

Aravalli stromatolites and biostratigraphy *Geol. Surv. India Misc. Publication* 44 : 128 - 133, 1980.

Amphibolites from granitic terrain, NW of Udaipur city, India. *Jour. Earth Science,* 8 : 35 - 43, 1980.

Nature of Udaipur phosphorite and its genetic implications. Proc. 4th International Workshop and Seminar on Phosphorite, *IGCP Project 156, Special Publication, Geol. Surv. India,* 17 : 79 - 104, 1984. (Jointly with M.S. Sisodia)

Sedimentary pyrite from Pb-Zn deposits of Zawar and Rajpura- Dariba region Sand its bearing on the genesis of base metal sulfides. In : A., Wauschkuhn *et al.* (Eds.) *Syngenesis and Epigenesis in the Formation of Mineral Deposits,* Springer Verlag, Heidelberg, 36 - 42, 1984.

Geological nature of Birmania phosphorite. *Jour. Indian Assoc. Sedimentologist,* 7 : 71 - 81, 1987 (Jointly with S.C. Mathur).

Phosphorites of Rajasthan. *Mem. Geol. Soc. India* 13:9-23 (Jointly with M.S. Sisodia)

Microbial activities and genesis of Aravalli phosphorite, Udaipur, Rajasthan, *Himalayan Geology* 13 : 39 - 52, 1989.

Geological analysis of part of Nagaur basin in the vicinity of Jodhpur city. In : S.K. Tandon; Charu; C. Pant and S.M. Casshyap (Eds.) *Sedimentary Basins of India,* Gyanodaya Prakashan, Nainital. 1991, pp. 64-73. (Jointly with J.C. Dubey, Bhanwara Ram).

The influence of magnesian ions during the formulation of stromatolitic phosphorites of Udaipur, Rajasthan, India. In : Notholt, A.V.G. and Jarvis, I. (Eds.) *Phosphorite Research and Development. Geol. Soc. London Spel. Publ.* 52 : 313-320. 1990 (Jointly with M.S. Sisodia).

Lithofacies evaluation and environmental interpretation of Birmania basin, District Jaisalmer, Western Rajasthan, India. In : R. Ahmed and A. M. Sheikh (Eds.) *Geology in South Asia-I,* Hydrocarbon Development Institute, Pakistan, Islamabad. 1994. (Jointly with S.C. Mathur).

Jivant Dhara Atur Nagraj - *Vilupta Saraswati ke Sandarbha Main* (Hindi) In : Devendra Singh Chauhan and Shishindra Vishvabandhu (Eds.) *Antahsalila* Vedic Saraswati Nadi Sodha Sansthan, Jodhpur 1995.

Deformation pattern in the Precambrian stromatolitic phosphorite of Sallopat area, Rajasthan. In : B.S. Paliwal (Ed.) *The Indian Precambrian.* Scientific Publishers (India), Jodhpur, 159-170, 1998 (Jointly with S.K. Trivedi).

Evolution of stromatolite bearing sedimentary sequences in the Precambrian Aravalli phosphorite deposits of Udaipur, Rajasthan, India. In : B.S. Paliwal (Ed.) *The Indian Precambrian.* Scientific Publishers (India), Jodhpur, 171-182, 1998. (Jointly with M.S. Sisodia).

Didwana-Rajod metiorite - an unusual chondrite. *Curr. Sci.* 73(6) : 499-502, 1997. (Jointly with B.S. Paliwal and V.K. Vaya).

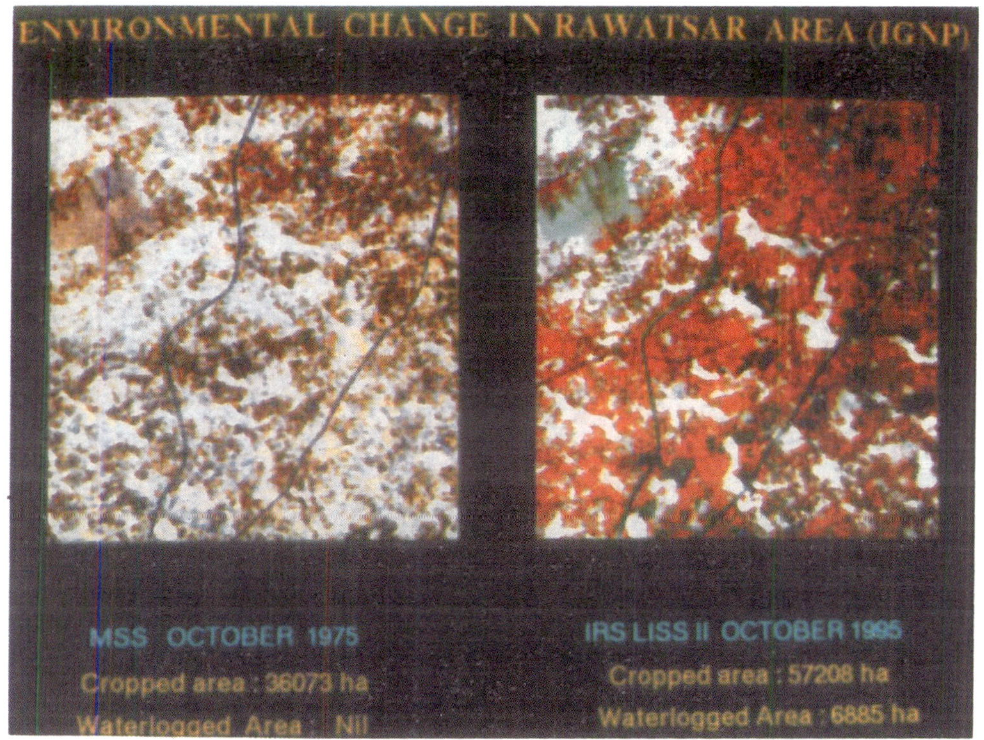

Plate 1. Environmental change in Rawatsar area, 1975-1995.

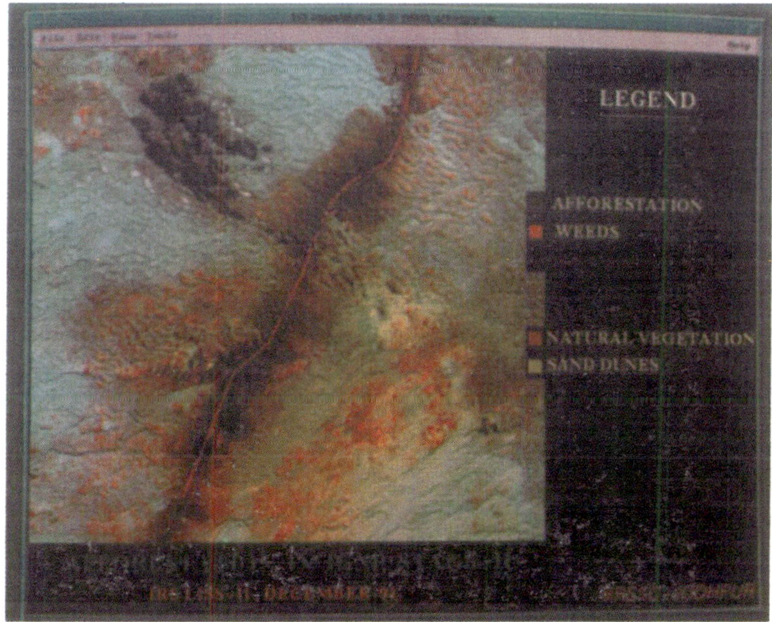

Plate 2. Afforestation and Weeds along the canal, 1991.

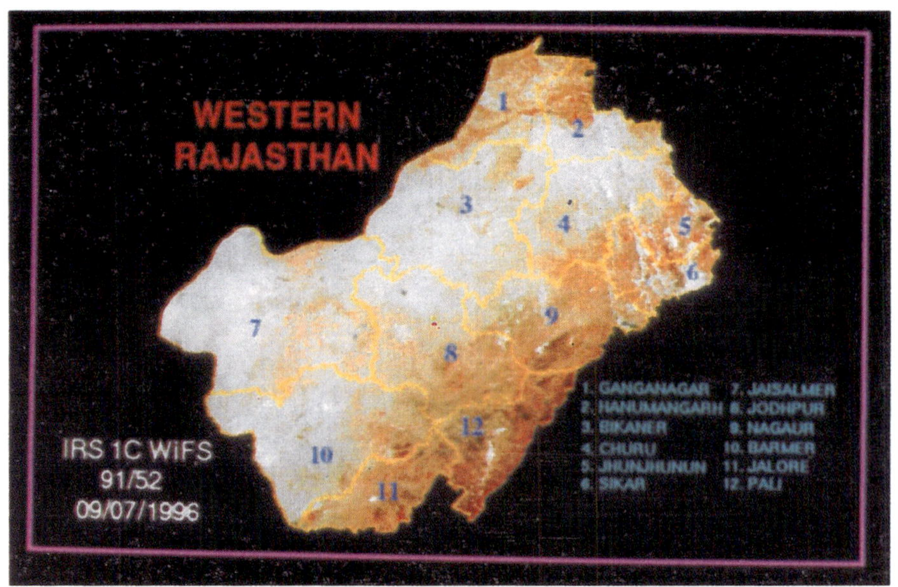

Plate 3. Vegetation in Western Rajasthan, July, 1996.

Plate 4. Sewan grassland in Nachana area, 1992.

CONTENTS

ieological Evolution of Northwestern India
.S. Paliwal (Ed.)
cientific Publishers (India), Jodhpur, 1999, pp. 1- 18.

MARWAR'S CONNECTION WITH THE HIMALAYA

K.S. Valdiya

Jawaharlal Nehru Centre for Advanced Scientific Research, Bangalore - 560 064

ABSTRACT

The ruggedly stupendous Himalaya and the sandy Marwar plains had close connections in the past. The Lesser Himalayan terrain in northern Pakistan and Jammu represent the northerly extension of the Marwar Basin of Rajasthan.The link was maintained for a long period, since the Vendian some 650 m.y. ago until about 3700 years B.P., with many breaks of variable duration. The Marwar Basin represents the northwestern submergent continental shelf of the Indian Shield. Obviously, the continental platform was extraordinarily wide, with a great thickness of sedimentary succession. As a consequence of Tertiary orogenic movements, the frontal part of this sedimentary prism got involved in the tectonic framework of the Himalaya. This caused the severence of Marwar's connection with the Himalaya for a short duration.

(ey words : Marwar Basin, Sedimentation, Lesser Himalaya, Western Rajasth
India.

INTRODUCTION

Those who have seen both the ruggedly stupendous Himalaya and the san ndulating Marwar plains would scarce believe that the two domains had cl onnections in the past. Indeed the Lesser Himalayan terrain in northern Pakistan a ammu represent the northerly extension of the Marwar Basin (Fig. 1). The link w naintained for a long period, since at least the Vendian some 650 m.y. ago until ab 700 years B.P., of course with many breaks of variable duration.

The Marwar Basin represents the northwestern submergent continental shelf of t ndian Shield. The platform extends, as geophysical probing and lithostratigraphi nalyses indicate, as far northwest as Hazara in northern Pakistan (Fig. 2). Obvious ıe continental platform was extraordinarily wide, with a great thickness of sedimenta ıccession. As a consequence of Tertiary orogenic movements, the frontal part of th edimentary prism got involved in the tectonic framework of the Himalaya. This caus ıe severence of Marwar's connection with the Himalaya, but only for a short duration.

DEPOSITION OF EVAPORITES AND TIME OF DESICCATION

Towards the end of the protracted period of the Purana sedimentation, widesprea ʼaporitic conditions developed on the shallow northwesterly continental shelf of t ıdian Shield (Fig. 1). This was particularly manifested in the partially barred Marw asin enclosed by the Lahore-Sargodha, Aravalli and Chohtan-Malani ridges (Fig. 3).

Cyanobacterial-built carbonate reefs and domes and sand bars, which a ıaracteristic of the Bilara and Nagaur Formations (Figs. 5 and 9) of the Marwar Grou ust have inhibited movements of currents and waves in the shallow sea. This led evelopment of evaporitic conditions under the arid climate which prevailed, as Ind ıd by then drifted into the equatorial belt (Virdi, 1990) (Fig. 4).

After the deposition of sands in the Jodhpur time (Chauhan *et al.*, 1992) carbonat ere laid down in the Bilara part of the basin in the south. Penecontemporaneously mo ıan 650-m thick Hanseran assemblage of halite, anhydrite, polyhalite and sylvi gether with subordinate magnesite and dolomite developed in the north-in t agaur-Ganganagar sector (Dasgupta *et al.*, 1988).

Very similar was the situation in the Salt Range in the Himalayn domain (Figs. 8) and 11) where nearly 2000 m thick succession of gypsum and salts of the Saline Seri alt Range Formation) was laid down (Gee, 1989). The common occurrence of pota ılts in the contemporaneous Marwar and Salt Range successions (Jones, 1970) on t milar Precambrian crystalline floor (Shah, 1977) is very significant. Doubtless, t ıallow platform with evaporitic condition prevailing extended (Fig. 6) from the Jodhpu agaur tract in the south through the Salt Range to the Abbotabad belt in Hazara (Vir)90) (Figs. 1 and 3), the Ramban-Sinohu area in Jammu (Jangpangi *et al.*, 1988) a haramkot-Ropri-Mandi belt in the Shali domain of Himachal (Srikantia and Sharm)72). Deep drilling at Kasganaj in the Ganga Basin revealed presence of gypsum laye . the Proterozoic sedimentary rocks forming the floor (Sastri *et al.*, 1977).

Conditions of aridity and resultant desiccation continued in the Early Cambrian, ɔrne out by salt-pseudomorphs in the shales of the uppermost Bhaganwala unit of t ıelam Group (Gee, 1989) in the Salt Range (Figs. 5 and 11) and in the Sirbu a anurgarh formations in the uppermost Vindhyan (Misra, 1969). India had by no oved to the 25-30°S latitudes belt (Virdi, 1990).

CESSATION OF SEDIMENTATION : PAN-AFRICAN DIASTROPHISM

At the end of the Early Cambrian around 525 m.y. ago, the whole of the India ndmass was affected by tectonic upheaval, which abruptly terminated sedimentati ıroughout the Peninsular Shield and the Lesser Himalayan domain (Valdiya, 199 he whole of Marwar, the Salt Range and the Lesser Himalayan domain extending fro azara to Bhutan and beyond was lifted above the sea and exposed to processes ʼosion. This went on until the sea returned through a narrow depression in the Ear ərmian, aflooding large tracts of the Lesser Himalaya and Marwar (e.g. Badhaura).

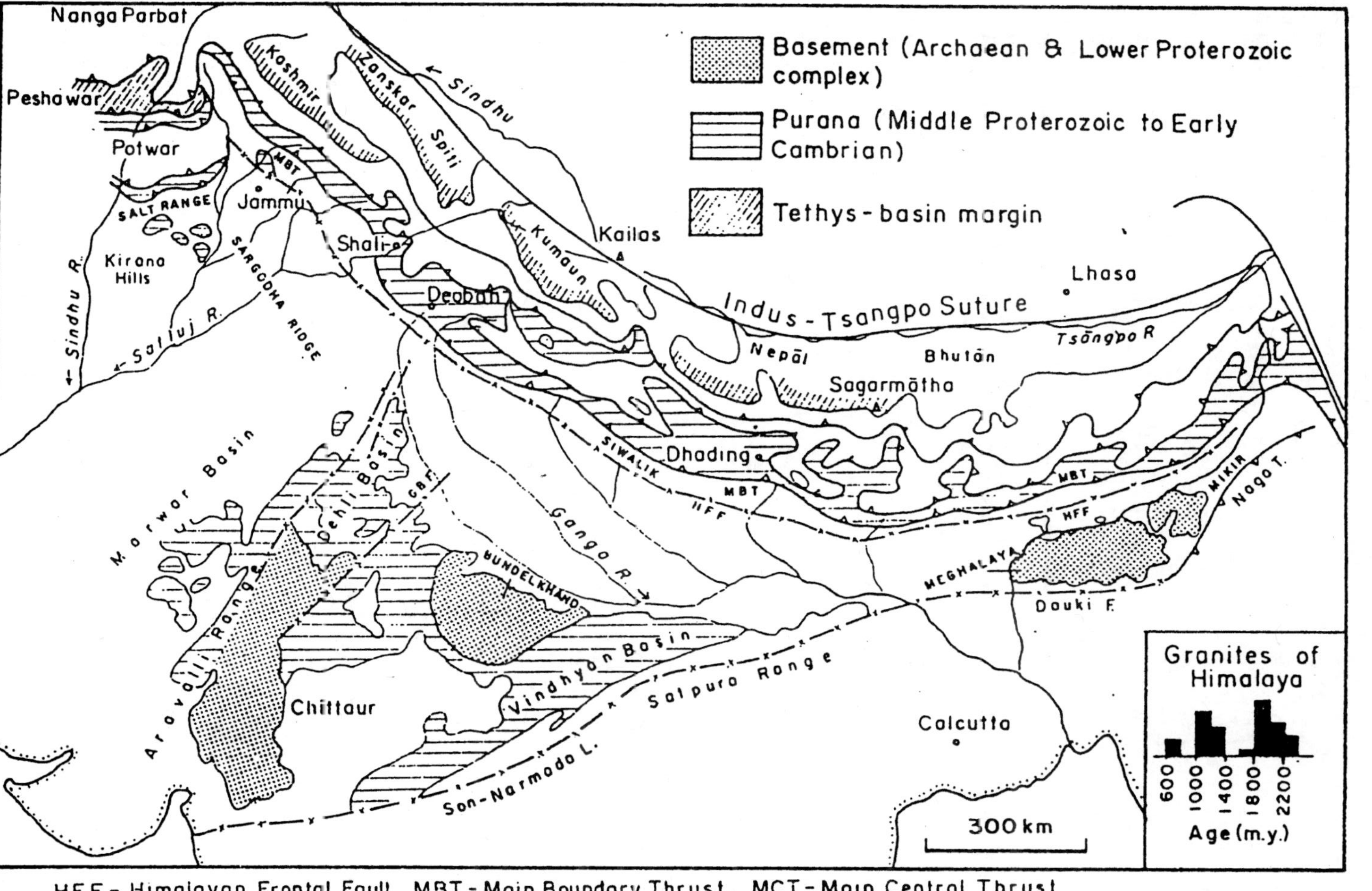

HFF - Himalayan Frontal Fault, MBT - Main Boundary Thrust, MCT - Main Central Thrust

Fig.1. Schematic map of the Himalaya and northern part of the Indian Shield, showing the extension of the Purana formations into the Himalaya (Valdiya, 1995)

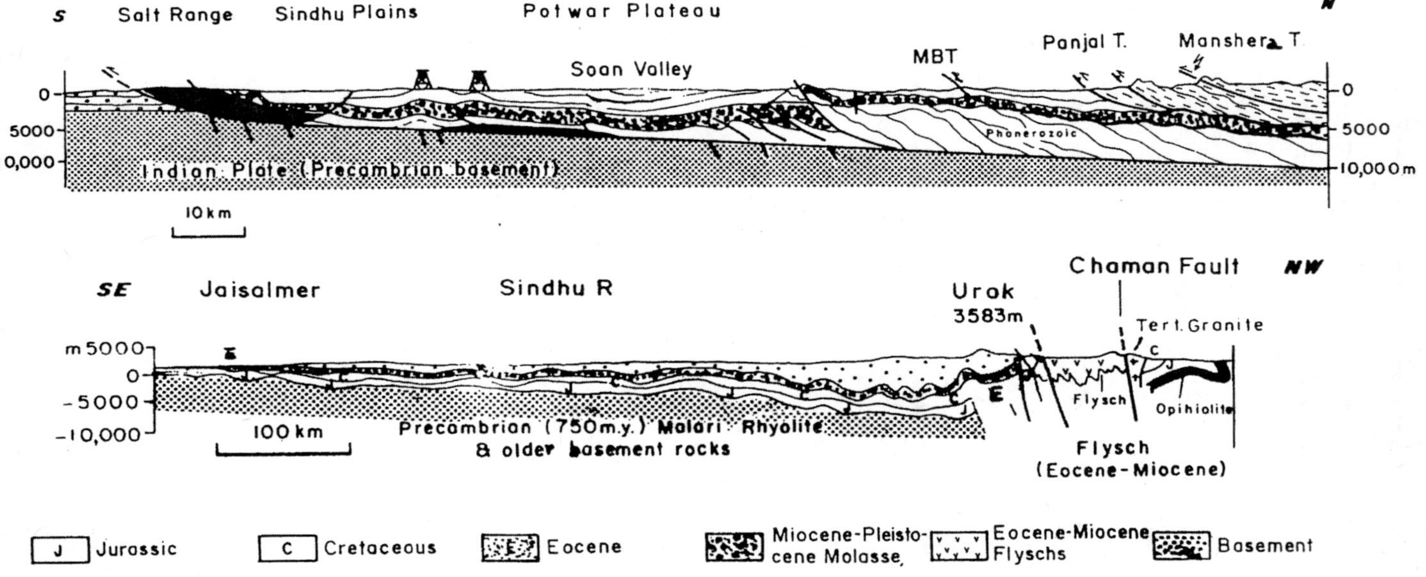

Fig.2. Cross sections across the northwestern continental shelf of the Indian Shield, showing the links of the Marwar Basin with the Himalayan province in northern Pakistan (Lillie et al., 1987, and Auden, 1957)

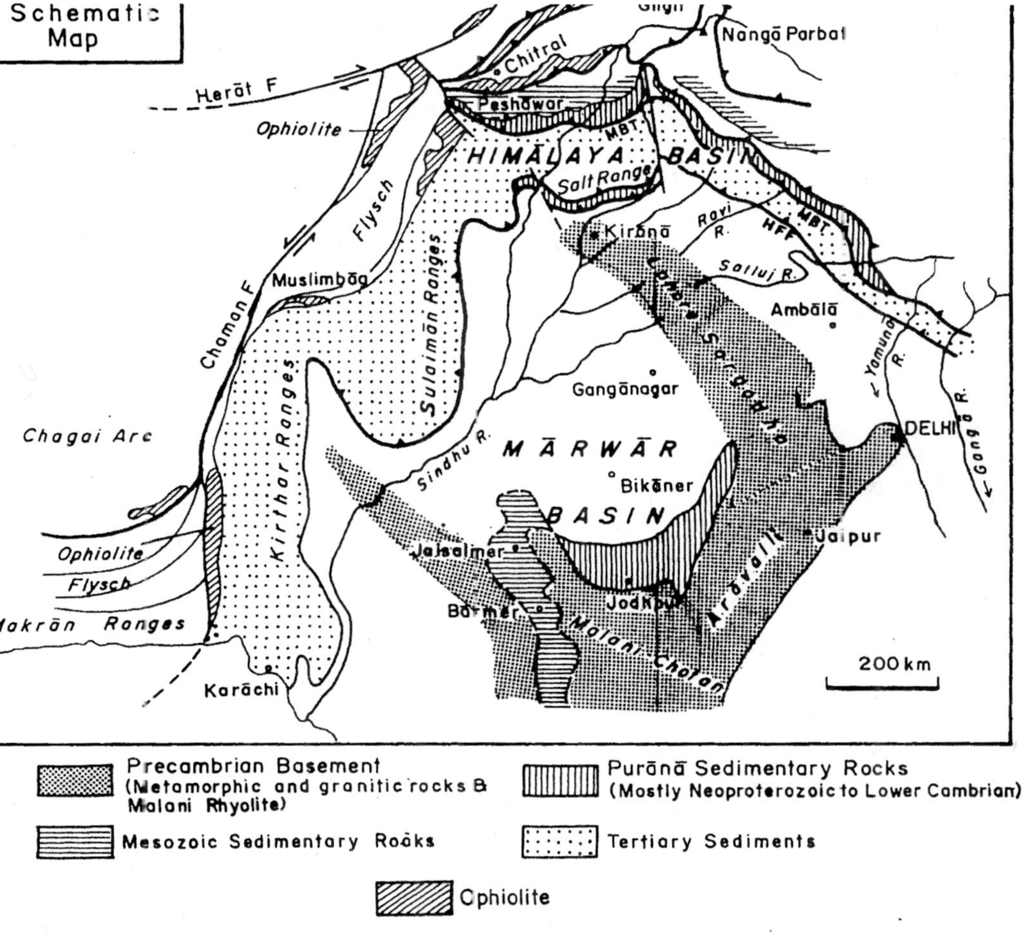

Fig. 3. Sketchmap of Marwar and adjoining sectors of the Himalaya, indicating close geological connections of the two domains (Based on Gansser, 1964; Valdiya, 1995, and Virdi, 1990)

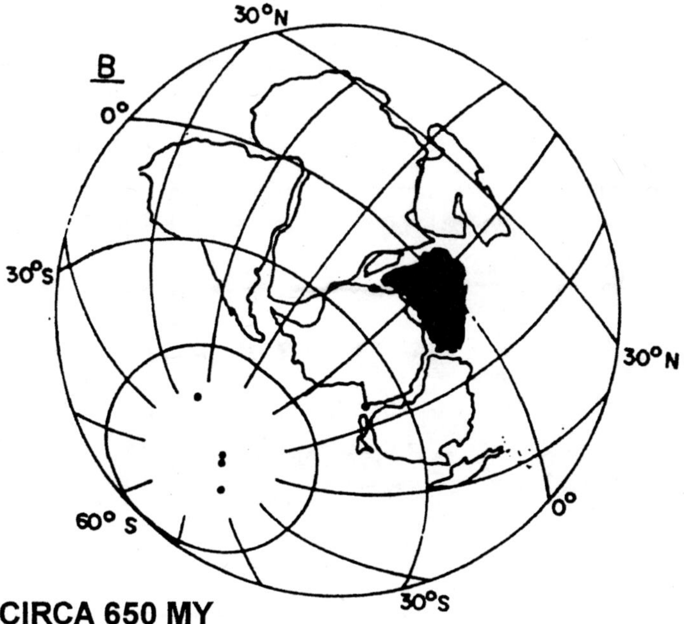

Fig. 4. Palaeoposition of India in the Vendian time (aftter Virdi, 1990)

s time the Gondwanaland was experiencing severe climatic refrigeration.

INTERLUDE OF GLACIATION

Glaciers covered large parts of India (Fig. 7A) as evident from deposits of tillites :h as the Talchir Boulder Beds in eastern India, the Bap Boulder Beds in Marwar, th bra Conglomerate in the Salt Range, and the Ganmachidam Conglomerate in Lahau g. 7B).

The glaciers that were nurtured by the Aravalli Range, moved northwestward eeping along the gravelly debris from the Aravalli and through Marwar - as far nortl the Salt Range 600 km away (Figs. 8 and 11). The presence of clasts of Malan yolite and Jalor Granite in the Tobra Conglomerate testifies to the Aravalli connectioı the Salt Range in the Early Permian time.

The period that followed witnessed practically similar sedimentation in the marin ılm encompassing the Jaisalmer Basin and Salt Range (Figs. 8, 9, 10 and 11) wever, the Bikaner-Nagaur and Barmer sectors had a little different history o liment deposition in the Mesozoic.

EARLY TERTIARY : FORMATION OF DEPOSITS OF FOSSIL FUELS

In the Palaeocene-Eocene time, the Marwar and Salt Range sectors went througl nost the same vicissitude of the tectonic movements and resultant sedimentatioı gs. 9, 10 and 11). The Bikaner and Barmer areas were flooded by sea water and th

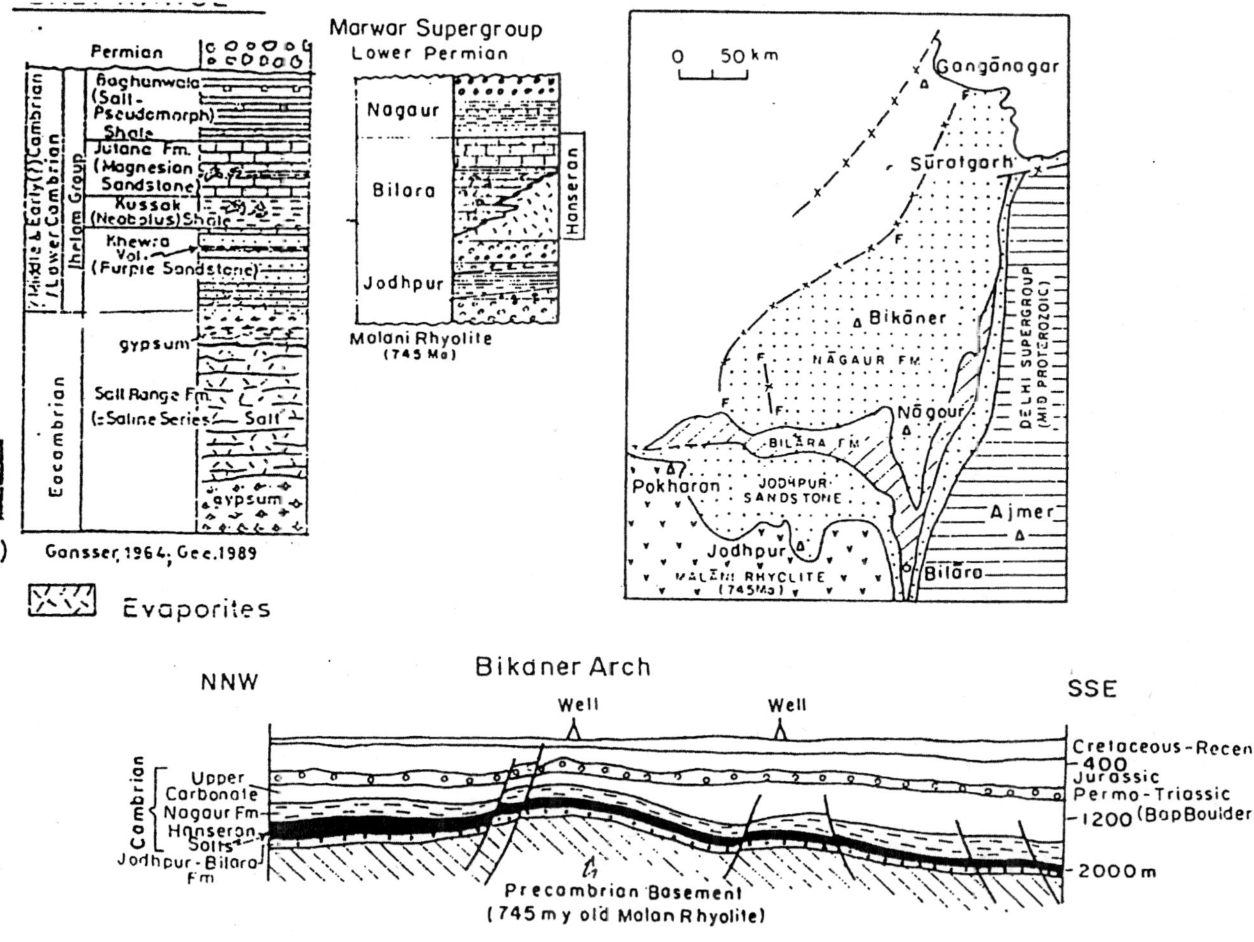

Fig. 5. Comparison of the Late Precambrian-Early Cambrian evaporite deposits of Marwar and the Salt Range of Pakistan. (From Valdiya, 1995, based on Dasgupta et al., 1988)

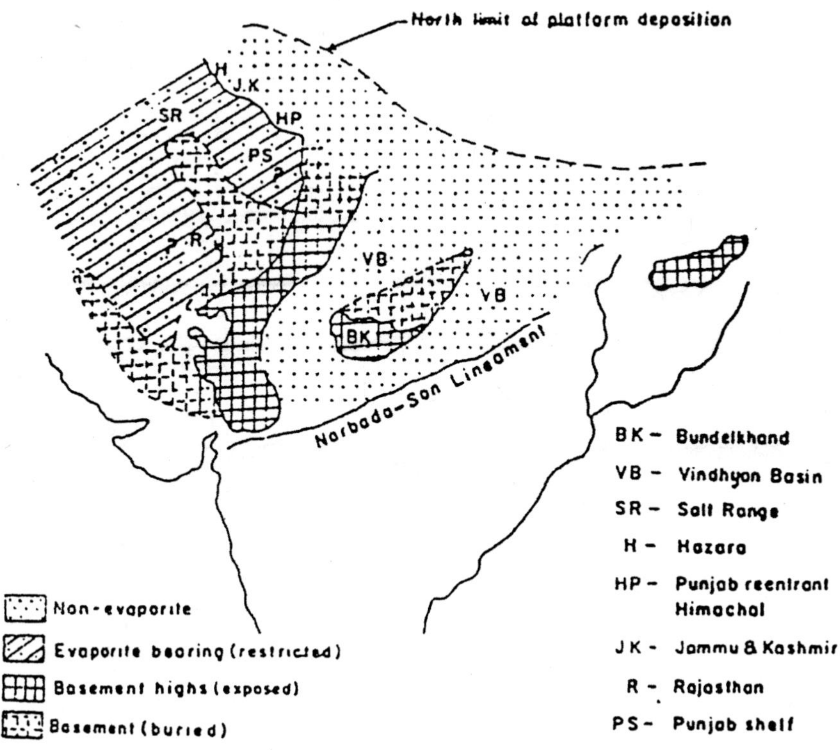

Fig. 6. Palaeogeographic map showing the extent of stratigraphic formations which bea: evaporite deposits (Virdi, 1990)

ntire southern front of the Lesser Himalaya sagged down, inviting marine invasion (ıe vast region extending from the Ranikot, Laki, Kirthar tracts in Sindh and tł aisalmer-Bikaner areas in Marwar, through the Salt Range-Hazara parts in norther 'akistan, and the Jammu-Subathu belt in western Himalaya.

Sediments were deposited on tidal flats and lagoons. Algae, bryozoans an oraminifers of a great variety grew prolifically. Under the peculiar biological an hysicochemical conditions that developed for the first time in the geological history (ıe Indian subcontinent, petroleum oil and gas were formed in the sediments (Fig. 12 'ne of the intriguing things, however, is the occurrence of petroleum in the Lat 'roterozoic Jodhpur and Bilara sediments of western Bikaner sub-basin, implying tha ıe oil and gas originated during the Precambrian time in this very region (Dasgupta an ulgauda, 1994).

Earlier in the Palaeocene, extensive deposits of lignite were formed, such as those ('alana (Bikaner, Nagaur and Barmer) in Marwar and Kalakot in Jammu. It seems tha ıe hinterlands, including the Aravalli, were covered with thick forests, and flooc 'ashed down plant debres to the tidal flats and lagoons, giving rise to lignite on the urial.

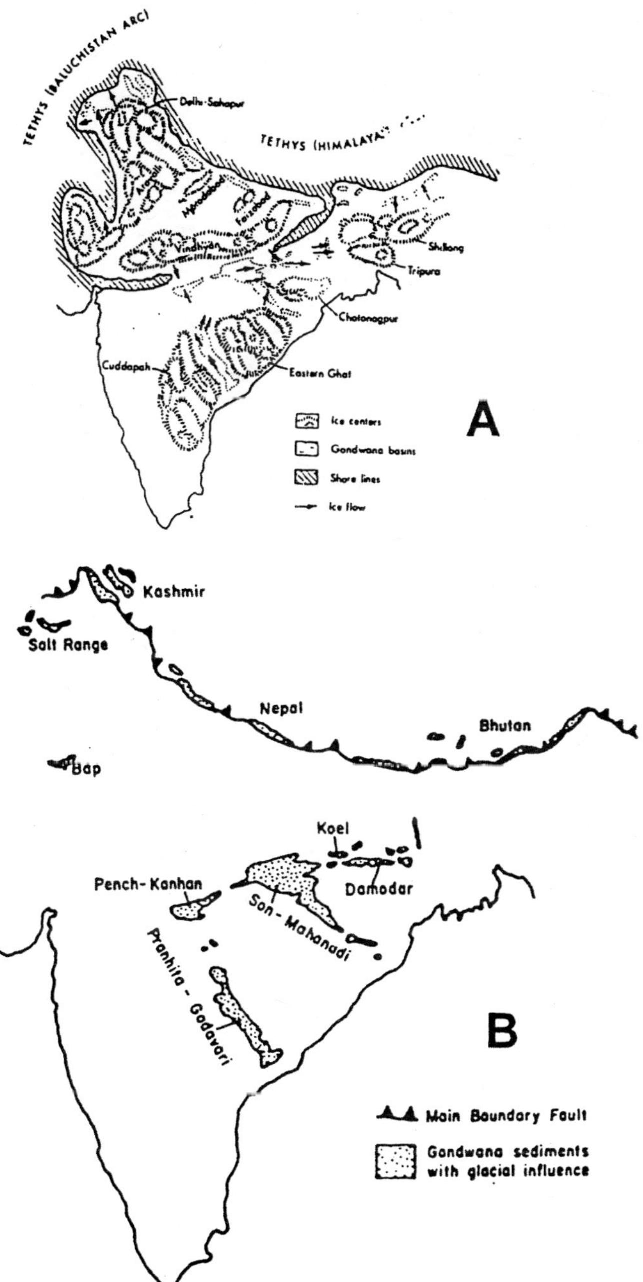

Fig. 7.A. Hilly tracts of the Gondwanic Indian Shield which must have nourished glaciers in the Lower Permian time.
B. Deposits of tillites emplaced by the Gondwana glaciers (Chatterjee and Hotton, 1986).

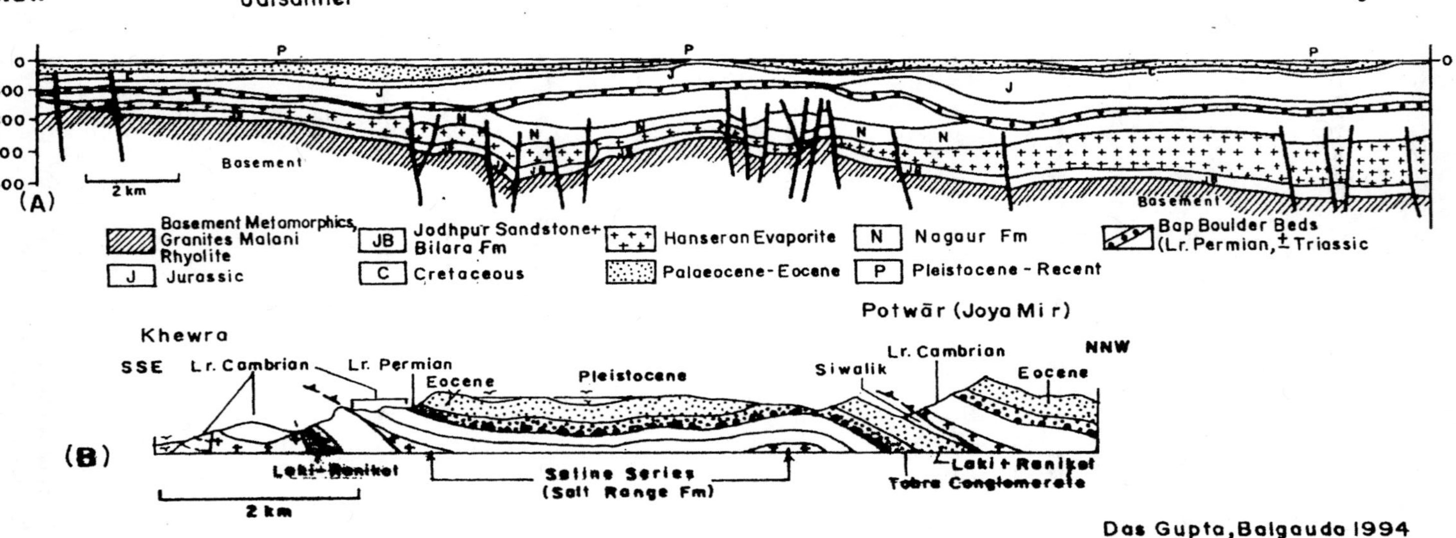

Fig. 8.A. Longitudinal profile of the Jaisalmer-Bikaner basin highlighting deposits of the Vendian evaporites, Lower Permian tillites and petroleum-bearing Lower Tertiary formations (After Dasgupta and Bulgauda, 1994) B. Cross section of th salt Range Potwar Plateau in Pakistan highlighting the above-mentioned stratigraphic horizons (After Gansser, 1964)

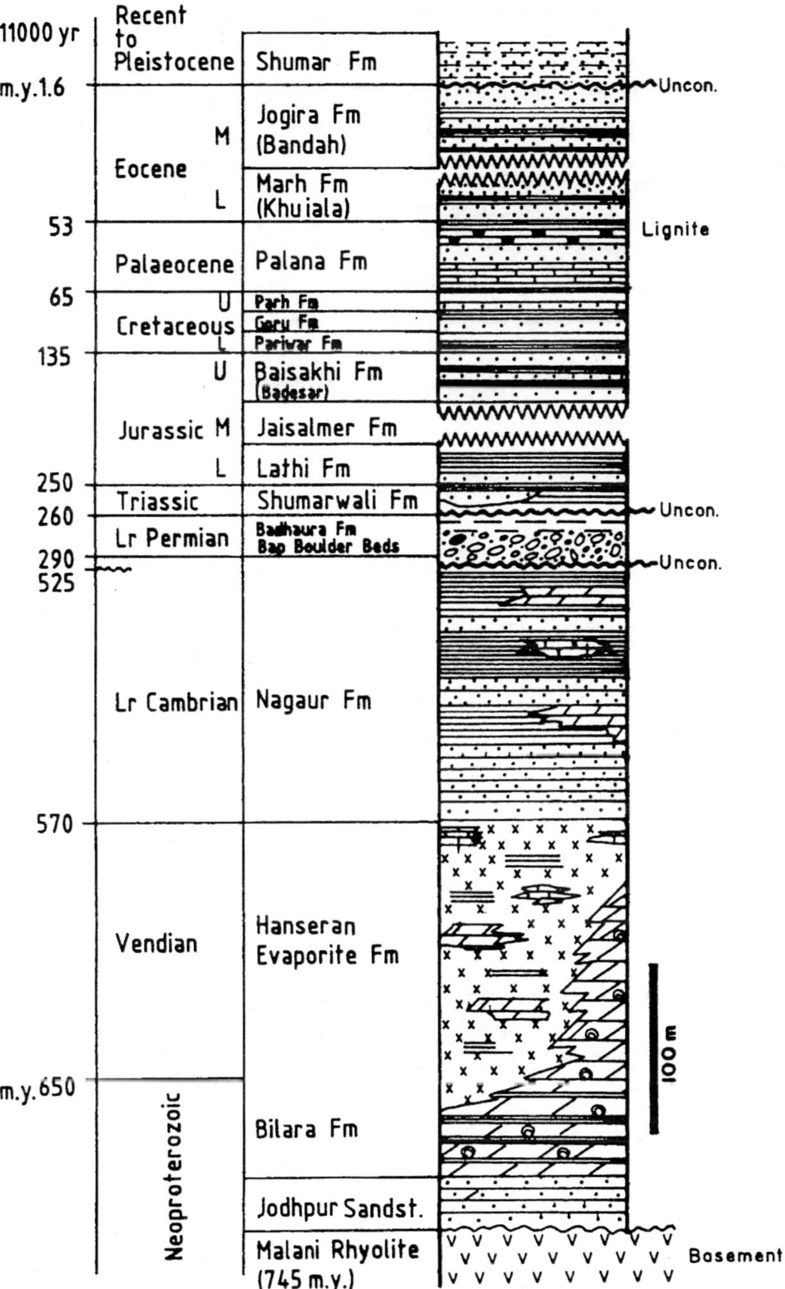

Fig. 9. Integrated lithostratigraphical column of the sediments of the Marwar basin (Based on Dasgupta et al., 1988; Dasgupta and Bulgauda, 1994, and Mohan and Sang[illegible] 1995).

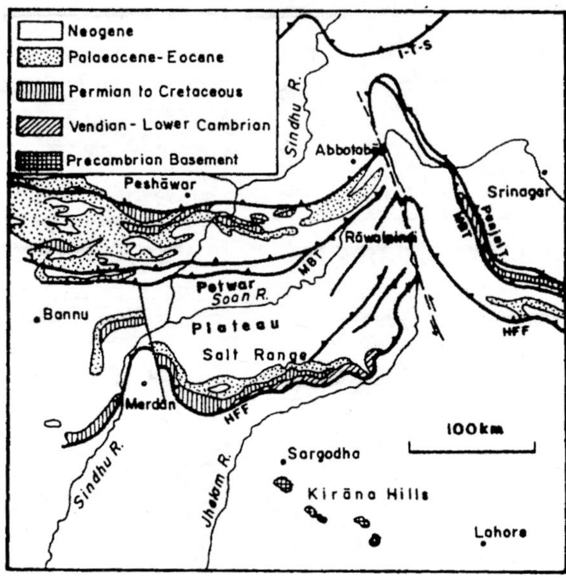

Fig. 10. Sketchmap of northern Pakistan Himalaya, showing the horizons that indicate its Marwar connections (After Lillie et al., 1987)

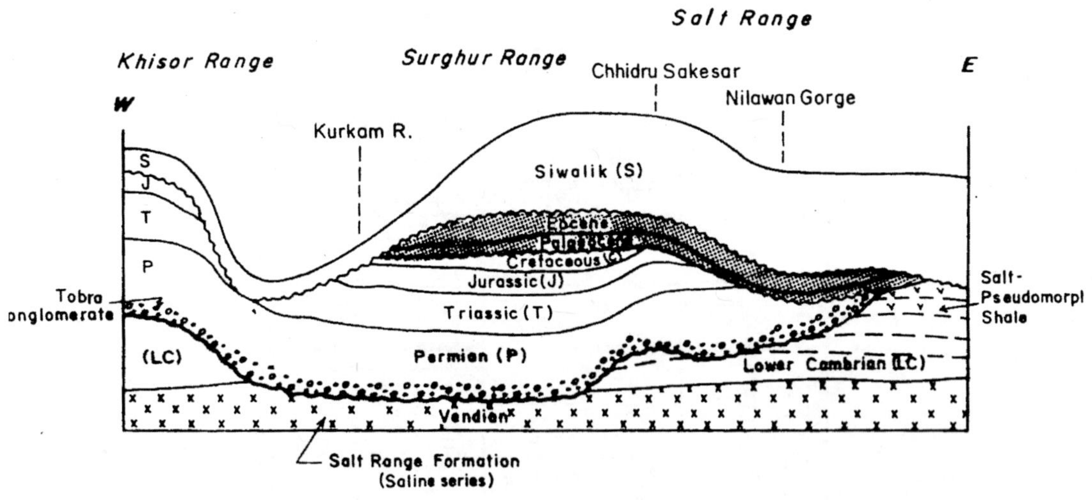

Fig. 11. Pattern of sedimentary succession in the Salt Range and adjoining Potwar Plateau in northern Pakistan (Gee, 1989)

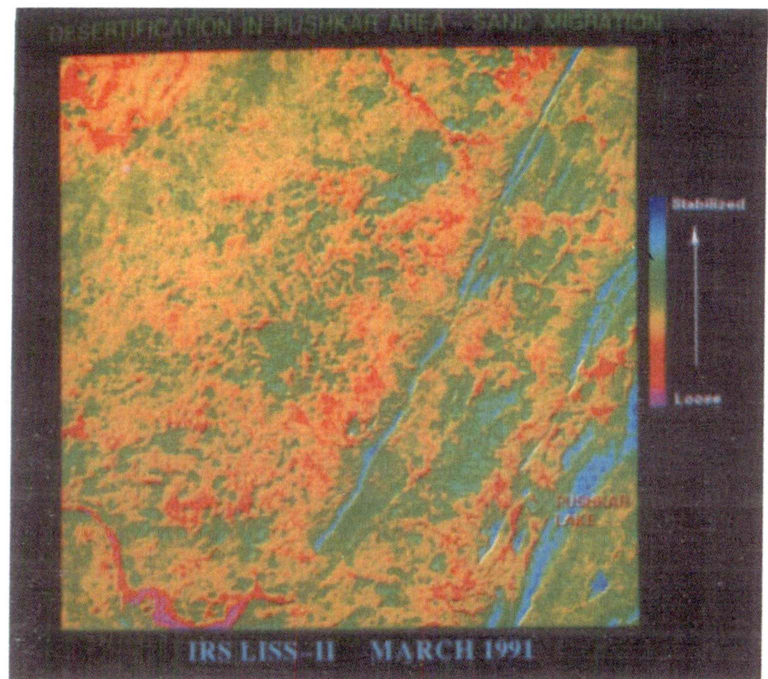

Plate 5. Desertification in Pushkar area, 1991.

Plate 6. Palaeo Channels in Kalti River Basin.

Plate 7. Limestone mining in Sanu area, 1993 & 1989.

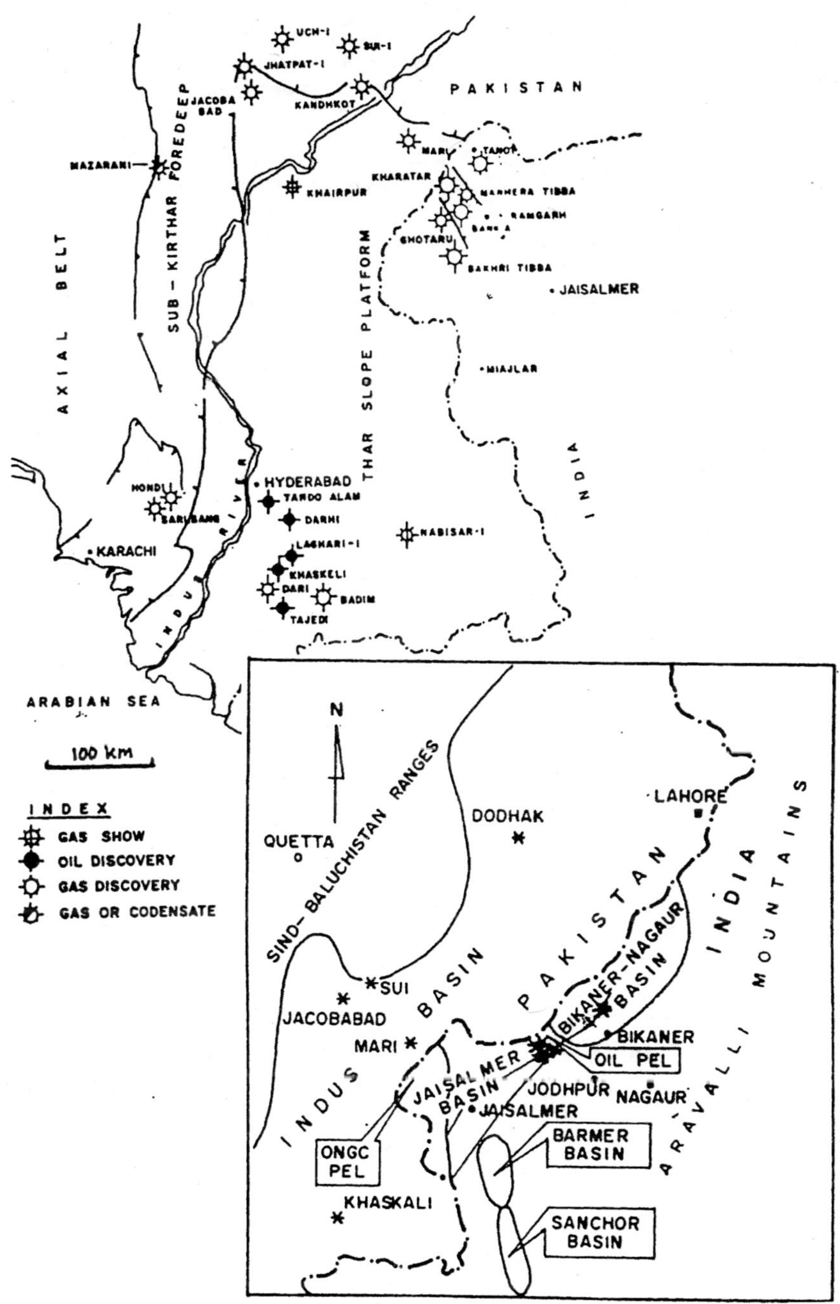

Fig. 12. Oil and gas wells in western Marwar and adjoining Sindh (Mohan and Sangai, 1995)

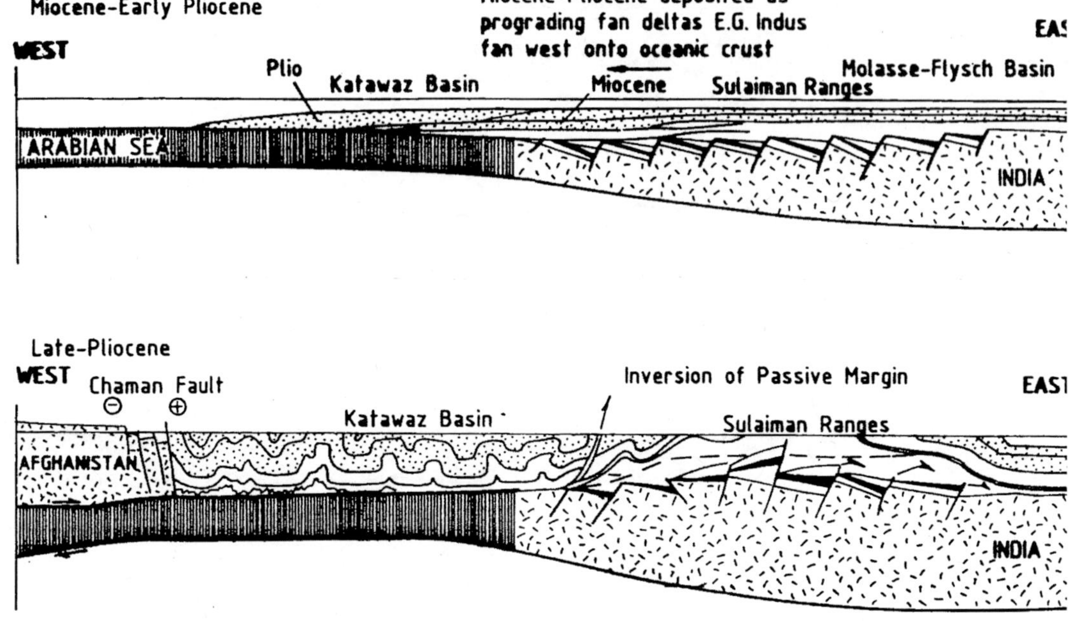

Fig. 13. Tectonic evolution of the Sindhu Basin and adjoining domains (Treloar and Izzat, 1993)

DRAINAGE REVERSAL AND TERRESTRIAL SEDIMENTATION

In the later part of Eocene, the sea suddenly-and finally-withdrew from the Sirma nd Marwar basins. The newly emerged land at the foot of the Himalaya became the s f vigorous sedimentation by rivers and streams descending from the rising mounta rovince. All through the long history of the Himalaya, from Proterozoic through Ea ertiary rivers flowed north and northwest from the Peninsular Shield. Suddenly t rainage was reversed, and rivers started flowing southwards in the northweste Iimalaya (Srivastava and Casshyap, 1983) and eastwards in southwestern Pakist Iimalaya (Waheed and Wells, 1990). That happened at the beginning of the Oligocer he rivers carried great volumes of detritus and dumped them in their channels, flo lains and deltas in an essentially terrestrial setting. The resultant piles are known ne Murree and Siwalik groups. Since there was little sediment contribution from t ravalli-which was presumably tectonically undisturbed and a greatly denud errain-no sediment accumulated in the Marwar front during the Murree-Siwal eriods.

Towards the close of the Pleistocene, the land in front of the Murree-Siwalik terra agged, even as the mountain province rose up higher and higher. The depression w ery rapidly filled up with sediments into what we know as the Indo-Gangetic Plai ncompassing the flood plains of the Sindhu, Ganga and Brahmaputra systems.

COMING OF THE SARASWATI

One of the major rivers draining the Himalayan province and building the In angetic Plains, was the Saraswati (Fig. 14). Formed by the confluence of rive amuna' and Satluj, this mighty river held profound sway over Marwar in the Rigve mes prior to the Mahabharat epoch. The Yamuna then flowed southwest through wh today the channel of the Ghaggar, and the Satluj used to take southeasterly course in the former at Shatrana, 25 km south of Patiala (Yashpal *et al.*, 1980). The Sarasw as 6 to 8 km wide downstream of the confluence. Building a vast flood-plain (of Hakr northwestern Marwar and adjoining Bahawalpur (in Pakistan) the Saraswati empti self through the channel of the Nara into the Gulf of Kachchh (Fig. 14).

The Saraswati had reestablished Marwar's link with Himalaya in the Pleistoce me. Some 40,000 years ago western Marwar was a much wetter region than it is tod id the Luni was quite a vigorous river made up of a network of perennial streams (K 995). It was in this well-watered and congenial land of the Saraswati and Luni that t one-age people established their settlements (Fig. 15) and developed Palaeolithic a esolithic cultures, particularly in the period Middle to Late Pleistocene (Dhir *et a*

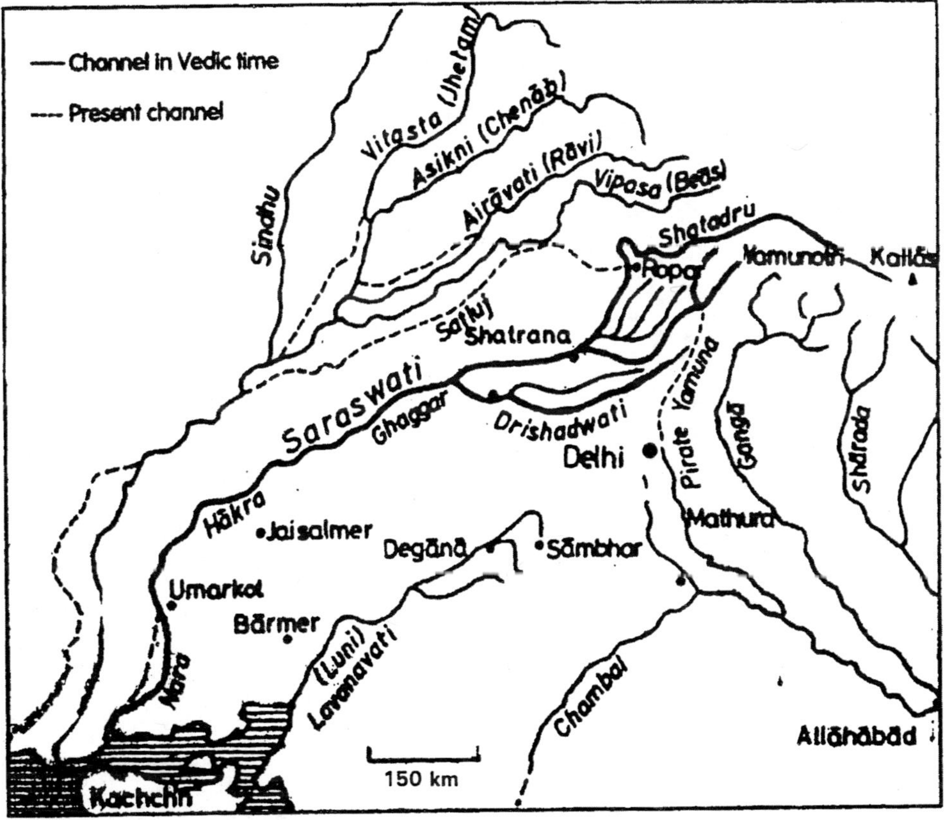

Fig. 14. Changing courses of the Saraswati River of the Rigveda time – before neotectonic movements in the Aravalli overtook the river system (After Kar, 1988)

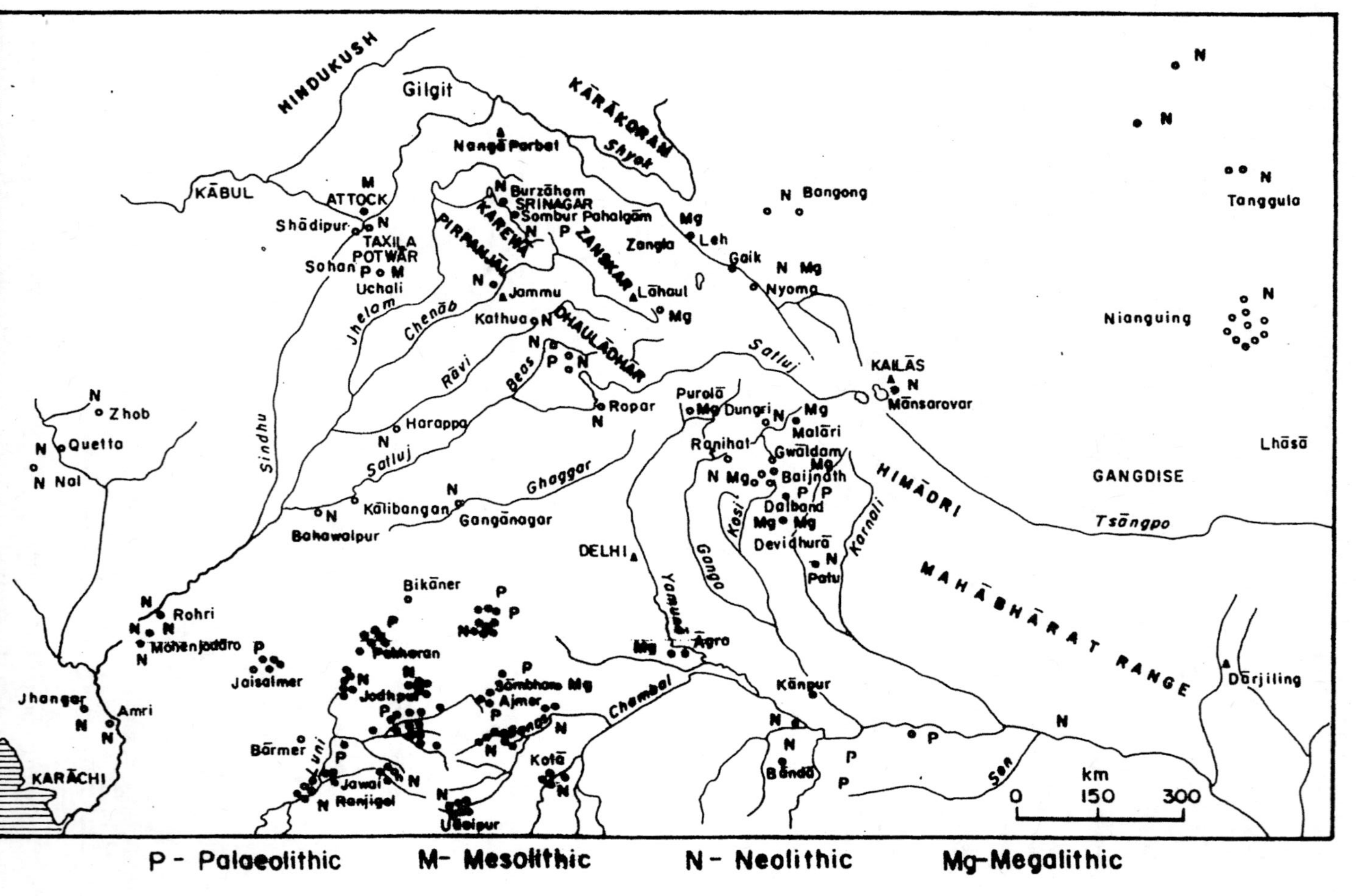

Fig. 15. Stone-age people lived extensively in western India and adjoining Himalayan terrains (compiled from various sources including Dhir et al., 1992; Misra et al., 1988)

92; Misra, 1995; Misra *et al.*, 1988). These people must have had connection with the ıtemporaries living in the Himalayan domain (Valdiya, 1993). Then the rainfa clined, the climate grew drier and the wind shifted finer alluvial sediments buildir nes. Compounding this environmental problem, the Aravalli started rising up. Th ·ced the Saraswati to migrate progressively westwards as evident from the numerou andoned channels (Fig. 14).

The drier period was punctuated with wetter speels-there were atleast two maj ·tter phases in the Quaternary period (Singhvi and Kar, 1992). The climate was qui ·t in the period 10,000 to 3,500 years B.P. (Singh *et al.*, 1974; Bryson and Swain 83). The amelioration of climatic condition was conducive to the prolific development .crolithic culture. Some time later, in the valleys of Saraswati and Drishadva urished the great Harappan civilization, 4,600 to 4,200 years B.P. (Bisht, 199 ·stified by remains of hundreds of settlements in the valleys.

As the Aravalli continued to rise, the Yamuna moved away eastwards, and w .ally pirated by a branch of the Chambal, a tributary of the Ganga. The Satluj, likwis grated progressively westwards. Abrupt uplift-presumably of the order of 20 to 30 m th possible subsidence of the ground to the west caused sharp westward deflection e Satluj at Ropar. The Satluj then became the tributary of the Sindhu. The Saraswa ıs left high and dry.

With the disappearance of the Saraswati collapsed the great civilization of tl ırappan people. Marwar's last link with the Himalaya was thus snapped.

ACKNOWLEDGEMENT

Financial assistance from the Council of Scientific and Industrial Research for tl thor's neotectonic studies in the Kumaun Himalaya is gratefully acknowledged.

REFERENCES

sht, R.S. 1993. Dhaulavira : New dimensions of Harappa Civilization, *Aajkal*, **49(6)** : 3-12.

yson, R.A. and Swaine, A.M. 1981. Holocene variations of monsoon rainfall in Rajasthan. *Qua Research.* **16** : 135- 145.

.auhan, D.S., Dubey, J.C. and Ram, B. 1991. Geological analysis of Nagaur basin in the vicini of Jodhpur City, In : S.K. Tandon *et al.* (Eds.) *Sedimentary Basins of India*, Gyanoda Prakashan, Nainital, pp. 63-73.

.atterjee, S. and Hotton, N. 1986. The palaeoposition of India, *Jour. Southeast Asian Ear Sciences*, **1** : 145-189.

ıir, R.P., Kar, A., Wadhwa, S.K., Rajaguru, S.N., Misra, V.N. and Singhvi, A.K. 1992. *Th Desert in Rajasthan*, *Geol. Soc. India*, Bangalore, 191 p.

.sgupta, U. and Bulgauda, S.S. 1994. An overview of the geology and hydrocarbon occurrences the western part of Bikaner-Nagaur Basin, *Ind. Jour. Petroleum Geol.*, **3** : 1-17.

.sgupta, S.P., Kumar, V., Chandra, R. and Jairam, M.S. 1988. A framework of the Nagau Ganganagar evaporite basin, *Indian Minerals*, **42** : 57-64.

.nsser, A. 1964. *The Geology of the Himalayas*, Interscience, New York, 289 pp.

e, E.R. 1989. Overview of geology and structure of Salt Range, with observations on relat areas of northern Pakistan, In : L.L. Malinconico and R.J. Lillie (Eds.) *Tectonics of Weste Himalayas, Geol. Soc. Am. Spel. Paper,* 232 : 95-110.

ngpangi, B.S., Kumar, G., Rathore, P.R. and Datta, S. 1986. Geology of the Autochthono

Folded Belt of Kashmir Himalaya. *Jour. Palaeont. Soc. India*, **31** : 39-59.

nes, C.L. 1970. Potash in halitic evaporite salt, western Pakistan, *U.S.G.S. Professional Paper* 700D, pp. 140-145.

ar, A. 1988. Evidence for neotectonism from the Indian desert, In : S. Singh and R.C. Tewa (Eds.) *Geomorphology and Evironment*, Allahabad Geographical Society, Allahabad, 30 310.

ar, Amal, 1995. Geomorphology of arid western India, In : S. Wadia *et al.* (Eds) *Quaterna Enviornments and Geoarchaeology of India*, *Geol. Soc. India*, Bangalore, 168-190.

atif, M.A., 1973. Partial extension of the evaporitic facies of the Salt Range to Hazara, Pakista *Nature*, **244** : 124-125.

llie, R.J., Johnson, G.D. and Yousuf, M. 1987. Structural development within the Himalaya foreland fold and thrust belt of Pakistan, In : C. Beaumont and A.J. Tankard (Eds *Sedimentary Basins and Basin-Forming Mechanisms*, *Canadian Soc. Petroleum Geol.*, *Mem* **12** : 379-392.

isra, R.C. 1969. The Vindhyan system, Presidential Addres, *Geology and Geography Sectio 56th Indian Science Congress,* Pawai Bombay : 1-32.

isra, V.N. 1995. Geoarchaeology of Thar Desert, northwest India, In : S. Wadia *et al.*, (Eds *Quaternary Environments and Geoarchaeology of India, Geol. Soc. India Spel. Pub* Bangalore, 210- 230.

isra, V.N., Rajaguru, S.N. and Raghavan, H. 1988. Late Middle Pleistocene environment ar Acheulian Culture around Didwana Rajasthan, In : *Palaeoclimatic and Palaeoenvironment Changes in Asia During the Last Four Million Years*, Indian National Science Academy, Ne Delhi, 93-106.

ohan, Shyam and Sangai, Pranaya, 1995. ONGC through Thar Desert : an exploration view, In *Proc. First International Petroleum Conference and Exhibition*, Vol. I, B.R. Publishing Cor Delhi, 133-145.

astri, V.V., Bhandari, L.L., Raju, A.T.R. and Datta, A.K. 1977. Tectonic framework ar subsurface stratigraphy of the Ganga Basin, *Jour. Geol. Soc. Ind.*, **12** : 222-233.

aah, S.M.I. 1977. Stratigraphy of Pakistan, *Mem. Geol. Surv. Pakistan*, **12** : 1-138.

ngh, G., Joshi, R.D., Chopra, S.K. and Singh, A.B. 1974. Late Quaternary history of vegetatio and climate of Rajasthan desert, India, *Philos. Trans. Royal Soc. London*, 267B : 467-501.

nghvi, A.K. and Kar, A. 1992.. *Thar Desert in Rajasthan*, *Geol. Soc. India, Spel. Publ.* Bangalor 191p.

rikantia, S.V. and Sharma, R.P. 1972. The Precambrian salt deposits of Himachal Prades Himalaya : Its occurrence, tectonics and correlation, *Himalayan Geology*, **2** : 222-228.

rivastava, V.K. and Casshyap, S.M. 1983. Evolution of pre- Siwalik Tertiary basin of Himach Himalaya, *Jour. Geol. Soc. Ind.*, **24** : 134-147.

reloar, P.J. and Izzat, C.N. 1993. Tectonics of the Himalayan collision between the Indina pla and the Afghan block : a synthesis, In : P.J. Treloar and M.P. Searle (Eds.) *Himalaya Tectonics*, *Geol. Soc. London*, 69-87.

aldiya, K.S. 1995. Proterozoic sedimentation and Pan-African geodynamic development in th Himalaya, *Precamb. Research*, **74** : 35-55.

rdi, N.S. 1990. Vendian-Lower Cambrian desiccation along the southern margin of the Prot Tethys, *Jour. Himalayan Geol.*, **1** : 103-114.

aheed, A. and Wells, N.A. 1990. Changes in palaeocurrents during the development of a oblique convergent plate boundary, Sulaiman fold belt, southwestern Himalay *Sedimentary Geol.*, **67** : 237-261.

ieological Evolution of Northwestern India
3.S. Paliwal (Ed.)
icientific Publishers (India), Jodhpur, 1999, pp. 19-33.

GEOLOGY OF THE REGION AROUND SIROHI TOWN, WESTERN RAJASTHAN - STORY OF NEOPROTEROZOIC EVOLUTION OF THE TRANS-ARAVALLI CRUST

A.B. Roy and K.K. Sharma*

Department of Geology, Mohanlal Sukhadia University, Udaipur 313 002
** Department of Geology, Governmnet College, Sirohi*

ABSTRACT

The rocks exposed around the Sirohi town and it's neighbourhood (around Sindreth) representing two distinct events are popularly classified as Sirohi Group and Sindreth Group. Present study indicates that rocks around Sirohi town show typical "thurst related" structures with an attendant thermal metamorphism and emplacement of highly fractionated bodies. The 1000 Ma diorite (gabbro) emplacement event in the Delhi fold belt, as well as in the region west of it (as around Sirohi) has been assumed to record the maximum age of the Sirohi orogeny. The succession of events which post dated the cratonisation (and peneplanation) include deposition of a volcano sedimentary sequence the Sindreth Group. The phase of activity has been dated at 780 Ma. The recorded geological history (based on available isotope data) of the Trans-Aravalli belt indicates that the Neoproterozoic processes were triggered off by the aborted attempt to split the Aravalli crust when the Rodinia super continent started to fragment for a new assembly of continental blocks, the Gondwana, around 1000 Ma ago.

Key words : Neoproterozoic, Trans-Aravalli Crust, Sirohi Group, Sindreth Grou Rajasthan

INTRODUCTION

In the region west of the Aravalli mountain range (a region popularly described he trans-Aravalli belt), we note a spectacular change in the geological scenario from pical mobile belt milieu to that of a stable intra-cratonic platformal setting in th st-west direction. Thus, Sirohi (Fig. 1) not only is a door step to the vast arid zon rdering the Thar desert (as described by Tod, 1909), its geological record is replete wit idence marking threshold of evolution from orogenesis to a phase of stable platform agmatism and sedimentation.

Another important aspect, disquieting though, about the geology of this part (jasthan, is the confusion that prevails over the stratigraphic status of the differer cks of the region (the neighbourhood of the Sirohi town). One basic reason for th nfusion is the scanty information that is available on the geology of the regio rticularly on the structural setting of the rocks. Primarily because of this, most of th rlier conclusions, on the geology of Sirohi are speculative in nature.

The only redeeming aspect about the geology of western Rajasthan is th cumulation of vast geochemical data on the volcanic rocks which show scattere currences all over the region (Pareek 1981; Gaur, 1984; Maheswari, 1983; Srivastava , 1989; Bhusan, 1995). The present authors worked in the Sirohi region with two fol jectives (1) to elucidate the structural history of the region, and (2) to establish relativ ratigraphy of different rocks based on studies on structure, metamorphisn lationship between granites and sedimentary units (part of which are metamorphosed d the nature of volcanic rocks (Figs. 2 and 3). The ultimate aim of the study is t derstand the Neoproterozoic (Plumb, 1990) processes which led to the evolution of th ans-Aravalli crust.

STRATIGRAPHIC CORRELATION : PROBLEM AND PERSPECTIVE

Hacket (1889) was the first to describe the geology of the Sirohi region. H troduced the term "Aravalli series" for the rocks, apparently thinking them to b uivalent to the rocks of the Udaipur region. He had also identified rocks which wer rlier described as the 'Malani beds' by Blanford (1877) from the Jodhpur regior dham (1893) in his classic book "Manual of the Geology of India" wrote about th eissic rocks of the earstwhile Sirohi State. Coulson (1933) classified the geology of th rohi region adopting terms introduced by Heron (1917) for the rocks of the Araval ountain region in the northeast. The stratigraphic succession of the Sirohi regio ected by him is given in Table 1. Coulson (1933) correlated the rocks of the Sirohi an ndreth (Fig.1) region to the Aravalli system of Archaean age (sensu Heron, 1935 owever he had expressed some doubt about the position of the Sindreth basalt.

Raja Rao *et al.* (1971), while describing the geology of the Sirohi town, were the firs express doubt about the correlation of the Sirohi rocks with the Aravalli "system" (e type area around Udaipur. It may be worth quoting here their judiciously expresse inion : *"The question must necessarily be kept open until the structural, metamorphi d igneous episodes of Sirohi is unravelled in greater detail. At this stage it is justifie ly to conclude that these rocks are pre-Erinpura granite in age"*. Gupta *et al.* (1980) wh oposed a revision of the lithostratigraphy of the Aravalli region, included the Sirol cks in their Delhi Supergroup (series and system have been replaced by Group an pergroup), constituting the upper part of the succession. These authors recognized tw oups : the lower Sirohi Group underlying the upper Sindreth Group (considered to b uivalent of the Punagarh Group of the Pali region). A time span from 2000 to 700 M s assigned for the entire Delhi cycle, which also included the Sirohi and Sindret oups.

'able 1. Straigraphic succession of the Sirohi State (Coulson, 1933)

'ost-Tertiary and Recent	Wind-blown sand, tallus, alluvium and conglomerate etc.
'ost Malani basic intrusive	Dolerites
⁄lalani system (?Purana)	Idar granite, porphyries, rhyolites, and dellenites.
'ost-Erinpura granite but pre-Malani asic intrusives	Dolerites, epidiorites, gabbors, pyroxenites, picrit basalts and sodalite-syenites.
'ost-Delhi but pre-Malani granitic intrusives	Erinpura granite and accompanying quartz ree pegmatites, and aplites.
)elhi and post-Delhi, but pre-Erinpura granite basic intrusives	Dolerites, epidiorites, amphibolites, homblencde a actinolite schists.
)elhi system-Ajabgarh series	3. Limestones and calc rocks with associated ba rocks 2. Mica schists and thin quartzite, with abunda associated basic rocks. 1. Quartzites.
	UNCONFORMITY
ıravalli system (Archaean)	Mica schist, phyllites, shales, crystalline limesto quartzites with contemporaneous tuffs and lavas.

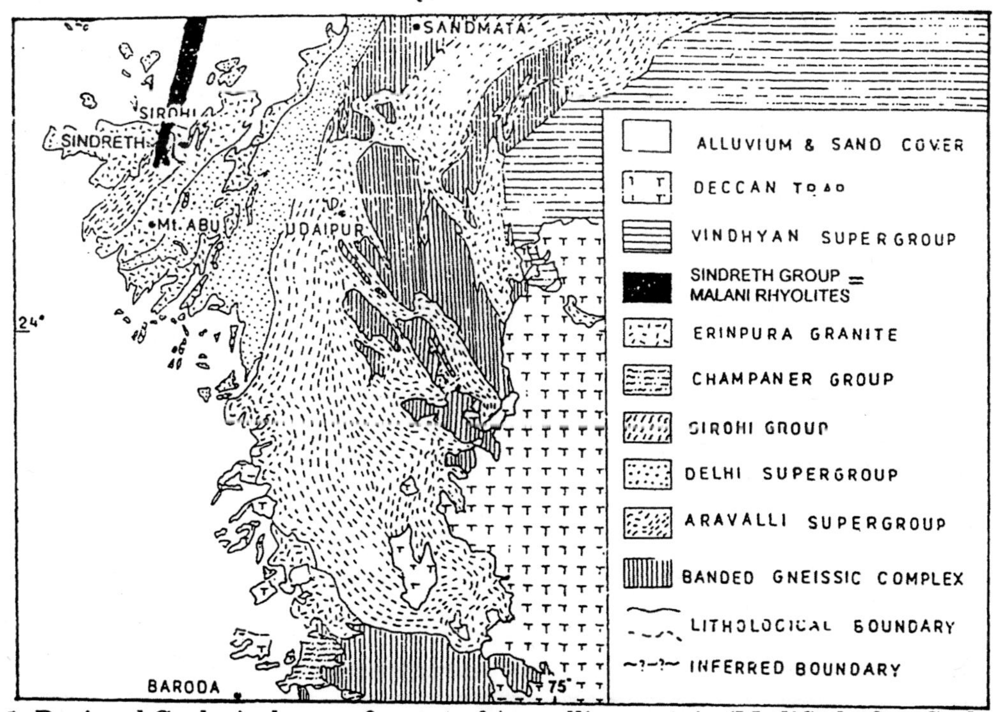

'ig. 1. Regional Geological map of a part of Aravalli mountain (Modified after Geologic Survey of India map, 1969)

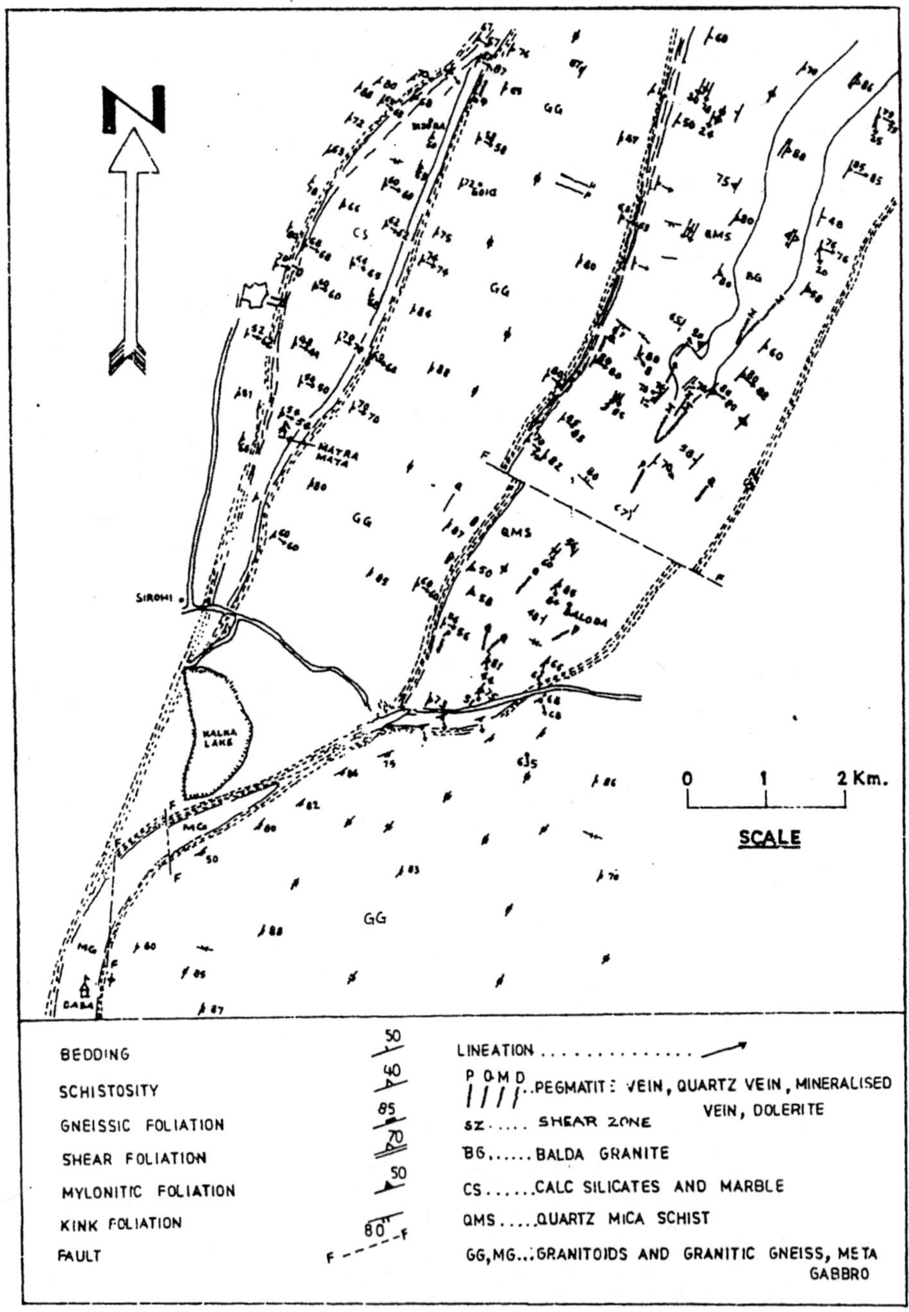

Fig. 2. Structural map of the Sirohi area

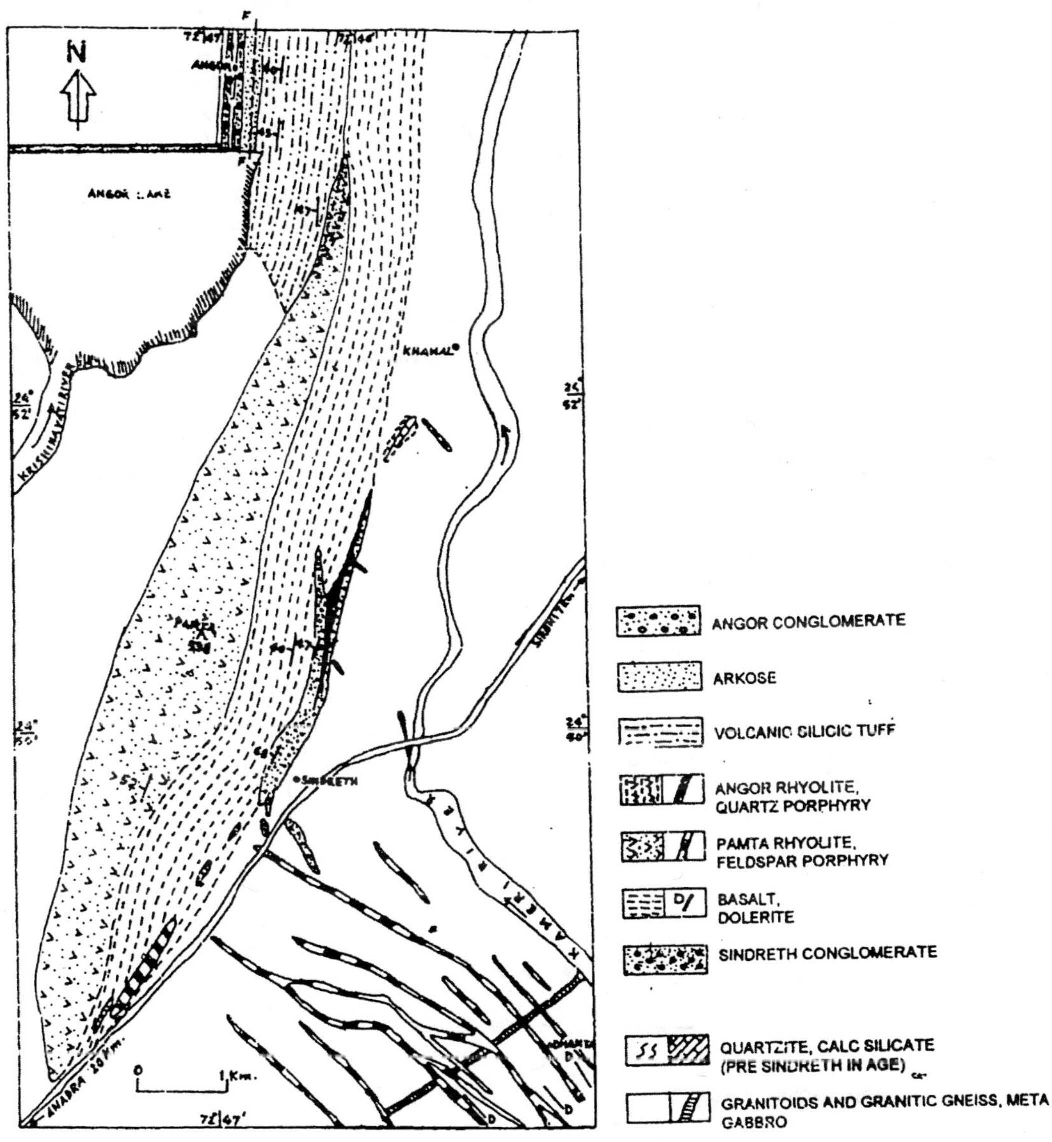

Fig. 3. Geological map of the area around Sindreth

Roy (1988, 1990) considered the Sirohi Group as separate tectonostratigraphic un unger to the Delhi Supergroup (Table 2). Bose (1989) raised questions on the status e Sirohi Group. Based on the study of conglomerate horizons Sinha-Roy *et al.* (199 oposed a stratigraphic model for the rocks of south- central Rajasthan. These autho aced the Sindreth rocks in the upper part of the Delhi Supergroup. The autho rrelated these rocks with those of the "South Delhi fold" belt. A surprising omission e model is the status of the Sirohi Group.

'able 2. Revised lithostratigraphic succession of the pre-Vindhyan rocks of Rajasthɛ and Gujarat (Roy, 1990)

		Age in Ga
	Vindhyan Supergroup	
	Malani rhyolite suite	0.75-
	Erinpura granite and Godra granite	0.75-0.90
	Champner Group	
	(Sirohi Group?)	
	Post-Delhi granites	1.45
	Ajabgarh Group	
Delhi Supergroup	Alwar Group	
	Rayanhala Group*	
	Post Aravalli granites	
	(Darwal and Amet granites)	1.9
	Upper Aravalli Group	
Aravalli Supergroup	Middle Aravalli Group	
	Lower Aravalli Group	
	Berach granite	(~2.6)
	Untala and Gingla granites	3.0
Mewar Gneiss and Granite	Pelitic gneiss, quartzite, marble, calc-silicates	
	Tonalitic biotite gneiss, amphiblite	3.5

Sensu Roy (1988), 'Raialo series' sensu Heron (1917, 1953)

STRATIGRAPHIC CORRELATION : THE QUESTION OF BASEMENT

Establishment of correct mutual relationship amongst various lithologic uni eveloped in an area is an essential prerequisite to understand geology and tecton volution of the belt. Most important stratigraphic relationship that led to establishme f correct stratigraphic order in most orogenic belts, and the Sirohi basin being ı xception, is the basement-cover relationship.

While describing the geology of the Sirohi State, Coulson (1933) considered tl ıetasedimentary as well as sedimentary-volcanic units as belonging to the 'Arava ystem' of supposedly Archaean age. No where any outcrop of basement rocks w: ecorded by him. Heron (1953), however, mapped outcrops of Banded Gneissic Compl considered to be the basement of the 'Aravalli system' by Heron, 1935, 1953) around K ıorth of the Sirohi town). He, however, commented on the difficulty in separating tl asement gneisses' from the 'intrusive' Erinpura granite.

From around the Sirohi town, Coulson (1933) described two different types ranites : (i) Erinpura granite, and (ii) Idar granite. The latter granite constitutes tl oungest Precambrian granite correlatable with the granites of Jalore and Siwana. ecent years, a separate Balda granite has been described, which is associated wi olframite mineralisation (Chattopadhyaya *et al.*, 1982).

However, none of the above granites is considered to form basement of the Siro asin. On the contrary, a clearly intrusive relationship is observed between the Bal anite and the Sirohi metasediments. Idar granite outcrops are outside the map ar ıt its intrusive relationship with its host rocks (which include Sirohi metasedimer ıd Erinpura-type granites) is unequivocal.

All the granite rocks which occur at the contact of the Sirohi metasediments are r ıly granitoid in character. On the other hand there are gneissic granites (which ha ıysical characteristics quite similar to that of the type pre-Aravalli basement gneiss ewar Gneiss of Roy, 1988) and mylonitic gneisses. During the present work duct .ear zones have been observed almost all along the contact of the granite gneisses ıase of Erinpura granite) of Coulson (1933) and the Sirohi metasediments. Not or at, a number of slices of mylonitised granite gneisses are present in t etasediments. Further unlike the Balda granite which produced a thermal aure ound it, there is no evidence of any contact metamorphism or "wall rock alteratio ıerever the contact has been mapped. The contact is as sharp as a knife's edge, and ghly tectonised. Thus, although virtually no other physical characteristics are observ a marker of unconformity, the above relationship may be construed as indicative ısement-cover relationship between the gneissic granites and the metasedimen ıother indirect relationship indicative of such a relationship, is the quartzose charact the Sirohi mica schist, which suggests granitic provenance. In view of the abo scription, it may be prudent to describe the granites earlier mapped as Erinpu anite (Coulson, 1933; Heron, 1953) as older Erinpura granite and newer Erinpu anite.

The essentially anorogenic origin of the Sindreth basin containig volcanic rocks a diments, is clear in the petrographic and structural description of the rocks. Thus the hardly any problem in deciphering all other rocks, the Sirohi Group and the Erinpu anites (both the older and younger types), as the basement rocks of the Sindre roup. Further confirmation comes from the Sindreth conglomerate, the basal unit of t oup, of which pebbles are drawn from granite, granite gneiss, mica schist, and ev rbonaceous phyllite.

STRATIGRAPHIC SUCCESSION OF COVER ROCKS

The two main cover rock units of the area are the Sirohi Group and the Sindre roup. Both these units differ from each other in terms of lithology, structural evoluti etamorphism and relationship to the granites.

The Sirohi Group is made up dominantly of psammo-pelitic schists (as quartz ica schists) and carbonates (dolomite marbles) grading to calc-silicate rocks. There ı almost total absence of rocks of larger terrigenous clasts (conglomerate, arkose, en quartz arenite), as well as volcanics. A strong penetrative deformation resulti om shearing and folding, characterises these rocks. The Sirohi rocks characteristica ıow development of LS fabric. A thermal metamorphic aureole has developed in t rohi rocks of the Balda granite body. Besides these, the development of quartz a ldspar "augen" along mylonitic foliation in the mylonitic granite gneisses, a velopment of garnet-biotite pair in the mica schists are indicative of relatively high] ndition corresponding to the upper green schist facies of metamorphism.

In strong contrast to all these, the rocks of the Sindreth Group are tota ınmetamorphosed and undeformed (except some depositional and post deposition aults and fault related drag folding) comprising a sequence of conglomerate, arkose aı ›imodal volcanic flows and volcanoclastic layers. Thus the separation of the two groups ʇuite easy in the field. The rocks of Sirohi Group show penetrative deformation aı netamorphism; by contrast the Sindreth rocks are unmetamorphosed and do not shc ıny trace of penetrative deformation.

Stratigraphic succession and regional tectonostratigraphic correlation of the roc ›f the area are presented in Tables 3 and 4.

'able 3. Stratigraphic succession of Sirohi Rocks

›indreth Group	Angor conglomerate Arkoses Volcanic silicic tuff Angor rhyolite Pamta rhyolite Basalt Sindreth conglomerate	 Quartz porphyry Feldspar porphyry Dolerite
›irohi Group	Balda granite Carbonaceous phyllite (local) Calc-silicates and marble Mica schist	
3asement rocks	Granite gneiss (older Erinpura granite) and associated meta-gabbro	

'able 4. Geochronology and Tectono-stratigraphy of the Precambrian rocks Sirohi-Pali-Ras Region

٩ge	Reference	Tectono-stratigraphy	Events
).78 Ga	Rathore (1994)	Sindreth Group = Malani Volcanics	Oldest age of the Malani volcanics fro Diri-Gurapratap Singh, Pali regio dating basin opening in the Sindre region
).83 Ga	Chaudhary et al. (1984)		Granite intrusion in Mt. Abu, Erinpu and Pali region and end of Sirohi cycle
l.0 Ga	Volpe and Macdougall (1990)	Sirohi Group	Opening of the Sirohi-Ras basin with tl intrusion of gabbro in granite gnei (Daba-Dhanta region) which is coev with the emplacement of gabbro-diori intrutisons in Delhi metasediments
l.48 Ga	Chaudhary et al. (1984)	Delhi Supergroup	Granite intrusion in the Del Supergroup of rocks (Roy, 1990), end Delhi orogenic cycle

REGIONAL CORRELATION

The Sirohi town is the virtual western limit beyond which there is no outcrop leformed and metamorphosed rocks. Geological map of Coulson (1933) shows continui the rocks correlatable to the Sirohi Group in the south in the Palanpur, district ıjarat. In the north, the outcrops become scarce, particularly beyond Jawai Rive ırther, linear outcrops of marble appear (the Ras marble). Other deform etasediments which occur in the north are the phyllites of Degana, the Makraı arbles, and the carbonaceous phyllites and minor carbonate rocks around Didwana. A ese outcrops which occur west of the main Delhi basin, are tentatively correlated wi e Sirohi Group on the basis of deformation characteristics, such as prenetrati iation and shear (thurst) related steep plunging lineation. This correlation, however ırely tentative, not based on strong field data.

Structural analysis of the Sirohi Group of rocks clearly brought out a pattern formation which is in strong contrast to the known structural style of the Delhi aı 'avalli Supergroups. (Roy and Das, 1985; Roy, 1991; Roy *et al.,* 1993; Paliwal, 198 ngh, 1988). The Delhi rocks leaving aside the complexities observed in the sou iswal, 1988), are characterised by coaxial folding of F_1 and F_2 on NE-SW ax perimposed by a third F_3 set at almost right angles. Nowhere in the Sirohi Group roc ch a deformation pattern is ovserved. When compared with the deformation pattern e Aravalli Supergroup of the type area (Naha and Halyburton, 1977; Roy, 1972, 199 ›y *et al.,* 1971, 1980; Paliwal, 1988), the structural pattern in the Sirohi Group appea ɔ simplistic with only one schistosity and one down dip lineation.

The rocks of the Sindreth Group are correlatble with those occurring in the Pali e north. The Sindreth basin extends north-south at Diri and Gurapratap Sin [aheswari, 1983), there is a thick sequence of conglomerate, basalt, rhyolite includiı trapotassic types. Basaltic rocks of the Sindreth Group showing pillow structures, a nilar to those of the Bombholai (Singh, 1983).

TECTONO STRATIGRAPHIC EVOLUTION

The history of crustal evolution of the Aravalli mountain is a story of repeat velopment of orogenic belts and cratonisation showing westward polarity. The orogen cles which began with rifting of the thick Archaean crust possibly during the earlie 'oterozoic (Roy *et al.* 1988, 1990; Sharma 1988, 1990; Sinha Roy *et al.,* 1993) aı ntinued through the Middle and Late Proterozoic, culminated with the closing of tl rohi and associated basins. The geological maps of the Aravalli mountain, showiı sposition of the different sedimentary-volcanic basins (Fig.1) clearly elucidate tl cture of westward migration of the basin axis with a corresponding shift of tl ctonothermal activity both in space and time.

The peaks of thermal activities in the Aravalli mountain belt have been dated at 00 Ma, (stabilization of the Archaean craton; Wiedenbeck *et al.*, 1995); *ca* 1800 ± 1 a (closing of the Aravalli orogeny; Choudhary *et al.* 1984, Roy, 1988, 1990); *ca* 1400 ± a (closing of the Delhi orogeny Roy, 1988, 1990); *ca* 900 to 850 Ma (Choudhary, *et a* 84; Volpe and Macdougall, 1990). The last event is related to the Sirohi cycle of tl 'avalli mountain building episode (Roy, 1988, 1990).

The deformation features and associated plutonism observed in the rocks of tl rohi Group are unique and do not match with those in the Delhi and Arava ıpergroups. The Sirohi Group showing a single phase of deformation, metamorphisı trusion of granite, represents the youngest of the three Proterozoic orogenic cycl hich built up the Aravalli mountain. The culmination of the Sirohi cycle actually mar e cratonisation of the Archaean-Proterozoic Aravalli platform. The subsequent phase e Neoproterozoic (Plumb, 1990) crust building processes in the trans-Aravalli be tnessed basin evolution and volcanicity (generally bimodal; Bose, 1989; Bhusha 95) under stable platformal condition.

At Sirohi, this transition from orogenic to stable cratonic condition is recorded e development of shallow intra-cratonic basin in which conglomerate, basalt, rhyoli d arkosic sandstones were deposited successively over deformed and metamorphos sement rocks. This phase has been correlated with the earliest Malani Group lcanics and sediments (the maximum age dated at around 780 Ma, Rathore, 1994; 7 a, Crawford and Composton 1971; Crawford, 1975).

The Sirohi Group must have predated the age of the Malani Group by about illion years. Confirmation for this comes from the 830 Ma/Sr age (Choudhary et 84) of the Erinpura granite (sensu stricto). The features in the granite such ignment of feldspar phenocrysts parallel to mylonite foliation with a correspondi eation parallel to the shear direction, suggest that the rock has undergone parti elting coevally with the ductile shearing of the protolith. The observed comm formantion history of the granite and the meatsediments may, therefore, lead rrelation of 830 Ma age as the minimum age of the Sirohi Group. This age has al en recorded from other areas north of this belt (Pali, Jhunjhunu and Tosha oudhary *et al.,* 1984). Incidently, it may be mentioned that this is also the belt ngsten mineralisation (rather exclusively) along the western fringe of the Arava ountain.

The younger dates recorded in the adjacent Abu granite (735 Ma, Crawford, 197 oudhary, *et al,* 1984) may be interpreted as the age of the younger thermal eve arking the phase of Malani volcanicity in the trans-Aravalli belt.

Although the lithological association does not strictly favour a rifting evolution e Sirohi basin, yet there are two features which indicate evolution of the Sirohi bas an "aulacogen" (used in the sense of Shatsky, 1964, quoted by Milanovsky, 198 rstly, there is a strict linearity of basin over a long distance; and secondly, t velopment of ductile shear zones at the contact of the sediment and the baseme cks (the Erinpura granite gneiss, *sensu lato*).

The term 'aulacogen' used by Shatsky (1964, quoted by Milanovsky, 1987), is rivation from the Greek word meaning structural zones evolved by furrow. T rrelation of Sirohi rocks with the Ras and Makrana metasediments (mainly marbl d the phyllite occuring around Degana, and Didwana and at other places) would me at the Sirohi basin evolved as a narrow linear graben similar to an aulacogen. T hological characteristics of this Sirohi-Ras-Makrana basin (showing typing absence cks of large clasts and volcanics) indicate that the rift shoulders were the terrains bdued topography. A necessary corollary of this interpretation is that the Abu blo hich now forms eastern shoulder of the inverted basin did not exist as a topograph gh at the time of evolution of the Sirohi basin.

The total absence of any mafic or felsic volcanic material in the Sirohi Group m ad to the conclusion that the faults which bounded the basin margins were probab allow features and did not penetrate deep into mantle or lower crust. In view of tl or geophysical information it is difficult to comment on the actual process of the riftii ıich led to opening of the Sirohi basin. However, some comments can be made on tl olutionary history of the Sirohi basin.

The strict parallelism of the Sirohi basin with the Delhi fold belt, and the young e of the Sirohi rocks in relation to the Delhi orogeny suggest possibility of the Siro sin as successor basin. However, as mentioned earlier, there is a total lack of lar ıstics in the Sirohi basin. This fact, therefore, militates against the suggestion th rohi basin is a typical successor basin.

The available geochronological data from Delhi fold belt indicate its diachrono story, (Bose, 1989; Volpe and Macdougall, 1990). The *ca* 1400 ± 50 Ma has been mark the closing age of the Delhi basin, and *ca* 850 Ma have been stated to be the age ıorogenic magmatism (Roy, 1988, 1990). Recent geochronological data of Volpe aı acdougall (1990), however, suggest that the anorogenic magmatism in Delhi fold be ıd initiated earlier, at around 1000 Ma before present when series of diorite bodi ɔng with gabbro and serpentinites were emplaced along the axial zone of the fold be ıe intrusions which post dated the folding and regional metamorphism of the Del ıpergroup (Roy, 1988) caused wide spread thermal metamorphism in the Delhi fo lt, more particularly in the southern region (Sharma, 1988).

In the area around the Sirohi town, a linear body of meta-gabbro occurs subparall the Sirohi basin. Deformation and the consequent retrogression of the gabbro bo ıring the closing of the Sirohi basin led to development of schistose amphibolite. Tl ırallelism of gabbro body with the Sirohi basin and the time relation of deformation aı etamorphism might lead to the correlation of the gabbro intrusion with the evolution rohi basin (Fig. 4). Presuming that the mafic plutonic activities in the Delhi fold be ıd in the Sirohi belt are synchronous, there lies a strong possibility that the opening e Sirohi basin is caused by the emplacement of mafic magma body (manifestation ıderplating) either at the crust-mantle interface or in the lower or middle crust. Th sumption satisfies one requirement of rifting (passive or active) that the crust ining which led to basin opening is an outcome of heat addition to the syste hattacharji and Koide, 1987).

Subsequent to the basin infilling and tectonic inversion accompanied mpressional deformation, a narrow linear uplifted block had formed similar to tho scribed from Siberian and North American platforms (Milanvosky, 1987) (Fig.4). T formation, metamorphism and granite emplacement features observed in the Siro roup appear typical of an 'aulacogen' undergoing inversion related deformation, *i* rust displacement, ductile shearing and related metamorphism (cf. Milanovsky, 1987

The Sirohi basin became thermally stabilized rifted lithosphere (cf. Morgan aı amberg, 1987) prior to the opening of Sindreth volcano-sedimentary basin. The Mala lcanicity as recorded at Sindreth as well as at Gurapratap Singh and Diri, Pali distri Iaheswari, 1983; Gaur, 1984; Srivastava *et al.*, 1989) is typically bimodal in characte edimentary association of the volcanics which includes conglomerate and arkose alo ith tuffs is suggestive of formation of a shallow basin.

The eastern margin of the Sindreth basin, marked by lenses of conglomerate ru most north-south. The basin clearly transects the NNE trending inverted basin of t

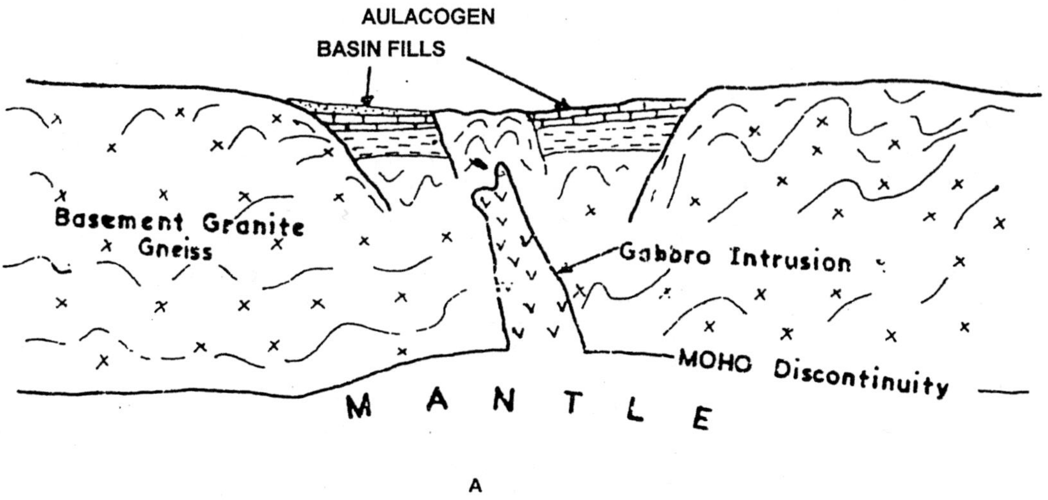

A

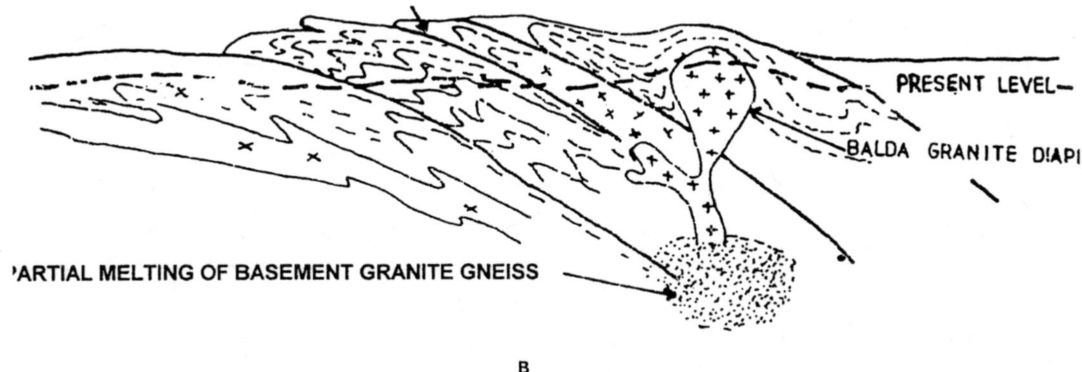

B

'ig. 4. (a) Evolution of the Sirohi basin as aulacogen subsequent to crustal thinning and intrus of gabbro body. (b) Tectonic inversion and thrust stacking of Sirohi metasediments. Part melting at depth resulted in diapiric rise of granite body (Balda granite)

;irohi Group. This, abrupt change in the structural grain from NNE to almost N-S, uggestive of rotation of the stress system which were operative at the time of Delhi a ;irohi orogenies respectively.

The evolution of dykes transverse to the Sindreth basin margin were possibly alo econdary rift fractures transverse to the main basin boundary (cf. Morgan and Rambe 987). The possible mode of Sindreth basin opening and the formation of dykes has be llustrated diagrammatically in Figure 5.

Interpretation of the available geochemical data and correlation of tectono-thern vents suggest a close link between the anorogenic mafic-ultramafic magmatism in t)elhi fold belt in the east and the opening of the Sirohi rift basins in the west. *Ca* 10

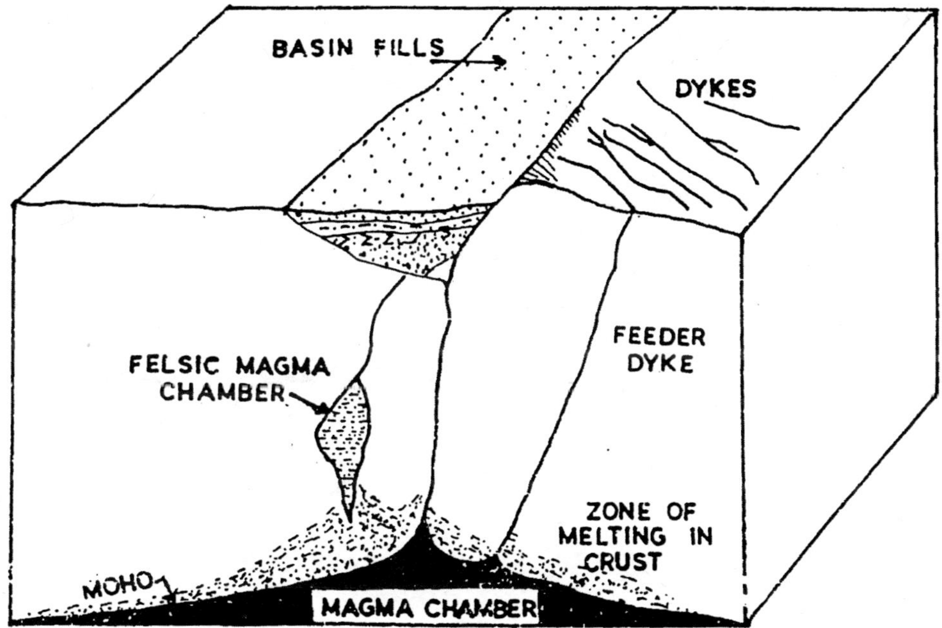

ig. 5. Evolution of the Sindreth basin due to rifting. Secondary rifts, transverse to basin, forme on the eastern shoulder which became site of dykes.

Ia age of the diorite rocks associated with mafic plutonics in the Delhi fold belt may b nterpreted as a phase related to an "unsuccessful" attempt to reopen the Delhi basin ctual basin opening took place in the west which also was active, i.e. did not lead t cean opening. The event may be linked with the rifting of the "Rodinia supercontinen McMenamin and McMenamin, 1990; Rogers *et al.,* 1995) which is believed to have bega t 1000 Ma and continued during much of the Neoproterozoic.

REFERENCES

hattacharji, S. and Koide, H. 1987. Theoritical and experimental studies of mantle upwellin penetrative magmatism crust. *Tectonophysics,* 143 : 13-23.

hushan, S.K. 1995. Late Proterozoic continetal growth : Implication from Geochemistry of ac magmatic events of west Indian craton. *Jour. Geol. Soc. India,* 34 : 339-353.

iswal, T.K. 1988. Polyphase deformation in Delhi rocks, SE of Amirgad, Banas Kantha Dist. Gujarat. *Mem. Geol. Soc. India,* 7 : 267-278.

lanford, W.T. 1877. Geological notes on the Great Indian Thar Desert between Sind ar Rajputana. *Rec. Geol. Surv. India* 10(1) : 10-21.

ose, U., 1989. Corrleation problems of the Proterozoic stratigraphy of Rajasthan. *Indio Minerals,* 43(3-4) : 183-193.

hattopadhyay, B., Mukhopadhyay, A.K., Singhai, R.K., Bhattacharjee, J. and Hore, M.K. 198 Post-Erinpura magmatism in Sirohi, Rajasthan and its bearing on tungsten mineralisatio Symp. on Metallogeny of the Precambrian IGCP-91, 115-132.

ıoudhary, A.K., Gopalan, K. and Anjaneya Sastry, C. 1984. Present status of the geochemistry the Precambrian rocks of Rajasthan. Tectonophysics, 105 : 131-140.

ulson, A.L., 1933. The geology of Sirohi State, Rajputana, *Mem. Geol. Surv. India,* 63(I) : 1-16

awford, A.R. and Compston, W., 1970. The age of the Vindhyan system of peninsular Indi *Quart. Jour. Geol. Soc. London* 125 : 351-372.

awford, A.R. 1970. The Precmbrian geochronology of Rajasthan and Bundelkhand, Northe India. *Can. Jour. Earth Sci.*, 7 : 91-110.

awford, A.R., 1975. Rb-Sr age determination for the Mount Abu granite and related rocks Gujarat. *Jour. Geol. Soc. India.* 16(1) : 20-28.

aur, C.P. 1984. Geochemistry of acid volcanics of Manihari, Pali dist., Rajasthan. Unpublish Ph.D. thesis, Raj. University.

ıpta, S.N. Arora, Y.K., Mathur, R.K. Iqbuluddin, Prasad, B. and Sahai, T.N., Sharma, S.B. 198 Lithostratigraphic map of Aravalli region, southern Rajasthan and northern Gujarat. *Ge Surv. India,* Calcuttta.

acket, C.A. 1889. Progress Report. *Rec. Geol. Surv. India,* 22 : 1-5.

eron, A.M. 1917. The geology of NE Rajputana and adjacent district. *Mem. Geol. Surv. Ind* 45(1) : 1-128.

eron, A.M. 1935. Synopsis of Pre-Vindyan Geology of Rajputana. *Trans Nat. Inst. Sci. India,* 1(: 17-33.

eron, A.M., 1953. Geology of central Rajputana. *Mem. Geol. Surv. India,* 79 : 1-339.

aheshwari, A., 1983. Geochemistry of volcanic rocks of Gurapratap Singh and Diri, Pali dis Rajasthan, *Unpublished Ph.D. thesis,* Raj. University.

cMenamin, M.A.S. and McMenamin, D.L.S., 1990. The emergence of animals : The Cambri breakthrough., Columbia Univ. Press, New York, 456 p.

ilanovsky, E.E., 1987. Rifting Evolution in geological history. *Tectonophysics,* 143 : 103-118.

organ, P. and Ramberg, I.B. 1987. Physical changes in the lithosphere associated with therm relaxation after rifting. *Tectonophysics.* 143 : 1-11.

aha, K. and Halyburton, R.V. 1977. Structural pattern and strain history of a superposed fo system in the Precambrian of central Rajasthan, India. Structural patterns in the "Ma Railo Syncline" central Rajasthan. *Precambrian Research,* 4 : 39-84.

dham, R.D. 1893. Manual of Geology of India, GSI, Calcutta, pp. 45.

liwal, B.S. 1988. Deformation pattern in the rocks of the Aravalli Supergroup around Udaip city. Rajasthan. *Mem. Geol. Soc. India* 7 : 153-168.

reek, H.S. 1981. Petrochemistry and petrogensis of the Malani igneous suite, India. *Geol. So Am. Bul.* 92(2).

umb, K.A. 1990. New Proteozoic Time scale. *Episodes* 14 : 139-140.

ja Rao, C.S., Poddar, B.C., Basu, K.K. and Dutta, A.K., 1971. Precambrian stratigraphy Rajasthan : A review. *Rec. Geol. Surv. India.* 101(2) : 60-79.

thore, S.S., 1994. Geochemistry of Malani volcanics and associated igneous rocks of southwe Rajasthan, India : Implications to crustal evolution. *Ph.D. thesis* M.S. University of Baro (Unpublished).

gers, J.J.W., Raphael, U. and Mohamed, S. 1995. Tectonic assembly of Gondwana. *Jou Geodynamics,* 19(1) : pp. 1-34.

y, A.B., Paliwal, B.S. and Goel, O.P., 1971. Superposed folding in the Aravalli rocks of the ty area around Udaipur, Rajasthan. *Jour. Geol. Soc. India,* 12 : 342-350.

›y, A.B., Nagori, D.K. Golani, P.R., Dhakar, S.P., and Choudhuri, R., 1980. Structural geomet: of rock phosphate bearing Aravalli rocks around Jhamarkotra mines area, Udaipur dis Rajasthan. *Indian Jour. Earth Sci.,* 7 : 217-231.

›y, A.B. and Das, A.R., 1985. A study of the time relations between movements, metamorpohis and granite emplacement in the Middle-Proterozoic Delhi Supergroup of Rajasthan. *Jou Geol. Soc. India,* 26 : 726-733.

›y, A.B., Sharma, B.L., Paliwal, B.S., Chauhan, N.K., Nagori, D.K., Golani, P.R., 199 Lithostratigraphy and tectonic evolution of the Aravalli Supergroup : a prtogeosynclin sequence. In : S.M. Casshyap (Ed.) *Rift Basins and Aulacogens,* Gyanodaya Prakasha Nainital, pp. 73-90.

›y, A.B. 1972. Significance of variable plunge of trend of small scale upright folds in the ty] Aravalli rocks around Udaipur, Rajasthan (Western India). *Geol. Soc. Am. Bull.* 83 1552-1556.

›y, A.B. 1988. Stratigraphic and tectonic framework of the Aravalli Mountain Range. *Mem. Ge(Soc. India.* 7 : 3-32.

›y, A.B. 1990. Evolution of the Precambrian crust of the Aravalli Mountain Range. *Developmen in Precambrian Geology*, Elsvier Pub., 8 : 327-348.

›y, A.B. 1991. Evolution of the Early Proterzoic Aravalli depositional basin. In : S.K. Tando C.C. Pant and S.M. Casshyap (Eds.) *Sedimentary Basins in India* : Tectonic Conte› Ganodaya Prakashan, Nainital, pp. 1-12.

ıarma, R.S., 1988. Patterns of metamorphism in the Precambrian rocks of the Aravalli mounta belt. *Mem. Geol. Soc. India,* 7 : 33-76.

ıarma, R.S. 1990. Metamorphic evolution of rocks from the Rajasthan craton, NW Indian shiel *Developments in Precambrian Geology,* 9 : 349-366.

ıatsky, N.S., 1964. On the Depressions of Donetz type. In : N.S. Shatsky (Ed.) Selected Work vol. 2, Nauka, Moscow, 544- 553.

ngh, S.P. 1983. Pillow lavas from Delhi Supergroup near Bombholai. Pali district, Rajastha *Jour. Geol. Soc. India*, 24 : 206-211.

ngh, S.P., 1988. Stratigraphy and sedimentation pattern in the Proterozoic Delhi Supergrou northwestern India. *Mem. Geol. Surv. India*, 7 : 193-206.

nha-Roy, S., Mohanty, M., Malhotra, G., Sharma, V.P. and Joshi, D.W., 1993. Conglomera horizons in south-central Rajasthan and their significance on Proterozoic stratigraphy aı tectonics of the Aravalli and Delhi fold belts. *Jour. Geol. Soc. India*, 41 : 331-350.

ivastava, R.K., Maheshwari, A., and Upadhyaya, R., 1989. Geochemistry of felsic volcanics fro Gurapratap Singh and Diri, Pali district, Rajasthan (Part I), Major elements. *Jour. Ge(Soc. India* 34 : 467-486.

›d, C., 1909. Travels in western Rajputana. Rajputana Gazetter, III(a) : 229 p.

›lpe, A.M. and Macdougall, J.D. 1990. Geochemistry and isotopic characteristic of mafic (Phul: Ophiolite) and related rocks in the Delhi Supergroup, Rajasthan, India : implications f rifting in the Proterozóic. *Precambrian Research*, 48 : 167-191.

'eidenbeck, M., Goswami, J.N. and Roy, A.B., 1996. Stabilization of the Aravalli craton Northwestern India at 2.5 Ga; An Ion Microprbe Zircon study. *Chemical Geology, Isoto, Geoscience,* 129 : 325-340.

eological Evolution of Northwestern India
.S. Paliwal (Ed.)
cientific Publishers (India), Jodhpur, 1999, pp. 34 - 60.

EVOLUTION AND GEOLOGICAL SET UP OF THE NAGAUR - GANGANAGAR BASIN, NORTH WESTERN RAJASTHAN

Virendra Kumar

Geological Survey of India, Western Region Jaipur-302004

ABSTRACT

The Nagaur-Ganganagar Basin lies in the western part of the Indian Shield. The Salt Range of Pakistan represents the northern part of this Basin, while the southern part occurring in India is named as Nagaur-Ganganagar Basin. The Basin came into existence by the rejuvenation of NNE SSW trending Archaean and Proterozoic lineaments. Initial rifting along these lineaments resulted in the interbasinal faults and formation of the horst and graben structures, during post-Delhi orogenic and later Malani igneous activites, between 745 ± 10 Ma to 600 ± 70 Ma. Though the sedimentation of the rocks of the Marwar Supergoup in the basin commenced under glacial conditions during Late Proterozoic time and soon after the climatic conditions became semi-arid to arid when the rest of the overlying sediments of the Jodhpur Group were deposited under near-shore marine platform conditions. The provenance of these arenaceous sediments was Malani lgneous Suite of rocks and Delhi metamorphites. The sediments of the Jodhpur Group and those of the Bhander Group (Upper Vindhyans) are similar and correlatable in time and space. The sedimentation in the Vindhyan Basin probably ceased after deposition of the Upper Bhander Sandstone (late Proterozoic) while it continued in the Nagaur-Ganganagar Basin during Eocambrian period to super saline conditions of evaporation leading to deposition of carbonate predominant Bilara Group in the southern parts and halite predominant Hanseran Evaporite Group in the northern part and a transitional sulphate predominant facies in between the two. However, during Lower Cambrian period the conditions of evaporation gradually changed and the youngest Nagaur Group of rocks, predaminantly red argillites, were deposited under shallow water in arid climatic conditions.

ey words : Nagaur-Ganganagar Basin, Geological Evolution, Marwar Supergro Rajasthan.

INTRODUCTION

The Nagaur-Ganganagar Basin (southern part of which has previously bee ›ferred to as Nagaur Basin by Khan and Sogani (1973) covers an area of about 100,0C ı. km in parts of Nagaur, Churu, Jaisalmer, Pali, Jodhpur, Bikaner, Hanumangarh an riganganagar districts of Rajasthan, parts of Bhatinda and Faridkot districts of Punja ıd parts of Sirsa district of Haryana.

The evolution of Nagaur-Ganganagar Basin is envisaged to be related with tł recambrian post-Delhi orogenic and later anorogenic activities between 745 ± 10 M.Y 1)0 ± 70 M.Y and completed with the deposition of Nagaur Group of rocks in the Ear] ambrian period.

Almost the entire area of the basin is covered by thick aeolian and/or alluvi: ›diments except for the southern part where scanty outcrops can be seen exposing tł ›ck formations disposed horizontally to sub-horizontally with a rolling dip of 5° to 1ɛ ue W and NW. Based on the information of the exposed area, data of dug well ıbewells and boreholes drilled for potash investigation (in all about 60,000 metres ‹ rilling in 72 bore holes in Rajasthan, Punjab and Haryana) a few bore holes drilled ŀ il and Natural Gas Corporation and Oil India, sub-surface stratigraphy of the basin hɛ ɜen built up, and correlated with the exposed stratigraphy. Further stratigraph ›rrelation has been envisaged across the Indo-Pak international border based c ublished data of Karampur oil well drilled on the 'foreland slope' of the Indian plate an ıe exposed geology of the Salt Range of Pakistan. Besides, correlation has also bee ttempted betweeen the Jodhpur Group of the Marwar Supergroup and the Upp‹ hander Sandstone of Vindhyan Supergroup occurring across the Aravalli mountai ınge.

GEOLOGICAL SETUP

The Nagaur-Ganganagar Basin forms the southern part of a large, elongate‹ symmetrical, intracratonic basin trending NNE SSW with slight westerly tilt. The Sa ange of Pakistan is inferred to constitute the northern extremity of the same basi /irendra Kumar, 1995 a and b; Virendra Kumar *et al.*, 1995). Dasgupta *et al.* (198‹ nd Jones (1970) are also of the similar opinion. Situated in the western part of tł ıdian Shield, it spans about 750 km from south to north and 300 km from east to wes he northern part of this basin fringing the northern edge of the Indian plate has bee rought into juxtaposition by thrusting with the Eocambrian sequence of the Salt Ranɡ f Pakistan (Fig. 1) The Precambrian rocks of the Western Indian Shield forms a stab rea marginal to the orogenic belt of Baluchistan geosyncline towards which rocks of tł ıdian Shield were thrust or overfolded (Krishnan, 1968). This Precambrian landma: ›rms the edge of the craton. Since the sediments were deposited on part of tł ›ntinental crust, the whole set-up may be described as 'foreland slope' of the risir ravalli range (Dasgupta *et al*, 1988).

The southern part of this large sedimentary basin, named Nagaur- Ganganag: asin (Fig 2), is bounded by (a) the Aravalli range in the east; (b) Delhi-Lahore/Sargodł

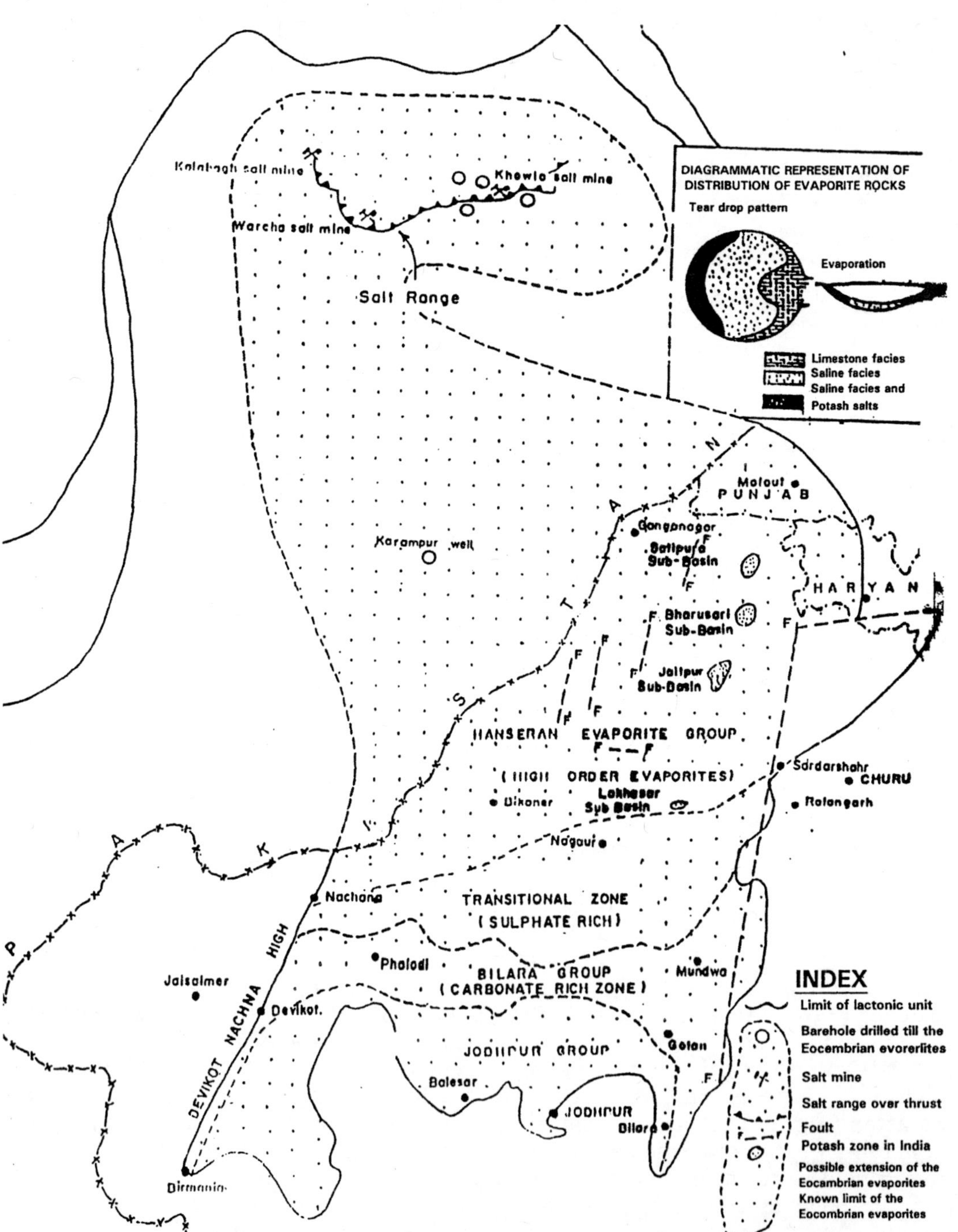

Fig. 1. Lithofacies variation map of Eocambrian Evaporite Basin through India and Pakistan

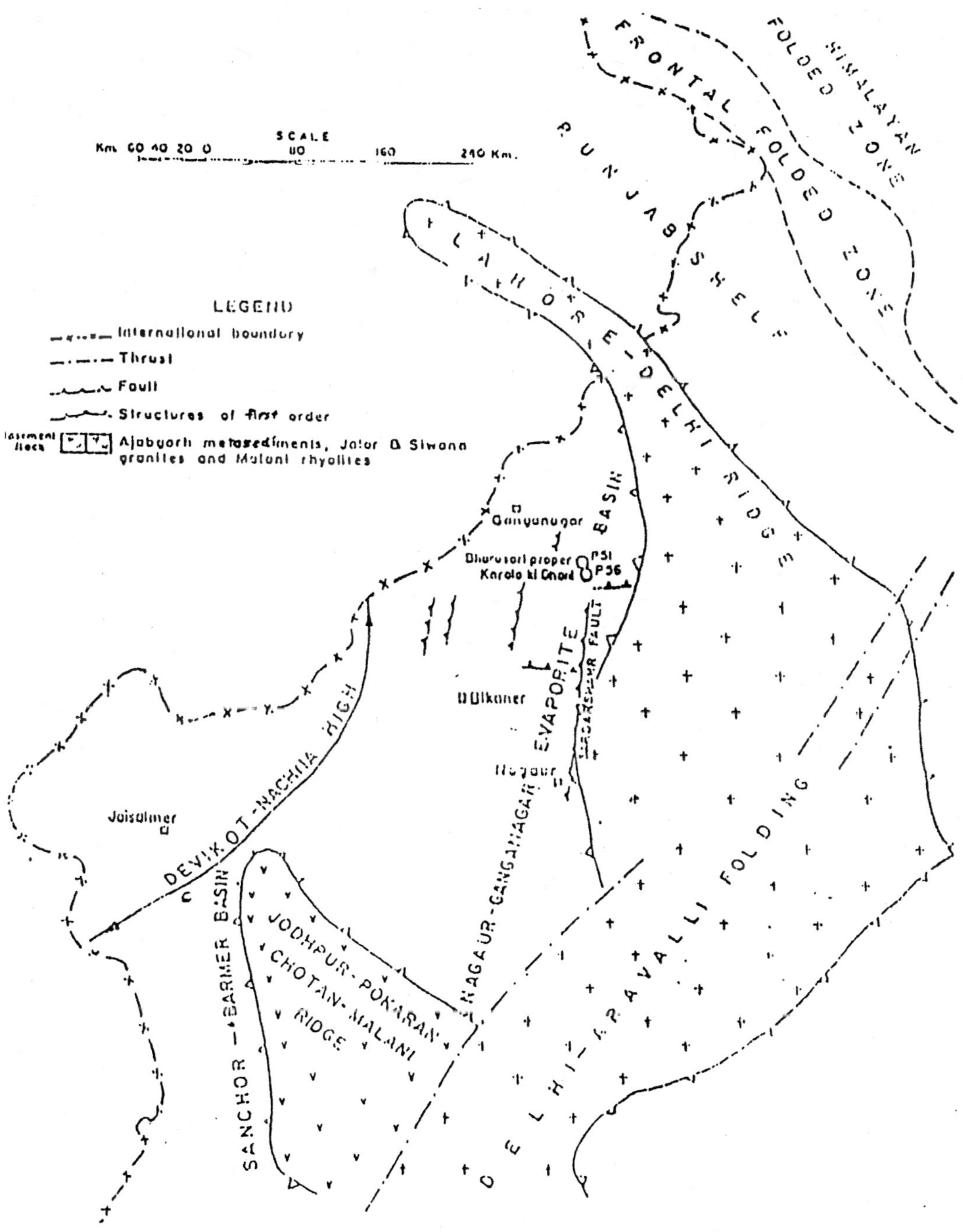

Fig. 2. Configurational setup of Nagaur-Ganganagar Evaporite Basin, Rajasthan.

ıb-surface high in the north and northwsest, (c) Devikot-Nachna subsurface high in tl ›uthwest separating it from the adjoining Jaisalmer Basin and (d) the Jodhpu okaran-Chohtan-Malani ridge in the south.

The northwestern part of the Nagaur-Ganganagar Basin continues across the Ind ak international border and appears to merge with the Indus Basin. The sub-surfa elhi-Lahore-Sargodha ridge surfaces in the form of Shahpur, Sangla and Kirana hills ıe Salt Range of Pakistan, where older metamorphites and Malani Igneous Suite ›cks are exposed.

The surface and subsurface information indicates that this basin contains a ve ıick pile of marine sediments of the Marwar Supergroup (Late Proterozoic to Ear ambrian) lying unconformably over the basement rocks of the Malani Igneous Suite aı · Delhi metamorphites. The Marwar Supergroup is unconformably overlain by ro ırmations of different ages at different places. These are Neobolus Bed (Low ambrian) in the Salt Range of Pakistan, while in Indian part Bap-Badhaura Formati ᵓermo-Carboniferous), Lathi Formation (Lower Jurassic) and Tertiary sediments but ıost of the places the Quaternary sediments directly rest over the rocks of the Marw upergroup. A generalised sequence is presented in Table 1.

BASEMENT ROCKS

The basement rocks of the Marwar Supergroup are essentially the metamorphit ; the Delhi Supergroup represented by massive quartzite, quartz veins, sericite schis ıyllite, slate and amphibolite intruded by post-Delhi quartz veins and post-Del ›lcano-plutonic rocks of the Malani Igneous Suite comprising rhyolite, tuff, rhyoli ›rphyry intruded by Jalore and Siwana granites. The effusives and intrusives a ·aversed by felsic, intermediate and mafic dykes. Some of the rhyolites of the Mala ;neous Suite have been dated at 745 ± 10 M.Y. by Crawford and Compston (1970) aı ılore-Siwana Granites at 600 ±70 M.Y. by Sharma et al. (1975).

MARWAR SUPERGROUP

The Marwar Supergroup (M.S.G.) lying unconformably above the basement rocl ›mprises (i) predominantly arenaceous Jodhpur Group and its time-equivalent Randl ormation (in Birmania area) at the base, followed with a gradational contact by (redominantly carbonates of Bilara Group (in the southern part of the basin), its tin ıuivalent lateral facies variants, Birmania Formation (in the Birmania area), Hanser vaporite Group (in rest of the Nagaur - Ganganagar basin) and Saline Series (arampur and the Salt Range areas of Pakistan), and finally (iii) the predominant ›ddish argillites of the Nagaur Group or its time equivalent Purple Sandstoı ormation (in Karampur and the Salt Range of Pakistan). However, the Nagaur Group ›cks or its time-equivalent sediments are missing in the southwestern extremity of tl asin (in Birmania area) due to deposition or probably subsequent denudation.

Table 1. A generalized stratigraphic sequence in Nagaur - Ganganagar Basin, Rajasthan, based on surface and subsurface data

Age	Supergroup/ Group	Formation	Rock Type	Thickness (m)
Recent to Pleistocene	Quaternary		Aeolian sand, alluvium, calcrete, gypcrete, kankar, clay, clacareous sandstone and siltstone	0-373
			Unconformity	
Post-Eocene		Mar	Poorly sorted ferruginous sandstone, gritty and gravely with subordinate clay and siltstone	50-110
			Unconformity	
Middle to Uper Eocene		Jogira Fuller's Earth	Shaly and marly limestone with foraminiferous white clayey marl - unfossiliferous Dirty brown impure limestone with broken shells of ostrea and foraminifera Fuller's earth with shale and shaly limestone as partings Foraminiferal limestone with a thin band of fuller's earth near the base Whitish fossiliferous marl with shale, limestone, white clayey limestone and siltstones.	116-121
			Angular Unconformity	
Upper Palaeocene to Lower Eocene		Marh	Ferruginous bands upper clay horizon Ferruginous sandstone, gritty sandstone and glass sand Middle clay horizon Ferruginous sandstone, gritty sandstone, grit, siltstone Lower clay horizon Ferruginous sandstone, gritty sandstone, grit, siltstone with leaf impressions	238-210
		Palana	Clay (light brown, grey to dirty white, sticky), limestone, greyish white to yellowish, rich in marine fossils)	40-120
			Non- Conformity	
			Shales (grey, carboniferous with lignite interbeds at places) and sandstone (fine grained, grey to yellowish and buff coloured, friable to compact)	
		Malout Conglomerate	Polymictic conglomerate and grit-calcareous, with interbedded buff coloured, calcareous sandstone. (with entrapped gas and carbonaceous flakes)	6 - 30

Age	Formation	Lithology	Thickness (m)
		—— Unconformity ——	
Lower Jurassic	Lathi	Conglomerate and arenites with subordinate argillites	70 - 280
		—— Unconformity ——	
Permo-Carboniferous	Badhaura	Coarse grained, brownish, yellowish, ferruginous sandstone, shale and siltstone with 10% to 15% lithic fragments of granite, rhyolite, biotite schist and dolerite containing permian fauna & flora.	25 - 30
	Bap Boulder	Rounded, subrounded, ellipsoidal and discoidal cobbles, pebbles of Jalore and Siwana granites, basalt, dolerite, phyllite, quartzite, gneisses, limestone, dolomites etc with a matrix of *Khaki* clays.	
		—— Unconformity ——	

		Developed in the southern part of the basin	Developed in south- western part of the basin	Developed in the central and northern parts of the basin
MAR WAR SUPER GROUP	Nagaur Group (75m to 500 m) (Early Camb- rian)	*Tunklian Formation* Grey sandstone, gritty sandstone and pebbly sandstone with subordinate brick red siltstone and claystone (3, 10-15 m)		Grey to dark micaceous and feldspathic, noncalcareous, coarse grained sandstone (3 - 20 m)
		Nagaur Formation : Brick red, purplish to brownish sandstone with greenish clayey blotches at places, siltstone shale and claystones and locally developed conglomerate (150-450 m)	Not developed	Maroon, brick red, purplish, reddish brown shale, clay, siltstone with greenish blotches and thin bads at places, and subordinate fine grained sandstone (50-150m)

—— Gradational Contact (in general) ——

				Formation	Unit	Lithology	Thickness (m)
Bilara Group (100 to 300 m)	*Pondlo Formation* Dolomite, dolomitic limestone, cherty dolomite, stromatolitic limestone and dolomite siliceous oolites and pisolites	*Birmania Formation* 7. Shale, limestone and cherty limestone	H	Lakhsar	E-16	Clay, claystone and anhydrite	1.14 - 70.3
		5. Chert, dolomite limestone and minor sandstone	A	Sataiyan	E-15	Halite (H_7) polyhalite clay and anhydrite	8.12 to 84.25

		sandstone and shales				dolemite	36.50
		4. Phosphorite and phosphatic limestone	S	Harsinghpura	E-13	Halite (H_6), polyhalite anhydrite/domomite	5.10 to 21.75
		3. Calcareous and pure quartzitic sandstone	E		E-12	clay with anhydrite and/or dolomite	2.20 to 88.45
		2. Calcareous and ferruginous sandstone	R	Kupli	E-11	Halite (H_5), polyhalite clay anhydrite/ dolomite	5.30 to 54.60
		1. Cherty limestone	A		E-10	Dolomite, dolomitec anhydrite and clays	1.45 to 93.95
(Eo-cam brian)	*Gotan Formation* Stromatolitic limestone dolomite, dolomitic limestone, chert and cherty dolomite.		N	Malkisar	E-9	Halite (H_4) polyhalite clays and anhydrite	1.35 to 86.60
			E		E-8	Anhydrite, dolomite clays	1.70 to 87.83
			V	Chattargarh	E-7	Halite (H_3) polyhalite clay and anhydrite	3.50 to 85.58
			A		E-6	Dolomite, anhydrite and clay	2.52 to 17.17
			P		E-5	Halite (H_2) polyhalite silvinite silvite langbeinite carsallite (Jr) with clay anhydrite and dolomits	15.80 to 138.95
			O				
	Dhanapa Formation Stromatolitic limestone dolomite, dolomitic limestone, chert and cherty colomits		R		E-4	Clay, dolomite and anhydrite	2.94 to 32.58
			I		E-3	Halite (H_1) poly hallite, silvinite/ silvite, Tr. of carsallite, clay, anhydrite and dolomite	16.11 to 60.45
			T				
			E		E-2	Anhydrite, dolomite and clays	0.00 to 8.83
			G	Lakhasar	E-1	Clay and anhydrite	1.40 to 260.11

			O U P	E-0	Clay, marl, lime stone, gypsum, anhydrite and dolomite with chert	0.00 to 90.00
	Gradational contact ——	Unconformable contact	——————		Gradational contact	
Jodhpur Group (Late Proterozoic)	Gribhakar Frmation Brick red to brownish siltstone, shale and sandstone, current bedded pisolitic with pellets				Reddish brown or buff and grey sandstone with purple to greenish shale, dolomitic limes tone with bands of chert and Jasper. Drilled maximum upto 423 m base not touched	
	Sonia Formation Siltstone shale, sandstone with sedimentary structures and salt pseudomorphs banded chert, jasper - dolomite	*Randha Formation* Sandstone, subordinate calcareous rocks and maroon shale				
	Pokaran Boulder Bed Boulders					
		——— Unconformity ———				
Precambrian	Malani Igneous Suite and Delhi Supergroup	Rhyolites, granites etc. and Metamorphites				

odhpur Group

The Pokaran Boulder Bed, Sonia Formation and Girbhakar Formation constituti ie Jodhpur Group lie unconformably over the eroded surface of the Malani Ignec iite and/or Delhi metamorphites.

The basal Pokaran Boulder Bed crops out north of Pokaran and extends southwar Ujlan, Barat-ka-Gaon, Uchpadra and finally to Hariyoli Nadi over a strike length of n (Bhushan, 1977). The Boulder Bed (3 m thick) consists of boulders, pebbles a bbles of Malani rhyolite, granite, rhyolite porphyry, syenite, quartzite etc. embedded ddish to brownish sandy or clayey matrix. It exists only in the southwestern fringe ie basin and has not been reported elsewhere. The assorted assemblage of glaciat oulders, cobbles and pebbles of different rock formations ranging from Precambrian arwar Supergroup suggests its glacial/fluvioglacial origin, and its stratigraphic positi lates it to Late Proterozoic glaciation period. The Palaeocurrent direction recorded ie overlying rocks of Sonia Formation indicates westerly provenance. The Mala asement high exists mearly 10 kms west of this boulder bed.

The Sonia Formation commences with a maroon shale with mega-ripples a eenish grey lenses of silty material and is overlain by creamish, buff and reddish, fi medium grained sandstone. The beds of this formation have exposed thickness of 12 id 5 m in the type sections of Garsuria and Sonia Hills, where they are dispos orizontally with rolling dips. Convolute laminations, salt pseudomorphs and pisoli ave been recorded from the rocks of this formation (Khan, 1973).

The contact with the overlying Girbhakar Formation is marked by a short break dimentation indicated by the presence of pebbles of Sonia Sandstone and che nstituting the basal pebbly horizon of the Girbhakar Sandstone.

The Girbhakar Sandstone comprises reddish to brownish, coarse to gritty micaceo indstone alternating with gritty and pebbly layers. Bands of reddish and light greeni iale usually alternate with the sandstone. It exhibits current bedding and nodul odies upto 5 cm in diameter (Pareek, 1984). The sandstones are characterized by t esence of primary sedimentary structures like ripple marks, cross-bedding, flute cas oove casts and rib - and - furrow structures.

The palaeocurrent direction of the Jodhpur Group of rocks as revealed from t dimentary structure is northwesterly to westerly, thereby indicating the provenance ie source rocks occurring towards southeast and east. It is further confirmed by study eavy minerals. A varied source for the sediments of the Jodhpur Group is indicated ie heavy minerals consisting of zircon, tourmaline, rutile, anatase, andalusite a ebeckite (Khan and Sogani, 1973). Zircon and riebeckite are inferred to be derived fro ie Malani Igneous Suite, tourmaline from pegmatites, andalusite and anatase fro etamorphites.

andha Formation

In the southwestern extremity of the basin (in Birmania area) Randha Formati ie time equivalent of Girbhakar Formation of Jodhpur Group is deposit nconformably above the Malani Igneous Suite of rocks. In this part of the basin eith ie older Sonia Formation time equivalent sediments were not deposited at all or e ave been erroded away. The Randha Formation comprises thinly bedded quartzi indstone with gritty layers, white to purple coloured quartzitic sandstone and maro iale.

HANSERAN EVAPORITE GROUP AND ITS TIME EQUIVALENTS

The rocks of Jodhpur Group are conformably overlain by the sediments of Hanserа 'aporite Group and its time equivalents lateral facies variant Bilara Group in most e Indian part of the basin (Nagaur-Ganagangar Basin), but in the southwestern part is basin (in Birmania area) the Birmania Formation time equivalent of Pond rmation (the youngest formation of Bilara Group) lie unconformably above the Randh rmation (time equivalent of Girbhakar, the youngest formation of Jodhpur Group).

However, the Saline Series of Pakistan in Karampur area and Salt Range occ conformably directly above the Precambrian metamorphites and/or Malani Igneo ite of rocks. All these time equivalent sediments are overlain conformably by th edominantly red argillites grouped under Nagaur Group in Indian part and Purp ndstone in Pakistan part of this large sedimentary basin, of course, in Birmania are W extremity of the basin) Nagaur Group of rocks are not present either due to nc position or due to subsequent denudation. All over the basin, these time equivale diments of the Marwar Supergroup occurring below the Nagaur Group or its tin uivalent Purple Sandstone are example of lateral facies variation caused by differe ysico - chemical environments of deposition.

The evaporite sequence of the Marwar Supergroup is an example of lateral faci riation caused by different physico-chemical environments of deposition. In th uthern part of the basin these sediments are represented by typical carbonate faci ouped under the Bilara Group. In the northern parts, upto Salt Range across th do-Pak international border, high order evaporites grouped under Hanseran Evapori oup and are represented by cyclic deposits of halite (most dominant constituen ntaining potash minerals (localised at places) alternating with clay, marl, limeston lomite and magnesite etc. These are grouped under Saline Series in the Pakistan pa between these two, a sulphate dominant (intermediate order of evaporites) zo mprising a sequence of marl, limestone, dolomite, gypsum and anhydrite exists in th agaur - Ganganagar basin and is referred to as transitional zone (Fig. 1). However, e southwestern extremity of the basin, in Birmania area, the carbonates (limestone lomite) host phosphorite also. Thus, the sedimentation during Eocambrian period to ace under shallow marine conditions in saline to supersaline conditions of evaporatio elding evaporites ranging from carbonate facies in the southern parts throug edominantly sulphate facies in the transitional zone to chlorides and potash beari cks in the northern parts (Fig. 1).

uthern part

In the southern part of the basin, the rocks of the Jodhpur Group are conformab erlain by the Bilara Group which comprises three dominantly calcareous ro rmations. The stratigraphic sequence of the group as worked out by Khan and Soga 973) is summarised in Table 1.

The outcrops of the Bilara Group are scanty and its presence has been confirm om the dug well data in the southern part of the basin. It occurs at Rol Qazia undwa, Dara, Bhakrod, Markori etc. The Bilara Group of rocks exhibit extensi velopment of biohermal, non-phosphatic stromatolite - bearing limestone/dolomite at mber of localities in a large tract extending from near Sojat in the south to Nagaur he north and Phalodi in the west. The beds are horizontally to subhorizontally dispos vith occasional higher dips due to slumping. It has maximum thickness of 300 m and inches out in the west.

The most common forms of stromatolites include *Collenia Pseudocolumnai Maslove. Colleniella* Karolyuk, *Concollenia* Maslov, *Croyotozoon Occidentalc* Dawsc *rregularia* Korolyuk and *Stratifera* Korolyuk with occasional *Oncolites* Pia and wi pread cryptoalgal laminates (Barman, 1980), *Weedia, Collenia, Cryptozoon* and alg odules (Khilnani, 1966).

Barman (1980 and 1987) found the stromatolites of the Bilara Group similar hose of the Salt Range, Pakistan and explained conspicuous difference between t tromatolites of the Bilara Group (Marwar Supergroup) and that of the main Vindhyan

Transitional Area

The stratigraphic sequence as observed in the transitioinal zone (between t arbonate dominant Bilaras in the south and the evaporites in the north) around Naga Badwasi area, where famous Nagaur Gypsum deposits are located, is summarised Table 2 (modified after Roy Chowdhury *et al.* 1965).

Table 2.

		Lithology	Age
		Desert sand at places with gypsite, also soil and kankar	Recent to
		Boulders, cobbles and pebbles of quartzite and chert.	Sub-Recen
MARWAR SUPERGROUP	Kuchera Khajwara series	Ferruginous grits, purple and grey, soft, porous sandstones with lenses, pockets, and lenticles of sandy gritty clays, sand, shale sequence with lignite occurrences.	Tertiary
		Unconformity	
	Nagaur Group	Interbedded red sandstone, sandy shale, siltstone	Lower Cambrian
		Red compact sandstone, medium to coarse grained, often gritty or pebbly	
		Pale, pink or red to dark red, compact, hard generally medium to fine grained sandstones, often interbedded with siltstones and shales; also soft, porous sandstone	
		Red and mottled siltstones, clay-stones and clays with interbedded silty sandstone.	
	Hanseran Evaporite Group	Greenish white to reddish vuggy non-gypseous limestones gypseous claystones, clays and siltstones with main gypsum bed and other subsidiary beds and bands of gypsum; also bands of greyish dolomite and sandstone	Eocambria
		Grey to dark gypseous dolomites, limestones and dolomitic limestones with interbedded claystones, clays gypsum and occasionally anhydrite	
		Unconformity	
		White to dark grey dolomites and grey to dark grey limestones and dolomitic limestones often cherty.	

›uthwestern Extremity (Birmania Area)

In the southwestern extremity of the basin the phosphate bearing and carbonɛ ›minant Eocambrian sequence is grouped under Birmania Formation lyi ıconformably above the Randha Sandstone Formation (time equivalent lithofac: ıriant of Girbhakar Formation of the Jodhpur Group). The Birmania Formati mprises a sequence of splintery needle shales (at the base), yellow cherty limestoı ey cherty limestone, dolomitic limestone, phosphorite, quartz arenite a lcareous-ferruginous sandstone (at the top). Except for the concentration ıosphorite, the Birmania Formation is similar to the Pondlo Formation of the Bilɛ roup (Table 1).

orthern Part

The Hanseran Evaporite Group subcropping in the northern part of the basin l: nformably above the Jodhpur Group with gradational contact as revealed by ›reholes drilled by Geological Survey of India, Pugal oil well by Oil and Natural G ›rporation, and Baghewala and Tabriwala oil wells by Oil India Ltd. The Hanser ⁄aporite Group comprises rhythmic cyclic deposits of halite containing potash minerɛ t localised places), alternating with clay, dolomite, anhydrite, magnesite, occasiona auberite, glauconite, rarely vanthofite etc. Potash minerals within the halite inclu ›lyhalite, sylvinite, sylvite, langbeinite and carnallite in decreasing order of abundan› ıe dolomite at places contain nonphosphatic stromatolites which include collen llleniella and stratifera forms of stromatolites (Ramchandra *et al.*, 1989). The thickne the Hanseran Evaporite Group ranges from 103.20 m to 650.15 m containi .mulative thickness of halite zones upto 488.50 m. A maximum of seven halite cycl ımbered H_1, H_2, ---, H_7 (from bottom to top) have been identified. The basin deepe wards the centre and in the northwest direction due to steep NNE - SSW trendi avity faults with their down thrown sides successively towards west. Thickness of t ansera Evaporite Group as well as the number of halite cycles increases from t argins of the basin towards its centre. The Hansera Evaporite Group has been divid to eight formations and 17 lithounits (Table 1) on the basis of lithological characte ıd order of superposition.

The Hanseran Evaporite Group present in the northern part is considered tiı ıuivalent lateral facies variant of the Bilara Group present in the southern part of t ısin on the following consideration (Virendra Kumar and Saha, 1985).

) Both, the Hanseran Evaporite Group and the Bilara Group occupy the saı stratigraphic position i.e., both being underlain by the Jodhpur Group and overlɛ by the Nagaur Group, the upper and the lower contacts being gradational, almc everywhere.

) The carbonate rocks (limestone, dolomite) present in both, the Bilara Group and t Hanseran Evaporite Group have yielded similar types of non- phospha stromatolites.

) The gradual variation in the rock assemblage of the sediments at the saı stratigraphic level within the basin is attributed to a gradual change physicochemical environment of deposition within the evaporite basin from the sou to the north. The distribution of the rock assemblage indicates that there would ha been a gradual change in the environment of deposition from the saline supersaline stage of evaporation from south to north during the time of th deposition.

The rocks of the Hanseran Evaporite Group were deposited under shallow ıoderately deep marine conditions under arid climatic conditions of evaporation ıdicated by the presence of algal stromatolites not only in the carbonates of the Bila roup but also within the dolomites of the Hanseran Evaporite Group. The rocks anseran Evaporite Group invariably associated with reddish, brownish and purpli ays and silty clays indicating oxidising shallow water conditions of their depositi nder arid climate.

It will not be out of place to mention here that the analogous evaporite sequen ouped under Saline Series continues further north through Karmapur area (320 k W of Nagaur area, India and 90 km ESE of Multan, Pakistan) upto the Salt Ran ormation in Pakistan.

The Karampur oil well drilled in the northwestern slopes of the Indian platform, akistan, has intersected evaporite sequence, 880 m thick, composed of rock salt wi olomite and anhydrite bands and clay interbeds resting unconformably directly on t ıetamorphic basement. This sequence is overlain by Purple Sandstone (160 r olomite sandstone (120 m) and clay with salt pseudomorphs (150 m) as reported harkov, (1984). At Karampur indications of 10 m zone with 12% K_2O have been notic 'aillard, 1992).

On the northern extremity of this large sedimentary basin 'Saline Series' exposed ıe Salt Range of Pakistan is correlated to the Hanseran Evaporite Group throu arampur Oil well section. The following succession (Krishnan, 1968) is present in t alt Range of Pakistan.

alt pseudomorph shales (ıpto 105 m)	Red to purple and greenish silty shales with ca or pseudomorph of salt crystals showing beddi planes, contain some gypsum.
[agnesian Sandstone (ıpto 75 m)	Well bedded, cream coloured, dolomitic sandston sandy dolomites and subordinate shales.
'eobolus Beds (upto 45 m)	Fossiliferous grey shales, sandy shales, sandston which may be micaceous, dolomitic and glauconit Pebbly bed at the base characterized by Neobol and other brachiopods as well as some trilobites.
urple Sandstone (75 - 140 m)	Fine grained, pink, purple and maroon sandston Lower part shales and flagstones.
aline Series (upto 450 m)	(c) Upper gypsum-dolomite with oil shale (a decomposed diabase, Khewra trap). (b) Pink, red and purple salt marl with beds of rc salt. (a) Lower gypsum-dolomite with Oil Shales.

In the Salt Range of Pakistan, Jones (1970) recorded that the lower part of tl aline Series consists of a basal gypsum layer with alternating salt marls and rock sa eds are well stratified, varying from a few metres to several hundred metres ıickness. There are at least ten major potash zones in the upper 75 m salt horiz erlying the beds of red impure salt.

From the above description it is clear that the evaporite sequence containing pota ılts exists in the entire basin from Nagaur - Ganganagar basinal part (named Hanser vaporite Group) in India through Saline Series in Karampur area, to the Salt Range akistan and is perfectly correlatable on the basis of their lithological sequence, order ıperposition and also with respect to the overlying rock sequence of Nagaur Group (tin ıuivalent Purple Sandstone in the Karampur area and the Salt Range of Pakistar any earlier workers like Swaminath *et al*, (1959), Pareek (1981) Wensink (1975) ha so correlated this sequence with the sequence developed at Salt Range, Pakistan.

agaur Group

In the entire evaporite basin extending from Nagaur-Ganganagar Basin in India cross the Indo-Pak international border through Karampur area upto Salt Range akistan the conditions of evaporation gradually changed to very shallow mari nditions and climate from arid to semi-arid, during the deposition of the Nagaur Grou nd its analogous Purple Sandstone.

In the southern part of the basin the rocks of the Nagaur Group are exposed, the e characterized by red, mottled red, purplish to brick-red clay-stone, siltstone ar ıbordinate sandstone with bands and blotches of greenish clay in the basal porti ouped under Nagaur Formation. It is overlain by Tunklian Formation comprising bri d claystone, siltstone, calcareous clay and gritty to pebbly sandstone which contai ebbles of chert, dolomite, quartz, quartzite, Malani rhyolite and granite.

In most of the exposed and sub-surface sections the contact between the underlyi ilara Group or Hansera Evaporite Group and the overlying Nagaur Group adational, except over a small exposed area along the Phalodi-Bap road section and amsar Pond near Phalodi where a nonconformity represented by a conglomerate exis etween the rocks of Nagaur Group. This conglomerate the Khichan conglomera easuring upto 5 m comprises of a wide assorted cobbles, pebbles, fragments and piec dolomite, chert, cherty dolomite, feldspar, quartz and sandstone cemented by pinki brownish sandy to calcareous matrix, (Pareek, 1984). In Pakistan area also the conta etween the Purple Sandstone (analogous to Nagaur Group) and the underlying Sali eries (analogous to the Hansera Evaporite Group) is gradational in nature.

The exposed Nagaur Group of rocks extend northwards from Merta Road undwa and Nagaur, and then takes an arcuate pattern, continuing westward up halodi and Pokaran. Data collected from the boreholes drilled for exploring gypsu gnite, groundwater, potash and oil by various agencies in this basin revealed that tl cks of Nagaur Group are present all over the Nagaur-Ganganagar Basin lying north e exposed Nagaurs. The arcuate nature of the contact with the underlying Bilara ansera Evaporite Group rocks indicates the receding shore line of the Nagau anganagar Basin in the southern part. In the north, across Indo-Pak border arampur-Salt Range area in Pakistan, Purple Sandstone (analogous to Nagaur Grou sandwitched between the underlying Saline Series and the overlying sequence eobolus Bed, Magensian Sandstone and Salt Pseudomorph Sequence of the Cambri ge.

On the basis of order of superposition, presence of stromatolites in Evapori ›quence of slightly advanced stage than those occurring in the Vindhyans, Eocambriɛ ;e has been assigned to the Evaporite Sequence sandwitched between the underlyiı ıte Proterozoic Jodhpur Group of rocks and overlying Lower Cambrian Nagaur Groı rocks which are analogous to the Purple Sandstone of the Salt Range of Pakistɛ 'irendra Kumar, 1995 a and b). Later on, Eocambrian age has been assigned to th 'aporite sequence by Peters *et al.* (1995) on the basis of their studies of age diagnost ıd source dependent biomarkers from the oil of Baghewala-well drilled in this basin l ıl India.

EVOLUTION OF NAGAUR - GANGANAGAR BASIN

The study, synthesis and evaluation of the basic geological data related agaur-Ganganagar Basin leads to conclude the following salient points about tl ılaeogeographic evolution of the Nagaur-Ganganagar Basin.

) The evidences of tectonic and thermal events in NW Rajasthan indicate that tl sedimentary basins, in general, and Nagaur - Ganganagar Basin, in particular, we developed by the rejuventation of NNE - SSW trending lineaments (e.g. tl Sardarshahar fault, Devikot - Nachna subsurface high etc.) of Archaean aı Proterozoic age. Evidences of plate tectonics indicate that at the time of developme of the Nagaur - Ganganagar basin India was situated in the Gondwana Land close Antarctica, Australia and Africa near the south pole.

) Initially, rejuvenation of NNE-SSW trending faults of older Precambrians took pla resulting in rift and graben structures. Large scale igneous activity represented l Malani Igneous Suite of rocks along these weak zones followed from 745 ± 10 Ma 600 ± 70 Ma (Crawford and Compston, 1970 and Sharma *et al.*, 1975).

The vulcanicity took place in three phases to complete the igneous cycle.

(i) The extrusive phase comprising tuff, welded tuff, rhyolite and rhyolite porphy with mafic bands.

(ii) The intrusive granite of the Jalore and Siwana type, and

(iii) The dykes of felsic, intermediate and mafic compositions. The extrusive pha was the most prolonged phase. The Malani outpourings were largely fissu controlled related to deep seated tectonic structures trending NNE-SS\ Outcrops of the Malani Igneous Suite of rocks and the subsurface Devik Nachana high (consisting mainly of Malani Igenous Suite of rocks) a oriented along these NNE-SSW trending lineament patterns. Some of tl rhyolites of Malani Igenous Suite have been dated 745 ± 10 Ma by Crowfo and Compston (1970) while Jalore Siwna Granites have been dated 600 ± ' Ma by Sharma *et al.* (1975). Hence, this period may be considered as the mc important period in the development of the Nagaur-Ganganagar Basin.

) The last cycle of igenous activity in Rajasthan marked by Malani Igenous Suite rocks range in age from 600 ± 70 to 745 ± 10 m.y. As the basement rocks of tl Marwar Supergroup are Malanis, the lower age limit of the Marwar Supergroı cannot be older than 600±70 m.y. It is also well established from field evidences th the Malani igneous activity was confined to the west of Aravalli Mountain Range. has not been reported to occur east of this range. Hence, it is inferred that during tl period of Malani igneous activity in this region, rocks of the Vindhyan Supergrou were deposited in the southern and eastern parts of Rajasthan. The deposition Vindhyans probably continued for some time even after completion of this last cyc of Precambrian igneous activity in Rajasthan. Concomitant with the deposition of tl Upper Bhander Sandstone in the southeastern part of the Rajasthan the rocks Jodhpur Group of the Marwar Supergroup were deposited in the Nagaur-Ganganag Basin. This view finds support from the correlation between the rocks of Jodhpu Group and the Upper Bhander Sandstone on the basis of similar colour, litholog nature of sediments, environment of deposition, almost comparable thickness sediments, structurally underformed and unmetamorphosed nature of the rocks. It also supported by the westward shifting hypothesis of geosyncline in this area in tl geological past. Thus, in the Vindhyan Basin sedimentation appeared to have cease after deposition of the Upper Bhander Sandstone during, Late Proterozoic perio while it continued in the Nagaur-Ganganagar Basin and further north upto Sa Range in Pakistan in the form of carbonates and evaporites. Based on the availab evidences, Late Proterozoic age has been assigned for the rocks of Jodhpur Group (c Paliwal, 1994). Thus, the lower age limit of the Marwar Supergroup is fixed at La Proterozoic.

) The sedimentation in the trans-Aravalli sedimentary basins commenced in tl Nagaur-Ganganagar Basin, while the adjoining parts of Jaisalmer and Barm basins remained the highland portion as depicted in palaeogeographic map (Fig. 3).

) In Nagaur-Ganganagar Basin, the sedimentation of Marwar Supergroup w probably initiated during the Late Proterozoic glaciation age and is represented l the Pokaran Boulder Bed at the base of Jodhpur Group. After deposition of th glacial/fluvioglacial boulder bed the climate of the region became semi- arid to arid evidenced by the overlying sediments. These sediments of the Sonia Formation of tl Jodhpur Group are predominantly yellowish brown, reddish brown to purplish ar arenaceous. Sedimentary structure like current bedding, ripple marks, convolu bedding, salt pseudomorphs and flute casts have also been recorded. Study of hea minerals and palaeocurrent of these sediments indicate that Delhi metamorphit and Malani Igneous Suite of rocks occurring to the south, southeast and east of tl Nagaur - Ganganagar Basin form the provenance. During the early period deposition of the sediments of the Jodhpur Group (i.e., Sonia Formation) the bas remained tectonically more stable than in the later part which is represented by tl older formation of the Jodhpur Group (i.e. Girbhakar Formation). This type tectonic behaviour of the basin is indicated by a short break in sedimentation aft deposition of Sonia Formation and is represented by a pebbly horizon (comprisi pebbles of Sonia Sandstone and chert) at the basal part of the overlying Girbhak Formation. During the deposition of Girbhakar Formation the basin remain tectonically unstable, as reflected in the nature of its sediments. The Girbhak Formation comprises reddish coloured, coarse to gritty micaceous sandsto alternating with gritty and pebbly layers. The bands of red and light green shal usually alternate with the sandstone. The tectonic instability and sedimentary bre mentioned above, are more pronouncedly reflected in the southwestern extremity the basin in Birmania area, where only Randha Formation is present and its tin equivalent sediments of the Girbhakar Formation are missing altogether. Instead a angular unconformity exists between the Randha Formation and the overlyi Birmania Formation. It implies that either the time equivalents of the Girbhak sediments were not deposited in the Birmania part of the basin or else they mig have deposited and were eroded before deposition of the Birmania Formation. T Jodhpur Group of rocks appear to have not deposited in deeper parts of the basi probably lying west of Dulmera high (as no borehole of potash investigation or Pug well of Oil and Natural Gas Corporation could even intersect the entire Evapori Sequence occurring stratigraphically above the Jodhpur Group). Also the Karamp oil well in Pakistan (located about 320 km NW of Nagaur) occurring further west a towards the central part of this large basin (of which Salt Range of Pakistan forn the northern part and Nagaur - Ganganagar basin forms the southern part), h interesected Evaporite Sequence (analogous to the Hanseran Evaporite Grou occurring unconformably above the older metamorphites.

) In the Nagaur - Ganganagar Basin the sediments indicate gradual change conditions of deposition from shallow marine to evaporitic conditions during the e of deposition of the rocks of Jodhpur Group. Such restricted basinal conditio prevailed all over the basin including the part occurring in the Salt Range Pakistan.

Due to physiogrphical conditions of the Nagaur-Ganganagar Basin, comprising t uthern and eastern marginal parts of the large asymmetrical basin whose long ax ends NNE - SSW, the evaporite sequence in the basin exhibits typical lateral lithofaci ariation. By the time evaporation conditions were set in the basin the sea had regresse om SSE to NNE. The distribution of various lithofacies of the evaporite sequenc ntrolled by the gradually regressing sea from SSW towards NNE and from ea wards west (i.e. towards the central part of the basin) is very well found in the Nagau anganagar Basin (Figs. 3,4,5 and 6). The low order evaporite facies were developed e southern part of the basin (Figs. 2 and 4) and are represented mainly by t rbonates (limestones and dolomites) associated with chert and stromatolites of Bila roup of rocks. This carbonate predominant facies gradually changes northwards redominantly sulphate facies represented by gypsum/anhydrite with associated che mestone and marl. It is in this lithofacies where famous Nagaur Gypsum deposit cated. This sulphate facies (moderate order saline evaporites) acting as transition ne between the low order evaporites (occurring in the south) and the high ord aporites occurring in the north and northwestern parts (i.e. towards the central par tends all along the eastern and northeastern margins of the basin where the low ord aporite facies (i.e. carbonate predominant) is not well represented as indicated by t tersections of potash in the boreholes. In Nagaur-Ganganagar Basin the Hanser vaporite Group comprises cyclic deposits of halite (most dominant constituen ntaining potash minerals, alternating in general with clay, dolomite, anhydrit agnesite etc. The potash minerals include polyhalite, sylvite/ sylvinite, langbeinite a aces of carnallite in decreasing order of abundance. However, in Saline Series of t alt Range in Pakistan a few more additional potash salts have been reported to occur at time equivalent Evaporite Sequence, Seven halite cycles are present agaur-Ganganagar Basin. Each cycle starts with low order evaporites (dolomite a nhydrite associated with argillites) and end up with a halite zone at the top. Thus, ea alite zone is underlain and overlain by an intervening zone comprising dolomite and/ nhydrite and/or associated clay with or without magnesite and other evaporite salt

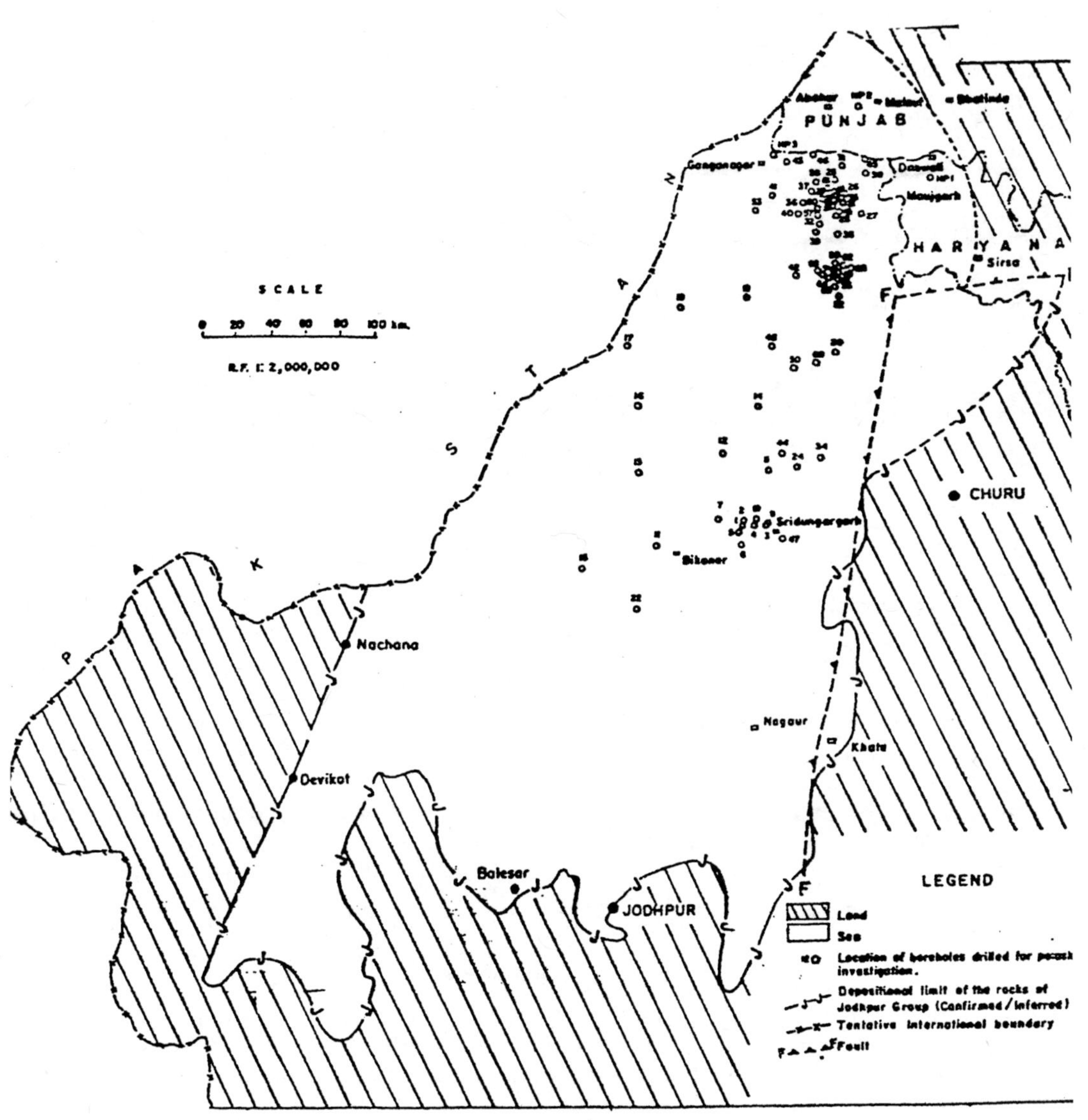

ig. 3. Palaeographic map of Late Proterozoic period marked by the start of Jodhpur Group northern and north-western parts of Rajasthan

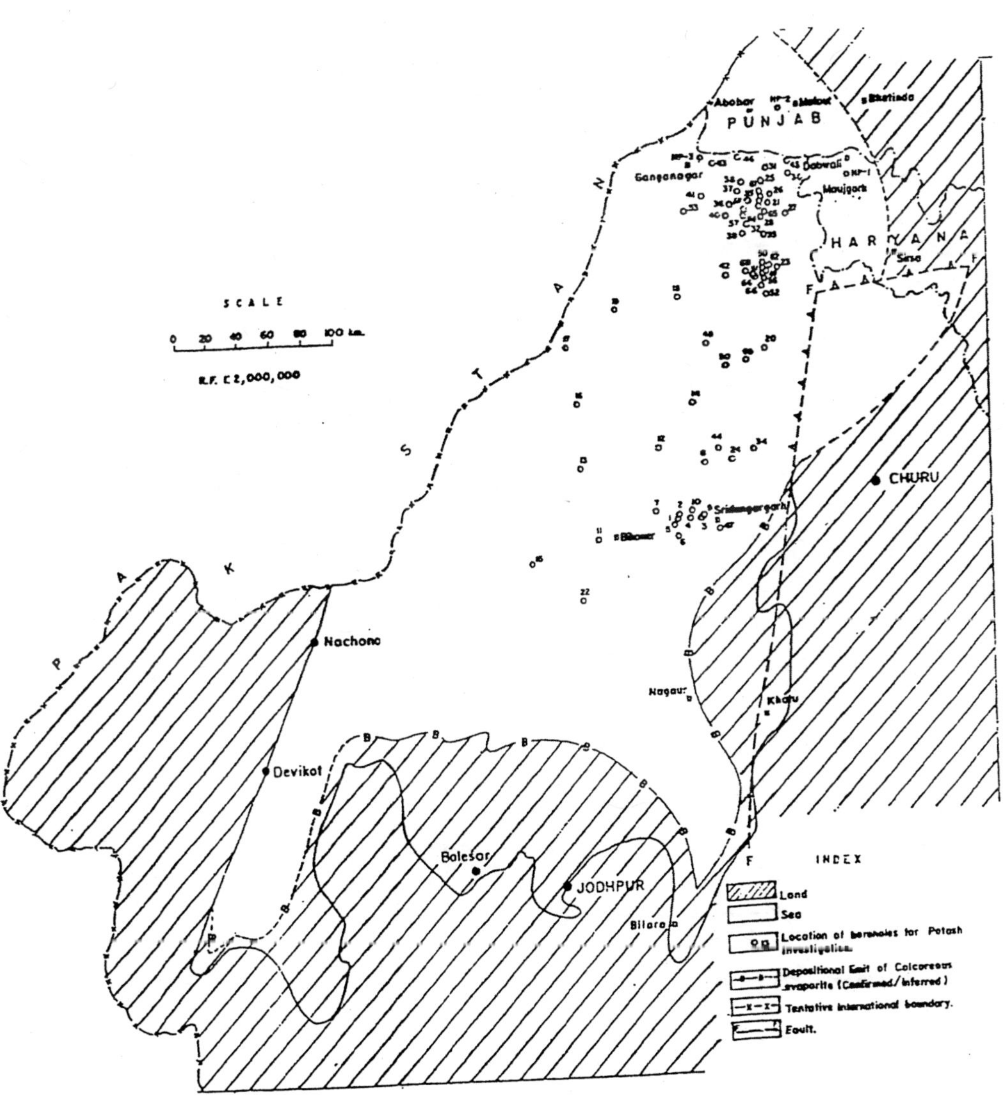

g. 4. Palaeogeograhic map of the Early Eocambrian period marked by the commencement of tl
Evaporite Sequnce in northern and north-western parts of Rajasthan

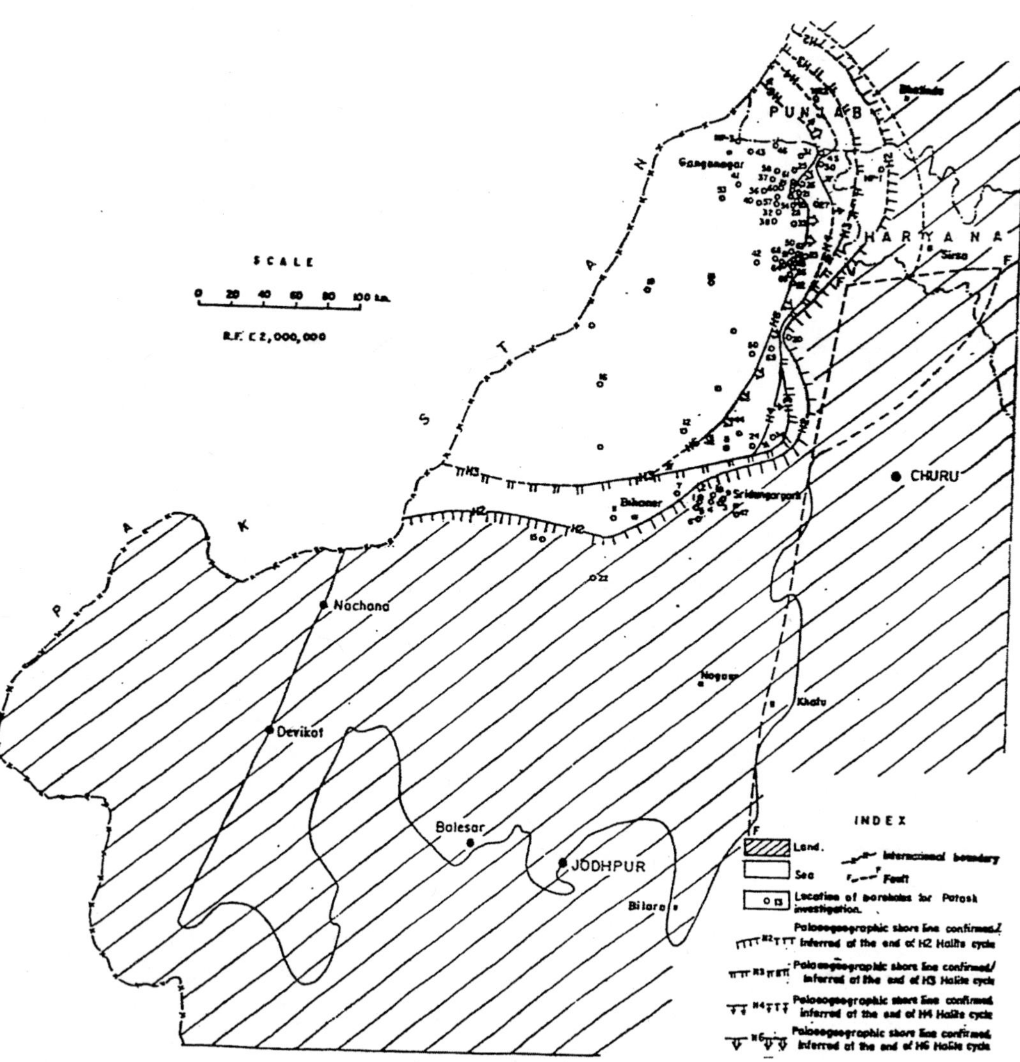

g. 5. Palaeogeographic map during the Eocambrian period (Marked by the end of H_2, H_3, H_4 a H_6 halite cyles) in the northern and north-western parts of Rajasthan.

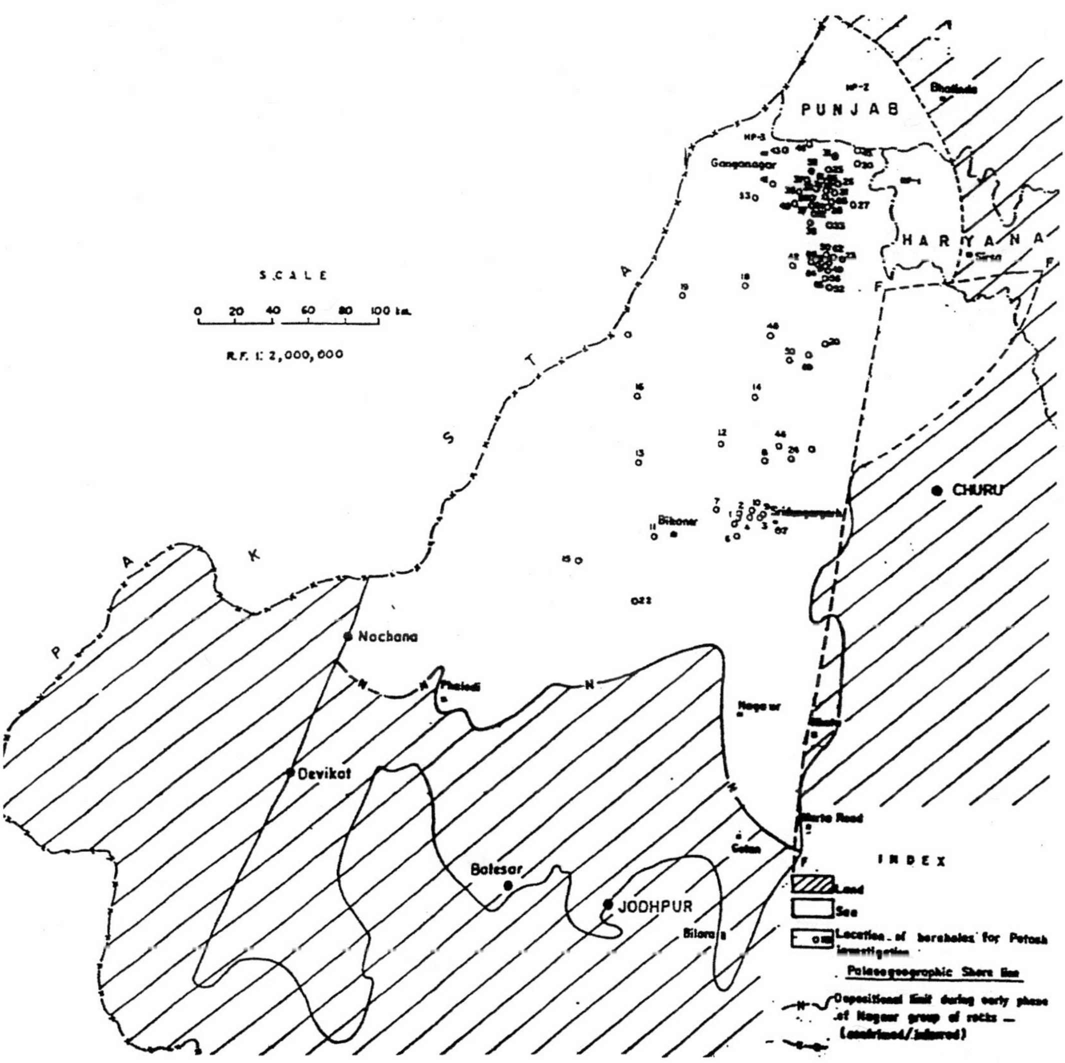

Fig. 6. Palaeogeographic map of the Early Cambrian (marked by the end of the Evaporite Sequence) in the northern and north-western parts of Rajasthan.

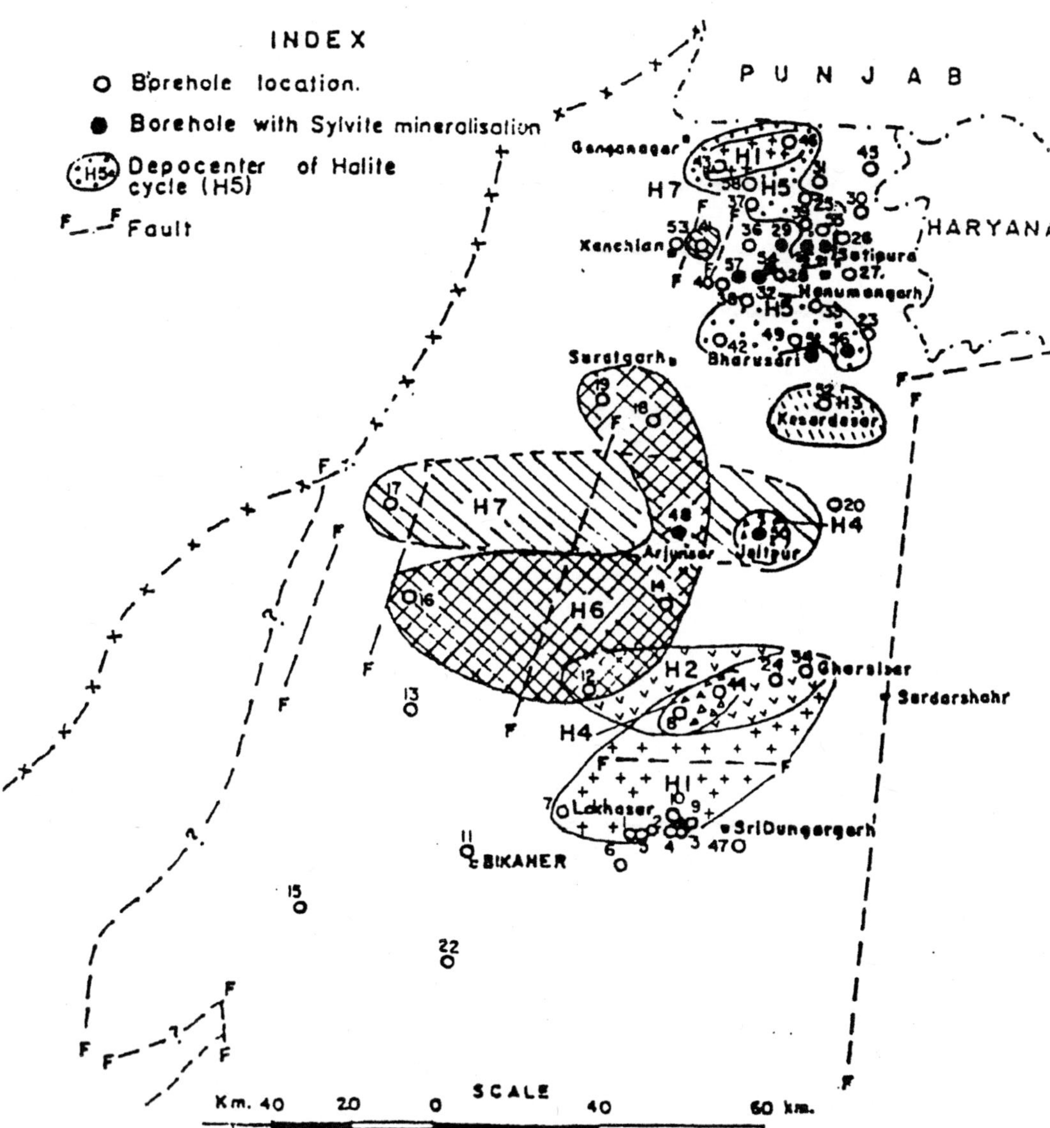

Fig. 7. Depocenters of various halite cycles in Nagaur-Ganganagar Basin, Rajasthan.

)uring the evaporite phase, the basin remained tectonically unstable as indicated by ›resence of several steeply dipping interfomational and intraformational faults (inferr rom subsurface data), (ii) frequent influxes of sea water as reflected by presence ıumber of halite cycles and sub-cycles in the evaporite sequence, (iii) fluctuations in t ›romine content is also indicative of sea water influxes, (iv) shifting of depocentres fferent halite cycles in space and time (Fig. 7) and above all (v) the presence of a hu ıickness of (652 m in Indian part and 880 m at Karampur oil well in Pakistа raporite sequence compels to invoke syndepositional sinking basin model : so as ;commodate this much huge thickness of evaporite sequence which has characteris atures of the deposits of shallow marine origin.

Based on the bromine content in the core samples seven halite cycles identified a grouped into three mega cycles viz, (i) the lower mega cycle encompassing the old vo halite cycles, namely H_1 and H_2 halite cycles, with Br-content from 20 ppm to 5 om; (ii) the middle mega cycle, including H_3, H_4, H_5 and H_6 halite cycles, with t r-content from 10 ppm to 780 ppm; and (iii) the uppermost mega cycle represented $_7$ halite cycle in which Br-content ranges from traces to 465 ppm. It indicates thı ajor tectonic disturbances-probably related to basinal faulting causing the basin eepen and major influxes of sea water. Further, the fluctuations in the bromine val id the presence of intervening low order evaporites between the seven halite cycl dicate that the three major tectonic events occurred in number of slow and sh npulses.

Mineral assemblage of the evaporite sequence and the variations in the bromi intent in the evaporite sequence indicate that evaporation at localised places/sub-basi as reached to the stage of desiccation leading to the development of sylvite, carnall c. (chloride minerals) in an otherwise sulphate rich system. This leads us to conclu ıat at certain localities closed or nearly closed basinal conditions prevailed in the bas hich allowed evaporation to the stage of desiccation.

) In Nagaur - Ganganagar Basin the condition for formation of evaporites gradua ceased due to further regression of sea toward NNE after deposition of H_7 halite cy pertaining to the third and last mega cycle (Fig. 6). However, when the H_7 hal cycle was being terminated in the Indian part of the basin (i.e. Nagaur - Ganganag Basin), in the Salt Range area of Pakistan basic igneous activity took place, indicated by the Khewra trap. In the Indian part of the basinal area (i.e., Nagau Ganganagar Basin) H_7 halite cycle seems to have been terminated by the end Eocambrian time due to onset of extremely arid climatic conditions and regression sea towards further NNE culminating the deposition of Hanseran Evaporite Grou This was followed by deposition of the rocks of the overlying Nagaur Group und shallow water condition in arid climate. In the Salt Range Formation of Pakistan, t Purple Sandstone lying above the Saline Series is correlatable with the rocks of t Nagaur Group. On the basis of the fossil contents of the overlying Neobolus Bed Salt Range Formation, Lower Cambrian age has been assigned to the underlyi Purple Sandstone (Ravindra Kumar, 1992). Since the Purple Sandstones a considered time equivalents of the rocks of Nagaur Group also.

ı) Culmination of Hanseran Evaporite Group in the Nagaur-Ganganagar Basin, locally marked by a period of nondeposition, erosion and weathering and represented by conglomerate bed in Phalodi-Bap area. This five metre thick Khich conglomerate bed forming basal part of the Nagaur Group contains an assort assemblage of cobbles, pebbles, fragments and pieces, including grains of dolomi feldspar and sandstone, cemented by pinkish to brownish sandy to calcareous matr Thin conglomerate bed unconformably overlies the Pondlo Dolomite of the Bila Group and is overlain by red mottled claystone, siltstone and fine grained friab sandstone.

The general reddish, purplish and brownish colours of the predominant ·gillaceous with subordinate arenaceous nature of the sediments of the older Nagaı ›rmation of the Nagaur Group along with the sedimentary structures (ripple mark ırrent bedding, cracks etc.) preserved in them indicate that these sediments we: ·posited in arid climatic environment under shallow water conditions. However, tl ·erlying Tunklian Formation developed in the marginal parts of the basin w: ·posited under tectonically disturbed conditions. It comprises reddish, brick r(ıystone, siltstone and calcareous clay and gritty to pebbly sandstone, which contaiı ·bbles of chert, dolomite, quartz, quartzite, Malani rhyolite and granite. The nature .e sediments (predominantly grey to buff feldspathic coarse sandstone, grit and pebble Tunklian Formation, further indicates that by the end of their deposition somewh: ımid climatic conditions prevailed in the basin.

) Till the development and filling up of the Nagaur-Ganganagar Basin by tl sediments of the Marwar Supergroup, the geographical position of this land porti(in the Gondwana Land was located close to Antarctica, Australia and Africa near tl south pole.

0) After compaction of the rocks of the Marwar Supergroup, the area was uplifted aı extensive erosion took place during Cambrian to Lower Carboniferous times. An i(cap covered the Aravalli, during the Upper Carboniferous. The glaciers moved fro the Aravalli towards north, northwest and west and carried with them the bouldeı cobbles and pebbles from the Aravallis, Delhis, Malanis and Marwars and deposit(them in the foot hill region extending from Pokaran-Phalodi area to the Upper Indı Basin through northwestern parts of Rajasthan. Bap Boulder Bed forming the glaci and fluvioglacial deposits were formed in Pokaran-Phalodi area and extending Tabariwala-Baghewala area in Rajasthan and to Talchir Boulder Bed in Pakista The Talchir Boulder Bed dominantly comprises boulders of Malani Igneous Suite rocks through out its stretch. However, in the Nagaur-Ganganagar Basin tl boulders, cobbles and pebbles are subrounded to ellipsoidal in shape exhibitiı glacial striation on their surface. These include rocks of Malani Igneous Suit pre-Delhi rocks and Delhi and Marwar Supergroups in decreasing order abundance. Pareek (1984) has correlated this Boulder Bed with that of the Umar area of Madhya Pradesh.

ACKNOWLEDGEMENT

The author wishes to record his sincere thanks to Dr. S. Sinha Roy, Deputy Direct ɘneral, G.S.I., W.R. and Shri U. Bose, Director-in-Charge, Operation Rajasthan f .eir encouragement and to Shri P.C. Bakliwal, Director for critical discussions aı -operation for finalising this paper.

REFERENCES

ıushan, S.K. 1977. A note on the stratigraphic position of Pokaran Boulder Bed. *Ind. Min.* 3 43-45.

ırman, G. 1980. An analysis of the Marwar Basin, Western Rajasthan in the light Stromatolites, Characteristics and Utility. *Geol. Surv. Ind. Misc. Publ.* 44 : 292- 297.

ırman, G. 1987. Stratigraphic position of Marwar Supergroup in the light of stromatoli studies. *Spl. Publ. Geol. Surv. Ind.* 11(1) : 72-80.

awford, A.R. and Compston, W. 1970. The age of the Vindhyan system of Peninsular Indi *Quart. Jour. Geol. Soc. London,* 125: 351-372.

ısgupta, S.K. 1975. A revision of Mesozoic-Tertiary stratigraphy of the Jaisalmer basi Rajasthan. *Ind. Jour. Earth. Soc.* 12 : 77-94.

ısgupta, S.P. Ramchandra, Jairam, M.S. and Kumar, V. 1984. The halite deposits of Bikane Ganganagar evaporite basin, Rajasthan. *Seminar on Salt industry in India*, Aug. 11,1 1984, Jaipur.

ısgupta, S.P., Ramchandra, Virendra Kumar and Jairam, M.S. 1988. A framework analysis Nagaur-Ganganagar evaporite basin, Rajasthan. *Ind. Min.* 42 : 57-64.

illard, Luc. 1992. *Potash in Rajasthan*. Appraisal of explorations status (India), GSI-MDPA.

nes, C.L. 1970. Potash in halitic evaporite salts, western Pakistan, *Geological Survay Researc U.S.G.S., Prof. Paper* 700(D) : 140-145.

ıan, E.A. 1973. Discovery of fossil brachiopods of Cambrian affinity from Trans-Araval Vindhyan sequence, near Jodhpur, Rajasthan. *Seminar on Recent Advances in the Geology Rajasthan and Gujarat*, Abst. 74-75.

ıan, E.A. and Sogani, P.C. 1973. Stratigraphy and sedimentation of Trans-Aravalli Vindhya basin, Western Rajasthan. *Seminar on Recent Advances in Geology of Rajasthan an Gujarat.* Abst. 103.

ıilnani, B.V. 1966. Algal stromatolitic structures of limestone of Trans-Aravalli Vindhyan Basi Western Rajasthan. *Proc. 53rd Ind. Sci. Cong. Asso.* Chandigarh Pt. III, Abst. Sec. V. 190.

ishnan, M.S. 1968. *Geology of India and Burma*, Vth Ed. Shankar Printers, Madras.

ıliwal, B.S. 1994. Tectonics of the post-Aravalli mountain building activity and its bearing on tl accumulation of sediments along the western flank of the Aravalli Range, Rajasthan Indi In Ahmed, R and Sheikh, A.M. (Eds.) *Geology in South Asia - II*. Proceedings of GEOSAS- February 23-27, 1992, Hydrocarbon Development Institute of Pakistan, Islamaba Pakistan. 52-60.

ıreek, H.S. 1981. Basin configuration and sedimentary stratigraphy of Western Rajasthan. *Jou Geol. Soc. Ind.* 22(11) : 517-527.

ıreek, H.S. 1984. Pre-Quaternary Geology and mineral resources of northwestern Rajastha *Geol. Surv. Ind. Mem.* 115 : 1-95.

ters, K.E., Clark, M.E., Dasgupta, U., Mc Caffeey, M.A. and Lee, C.Y. 1995. Recognition of Infra Cambrian source rock band on boikarbers in the Baghewala-1 *Oil, India AAP Bulletion,* 79(10) : 1481-1494.

ımchandra, Jairam, M.S., Virendra Kumar and Dasgupta, S.P. 1989. Occurrences stromatolites in Sri Ganganagar-Nagaur Evaporite Basin, Rajasthan *Rec. Geol. Surv. In* 115(7 & 8) : 49-52.

ıvindra Kumar. 1986. *Fundamentals of Historical Geology and Stratigraphy of India*, Wil Eastern Ltd., New Delhi.

ıy, Chowdhury, M.K., Sogani, P.C. Chandak, G.J. and Mehra, S.L. 1965. *Nagaur Gypsum. Bu Geol. Surv. India Sr. A, Econ. Geol.* 24 p.

ıarma, K.K., Nagapal, M.K., and Jhingran, A.G. 1975. Fission track geochronology of Ear Precambrian, Rajasthan, India. *Chayanica Geologica,* 1(2) : 196-211.

vaminath, J., Krishanamurthy, J.G., Verma, K.K. and Chandak, G.J. 1959. General geology the Jaisalmer area, Rajasthan, Western India. *Proc. Symp. Develop. Petrol. Resource As and the Far East Mineral Resources Development Series*, Bangkok, 10:154-155.

rendra Kumar and Saha, C.R. 1989. Stratigraphy of Marwar Supergroup with special referen to Hanseran Evaporite Group in Nagaur Ganganagar evaporite basin, Rajasthan. Abst. *Xi Indian Colloquim on Micropalaeontology and Stratigraphy, Lucknow,* Jan. 23-25, 1989.

rendra Kumar 1995a. Eocambrian sedimentation in Nagaur-Ganganagar Evaporite Basi Rajasthan. Abst. Vol. XI Convention, *Ind. Assoc. Sed. Roorkee*, Jan. 27-29, 1995 : 42-43.

rendra Kumar 1995b. Marwar Super Group - The missing stratigraphic link in the Weste parts of Indian Shield. Abst. Vol. *National Seminar on Recent Researches in Geology Western India*. Oct. 29-31, p. 37.

rendra Kumar, Misra, A.K. and Bohra, S.K. 1995. Prospective locales of primary gypsu deposits in Rajasthan. Abst. Vol. *National Seminar on Earth Resources, Industri Development and Environmental Issues*. March 20-22, 1995, Jaipur.

ensink, H. 1975. The structural history of the Indian-Pakistan Subcontinent during tl Phanerozoic. *"Progress in Geodynamic"*, *Royal Netherlands Academy of Arts and Scienc Amsterdam*, 193-207.

ıarkov, M.A. 1984. *Palaeozoic Salt Bearing Formation Of The World.* Springer verlag, Berlin.

əological Evolution of Northwestern India
S. Paliwal (Ed.)
:ientific Publishers (India), Jodhpur, 1999, pp. 61 - 68.

Rb-Sr AGE OF PERALKALINE GRANITES AND ASSOCIATED PERALKALINE VOLCANICS OF BARMER DISTRICT, RAJASTHAN

S.S. Rathore and T.R. Venkatesan*

Geochemistry Division, KDMIPE, ONGC, Dehradun 248 195
**Physical Research Laboratory, Navrangpura, Ahmedabad 380 009*

ABSTRACT

A new Rb-Sr age of 698 ± 10 Ma has been obtained for peralkaline granites belonging to the Malani Igneous Province of Rajasthan. The initial $^{87}Sr/^{86}Sr$ ratio (0.7067 ± 0.0041) of these granites indicates a crustal source of their magma. These granites and the associated peralkaline volcanics are coeval and cogenetic. Present study suggests that the two types of granites, i.e. peraluminous (Jalore type) and peralkaline (Siwana type) granites, are not contemporary and represent two different magmatic activities at a short interval.

.ey words : Peralkaline Granites, Volcanics, Rb-Sr Age, Barmer, Rajasthan.

INTRODUCTION

The peralkaline granites, commonly referred to as Siwana type granites, form pa ' the Malani Igneous Province (*MIP*) of Rajasthan (Chandrasekaran and Srivastav ɔ92). The Malani Igneous Province, covering an area of about 50,000 sq. km in westeı ıd southern Rajasthan to the west of the Aravalli mountain range, marks the younge roterozoic tectonomagmatic event in the western part of the Indian shield.

The province consists dominantly of felsic volcanics and two types of granites — ıbsolvus hornblende-biotite bearing Jalore type granite and a hypersolvus peralkaliı iwana type grnaite. The volcanics overlie the Proterozoic metasediments of uncerta ʒe (pre-Delhi or Delhi Supergroup) and underlie the Trans - Aravalli Vindhyaı rivastava, 1988). Besides the felsic volcanics and the granites the Malani Igneoı 'ovince also contains minor amount of basic, alkaline and intermediate alkaline rocks.

It is generally assumed that the Jalore and Siwana type granites represent tl .rent stock of the Malani Rhyolites and all the different igneous rocks of the provin long to a single Proterozoic event dated at 745 ± 10 Ma by Crawford and Compst(970). The view has been supported in the recent years by Pareek (1981), Kochh: 984) and Bhusan (1985, 1989). However, the detailed geochemical studies carried o ɔm different parts of the province by Srivastava (1988), Srivastava et al. (1989a, b) aı ochronological studies by Rathore (1995) and Rathore *et al.* (1996) have revealed th fferent rocks of the Malani province represent polyphase igneous activity. Bhusan aı ngupta (1979) have also reported four phases of Malani activity from Barmer distri(ıe first phase is represented by rhyolites and porphyry flows, while the second phase coarse grained biotite granite intruding into the first phase. The third phase includ th extrusive and intrusive rocks of acid and intermediate composition and the la ıase is of intrusive basic rocks. Recently Chandrasekaran and Srivastava (1992) ha so reported occurrence of polyphase igneous activity in Malani Igneous Province, bas multivariate statistical analysis.

The most important rock types of Malani Igneous Province are the granites whi(e mainly exposed around Jalore and Siwana areas of the province. The Jalore Grani peraluminous, while the Siwana Granite is peralkaline in nature. Kochhar (1984) aı ivastava et al. (1988) have considered these granites as anorogenic or A-type granite ter detailed field work in the area, Bhusan (1991) has identified fifteen plutons anite/granitoids in the Malani Igneous Province covering cumulatively an area of aboı 00 sq. km. Out of these, six ring dykes and a pluton, covering an area of about 1100 s n. are composed of riebeckite/ aegirine granite (Siwana Granite) and seven pluton vering a total area of about 7600 sq. km. are of biotite granite (Jalore Granit ırther, he has also identified two small plutons of hornblende granite, covering an ar about 70 sq. km and named it as Malani Granite.

The Jalore Granite is conspicuous by its pink colour and ranges in composition fro gmatitic granite to granodiorite. Discontinuous outcrops of the Siwana Granite occı an elliptical pattern (ring shaped) around Siwana, covering an area of about 290 s n. The peralkaline ashes, tuffs and volcanics (pantellerites and comendites etc.) a so associated with these granites (Fig. 1).

The peralkaline magmatism has attracted attention of the earth scientists wor er for a variety of reasons. The most important amongst them are their tecton gnificance and mode of occurrence. The peralkaline magmatism can occur during pr ting (e.g. epeirogenic doming), initial rifting (development of linear fractures aı ginning of crustal attenuation), continued rifting (extensive crustal attenuation leaviı tle or no sialic crust on the rift floor) and frequently displayed as sub-volcanic riı ructures (Bowden, 1974). Each event appears to be characterized by particular suit ıd varying proportions of peralkaline siliceous and associated rocks. Normall ralkaline volcanics are also associated with the peralkaline granites.

Comendites frequently occur in sub-volcanic ring complexes associated wi n-peralkaline volcanics. Comenditic associations are typical products of epeirogen ming exemplified by the elevated domes of vitreous to granophyric peralkaline aı

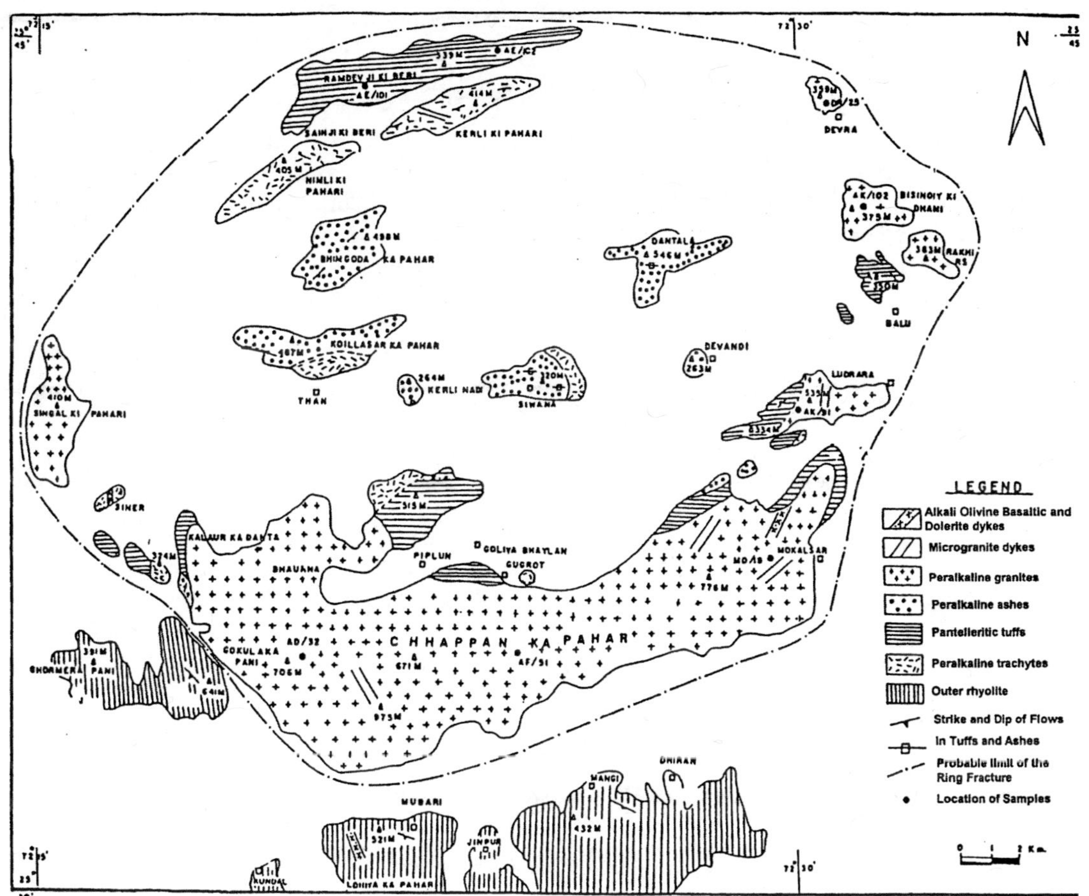

Fig. 1. Geological map of the Siwana area.

lkaline lavas in Tibesti (Vincent, 1970). In rift valleys, comendites are more close nked with peralkaline trachytes and pantellerites. However, the proportion omendite becomes diminished as the rift evolves and in well developed rift systems (e. thiopia), where crustal attenuation is more extensive, comendites are virtually abse Mohr. 1970).

Pantellerites are related to the development of rifting. The Ethiopian rift syste onsisting of the main Ethiopian rift and the Danakil depression provides good examp f pantelleritic volcanism and its relationship with the extent of rifting (Tazieff *et a* 969; Barberi *et al.,* 1970).

The intrusive nature of Siwana Granite was first indicated by La Touche (1902) an oulson (1933). Subsequently, various theories have been put forward for the ri tructure and emplacement of the peralkaline granites. Mukherjee (1958) suggested oubly pluging syncline for a subsidence structure. Pascoe (1950) speculated that t rcular plan at Siwana has a connection with the roots of a volcano. Murthy (1962) f he first time interpreted this as a feature of cauldron subsidence, having a similarity ıe Nigerian ring complex. Narayan Das et al. (1978) related the alkali magmatism 'ans-Aravalli region to crustal dislocations. Bhusan (1984) also suggested that t lkaline granites of Siwana and Barmer have been emplaced as ring dykes due ıuldron subsidence. Kochhar (1984) attributed this structure to an intra-cratonic h ɔot activity. Bhusan and Mohanty (1988) suggested that the Siwana granite h ıtruded along a collapse structure as a low-angle cone intrusion. Further, th ıvisaged that the peralkaline granites and associated peralkaline volcanics at Siwa ıdicate the zone of active rifting in a distinct late Proterozoic crustal regime. Srivasta .988) correlated sub-volcanic ring complexes to a period of crustal upwarping before t ɔmmencement of Gondwana rifting. We present in this paper a new emplacement a ır the peralkaline granites and associated peralkaline volcanics around Siwana armer district.

EXPERIMENTAL

Rb and Sr concentrations and their isotopic composition were determined usi otope dilution technique at the Keshava Deva Malaviya Institute of Petroleu xploration (KDMIPE), Dehradun, on VG 354 thermal ionization mass spectromet rovided with a computer control for data acquisition and reduction. About 100 - 150 r ; fresh whole rock samples, powdered to 200-230 mesh size, were treated with HF NO_3 + $HClO_4$ mixtures overnight in pressure digestion bomb at 150°C. Aft ɾaporation of the acid mixtures, the residues were dissolved in 6N HCl and dried befo nally making the solution in 2.5N HCl. The samples were spiked with high purity ^{87}] ıd ^{84}Sr spikes.

The errors in the mass spectrometric determinations of ^{87}Rb and ^{86}Sr are estimat ı be within ± 0.5% leading to a random error of not more than ± 1% for their ratios. ^{87}] ıd ^{86}Sr concentrations were calculated by isotope dilution technique. The $^{87}Sr/^{86}$ ıtios were corrected for mass fractionation assuming $^{86}Sr/^{88}Sr$ = 0.1194 in the samp he analysis of the NBS 987 standard, made during the course of this study gave a me: ılue of 0.710219 ± 0.000058 (2σ):

RESULTS AND DISCUSSION

Six whole rock samples of peralkaline granites were analyzed for Rb - Sr studi(ut of these three samples, viz. A/D/32, A/F/51 and MO/19, were collected from ma iwana granite pluton (Chappan-Ka-Pahar) while the remaining three samples, i /K/91, A/K/102 and DR/25, were collected from isolated hills near villages Ludraı ishnoiy-Ki-Dhani and Dewra, respectively on the north-eastern extremity of the Siwaı ng structure (Fig.1). The analytical data are given in Table 1. The sample show go ıdiogenic enrichment and mutual spread in their Rb/Sr ratios (4.9 to 29.4). In t ɔnventional Sr evolution diagram (Fig. 2), the six samples show a good line ɘlationship. This line interpreted as an isochron corresponding to an age of 698 ± 10 N ith an initial ^{87}Sr ^{86}Sr ratio of 0.7067 ± 0.0041. MSWD (0.85) of the fit indicates th ıe samples have remained in a closed system after their formation and meet t ochron conditions very closely.

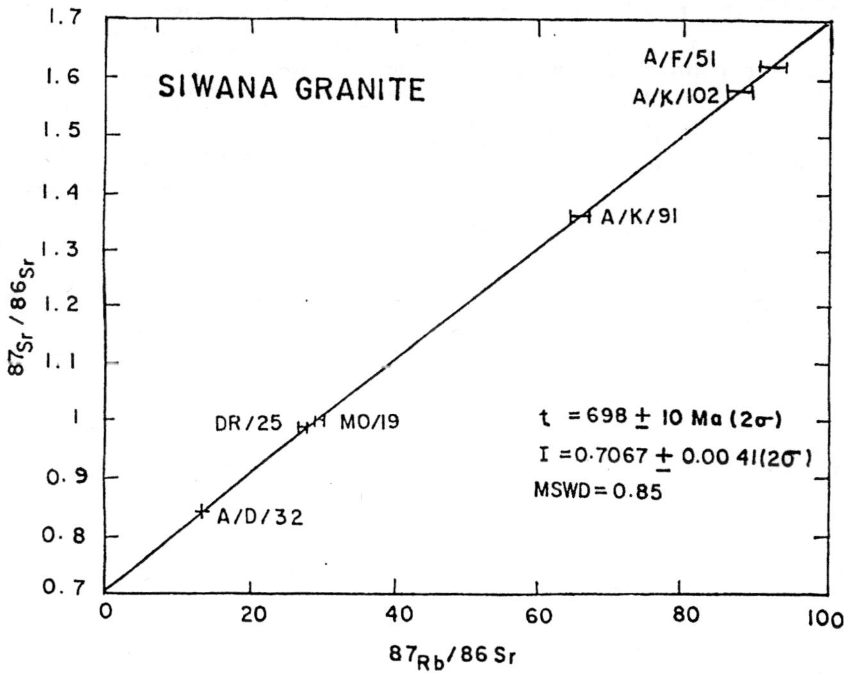

Fig. 2. Conventional Sr evolution diagram of Siwana Granite

'able 1. Rb-Sr isotopic data of peralkaline granites nad peralkaline volcanics of Barm district, Rajasthan

ample No.	^{87}Rb (ppm)	^{86}Sr (ppm)	^{87}Rb ^{86}Sr+ (atomic)	^{87}Sr / ^{86}Sr ± 2 σ
\/D/32	44.21	3.05	13.68	0.8411 ± 0.00021
)R/25	83.05	2.98	27.55	0.9903 ± 0.00140
10/19	46.67	1.55	29.76	1.0010 ± 0.002
\/E/101*	102.19	1.60	63.19	1.3390 ± 0.00072
\/K/91	111.40	1.67	65.81	1.3605 ± 0.00004
\/K/102	71.57	0.80	88.10	1.5799 ± 0.00066
\/E/102*	145.46	1.58	91.11	1.5859 ± 0.00059
\/F/51	129.11	1.38	92.48	1.6182 ± 0.00074

The age of 698 ± 10 Ma, interpreted as an emplacement time of peralkali ranites, is definitely lower than the age of 727 ± 8 Ma obtained for the Jalore Granite athore (1995). Present study, therefore, suggests that the two types of granites, i eraluminous (Jalore type) and peralkaline (Siwana type) granites, are not contempora s hitherto believed and instead represent two different magmatic activities at a sh nterval. The initial Sr ratio (0.706 ± 0.0041) of the Siwana Granite indicates a crus ource of their magma (Faure and Hurley, 1963).

The peralkaline volcanics i.e., soda-trachytes and pantellerites also occur i ;sociation with the peralkaline granites (Fig. 1). To understand their temporal an :netic relationship with the peralkaline granites, two pantellerites (i.e. A/E/101 an 'E/102) collected from northern extremity of the Siwana ring structure were analyze r Rb-Sr studies. The analytical data are included in Table 1. When these samples a gressed with the Siwana granites, the pooled isochron defines an age of 693 ± 8 M ith an initial Sr ratio of 0.7079 ± 0.0038 (Fig. 3). The MSWD of the fit is excellent a 92. This age and initial ratio are not different from the age and initial ratio of th anites alone, thus, suggesting contemporaneity and cogenetic nature of thes :ralkaline volcanics with that of the granites.

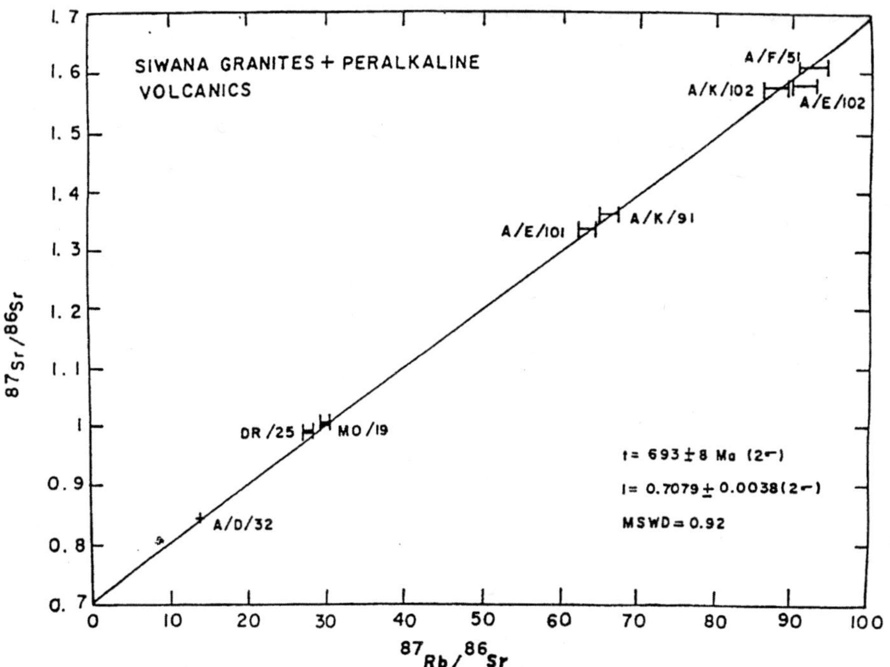

Fig. 3. Convential Sr evolution diagram of Siwana granites and prealkaline volcanics.

CONCLUSIONS

The peralkaline granites (Siwana type) from Barmer district give a whole roc)-Sr isochron age of 698 ± 10 Ma. These granites are slightly younger than th :raluminous granites (Jalore type) of the Malani Igneous Province dated at 728 ± 8 M :athore, 1995) suggesting two different magmatic activities at a short interval. Th itial Sr ratio of these granites indicate a crustal source of their magma. Further, th :ralkaline granites and associated peralkaline volcanics are coeval and cogenetic i ıture as indicated by the pooled isochron of these rocks.

ACKNOWLEDGEMENTS

I dedicate this work to late Prof. R.K. Srivastava who pioneered the geochemic ırk on Malani Igenous Province more than two decades ago. The authors are grateful t e management of Oil and Natural Gas Corporation Limited and the Director, Physic search Laboratory, Ahmedabad for according permission to publish this pape icouragement received from Sri Kuldeep Chandra, Regional Director and Hea)MIPE and Sri K.N. Misra (GM, Chemistry) is gratefully, acknowledged. We thank D N. Prabhu for critically going through the manuscript.

REFERENCES

rberi, F., Borsi, S., Ferrara, G., Marinelli, G. and Varet, J. 1970. Relations between tectoni and magmatology in the northern Danakil depression (Ethiopia). *Phil. Trans. R. Soc. Lond* 267(A) : 293-311.

ιusan, S.K. and Sengupta, R. 1979. *The Malani project : Geological Survey of India, Wester Region, Annual Programme for 1979-80*. 57-66.

ιusan, S.K. 1991. Granitoids of Malani igneous complex, western Rajasthan, India. *Jour. Eart Sci.* 18(3-4) : 1840-194.

ιusan, S.K. 1984. Classification of Malani Igneous Suite. *Symposium on Three decades Developments in Petrology, Mineralogy and Petrochemistry in India. Geol. Surv. Ind., Spe Publ.,* 12 : 199-205.

ιusan, S.K. 1985. Malani volcanism in western Rajasthan. *Ind. Jour. Earth Sic.* 12(1) : 58-71.

ιusan, S.K. and Mohanty, M. 1988. Mechanics of intrusion and geochemistry of alkaline granit from Siwana, Barmer district, *Ind. Jour. Earth. Sic.* 15(2) : 103-115.

ιusan, S.K. 1989. Mineral Chemistry and petrogenetic aspects of Malani volcanics, Wester Rajasthan. *Ind. Miner.,* 43(3&4) : 325-338.

wden, P. 1974. Oversaturated alkaline rocks : granites, pantellerites and comendites. I Sorenson, H. (Ed.) *The alkaline rocks*. John Wiley and Sons, London, 109-123.

ulson, A.R. 1933. The geology of the Sirohi State, Rajputana. *Mem. Geol. Surv. India* 63(1) 83-94.

ιandrasekaran, V. and Srivastava, R.K. 1992. Multivariate statistical analysis of polypha igenous rocks of the Malani igenous province with special reference to Sarnu Dandali are western Rajasthan. *Jour. Geol. Soc. India* 40 : 217-233.

awford, A.R. and Compston, W. 1970. The age of Vindhyan System of Peninsular India. *Qua Jour. Geol. Soc., Lond.* 125 : 351-372.

ure, G. and Hurley, P.M. 1963. The isotopic composition of strontium in oceanic and continent basalts : Application to the origin of igneous rocks *Jour. Petrol.,* 4 : 31-50.

ιchhar, N. 1984. Malani igneous suite : Hot spot magmatism and cratonization of the northe parts of the Indian shield. *Jour. Geol. Soc. India* 25(2) : 155-161.

touche, T.D. 1902. Geology of western Rajputana. *Mem. Geol. Surv. India* 35 : 1-16.

ıhr, P.A. 1970. Volcanic composition in relation to tectonics in the Ethiopian rift system : preliminary investigation. *Bull. Volcan.,* 34 : 141-157.

ıkherjee, A.B. 1958. Structure of Malani rhyolite and a part of the Siwana granite, arour Siwana, western Rajasthan. *Quart. Jour. Geol. Min. Soc. India* 30 : 158-161.

ırthy, M.V.N. 1962. The significance of the ring pattern of Siwana granite bosses in Weste Rajasthan. *Ind. Miner.* 16(2) : 297-298.

ırayan Das, G.R., Bagchi, A.K., Chaub, D.N., Sharma, C.V. and Navaneethan, K.V. 1978. Ra metal content, geology and tectonic setting of the alkaline complexes across tl trans-Aravalli region, Rajasthan. *Recent Researches in Geology*. 7, Hindustan Publishi Corpn., New Delhi.

ıreek, H.S. 1981. Petrochemistry and Petrogenesis of the Malani Igneous Suite, India. *Geol. Sc Amer. Bull.* 92(2) : 206-273.

ıscoe, E.H. 1950. *A Manual of Geology of India and Burma*. Govt. of India Press, New Dell 480-482.

ıthore, S.S. 1995. Geochronological studies of Malani volcanics and associated igneous rocks southwest Rajasthan, India : Implication to crustal evolution. *Unpublished. Ph. D. Thesı* M.S. University, Baroda.

ıthore, S.S., Venkatesan, T.R. and Srivastava, R.K. 1996. Rb-Sr and Ar-Ar systematics of Mala volcanic rocks of southwest Rajasthan : Evidence for a younger post-crystallization therm event. *Proc. Indian Acad. Sci. (Earth Planet. Sci.)* 105(2) : 131-141.

ivastava, R.K. 1988. Magmatism in the Aravalli mountain range and its environs. *Mem. Ge Soc. India*, 7 : 77-93.

ivastava, R.K., Yadav, A., Ashiya, I.D. and Chawde, M.P. 1988. Geochemistry of the peralkali silicic rocks from the Malani volcanic province, Rajasthan, India. *INIGCP 217 Workshop Proterozic rocks of India*, Calcutta. 106-107.

ivastava, R.K., Maheshwari, A. and Upadhaya, R. 1989a. Geochemistry of felsic volcanics fro Gurapratap Singh and Diri, Pali district, Rajasthan (Part - I, Major Elements). *Jour. Ge Soc. India* 34 : 467-486.

ivastava, R.K., Maheshwari, A. and Upadhaya, R. 1989b. Geochemistry of felsic volcanics fro Gurapratap Singh and Diri, Pali district, Rajasthan (Part-II, Trace Elements). *Jour. Ge Soc. India* 34 : 617-631.

ızieff, H., Marinelli, G., Barberi, F. and Varet, J. 1969. Geologie de I'Afar Septentrion Premiere expedition du CNRS- France et du CNR-Italie (December 6-7 - Fevrier 68). *Bu Volcan* 33 : 1039-1072.

ncent, P.M. 1970. The evolution of the Tibesti volcanic province, eastern Sahara, In *'Afric Magmatism and Tectonics* (T.N. Cliffort and I.G. Gass. Eds.), Oliver and Boyd, Edinburg 301-319.

ıdav, A. 1988. Geochemistry of Siwana granites and associated rhyolites. *Unpublished. Ph. Thesis,* Rajasthan University, Jaipur.

eological Evolution of Northwestern India
.S. Paliwal (Ed.)
:ientific Publishers (India), Jodhpur, 1999, pp. 69 - 82.

PETROLOGY OF AMPHIBOLITES FROM THE BANDED GNEISSIC COMPLEX OF NASIRABAD REGION, AJMER DISTRICT, RAJASTHAN

D.K. Khandelwal, and M.K. Pandya*

Department of Geology, Govt. Dungar College, Bikaner, Rajasthan 334 003
**Department of Geology, M.L. Sukhadia University, Udaipur, Rajasthan 313 001*

ABSTRACT

The amphibolites of the Nasirabad region constitute an important lithological unit, occurring as patches, lenses and linear bodies of variable dimensions in the high grade gneissic terrain of the Early Precambrian formations of central Rajasthan. Field relationship and petrographic characters indicate the presence of massive and stripped amphibolites. The close resemblance of these amphibolites in chemical behaviour with ortho amphibolites and some mafic rocks indicate their igneous parentage. The field and chemical evidences indicate the intrusive and extrusive nature of these amphibolites. The study reveals a probable oceanic tholeiitic basalts to be the progenitor of the amphibolites. Subsequent metamorphism ranging from almandine - amphibolites to granulite facies conditions have resulted in the formation of present amphibolities.

ey words : Petrology, Amphibalites, Banded Gneissic Complex, Nasirabad, Rajasthar

INTRODUCTION

The Banded Gneissic Complex of Central Rajasthan (Heron, 1953) holds promis or the better understanding of the evolution of the North-West Indian craton, yet epresents the debatable formation in the Precambrian Geology of Rajasthan. Vario orkers have reported and studied some of the metabasic rocks from the Centr ajasthan (cf. Khandelwal and Pandya, 1988).

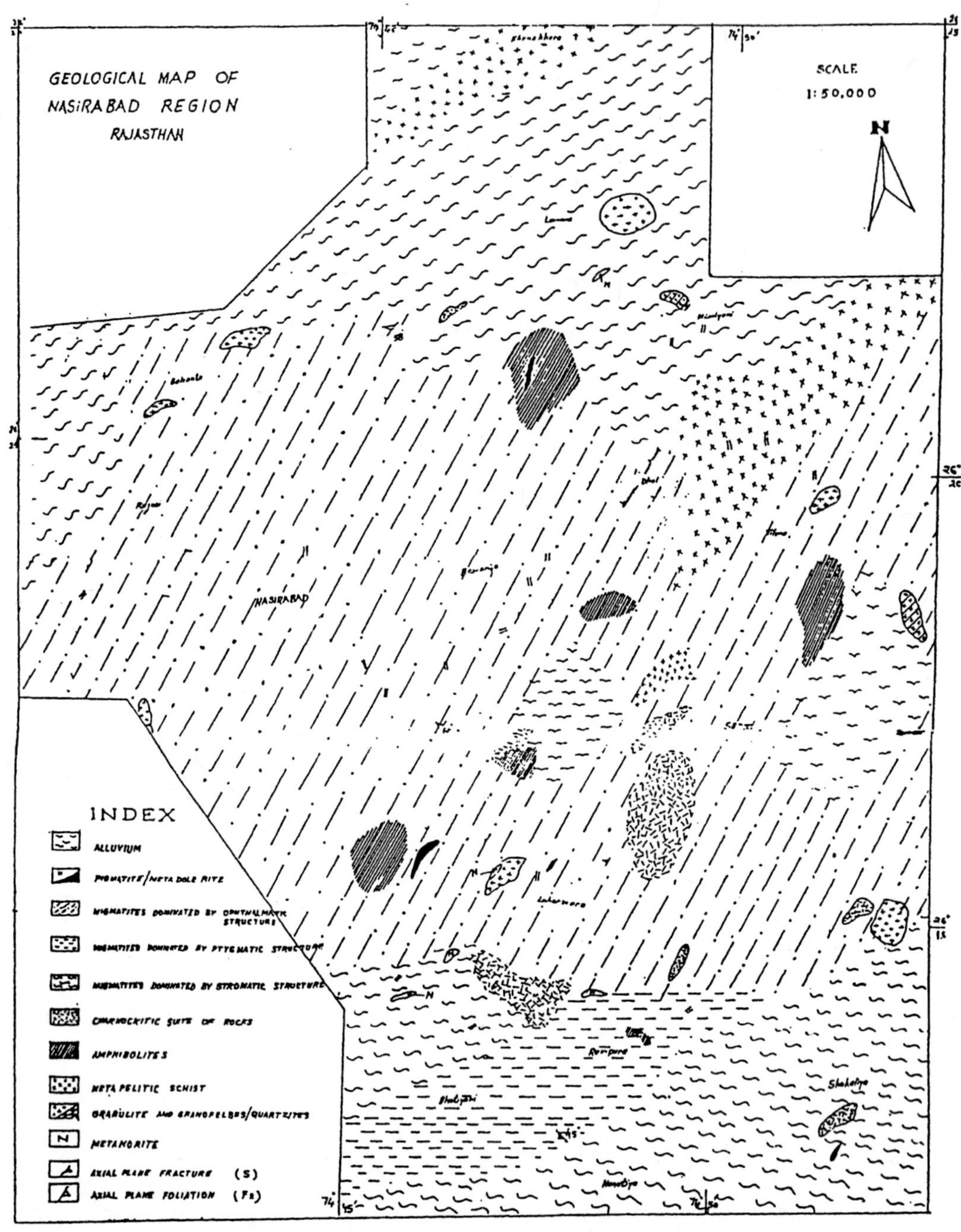

Fig. 1. Geological Map of Nasirabad Region Rajasthan

The objectives of this study are : (a) geochemical characterisation of tl ıphibolites, (b) an application of the geochemical data to determine the parentage aı to determine the tectonic setting of emplacement and magmatic source and the gra(metamorphism. The study is based on geological setting of various rocks and tl ochemical analyses of samples collected from Nasirabad region, Ajmer district.

GEOLOGICAL SETTING

Field work in the investigated region reveals that the Banded Gneissic Complex e area is represented by a large variety of interesting assembleage of deep seat(ιtazonal rocks viz. metanorites, metahyperties, metasedimentary rocks, amphibolite arnockite suite of rocks and a large variety of migmatites and associated pegmatit ιd aplites (Fig. 1) Migmatisation has affected almost all the rock types except tl etadolerite dykes which represent the youngest instrusives. Mylonites occurring only :ew places, represent the latest cataclastic event in this region (Khandelwal, 1992).

The amphibolites of varying minerology have a widespread occurrence around tl llage of Loharwara, Tilana and Bewanja. Field relationship and megascopic characte dicate two types of amphibolites : (i) massive and granular and (ii) stripped or band(ɑphibolites. The massive amphibolites are dark, greenish black, hard and compa cks with more or less granular habit, while the striped amphibolites are gre(loured, foliated rocks characterised by alternate dark and light bands, where in gener e dark bands are wider than light bands. Massive variety of amphibolites occ nerally in low lying plains as lenticular outcrops of varying dimensions within tl ιeissic terrain and show their transformation at various stages from the old etabasic rocks of the area. The stripped amphibolites occur either in strike ridges ιtches of variable dimensions following the foliation planes of the surrounding gneiss ιd are co-folded at places within them. Rafts of both the varieties of amphibolites .riable dimensions are also seen within the migmatitic gneisses of the area.

PETROGRAPHY

The massive and granular amphibolites are medium to coarse grained roc .aracterised by granoblastic texture. Different stages of transformation of old etabasic rocks like metanorite and metahyperties are observed and in such cases reli ıhitic to sub-ophitic textures are commonly observed in massive amphibolite evelopment of granular hornblende at the expense of pyroxenes are also observed. T ineral assemblages are more variable and the cataclastic effects are less dominating em as compared to the striped amphibolites.

The stripped amphibolites are medium grained, green coloured, foliated roc ιaracterised by dark and light colour bands showing parallel arrangements of elongat ismatic hornblende grains and plagioclase laths. These are charecterised by their ev(ained xenomorphic texture and distinct mineral assemblage (hornblande-plagoclas ıidote-quartz-sphene-magnetite-garnet). Effect of polymetamorphism is observed in sections of these rocks evidenced by the presence of streaks of cataclası anulation and epidotization along fractures and development of schistosity. Evidenc migmatisation on and mylonitization are commonly observed in them.

The major mineral constituents of the amphibolites of the region are hornblend agioclase, epidote, quartz, pyroxene and garnet; biotite occur in subordinate amou hereas sphene, zircon, apatite, magnetite and ilmenite are the important accesso inerals in these rocks.

Adopting Cannon's (1963) classification these amphibolites have been named l efexing the diagnostic minerals present in them. Following types of the amphibolit e established in Nasirabad region :

1. Amphibolites (Sensu Stricto), 2. Amphibolites (Sensu Stricto) with reli roxene, 3. Amphibolites (sensu lato), 4. Plagioclase amphibolites, 5. Clinopyroxe iopside) amphibolites, 6. Epidote amphibolites, 7. Garnet amphibolites, 8. Magneti nphibolites, 9. Quartz amphibolites.

A detailed account of various constituent minerals in these amphibolites is given able 1.

ible 1. Modal Analyses of Amphibolites from Nasirabad Region

	1	2	3	4	5	6	7	8
INERALS								
ornblende	78.3	55.3	60.2	63.7	57.8	55.3	49.7	55.6
agioclase	20.2	24.4	19.2	28.8	25.3	18.3	29.8	16.6
uartz	–	18.4	18.9	4.4	6.2	12.0	8.6	19.7
rthopyroxene	–	Tr.	–	Tr.	1.3	–	–	–
inopyroxene	Tr.	Tr.	–	–	8.5	3.7	2.1	2.8
otite	–	0.5	Tr.	1.2	Tr.	Tr.	0.9	2.4
arnet	–	0.6	Tr.	0.8	–	–	7.1	Tr.
idote	–	–	0.5	–	Tr.	8.9	Tr.	–
paques	1.4	0.8	1.1	1.0	0.8	1.3	1.6	0.8
thers	Tr.	Tr.	–	Tr.	–	Tr.	Tr.	Tr.

Amphibolite (nearer to hornblendite), near Rampura villag (2) Amphiibolite (sensu-stricto), 1 km south wera village. (3) Amphbolite (sensu-stricto), 1 km east of Bawanja village. (4) Plagioclase amphibolite (senu lat ar Ramsar village. (5) Pyroxene amphibolite (sensu lato), Loharwara village. (6) Epidote amphibolite (sensu lat m SE of Nasirabad. (7) Garnet amphibolite, 1 km North west of Ramsar village. (8) Quartz amphibolite, 1 k st of Loharwara village.

Hornblende occures in prismatic columnar laths which are devoid of inclusion ar ow well developed faces with perfect cleavages. In some cases, however, fine specks ached iron ores are seen arranged along the cleavage traces. The massive amphibolit ntain brownish hornblende, which indicates their igneous parentage (Deer 193 hereas striped amphibolites contain green hornblende. Plagioclase occurs as laths rying sizes, twinned on albite, peri-cline and combined carlsbad-albite law. In massi anular amphibolites the anorthite content of the plagiclase is comparatively high an in the stripped amphibolites. In general, the polygonal grains of quartz of varyi mensions are commonly associated with hornblende and plagioclase in the massi anular amphibolites whereas the elongated nature of quartz is observed in stripp nphibolites. Undulose extinction and crushing of quartz grains indicate mylonitizati these rocks. Diopside is the dominating pyroxene in amphibolites of the area. Vario ages of their transformation ranging from partial to complete transformation ɔrnblende is observed. Small spongy garnet grains are commonly observed. Inclusio ˈplagioclase, hornblende and quartz are seen in some cases, indicating the formation ırnet later than these minerals. Small colourless and nonspongy garnet at tim rmeds around the hornblende grains and which sometimes retains the form ɔrnblende; subherdral grains of variable sizes of epidote are also seen. Large grains ɔidote retain the form the cleavages of hornblende. Small granules of sphene either olated grains or as layers in association with the epidote, hornblende and opaqu ınstitute the important minor minerals in these amphibolites.

ıble 2. Major and Trace Element Analyses of Amphibolites of Nasirabad Region

	1	2	3	4	5	6	7	8	9	10
ꞮAJOR ELEMENTS :										
O_2	48.57	49.25	50.78	50.12	52.76	52.72	53.15	47.77	52.76	54.19
.O_2	1.32	0.87	1.43	1.21	0.97	1.32	1.32	1.61	0.92	1.23
l_2O_3	13.89	14.27	14.23	14.14	13.82	14.41	14.17	13.47	14.37	14.72
$_2O_3$	3.81	3.72	2.85	3.42	2.98	1.96	2.96	4.21	3.42	2.92
O	9.75	10.04	10.71	10.25	8.75	8.93	10.07	8.71	7.78	–
nO	0.28	0.16	0.21	0.76	0.32	0.12	0.29	0.31	0.25	0.25
gO	7.85	8.20	6.82	6.23	6.18	6.82	5.98	7.72	5.97	4.93
ıO	10.52	10.81	8.75	9.71	9.79	9.52	8.79	10.57	9.24	7.54
a_2O	1.21	1.12	1.81	0.92	1.75	1.85	1.92	2.01	1.79	2.67
$_2O$	0.91	0.88	0.76	0.81	1.02	0.91	0.78	0.87	1.43	1.61
O_5	0.16	0.21	0.27	0.24	0.19	0.24	0.16	0.23	0.24	0.22
ꞰACE ELEMENTS :										
i	13.5	132	85	90	52	140	70	92	112	70
ɔ	50	60	50	40	55	60	30	50	45	35
r	220	240	260	110	165	230	85	325	130	120
ɔ	15	16	13	13	12	22	13	10	20	11
	200	210	170	130	120	320	110	160	150	100
b	7	9	9	8	7	12	8	12	13	9
	31	34	28	22	25	35	17	25	25	22
	280	120	215	180	200	190	350	260	230	210
	130	70	170	90	85	140	145	180	70	80

Massive amphibolite, 1 km east of Bewanja village. (2) Massive amphibolite (sensu stricto), 1 km south of Lawe lage. (3) Stripped amphibolite, Ramsar village. (4) Stripped amphibolite, Lawera village. (5) Epidote amphiboli ‹m South-east of Nasirabad. (6) Massive amphibolite, Lawera village. (7) Plagioclase amphibolite, near Rams lage. (8) Pyroxene amphibolite, Loherwara village. (9) Garnet amphibolite, 1 km North-west of Ramsar villa)) Quartz amphibolite, 1 km West of Loherwara village.

GEOCHEMISTRY AND PETROGENESIS

Major and trace element analyses of amphibolites from Nasirabad region are give Table 2. The cation percentage, Niggli values, oxide ratio and discriminant functi c. of the investigated amphibolites are tabulated in Table 3.

able 3. CIPW Norms, Cation percentage, Niggli values, oxidation ratio, different index and important trace element ratios for the amphibolites of Nasirab region, Rajasthan.

	1	2	3	4	5	6	7	8	9	10
IPW NORMS										
Q	4.38	2.59	5.67	7.80	6.44	6.63	9.28	—	7.04	7.6
Or	5.38	5.20	4.49	4.79	6.03	5.38	4.61	5.0	8.46	9.5
Ab	10.23	9.46	15.29	7.78	14.81	15.64	16.23	16.77	15.13	22.5
An	29.74	31.27	28.34	34.74	26.79	28.29	27.7	25.3	26.91	23.3
Di	17.34	17.16	10.94	9.82	16.99	14.81	12.22	20.86	14.28	10.3
Hy	21.00	26.3	25.6	26.01	22.37	21.96	22.22	18.72	20.0	17.5
Ol	—	—	—	—	—	—	—	2.42	—	—
Mt	5.52	5.39	4.13	4.96	4.32	2.84	4.29	6.03	4.96	4.2
Il	6.31	1.65	2.76	2.3	1.84	2.51	2.51	3.04	1.75	2.3
Ap	0.38	0.5	0.64	0.57	0.45	0.57	0.38	0.67	0.57	0.5
ATION PERCENTAGE										
Si	51.73	52.5	55.36	54.70	56.29	56.56	57.72	57.57	57.19	59.7
Ti	2.65	0.69	1.17	0.99	0.77	1.06	1.07	1.30	0.75	1.0
Al	8.70	8.95	9.12	9.72	8.67	9.09	0.05	8.55	9.16	9.5
Fe+++	1.52	1.49	1.16	1.39	1.19	0.79	1.20	1.70	1.39	1.2
Fe ++	8.65	8.92	9.05	9.74	9.11	7.82	8.08	9.06	7.87	7.1
Mn	0.25	0.14	0.19	0.24	0.29	0.11	0.26	0.28	0.23	0.2
Mg	12.54	13.11	11.15	10.2	9.89	10.97	9.74	12.50	9.71	8.1
Ca	12.00	12.35	10.22	11.35	11.19	10.94	10.23	12.22	10.73	8.9
Na	1.25	1.15	1.91	0.97	1.81	1.92	2.02	2.10	1.88	2.8
K	0.62	0.6	0.53	0.56	0.69	0.62	0.54	0.6	0.99	1.1
P	0.07	0.09	0.12	0.11	0.08	0.12	0.07	0.10	0.11	0.1
IIGGLI VALUES										
al	18.49	18.56	20.50	21.32	19.69	21.44	21.38	17.55	21.14	23.6
fm	52.04	52.18	51.06	50.40	49.21	48.29	48.42	51.82	47.49	44.4
c	25.51	25.62	52.97	24.91	25.4	25.80	24.16	25.09	24.76	22.0
alk	3.96	3.64	5.48	3.37	5.68	4.48	6.04	5.54	6.61	9.8
si	109.92	108.93	124.38	119.98	127.80	133.33	136.33	105.83	131.94	147.
ti	5.63	1.44	2.63	2.17	1.76	2.50	2.54	2.67	1.72	2.5
p	0.15	0.20	0.28	0.24	0.19	0.26	0.17	0.21	0.25	0.2
k	0.03	0.34	0.22	0.37	0.28	0.33	0.21	0.22	0.34	0.2
mg	0.51	0.52	0.49	0.44	0.46	0.54	0.41	0.49	0.47	0.4
UGUMURA'S INDEX	38.49	40.04	38.22	42.69	39.19	39.57	39.86	32.94	38.05	34.5
IFFERENTIAL INDEX	19.99	17.25	25.45	20.37	27.28	27.65	30.12	21.77	30.63	39.7
Fe_2O_3 / FeO	0.39	0.37	0.28	0.32	0.29	0.22	0.38	0.42	0.39	0.3
RACE ELEMENT RATIOS										
Nb/Y	0.23	0.26	0.32	0.36	0.28	0.34	0.47	0.48	0.52	0.4
Y/Nb	4.43	3.77	8.11	2.75	3.57	2.92	2.12	0.08	1.92	2.4
Cr/Ni	1.63	1.82	3.06	1.22	3.17	1.64	1.21	0.53	1.16	1.7
Co/Ni	0.37	0.18	0.59	0.44	1.06	0.43	0.43	0.54	0.40	0.5
Rb/Sr	0.075	0.076	0.076	0.100	0.104	0.068	0.118	0.062	0.130	0.11
Zr/ P_2O_5	0.08	0.33	0.06	0.04	0.04	0.06	0.09	0.08	0.03	0.0

Though the amphibolites of the investigated region do not show much variation ii e chemical behaviour, it has been observed that the salient petrographic dissimilaritie e reflected in their chemical composition. The stripped amphibolites are rich ii leased iron ore derived from the hornblende and have more Fe_2O_3 in comparison to th assive amphibolites, likewise the epidote and garnet amphibolities have more Mn(ntents as compared to other amphibolites. However, the bulk chemistry of the massiv anular amphibolites of the region is more or less similar to the metanorites an etahyperites from which they were transformed.

The plots of major and trace elements of amphibolites in various binary and ternar agrams suggested by various workers, indicate the igneous parentage (Leake, 1964), o ese amphibolites.

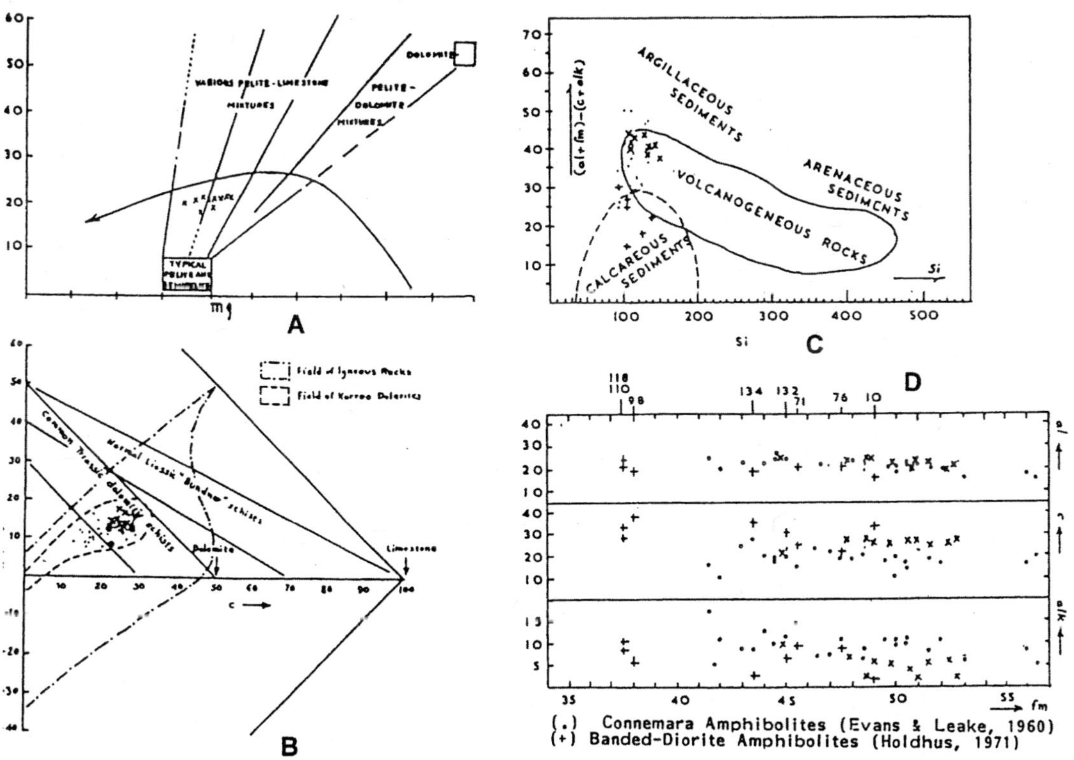

Fig. 2. Phase diagrams showing field of amphibolites of the study area

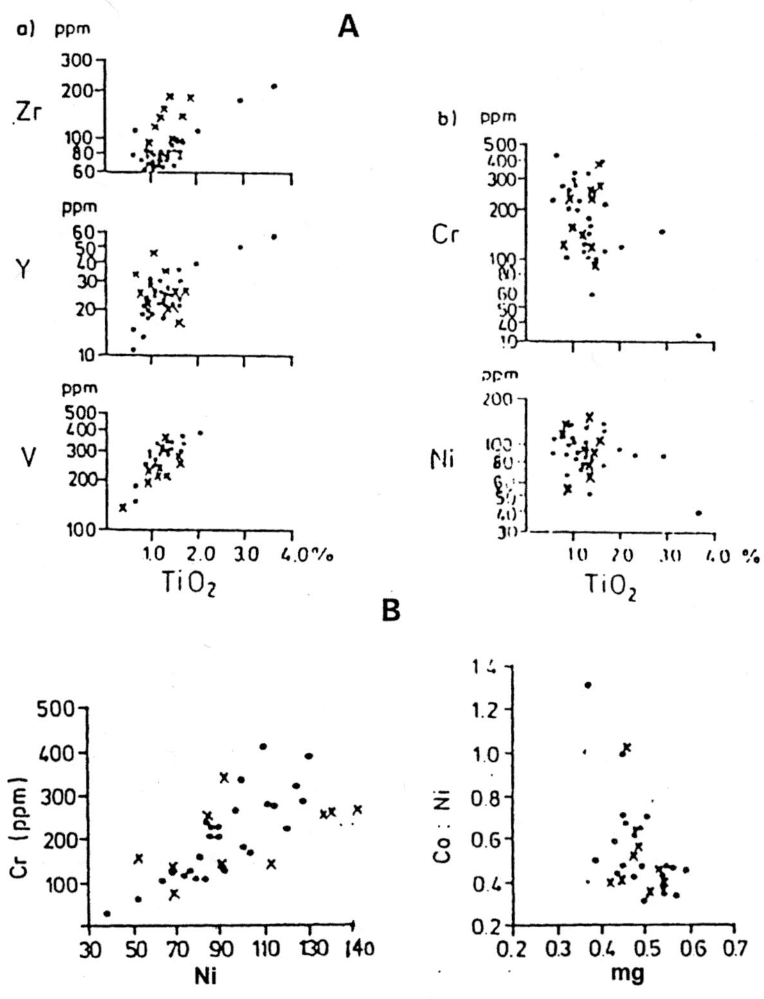

Fig. 3A. Plots showing relationship amongst Zr, Y, V, Cr and Ni with TiO_2 and 3B - Plots showing relationship of Cr v/s Ni; Co : Ni v/s mg.

The plots of the amphibolites in Niggli c vs mg diagram (Fig. 2a) and Niggli (al-all c (Fig. 2b) display a trend which is comparable with the middle stage differentatic end of Karoo dolerites. It is significant that plots of the studied amphibolites in th ggli (al+fm) - (c+alk) against Si values (Simonen, 1953) falls in the field of volcanogen cks (Fig. 2c). The Niggli c and alk and al vs fm values for the amphibolites, which ar uch comparable to striped amphibolites of Connemara region (Evans and Leake, 1960 dicate again the igneous parentage. (Fig. 2d). The trends of TiO_2 v/s Ni, Cr, V and ig. 3a) and Ni vs Cr and Niggli mg v/s Co : Ni (Fig. 3b) also show remarkab nilarities with the trends of similar rocks of the Amet area (Kataria, 1988). Th neous parantage is further evident in the ACF diagrams proposed by Orville, 1969 (Fi) and Walker, 1960 (Fig. 4b).

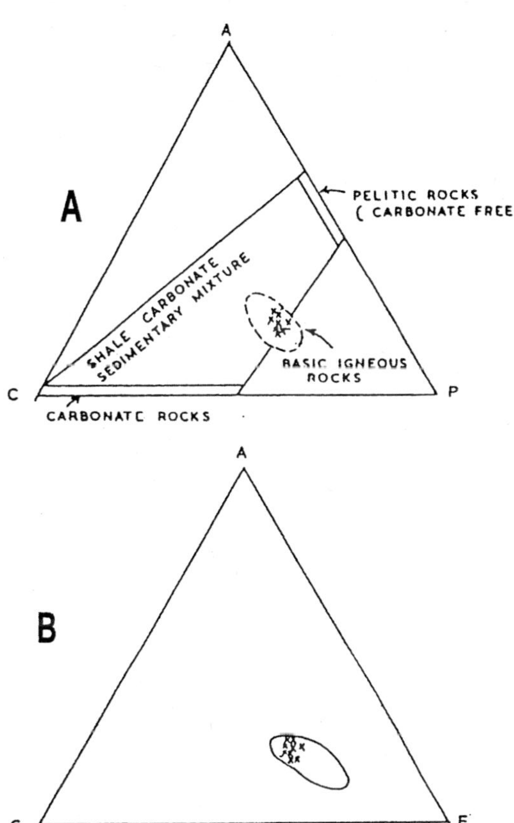

Fig. 4. ACF diagrams showing field of amphibolites of the study area, A- after Orville, 1969 and B - after Walker, 1960.

Thus it may be concluded that th amphibolites of Nasirabad region are o igneous parentage.

MAGMA TYPE, POSSIBLE TECTONIC SETTING AND METAMORPHISM

The amphibolites of the region unde investigation when plotted in variou binary diagrams, separating differen basalt fields like slica v/s alkalies (Kuno 1959; McDonald and Katsura, 1964 Irvine and Baragar, 1971) (Fig. 5a) alumina content and normativ plagioclase composition (100 An/Ab + Ab - Ne) after Irvine and Baragar, 1971 (Fig 5b), ratio of Nb and Y, Ti)2 - Zr/P_2O_5 an Nb/Y - Zr/P_2O_5 (Fig. 5c) used b Winchester, 1976, dividing the fields o tholeiitic basalt with calc-alkaline an alkaline basalts, clearly indicates th tholeiitic nature of the amphibolites fron Nasirabad region. Amphibolites of th region when compared with the simila rocks of other region for Zr/TiO_2 vs Nb/Y diagram (after Floyed and Winchester 1978) indicates the tholeiitic affinity. Th amphibolites when plotted in Y-T-(diagram (Fig. 6) proposed by Davies et al 1979, indicte that these rocks follow th Archaean tholeiitic trend, which ca easily be from the calc-alkaline rocks.

Pearce and Cann (1973) used Ti-Zr-Y and Ti-Zr-Sr diagrams to distinguish "Withi ate basalts" (Oceanic, island and continental basalts) from ocean floor basalts an w-k island arc tholeiites, using Ti-Zr-Y diagrams (Fig. 7a). Amphibolites of Nasiraba gion in the field of within plate basalts (Oceanic, island and continental) and field c ean floor basalt). Similarly in Ti-Zr-Sr diagram (Fig. 7b) these amphibolites plot i ean floor basalt field.

The oceanic tholeiitic affinity of the amphibolites of Nasirabad region is als ıpported when chemical compoisition of these is compared with the oceanic tholeiiti nphibolities of Amet area (Kataria et al. 1988). Oceanic tholeiites are characterised b out 50% silica, and low K_2O contents of crustal elements as compared to continent oleiites (Hess and Poldervaart, 1968; and Engel and Engel 1965). Average of silic 1.21%), K_2O (0.99) and some trace elements of present amphibolites like Sr=16 =26.4, Zr=116, Rb=14.5/ Nb=9.4, Cr = 188.5, Ni=97.8, V=223.5 ppm when compare ith oceanic tholeiitic composition it appears that they are having the oceanic affinity.

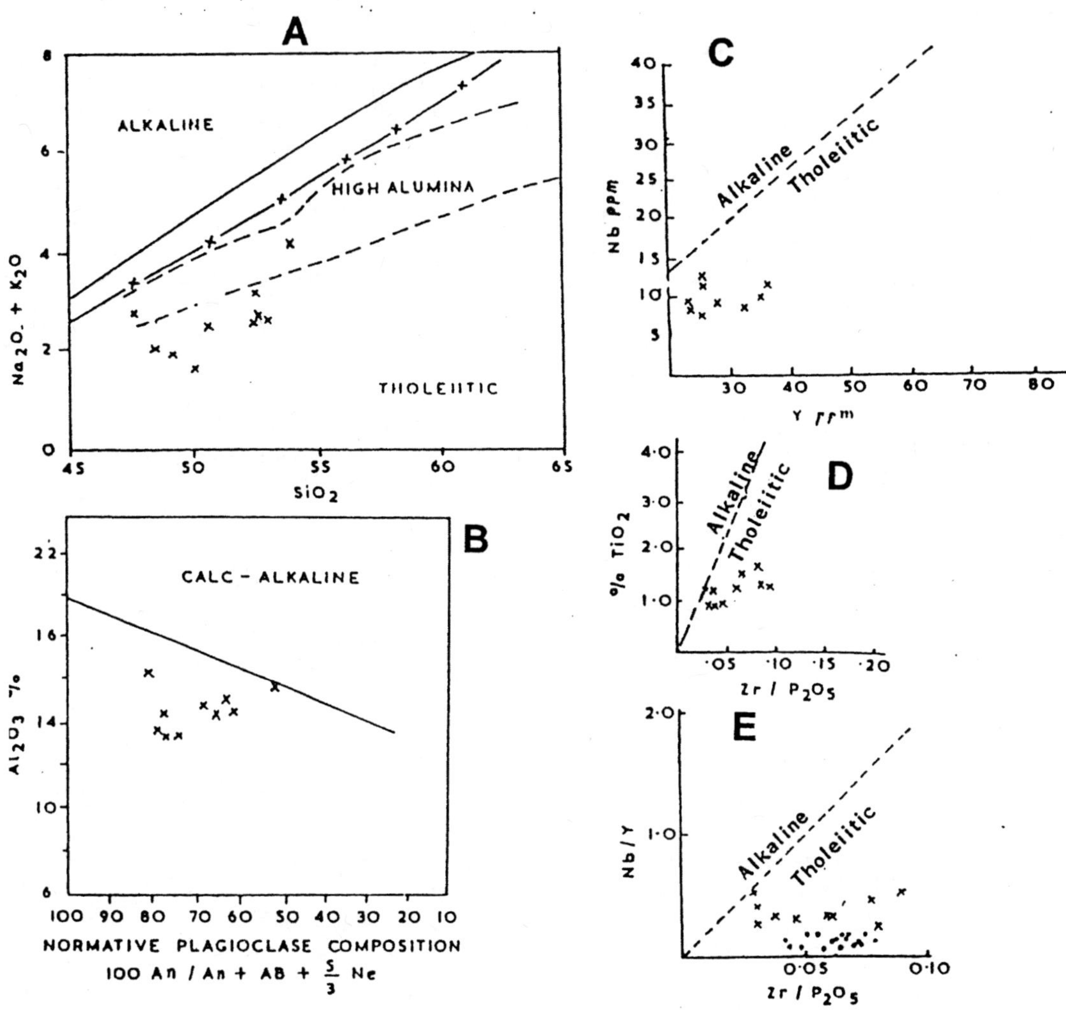

Fig. 5 (A-E). Binary diagrams showing different fields of baslts.

The "θ" index after Sugimura (1968) for the studied amphibolites is more than 3 .d values of Fe_2O_3/FeO ratio which match with theoleiitic series (Kuno et al. 195' dicate their tholeiitic affinity.

It is observed that the amphibolites of the investigated region plots in ocean oleiitic field delineated from various diagrams proposed by Floyed and Winchestε 979) namely TiO_2 v/s Zr/P_2O_5, Zr/P_2O_5 v/s Nb/Y and Y/Nb v/s TiO_2.

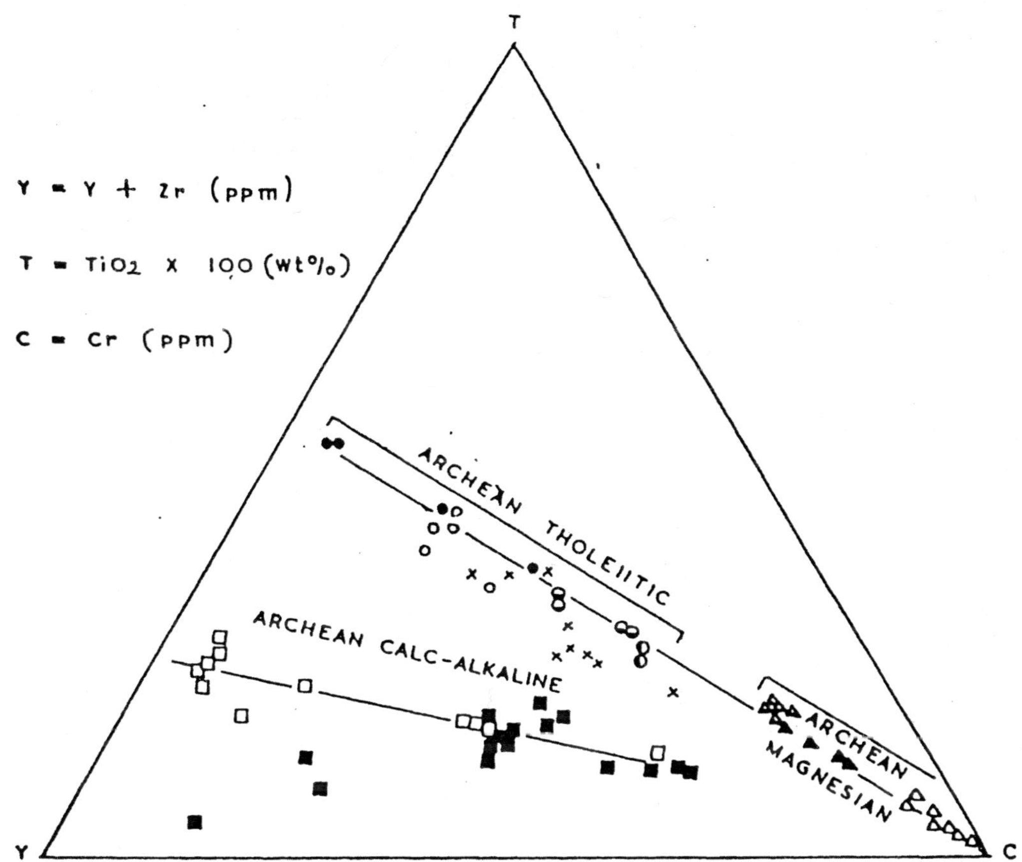

Fig. 6. A field of amphibolites of the study area in Y-F-C diagram

Field evidences support the probability that the stripped amphibolites could be tl :trusive mafic rocks whereas the massive amphibolites are formed instrusive rocks. Tl assive amphibolites show their trasnformation from metanorites and metahyperit hich occur as relict patches in the gneissic courtry whereas the stripped amphibolit ually cover vast areas and as occur parallel bodies along the foliation planes of tl ieissic rocks.

The P-T conditions, deduced from associated gneissic rocks, granulites, charnocki ite of rocks etc., mineral assemblages of these amphibolites and the plots in AC agrams (Fig. 4a and b) suggest that these rocks can be assigned for the most part to tl mandine - amphibolites facies and partly to the lower granulite facies conditions etamorphism.

It may be concluded that the present amplibolites appear to be the metamorphos oduct of oceanic tholeiitic basaltic rocks.

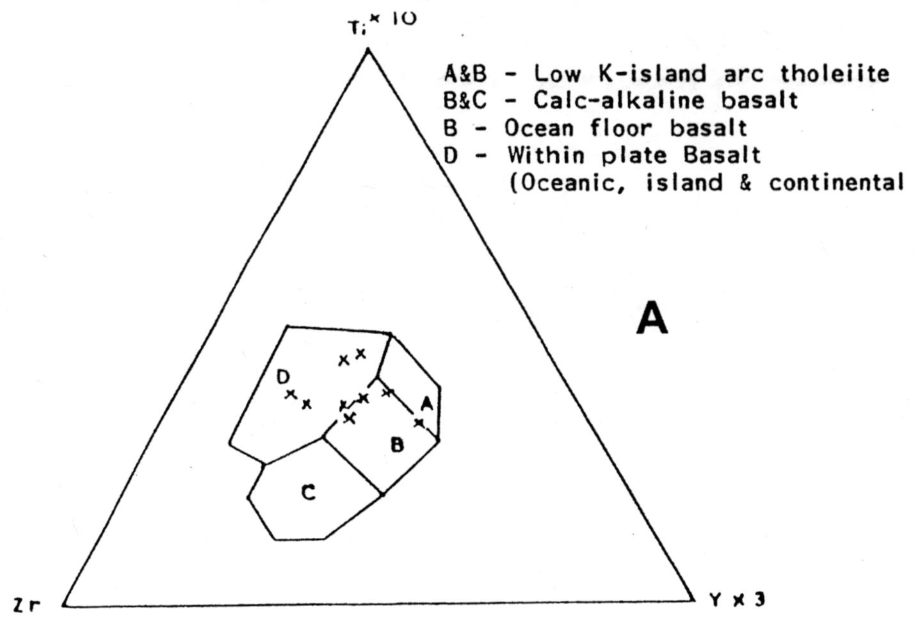

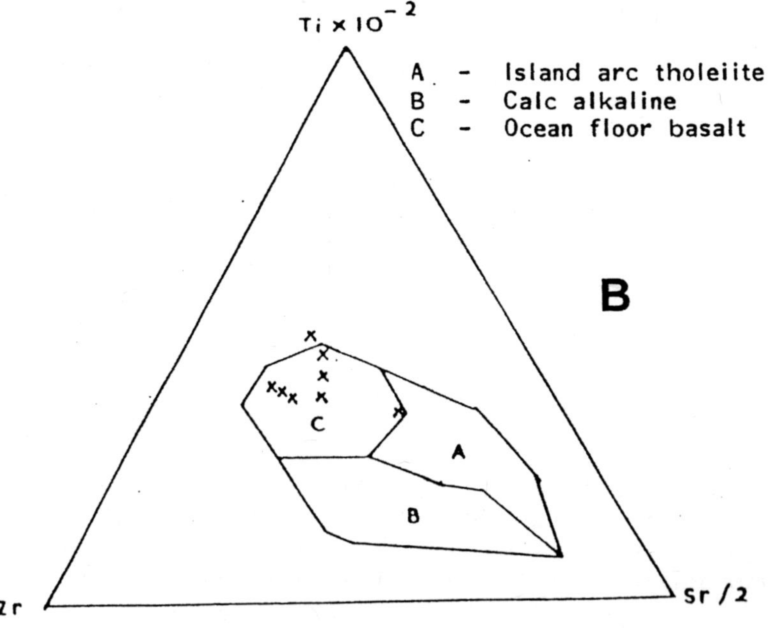

Fig. 7. A- Ti-Zr-Y diagram and B - Ti-Zr-Sr diagram indicating amphibolite plots in oceanic floor basalt field (cf. Pearce and Cann, 1973)

REFERENCES

annon, R.T. 1963. Classification of amphibolites. *Geol. Soc. Am. Bull.* 74. pp. 1087-1088.

avies, J.F., Grant, R.W.E. and Whitehead, R.E.S., 1979 : Immobile trace elements and Archaeε volcanic stratigraphy in the Timmins minning area, Ontario *Can. Jour. Earth. Sci.* 16 (2 305-311.

eer, W.A. 1938. The composition and paragenesis of hornblendes of the Glen Tile Comple Perthshire, Mi. Mag. 25 : 56-74.

ngle, A.E.J. and Engle, C.G. 1965. Chemical characteristic of oceanic basalt and the upp mantle. *Bull. Geol. Soc. Amer.* 76 : 719-733.

vans, E.W. and Leake, B.E. 1960. The composition and origin of the striped amphibolites Connemara, Ireland. *Jour. Petrol* 1 : 337-363.

oyed, P.A. and Winchester, J.A. 1975. Magma type and tectonic setting discremination usi immobile elements. *Earth Planet Soi. Lett.* 27 : 211-218.

oyed, P.A. and Winchester, J.A. 1978. Identification and discrimination of altered aı metamorphosed volcanic rocks using immobile elements. *Chem. Geol.* 21 : 291-306.

eron, A.M., 1953. Geology of Central Rajputana. *Mem. Geol. Surv. Ind.* 79 : 1-389.

ess, H.H. and Poldervaart, A., 1968. *Basalts.* J. Willey and Sons. 483-835.

vine, T.N. and Baragar, W.R.A. 1971. A guide to the chemical classification of the comm volcanic rocks. *Can. Jour. Earth. Sci.* 8 : 29-40.

ataria, P., Choudhari, M.W. and Althaus, E. 1988. Petro- chemistry of Amphibolites from tl Banded Gneissic Complex of Amet, Rajasthan, North Western India. *Chem. Erde* 48:89-111

handelwal, D.K. 1992. Petrology of Early Precambrian rocks around Nasirabad, Ajmer Distri Rajasthan, *unpublished Ph.D. Thesis,* M.L. Sukhadia University, 371 p.

handelwal, N.M. and Pandya, M.K. 1988. Petrochemistry and genesis of the Precambriε amphibolites from Masuda-Rampura region of Central Rajasthan. *Precambrian of ti Aravalli Mountain, Rajasthan, India. Mem. J. Geol. Soc. India.* 7 : 296-306.

uno, H. 1959. Origin of Cenozoic petrographic provinces of Japan and surrounding areas. *Bu Volcanol. Ser.* 2, 20 : 37- 76.

uno, H. Yamasaki, K. *et al.,* 1957. Differentiation of Hawaiian magma, *Jap. Jour. Geol. aı Geog.* 28 : 179- 218.

eake, B.E. 1964. The chemical distinction between ortho and para amphibolites. *Jour. Petr* 5:238-254

ac Donald, C.A. and Katsura, T.F. 1964. Chemical composition of Hawaiian Lavas, *Jour. Petr* 5 : 82-133.

rville, P.M. 1969. A model for metamorphic differentiation origin of thin layered amphibolite *Am. Jour. Sci.* 267 : 64-86.

earce, J.A. and Cann, J.R. 1973. Tectonic setting of basic volcanic rocks determined, using tra element analyses, *Earth Planet Sci. Lett.* 12 : 339-349.

monen, A., 1953. Stratigraphy and sedimentation of Sevecofennidic, early Archaeε supracrustal rocks in southwestern Finland. *Bull. Comm. Geol.* Finland 160 : 1-64.

ugimura, A. 1968. Spatial relations of basaltic magmas in island arcs. In : H.H. Hess and Poldervaart (Eds.) *Basalt,* J. Willey and Sons. 483 - 485.

alker, F. and Poldervaart, A. 1949. Karoo dolerites of the Union of South Africa, *Geol. Soc. A*
Bull. 60 : 591-706.

inchester, J.A., 1976. Different Monian amphibolite suites in northern Ross-shire. *Scott. Jou*
Geol. 12(3) : 187- 204.

›logical Evolution of Northwestern India
. *Paliwal (Ed.)*
›ntific Publishers (India), Jodhpur, 1999, pp. 83 - 93.

DEFORMATIONAL HISTORY OF THE PRECAMBRIAN METASEDIMENTS AROUND SHIVGANJ AND SUMERPUR, SOUTH WESTERN RAJASTHAN

G.R. Utta* and S.P. Sychanthavong

**Department of Mines and Geology, Jaisalmer - 342 001*
Department of Geology, M.S. University of Baroda, Vadodara - 390 002

ABSTRACT

The Precambrian metasediments exposed around Shivganj and Sumerpur area belong to Sirohi and Sindreth Groups. Earlier these rocks were considered as Aravallis but recent workers have assigned much younger age to them. The precise stratigraphic position vis-a-vis Delhi Supergroup is not yet finally decided. Recent studies carried out in the area have revealed an interesting structural and metamorphic history. The analysis of structural elements has shown three phases of folding episodes in the area. The F_1 and F_2 are more or less identical in geometry trending in the NE-SW direction whereas the F_3 is recorded to be open NW-SE flexures. An earlier NE-SW trending regional shear zone appears to be genetically related to F_2 deformation. The younger set of fractures trending NW-SE is recorded to be associated with the F_3 deformation. The regional metamorphism was synchronous with F_1 and F_2 whereas the emplacement of the Erinpura Granite caused a distinct contact metamorphism.

›y words : Deformation, Precambrian Metasediments, Shivganj, Sirohi Grou
Sumerpur, Southwestern, Rajasthan

INTRODUCTION

The study area located in the southwestern Rajasthan in the Sirohi and Pa ›tricts (Fig. 1) comprises calc schist, calc gneisses, crystalline limestone (marble) wi artzitic bands garnetiferous staurolite, silimanite-biolite schists and gneisses

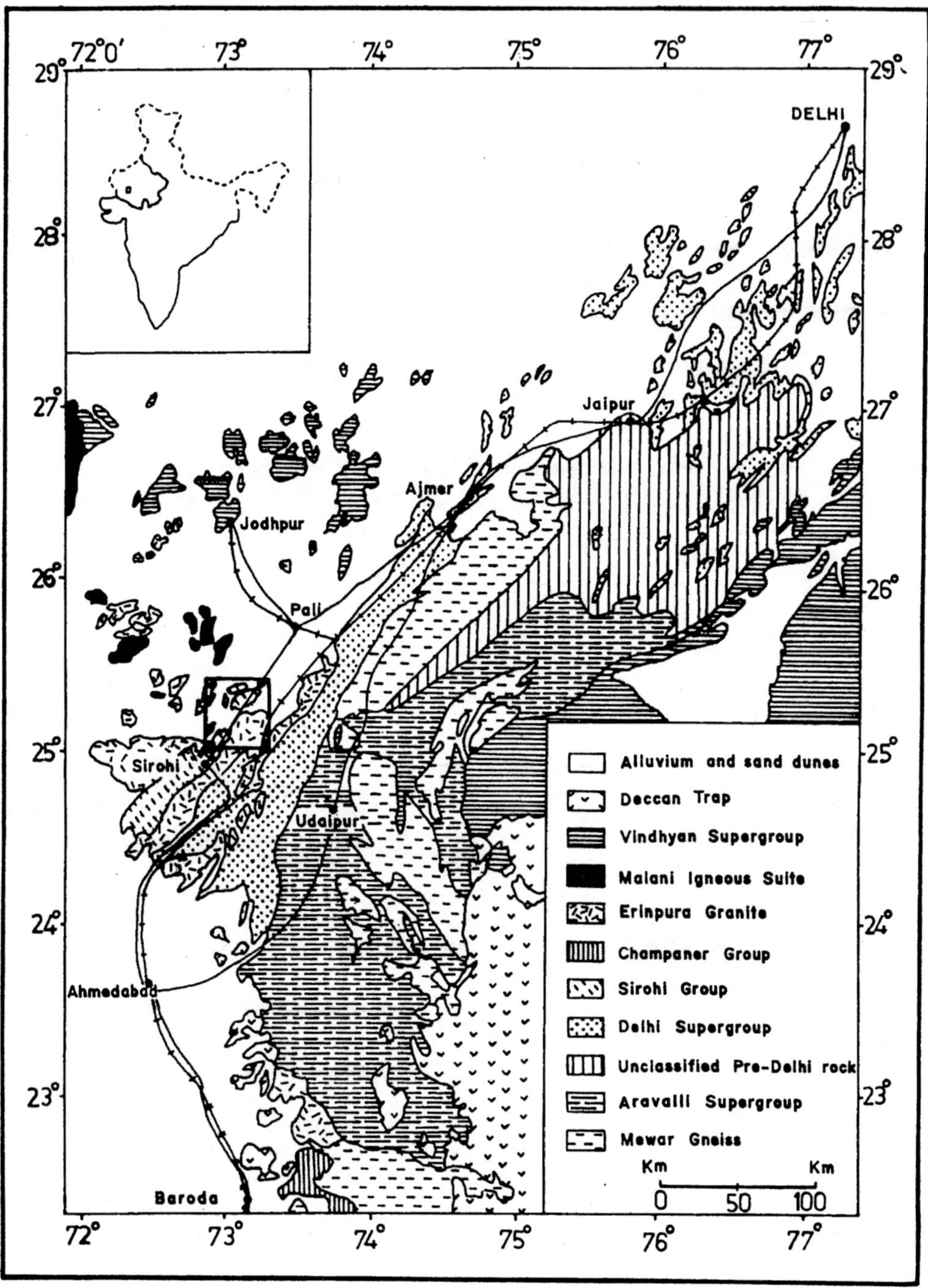

Fig. 1. Regional Geological Map of Southern Rajasthan (modified from Roy, 1988).

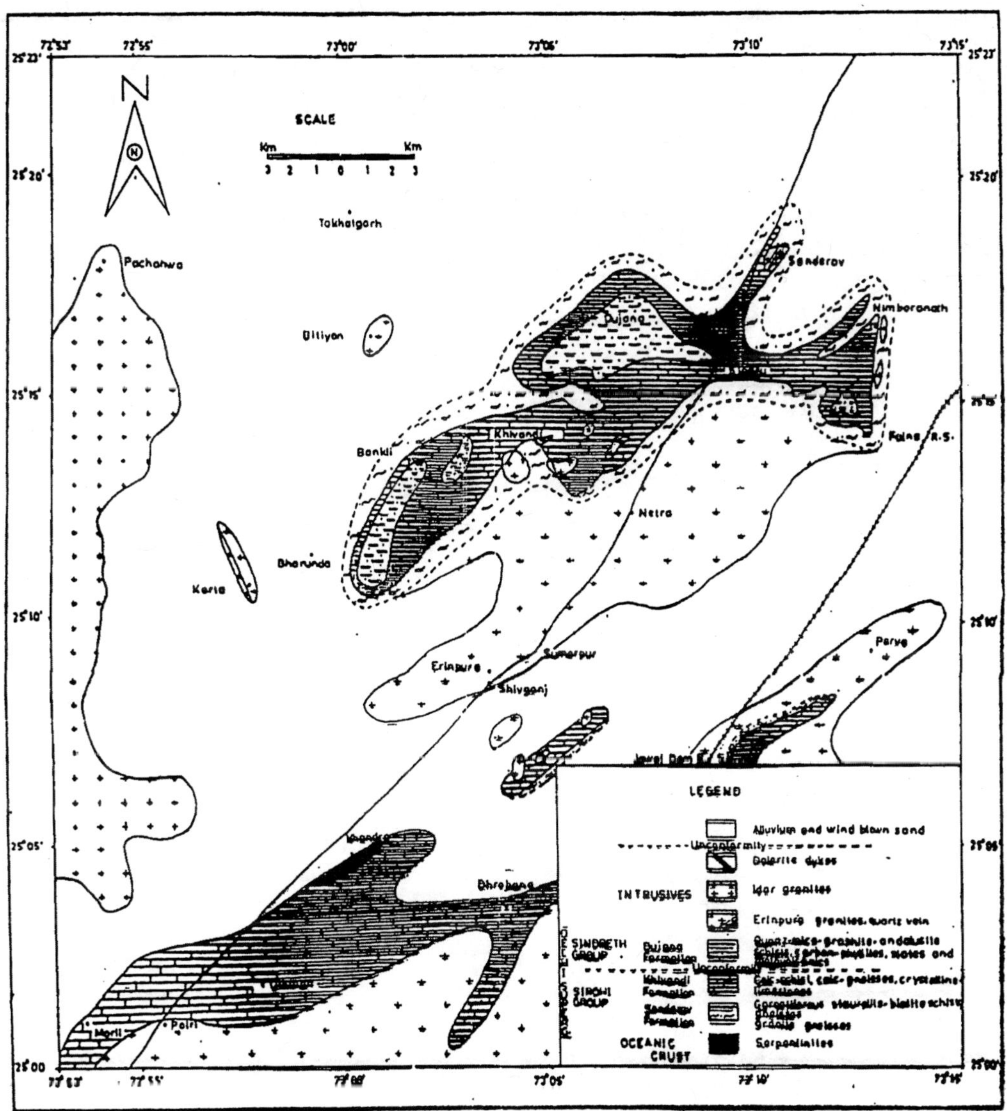

Fig. 2. Lithological Map of Shivganj-Sumerpur area.

nderav Formation of Sirohi Group. The Sirohi Group is unconformably overlain] ıjana Formation of Sindreth Group which comprises carbon phyllites, slates, quartzi d metavolcanics at the base. The intrusive Erinpura Granite and dolerite dykes a ɜn exposed near Jawai Dam, Sumerpur, Netra and Khandra areas. The basement ov ıich Sirohi Group of rocks are resting is part and parcel of oceanic crust. Baseme: nprises ortho amphibolite, meta gabbros, serpentinized ultramafics (Sychanthavoı d Desai - 1977, Sychanthavong and Merh - 1981). The green schist faci ɛtamorphism (chlorite-muscovite zone) to middle amphibolite (garnet-staurolite zon s been recorded in the area.

Based on the lithological similarity and field relation, earlier workers called the ɔks to be Aravallis (Coulson, 1933, Heron, 1953) while recent workers lil ·chanthavong and Desai (1977), Sychanthavong and Merh (1981) and Gupta *et c* 980, 1992) assigned a much younger age to these metasediments and have put them ·lhi Supergroup. Roy (1988), however, has designated these rocks a separate entity e Sirohi Group.

The rocks of the study area have been included in the Sirohi Group witho volving into the controversy whether they belong to the Delhi Supergroup or form parate stratigraphic unit overlying it (Fig. 2). On the basis of detailed field relatio ɔoratory studies, metamorphism and radiometric age data (Choudhary *et al.* 198 ·awford, 1970) the following stratigraphy of the area has been proposed (Table 1).

ıble 1. Stratigraphic succession of the metasedimentary rocks around Shivganj aı Sumerpur area.

Recent				Alluvium and wind blown sand
			Unconformity	
P R O T E R O Z O I C	D E L H I S U P E R G R O U P	Intrusive		Dolerite dykes Idar Granite, Erinpu Granite
		Sinderth Group	Dujana Formation	Cherty quartzite, slates, carbon phyllite quartz-mica graphite, andalusite schist ar metavolcanics
		Unconformity		
		Sirohi Group	Khivandi Formation	Crysatalline limestone (marble)
				Calc gneisses with bands and lenses quartzites calc schist etc.
			Sanderav Formation	Garnetiferous staurolite biotite schist ar gneisses (+ silimanite) granite gneisses
		Basement Rocks (Oceanic Crust)		Serpentinites, ortho amphibolite and me gabbros

GEOLOGICAL SETTING OF THE AREA

ıe Basement Rocks

The basement rocks of Sirohi Group comprise ortho amphibolites, meta gabbro rpentinised ultramafics and granitic gneisses which are part and parcel of the ocean ıst (Sychanthavong and Desai, 1977, Sychanthavong and Merh, 1981). Greenis loured serpentinised ultramafic arocks are exposed in the area near village Khandra.

ıe Sirohi Group

The Sirohi Group is classified in lower Sanderav Formation and upper Khivan ·rmation. The Sanderav Formation consists of garnetiferous staurolite-biotite schis

.d gneisses (at places silimanite bearing), migmatites and sheard granitic gneisses. T pe area for this formation covers Sanderav, Nimboranath, Dhrobana villages and t ıvarpahar hill area. The Ṡanderav Formation is directly overlain by Khivan ırmation (Gupta *et al.* 1980, 1992). The Khivandi Formation consists of calcareo etasediments with intercalations of calc schists, calc-gneisses, crystalline limestor ırble and bands and lenses of quartzites. This formation is exposed near Khivan mboranath, Sindru, Uthman and Morli villages. The Mineral components of calcareo :ks are calcite (dolomite), plagioclase, feldspar, diopside, actinolite, quartz and epidot

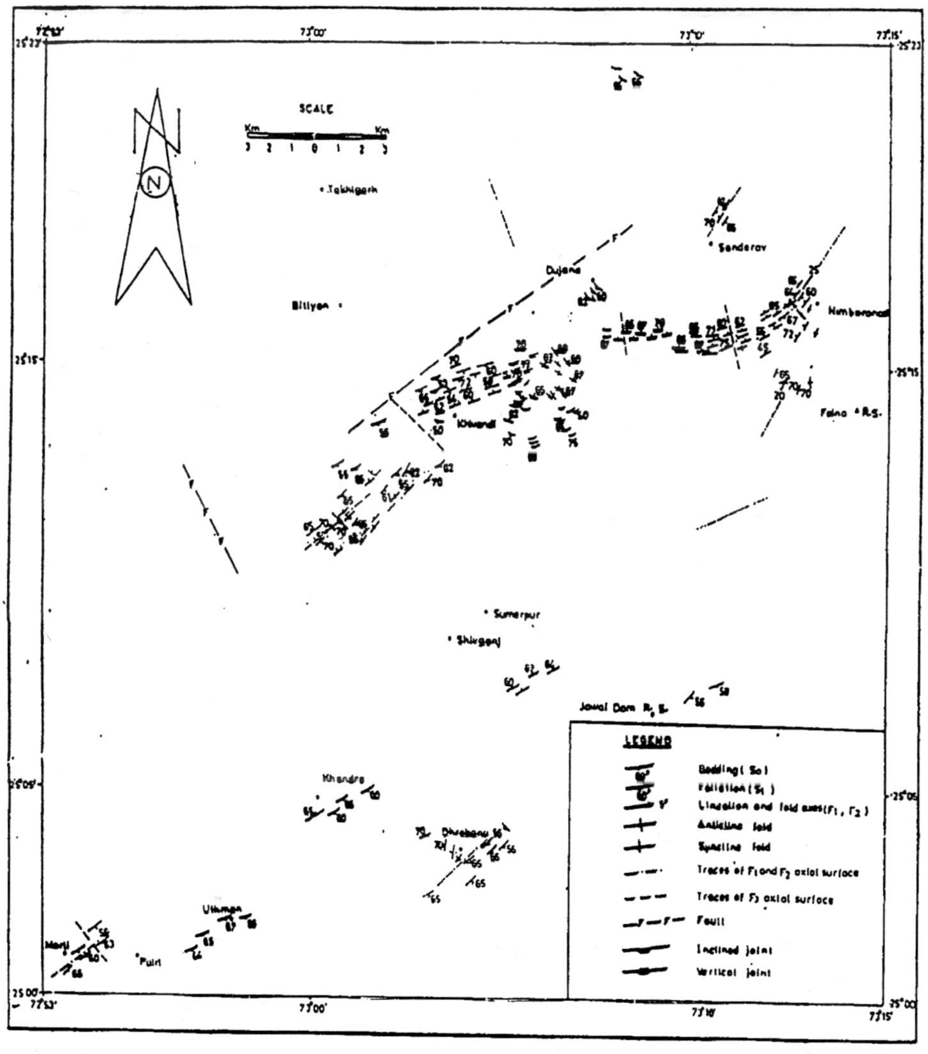

Fig. 3. Structural Map of Shivganj - Sumerpur area.

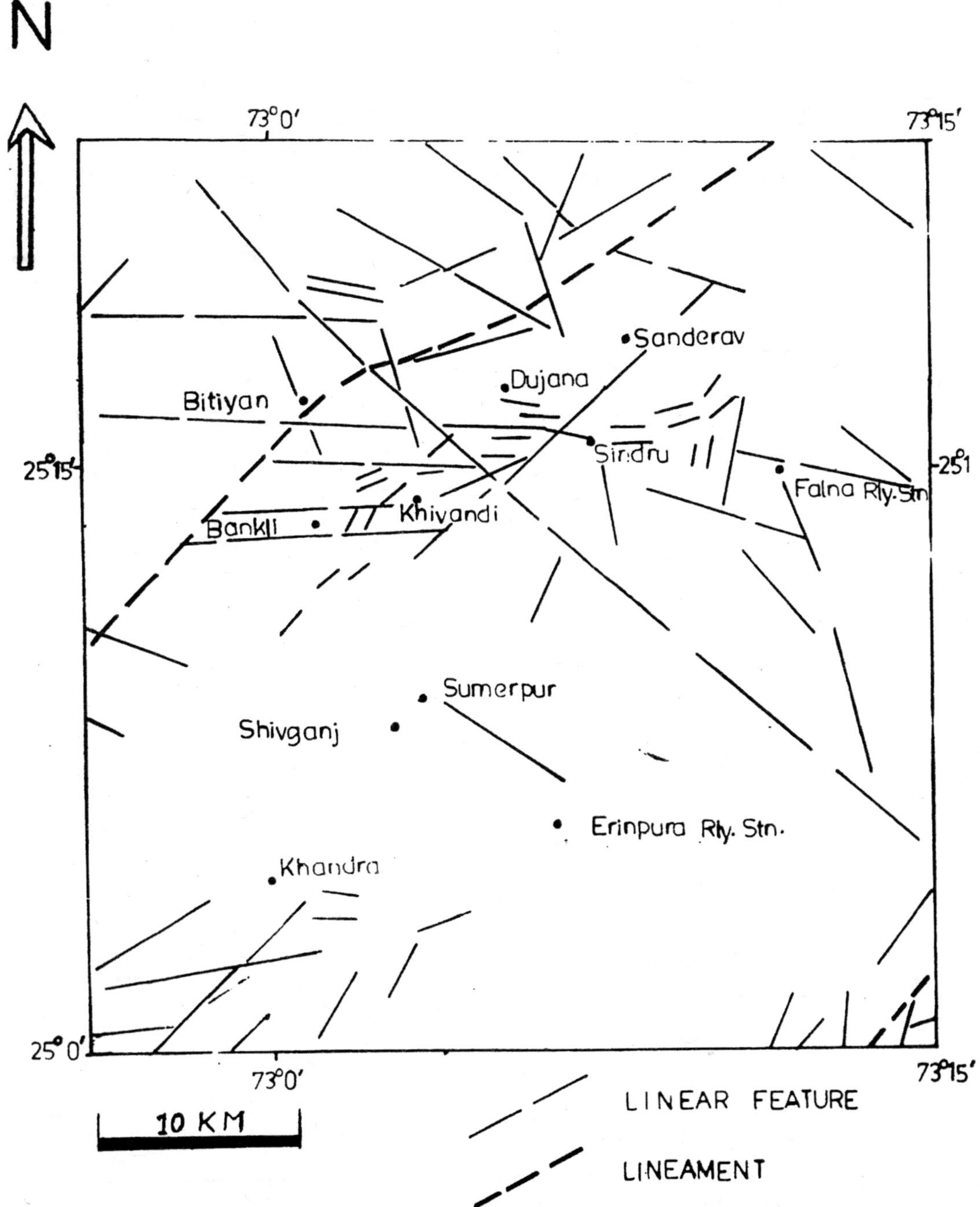

Fig. 4. Satellite Imagery Analysis Map of Shivganj-Sumerpur area (IRS-FCC data).

ıe Sindreth Group

The Sindreth Group is seen to rest unconformably over the Sirohi Group. It presented by a single formation - the Dujana Formation. The lower most member is formation is a hornfelsic derevative of metavolcanics, rich in hornblende, plagiocla dspar with quartz, sphene and epidote. The upper member of this formation compris artz-mica, graphite andalusite (chistolite) schist, carbonphyllite, slates and cher artzite. These rock types are exposed near Dujana, Koshelav and Morli Villages.

trusive Rocks

The Erinpura Granite showing an age of 830 m.y. (Choudhary *et al.* 1984) to 9: y. (Crawford 1975) is the prominent instrusive rock in the study area. It is general eyish-white in colour and medium to coarse grained. Besides Erinpura Granite tl her intrusives are dolerite dykes most of them are seen trending in NW-SE direction.

DEFORMATIONAL HISTORY

In the present investigation structural data were generated with the help of fie apping on 1.50,000 scale (Figs. 2 and 3) and analysis of satellite imagery (IRS-FC ta). The tracks of linear features from IRS- FCC are shown in Fig. 4. The structur story and deformational pattern has been worked out by carrying out geometric alysis of the bedding and foliation by preparing π and β diagrams with the help hmidt's equal area projection (Fig. 5).

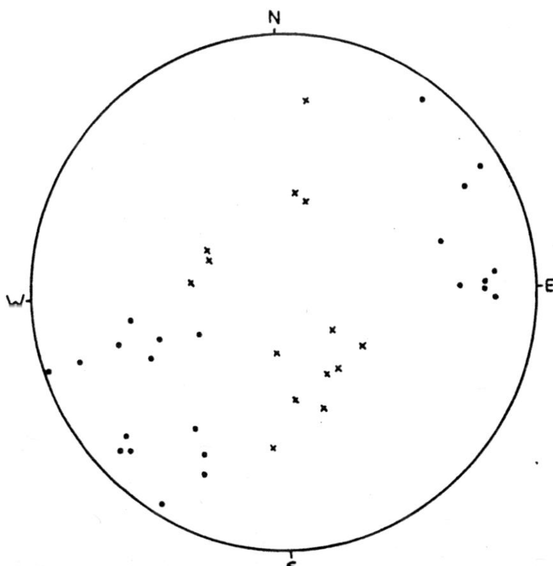

Fig. 5. Synoptic diagram of the study area showing orientation of F_2 and F_3 fold axes from the entire area. Dotts - Second generation fold axex and Cross - Third generatior fold axes.

The study has revealed convencing evidence in support of three phases of foldi ents in the Sirohi Group of rocks. Earlier two folding episodes in the Sirohi Group ocks are correlatable with the F_1 and F_2 folds of the Delhi Supergroup (Fig. 6).

Fig. 6. Photograph showing superimposition of F_1 and F_2 folds in the study area.

rst *Deformation* (**F_1 Folding**)

The earliest folds are tight, isoclinal folds (antiforms and synforms). They vary end between NE-SW to NNE-SSW (Fig. 7).

The most prominent planar structure (S_1) axial planar to F_1 folds is well develop schists and gneisses. These planar structures have been involved in F_2 and F_3 foldi hibiting discrete crenulation cleavage. Linear structures related to F_1 comprise min ld axes and cleavage bedding intersection.

econd *Deformation* (**F_2 Folding**)

The second generation folds (F_2) have deformed the earliest (F_1) folds. In the stu ea, the second folds have been observed in both microscopic as well as mesoscopic sca he upright folds in metavolcanics of Kavarpahar exhibit sharp hinges (Fig. 8).

The crenulation cleavage axial planar to these folds are seen at a number of plac he L_2 lineation related to F_2 folding consists of (1) minor fold axes and puckers edding (S_0) in the Sindreth Group of rocks and on S_1 planes in the Sirohi Group of roc) the intersection of S_1 and S_2 forming a lineation developed parallel to the F_2 fold axi

Fig. 7. Photograph showing the nature of the F_1 folds in the study area.

ird Deformation **(F_3 Folding)**

The F_3 folds, have developed on the limbs of F_2 folds. These folds trend in a NW-S
rection and are open upright in nature and form kink bands. These folds are seen we
veloped on the limbs of F_1 and F_2 folds and are accompanied by the development
ial plane fractures (S_3) trending in WNW-ESE direction.

STRUCTURAL ANALYSIS

The Stereographic analysis of various planar structures i.e. bedding and foliatic
as carried out by preparing π and β - diagrams using Ramsay's (1967) concepts
nderstand the over all geometry of major structures of the area. For this purpose tl
ea was divided into 5 sub-areas and for each sub-area a more or less indentical pictu
F_1, F_2 and F_3 relationship was observed. The axes of F_1, F_2 and F_3 of all the sub are
ve been plotted together in synoptic - diagrams and an overall and concise picture
e fold behaviour was evolved. It is seen that the axial planes of F_2 folds are more
ss vertical all over the area, though they do show a variation in the trend varyi
tween NNE-SSW and ENE-WSW due to F_3 superposition. F_2 axes however sho
nsiderable scattering and variation in plunge values between 2° to 60°. Most of the
lds plunge in the SW direction. The plunge values vary from moderate to almost su
rizontal. Very gentle plunge in NE to E directions is also seen.

Fig. 8. Second generation upright fold (F_2) in metavalconics at Kavarpahar in the study area.

CONCLUSION

The metasediments of the study area belong to the Sirohi and Sindreth Groups ıe Delhi Supergroup. The E-W Aravalli folding (cf. Paliwal, 1988) is not observed aı here in the area. Corresponding to Delhi folding the Sirohi Group of rocks show thr ıases of folding viz. F_1, F_2 and F_3 while rocks belonging to the Sindreth Group shc ıly last two folding episodes viz. F_2 and F_3.

ACKNOWLEDGMENT

Authors are thankful to Professor S.S. Merh of M.S. University of Baroda for tl ıidance rendered during the present work. Authors are also grateful to Professor B. aliwal of J.N.V. University, Jodhpur for offering valuable suggestions during final dra the manuscript.

REFERENCES

houdhary, A.K., Gopalan, K. and Sastry, C.A. 1984. Present Status of the geochronology of t Precambrian rocks of Rajasthan, *Tectonophysics,* 105 : 131-140.

ulson, A.L. 1953. The Geology of Sirohi State, Rajasthan, *Mem. Geol. Surv. India,* 53(1) : 1-22(

awford, A.R. 1975. Rb-Sr age determination of Mount Abu Granite and related rocks of Gujar *Jour. Geol. Soc. India,* 16 : 20-28.

ıpta, S.N., Arora, Y.K., Mathur, R.K., Iqballudin, Prasad, B., Sahil, T.N. and Sharma, S.B. 198 Lithostratigraphic map of Aravalli region, *Geol. Surv. India,* Calcutta.

ıpta, S.N., Mathur, R.K. and Arora, Y.K., Lithostratigraphy of Proterozoic rocks of Rajasth and Gujarat-Review. 1992. *Rec. Geol. Surv. India* 115 : 63-85.

ron, A.M. 1953. Geology of Central Rajputana, *Mem. Geol. Surv. India,* 79 : 1-389.

Touche, T.D. 1902. Geology of Western Rajputana, *Mem. Geol. Surv. India,* 35(1) : 78-93.

liwal, B.S. 1988. Deformation pattern in the rocks of the Aravalli Supergroup around Udaip city, Rajasthan. *Geol. Soc. India Mem.* 7 : 153-168.

ımsay, J.G. 1967. *Folding and Fracturing of Rocks.* McGraw- Hill, New York, 568p.

y, A.B. 1988. Stratigraphic and Tectonic Framework of the Aravalli Mountain Range. In : R A.B. (Ed.) *Precambrian of the Aravalli Mountain, Rajasthan, India, Mem. Geol. Soc. Ind* 7:3-32.

chanthavong, S.P. and Desai, S.D. 1977. Protoplate tectonics controlling the Precambri deformations and metallogenatic epochs of north western peninsular India, *Min. S Engngs,* 9 : 218-237.

chandthavong, S.P. and Merh, S.S. 1981. Protoplate tectonics : the energetic model for t structural, metamorphic and igneous evolution of the Precambrian rocks, north weste peninsular India, *Proc. SYMPET, Spel., Publ. Geol. Surv. India.* 419-457.

eological Evolution of Northwestern India
.S. Paliwal (Ed.)
cientific Publishers (India), Jodhpur, 1999, pp. 94 - 102.

Nb, Zr, Y, REE RICH ACID DYKE ROCKS FROM THE PIPLUN AREA OF SIWANA RING COMPLEX, WESTERN RAJASTHAN, INDIA

G. Vallinayagam

Department of Geology, Kurukshetra University, Kurukshetra 136 119

ABSTRACTS

The Siwana Ring Complex comprises acid volcano - plutonic rocks with minor amount of basics. The acid dyke rocks present here include microgranite, rhyolite, trachydacite, trachyandesite which cut across granites as well as rhyolites. They trend NE-SW and extend upto 50 - 70 m with a width ranging from a few cm to 2 m. The contact between the dyke rocks and the granites is sharp. The dyke rocks are fine to medium grained, vary in colour and consist of quartz, alkali feldspar, alkali amphibole, alkali pyroxene as essential minerals. The accessory minerals are hematite, ilmenite, zircon, monazite and rutile. They are high in SiO_2, Na_2O+K_2O, and total iron, low in MgO, CaO, Co, and Cr. They are anomalously high in Nb, Ta, Zr, Hf, U, Th, Y, Ga, La, Ce and HPU as compared to the normal acid rocks. The enrichment pattern may attribute the hydrothermal alteration process or high content of the metals in the host rocks. They have probably emplaced along the ring fractures of Siwana. These acid dyke rocks may be a possible target for detail rare metal exploration.

ey words : Siwana Ring Complex, REE, Acid Dykes, Piplun, Rajasthan.

INTRODUCTION

The rocks of Siwana include the volcano-plutonics of acidic and basic componen hey belong to the Late Proterozoic tectono thermal event of Malani Igneous Suite in tł orth western part of India. The literature survey (e.g. La Touche, 1902; Crawford an ompston, 1970; Bhushan and Mohanty, 1988; Bhushan, 1995; Eby and Kochhar, 199 ıows their regional tectonic, genetic and mineralization aspects. Narayan Das *et a*

978) inferred that Siwana microgranites contain high Y and Zr content. Kochhar (198 entified the Siwana, Jalore, Tosham granites of Malani Igneous Suite as High He roducing (HHP) granite based on the study of the surface abundance of U, Th, K ar elected trace elements. Based on certain pathfinder elements and hydrotherm teration pattern, Kochhar (1992) reported mineralization associated with A-ty alani magmatism. Baskar and Sharma (1994) classified Siwana granites as rare met anitoid of agpaitic type based on petromineralogical and geochemical data. Recentl ain *et al.* (1996) recorded a number of peralkaline, agpaitic granitoid dykes from tl stern half of the Siwana Ring Complex rich in rare metals and rare earth element he present note for the first time highlights the occurrence of the rare metal acid dyl cks from southcentral portion of the Siwana Ring Complex.

GEOLOGICAL SETTING

Geological mapping (on 1:25000 scale) of Piplun area (topographic sheet No. 45C\ Siwana Ring Complex reveals the rock units like basalt, gabbro, trachyte, rhyolit rious types of granites and dyke rocks (Fig. 1). No older rocks are exposed in the stu ea. The rock units encountered in the area can braodly be grouped into the followi ajor phases :-

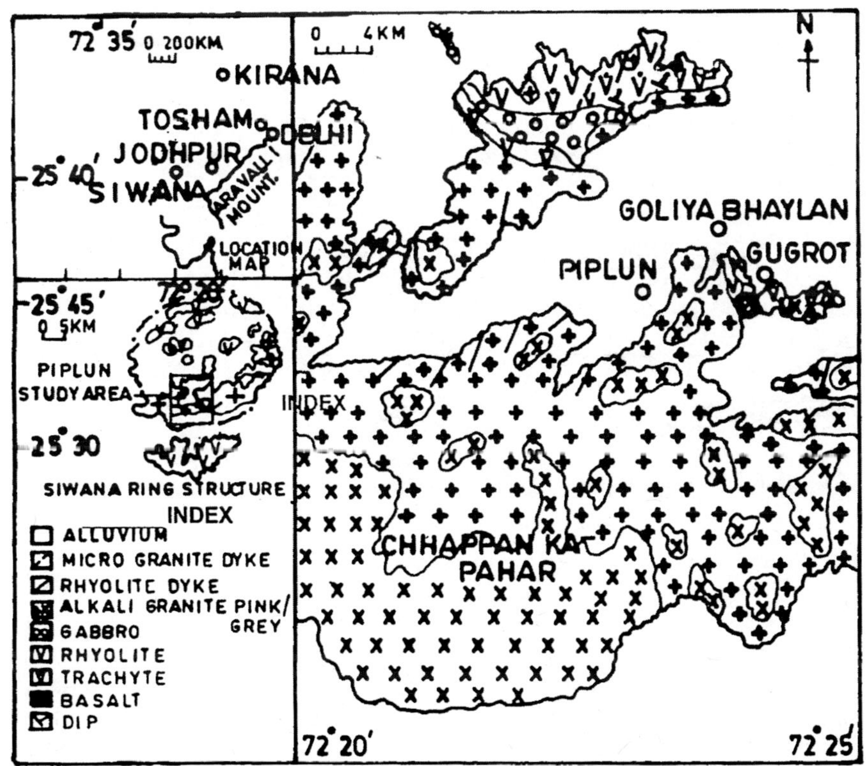

Fig. 1. Geochemical map of Piplun area, Siwana, Rajasthan.

Extrusive phase : Basalt, trachyte, rhyolite, welded tuff.

Intrusive phase : Gabbro, various types of granites (dark gray, gray, pink).

Dyke phase : Trachydacite, Trachyandesite, rhyolite, microgranite.

The basalt occurs as flow associated with rhyolite and welded tuff. Gabbro occurs . trusive body cut by granitic veins. The gabbroic enclaves occur in trachyte. The grani trusion into acid volcanics indicates the volcano-plutonic association and tl ıbvolcanic nature of the granite. It also suggests the fracturing of the earth crust durii ıeir emplacement. The contact between the granite and the acid volcanic rocks is sha ıd inclined towards the center of the Siwana Ring Complex.

These acid volcano-plutonic rocks are cut across by numerous dykes of microgranit ıyolite, trachydacite, trachyañdeste and trend NE-SW. The dyke rocks are compa ırd and porphyritic (pink orthoclase phenocryst) as well as nonporphyritic (fine graine ·een, grey, pink in colour with iron oxide encrustation on the surface of the dyke roc ıe to the hyrothermal alteration process. The dykes extend upto 50-70 m (at places up km) with width ranging from a few cm to 2 m. The contact between the dyke and tl ost (granite and rhyolite) rocks is sharp. Near the contact the minerals are fine grain ıd towards the centre they become coarse. At places, light and dark layers are observ ithin these dykes which are represented by felsic and mafic minerals. The xenoliths ·anite and rhyolite are also noted in the dyke rocks. Swelling and pinching are tl ıaracteristic features of the dykes.

PETROGRAPHY

Under the microscope, the microgranite dyke rock is fine to medium granied aı ontains (modal composition) quartz (24%), alkali feldspar (perthite, orthoclase) (39% rfvedsonite (10%), aegirine (4%) and riebeckite (3%) as essential minerals. T ccessories are hematite, ilmenite, zircon, monazite and rutile. Perthite and orthocla ccur as phenocrysts, subhedral and contain fine arfvedsonite, riebeckite inclusior erthite altered to sericite and kaolin. Sericite shows high order interference colou nder the crossed nicols. Kaolin appears light grey under polarised light and isotroı nder crossed nicols. The orthoclase is altered to Kaolin. Quartz occurs as phenocrys nhedral crystals are disseminated in the microcrystalline groundmass and show wa xtinction. The intergrowth of alkali feldspar and quartz which gives rise to granophyı xture is also observed. The arfvedsonite and riebeckite inclusions in quartz a ommon. The arfvedsonite and aegirine also occur as phenocrysts and fine crystals. T ebeckite is needle shape (accicular) crystals, blue in colour and is intergrown wi rfvedsonite. Few minute crystals of reddish brown aenigmatite are also encounter long with arfvedsonite.

The abundant accessories are hematite (7%), zircon (5%), ilmenite (4%) in quartz lsdpathic groundmass as well as in mafics. The hematite is red in colour and occurs henocrysts and fine anhedral crystals in the groundmass. The zircon is identified by rismatic form, absence of cleavge and high relief. It is included in the perthite and al ı the groundmass. The ilmenite is silver grey in colour, anhedral in shape, embedded ıe groundmass as well as intergrown with arfvedsonite. Ilmenite at plac ydrothermally altered to leucoxene which is identified by its white internal reflectic ght yellow colour, small crystals of monazite are also observed with quartz an oundmass. The rhyolite and trachydacite dykes show cryptocrystalline and trachyt: xture. They contain similar mineralogy of microgranite which suggest comagmat: ıture of the suite.

GEOCHEMISTRY AND MINERALISATION

Based on the above studies, the major, trace and rare earth elements of differer 'ke rocks were analysed (Table 2) by XRF, ICP-MS at National Geophysical Researc stitute Hyderabad. Volcanics are classified by TAS diagram (Lebas *et al.* 1986) and th utonics by QAP diagram (Streckeisen, 1967). They are high in SiO_2, Na_2O+K_2O, an tal iron, low in MgO, CaO, Co, and Cr and anomalously high in Nb, Ta, Zr, Hf, U, Tl Ga, La, Ce and HPU. They have acmite in norm and the agpaitic index 1 indicates th ır alkalinity of the rocks. The Differentiation Index is 76-89 and the Oxidation Ratio i :-83. All rocks were plotted in different variation diagrams to assess their chemic: ıriability, tectonic environment and the nature and evolution of magma. They are a ıder (Table 1)

ıble 1.

[> Wt. % of Oxides (Thornton and Tuttle, 1960)	Silica, alumina, soda, potash and iron show scatteı ing of data, P_2O_5, TiO_2, MgO and CaO define lineɛ trend which suggests non-orogenic environment.
›g 10 K_2O / MgO > SiO_2 diagram .oges and Greenbrg, 1981)	Plot in alkali granite field
FM diagram (Irvine and Barager, 1971)	Plot in tholeiitic field
olar A/CNK > A/NK diagram 1aniar and Piccoli, 1989)	Plot in peralkaline field (Fig. 2)
l > R2 diagram (Batchelor and Bowen, 1985)	Plot in anorogenic field
›g (Y + Nb) > Log Rb and Log Y > Log Nb 'earce et al. 1984)	Plot within plate granite field (Fig. 3)
/Rb diagram (Shaw, 1968)	Plot in main trend and cluster towards pH tren which may indicate the metasomatism of rock
ormalized REE diagram	Show high REE contents, enriched in LREE wit marked Eu anomaly (Fig. 4). Ce/Yb 9.85 - 13.26
'imordial mantle normalized trace element diagram	Peaks of U, Th, Nb, La, Ce, Hf, Zr, and through of B: Sr, Ti (Fig. 5) Th/U 2.8-4.5; Zr/Hf 32-41

The dyke rocks are very low in Ba/Rb (0.1-0.4), K/Rb (81), TiO_2/Ta (0.01) ratic 'able 2) and show negative correlation in Rb versus Sr plot. These in tern support tl 'ogressive differentiation of their host magma which may have caused tl ineralization of these dyke rocks. On Rb-Ba-Sr ternary diagram (Biste, 1979), the rock ot in the field of strongly differentiated granites. In K\Rb ratio versus Rb/Sr ratio plo .ey fall in the mineralized granite field.

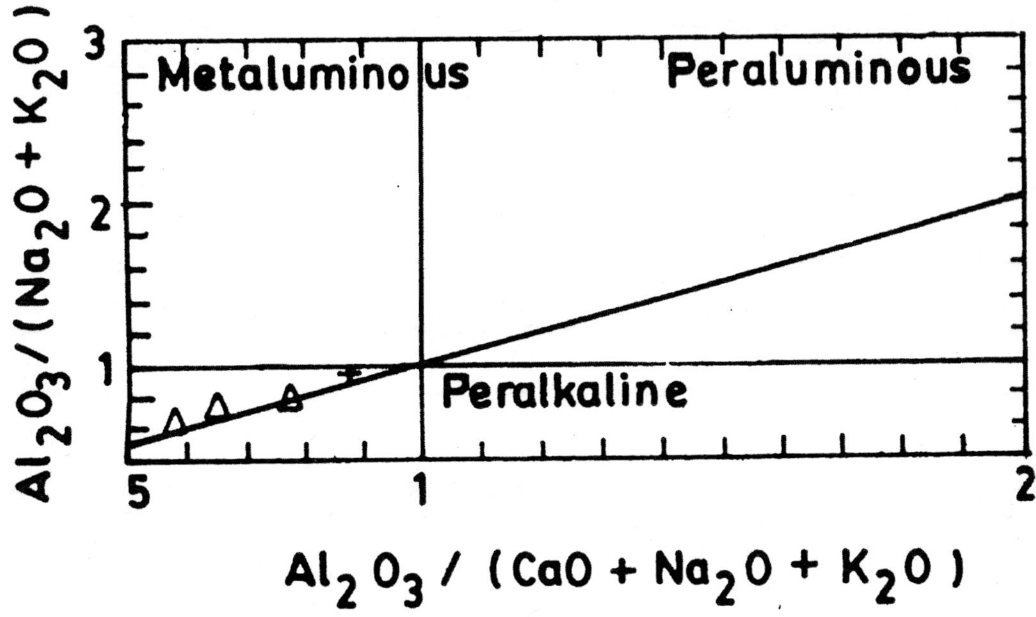

ig. 2. Molar Al_2O_3/ (CaO + Na_2O + K_2) - Al_2O_3 / Na_2O+K_2) diagram (after Maniar and Pic(1989) + Micrograinte, Δ rhyolite.

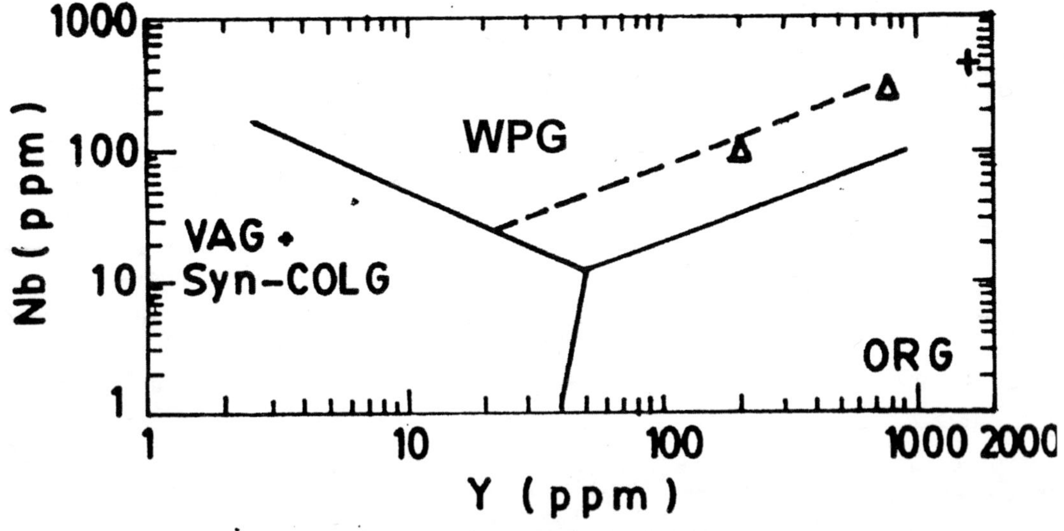

Fig. 3. Log (Y-Nb) (after Pearce et al. 1984) diagram. Symbols same as that of Fig. 2

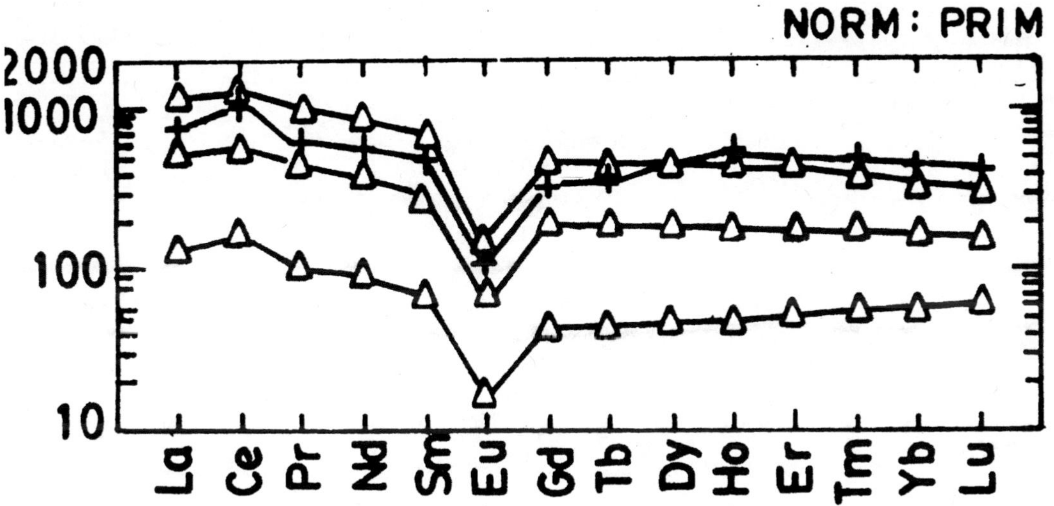

Fig. 4. Normalized (PRIM) REE plot. Symblos same as that of Fig. 2.

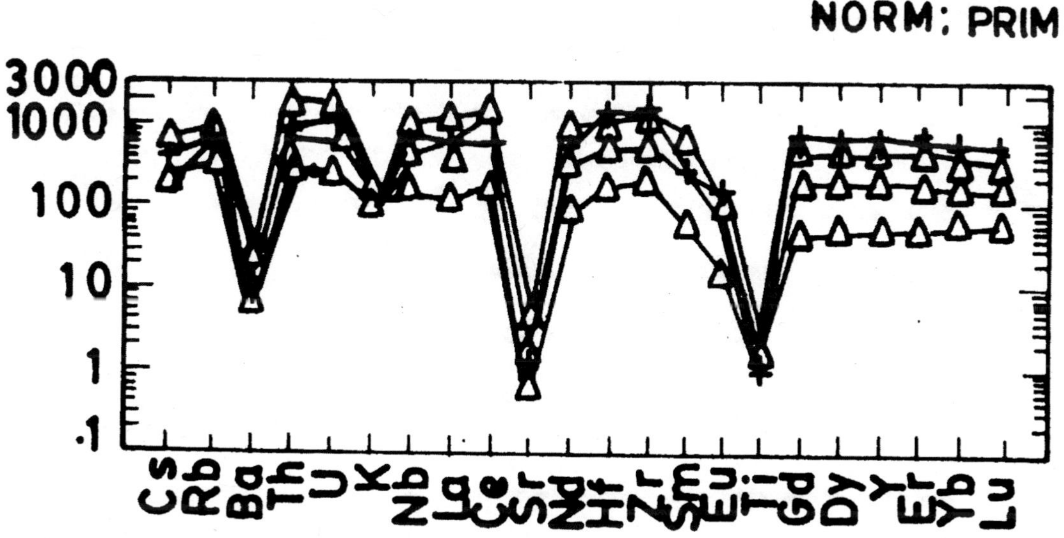

Fig. 5. Normalized (PRIM) trace element plot. symbols same as that of Fig. 2.

ble 2. Geochemical data of acid dyke rocks, Siwana Ring Complex, Rajasthan

mple	MIA	P1	C3	GWS
ck type	Mi gry dy	Rhy dyke	Rhy dyke	Rhy dyke
t. %				
O_2	69.05	73.69	73.19	73.05
O_2	0.21	0.26	0.27	0.28
$_2O_2$	11.10	8.97	7.39	11.08
$_2O_2$	7.49	6.02	8.45	5.45
nO	0.14	0.04	0.01	0.04
gO	0.27	0.14	0.22	0.29
O	0.61	0.89	0.75	0.27
a_2O	4.55	4.35	4.93	5.09
O	3.71	4.60	3.04	4.94
O_5	0.17	0.03	–	0.02
tal	97.30	98.99	98.25	100.51
m				
	—	2	2	2
	1	—	—	—
	1	1	1	1
	6	3	5	2
	1	12	6	18
	181	5	12	6
	330	93	175	27
	759	499	816	283
	30798	38186	25326	41008
	365	426	308	231
	2.75	3.38	1.70	2.26
	122	47	115	57
	27	39	74	11
	61	49	57	49
	54.33	20.90	52.81	7.20
	446.7	281.4	857.4	112.7
	358.90	185.44	308.25	50.83
	11545	8425	11162	2107
	1259	1559	1619	1679
	1883	802	1853	203
	83.05	68.44	162.16	27.27
	28.89	15.22	36.00	6.05

The dyke rocks are anomalously high in Nb (112-657 ppm), Ta (7-54 ppm), Z
167-11545 ppm), Hf (82-718 ppm), U (6-36 ppm), Th (27- 162 ppm), Y (203-1853 ppm
a (48-61 ppm), La (82-718 ppm), Ce (268-2026 ppm) and Heat Production Unit (Ashw
al., 1987) (HPU) 4-21 uWm^{-3} as compared to normal acid rocks. The unusally hig
undance of rare metals in dyke rocks is due to their high content in host rocks and th
evalence of hydrothermal alteration process for which these suite bear the eviden
allinayagam, 1988). The ring fractures of Siwana area also probably acted as channe
r the anomalous concentration of these metals. Hence these acid dyke rocks may be
ssible target for detail rare metal exploration.

ACKNOWLEDGEMENT

The present work was supported by DST SERC New Delhi grant (No
S/23/163/92). The author is grateful to Dr. V. Balaram and Dr. P.K. Govil, Nation
eophysical Research Institute, Hyderabad for the chemical analyses.

REFERENCES

hwal, L.D., Morgan, P., Kelly, S.A. and Preicival, G.A. 1987. Heat production in an Archea crustal profile and implication for heat flow and mobilization of heat producing element *Earth Planet. Sci. Lett.,* 85 : 439-450.

iskar, R. and Sharma, R. 1994. An assessment of the rare metal potential of the granitoids Siwana, Jalore, Jhunjhunu and Tosham, North Western Peninsular India. *Curr. Sic.*, 66 67-69.

itchelor, R.A. and Bowen, P. 1985. Petrogenetic interpretation of granitoid rock series usi multicationic parameters. *Chem. Geol.*, 48 : 43-55.

ste, M. 1979. Die Anwendung geochemischer Indikatoren auf die Zinn - Hoffigkeit herzynisch granite in Sud - Sardinien. *Berl. Geoweiss*, Abh., 18 : 1.

iushan, S.K. 1995. Late Proterozoic continental growth : Implications from geochemistry of ac magmatic events of west Indian craton, Rajasthan. *Mem. Geol. Soc. India* 34 : 339-355.

iushan, S.K. and Mohanty, M. 1988. Mechanics of intrusion and geochemistry of alkali granites from Siwana, Barmer, Distt., Rajasthan. *Ind. Jour. Earth Sci.*, 15 : 103-115.

awford, A.R. and Compston, W. 1970. The age of Vindhyan System. *Qurt. Jour. Geol. Sc London,* 125 : 351-371.

y, G.N. and Kochhar, N. 1990. Geochemistry and petrogenesis of the Malani Igneous Suit North Peninsular India. *Jour. Geol. Soc. India* 36 : 109-130.

rine, T.N. and Barager, W.R.A. 1971. A guide to the chemical classfication of the comm volcanic rocks. *Can. Jour. Earth Sci.,* 8 : 523-548.

in, R.B., Miglani, T.S., Kumar, S. Swarnkar, B.M. and Singh, R. 1996. Rare metal and ra earth rich peralkaline, agpaitic granitoid dykes of Siwana Ring Complex, District Barme Rajasthan. *Curr. Sci.* 70 : 854-858.

ochhar, N. 1989. High heat producing granites of the Malani Igneous Suite, northern Peninsul India. *Ind. Minerals*, 43 : 339-346.

chhar, N. 1992. Mineralization associated with A-type Malani magmatism, northwe Peninsular India. In : S.C. Sarkar (Editor) *Metallogeny related to tectonics of the Proterozc mobile belts* : Oxford and IBH Publishing Co. Pvt. Ltd. pp. 199-224.

Touche, T.D. 1902. Geology of western Rajputana. *Mem. Geol. Surv. India,* 35 : 116.

ɔas, M.J., Lematre, R.W., Streckeisen, A. and Zanetin, B. 1986. A chemical classification volcanic rocks based on the total alkali-silica diagram. *Jour. Petrol.*, 27 : 745-750.

.niar, P.D. and Piccoli, P.M. 1989. Tectonic discrimination of granitoids. *Geol. Soc. Am. Bu* 101 : 635-643.

rayan Das, G.R., Bagchi,A.K., Chaube, D.N., Sharma, C.V. and Navaneetham, K.V. 1978. Ra metal content, Geology and tectonic setting of the alkaline complexes across the Tran Aravalli region, Rajasthan. In : *Recent Research in Geology*, Hindusthan Publishi Corporation, Delhi, 7 : 201-217.

ırce, J.A., Harris, N.B.W. and Tindle, A.G. 1984. Trace element discrimination diagrams for tl tectonic interpretation of granitic rocks. *Jour. Petrol.*, 25 : 956-983.

ȝers, J.J.W. and Greenberg, J.K. 1981. Trace element in continental margin magmatism : ı III, Alkali granites and their relationship to cratonisation : Summary. *Geol. Soc. Am. Bul* 92 : 6-9.

ıw, D.M. 1968. A review of K-Rb fractionation trends by covariance analysis. *N. Jb. Minerc Abh.,* 109 : 144-240.

ɔrnton, C.P. and Tuttle, O.F. 1960. Chemistry of igenous rocks. pt. 1, Differentiation Index. *Ar Jour. Sci.,* 258 : 664-684.

llinayagam, G. 1988. Geology and Geochemistry of alkali granites and the associated ac volcanics around Mokalsar, Distt. Barmer, W. Rajasthan, India and their bearing on the ri tectonics. Unpublished. *Ph.D. Thesis,* Punjab University, Chandigargh. 1-113.

ological Evolution of Northwestern India
S. Paliwal (Ed.)
ientific Publishers (India), Jodhpur, 1999, pp. 103 - 108.

GEOCHEMISTRY AND PETROGENESIS OF ALKALI RHYOLITES FROM MANDOR, JODHPUR DISTRICT, WESTERN RAJASTHAN

Vinod Agarwal, Anil Maheshwari* and Mirza Mubeen Baig

Department of Geology, Mohan Lal Sukhadia University, Udaipur - 313 001
**Department of Geology, University of Rajasthan, Jaipur 302 004*

ABSTRACT

The alkali rhyolites from Mandor area are chemically similar to A-type granitoids and were generated by partial melting of curstal rocks under the influence of rise of mantle under an extensional tectonic regime.

ey words : Malani Igneous Suite, Geochemistry, Petrogenesis, Alkali Rhyolite Mandor, Jodhpur

INTRODUCTION

The Malani Igneous Suite of the Precambrian age in western and southwester ajasthan represents the largest felsic suite in India. Malani Igneous Suite covers a ea of 50,000 sq. kms. and occurs in scattered outcrops from south of Sirohi to Pokara d from east of Jodhpur to the edge of the Thar desert. The suite is characterized b eponderance of rhyolites over intermediate and basic rocks. The prese mmunication describes in brief the geochemistry including the oxygen isoto mposition of the Malani rhyolites exposed around Mandor, Jodhpur district (latitu 6°15′ - 26°25′ and longitude 72°55′ - 73°5′).

GEOLOGY OF THE AREA

The volcanics studied are represented by a number of hills trending NE-SW and a nconformably overlain by the sandstone of the Marwar Supergroup (Fig.1). Th

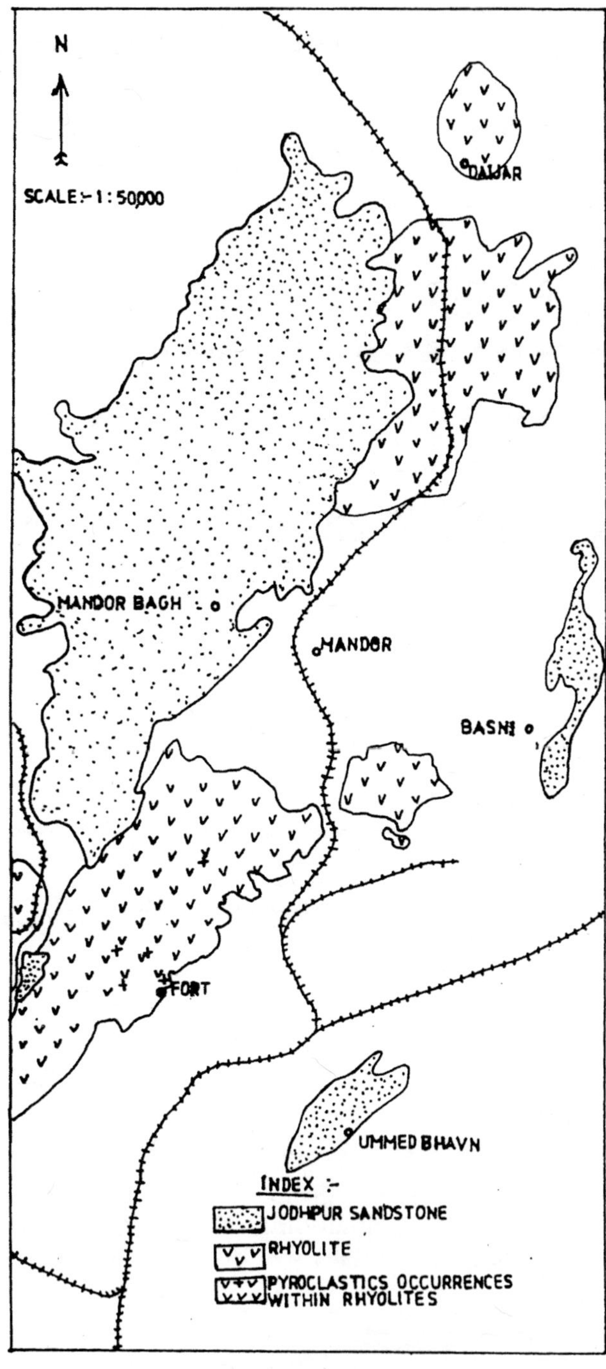

Fig. 1. Geological map of the Mandor area, Jodhpur

preliminary account of thes volcanics of the Jodhpur area wa provided by La Touche (190 Pareek (1981) and Paliwal (1994).

The investigated area i represented by rhyolitic lavas an ignimbrites. The preponderance c rhyolites over ignibrites is we observed in the area. The rock generally show well marke columnar jointing. The rhyolite are brown, green and grey i colour, at places typically flo banded, with occasion phenocrysts of quartz and alka feldspar, are set in a groundmas of micro to cryptocrystallin granular in nature and spherulite of quartzo-feldspathic materia The phenocrysts are generall unaltered.

The ignimbrites ar characerised by a variety c colours (i.e. grey, brown, reddis brown etc.), presence of devitrifie glass shards, angular phenocryst of quartz and feldspar an collapsed pumice fragment Volcanic breccia has also bee reported at places.

GEOCHEMISTRY AND PETROGENESIS

The major and trac elements were analysed at th Department of Geology, M.L.S University, Udaipur by usin Spectrophotometer and Atomi Absorption Spectrophotomete The oxygen isotope studies wer carried out at LABISE, Brasil an the values have been reported i the standard notations.

The chemical data and CIPW norms of the representative rhyolitic rocks have bee
ovided in Table 1. The rhyolitic rocks are charecterized by high silica contents, hig
vel of alkalis with the dominance of K_2O over Na_2O, low content of MgO and CaO an
pletion of Co and Ni. These chemical charecteristics are diagnostic of A-tyı
norogenic) granitoids (Loiselle *et al.*, 1979; Whalen *et al.*, 1987).

ıble 1. Chemical composition and normative mineralogy (CIPW) of alkali rhyolites froı Mandor, Jodhpur district. Oxides and norms (weight %), trace elements (ppm).

	MB 75	MB 52	MB 108	MB 82	MB 109
SiO_2	73.64	73.66	73.88	74.14	73.14
TiO_2	0.05	0.14	0.05	0.04	0.10
Al_2O_3	14.03	13.28	14.23	13.77	14.44
Fe_2O_3	1.46	2.64	1.46	1.60	1.84
FeO	0.08	0.04	0.08	0.04	0.08
MnO	0.03	0.03	0.02	0.01	0.03
MgO	0.21	0.10	0.01	0.08	0.02
CaO	0.08	0.15	0.39	0.0	0.33
Na_2O	2.55	0.25	2.26	2.52	3.54
K_2O	5.48	7.45	5.74	5.58	5.16
P_2O_5	0.01	7.45	0.01	0.00	0.03
IPW norms					
Q	37.39	43.33	37.94	38.01	32.15
Or	32.29	44.03	33.92	32.98	30.49
Ab	21.58	2.12	19.12	21.32	29.95
An	0.33	0.48	1.87	0.0	1.44
C	3.78	4.63	3.61	3.58	2.50
Hy	0.52	0.25	0.02	0.20	0.05
ace elements (ppm)					
Rb	271	172	253	260	159
Sr	24	9	29	22	35
Ba	136	38	312	79	771
Li	18	1	4	17	0
Ni	6	10	6	5	13
Cr	14	10	20	22	26
Co	8	8	13	8	8
Cu	7	7	6	5	8
Zn	35	33	47	25	41
$\alpha^{18}O$	+9.42				

The rhyolites in the TAS diagram (Le Bas *et al.*, 1986) plot in the alkali field (Fig.
the sense of Miyashiro (1978). The rhyolites are classified as high-K volcanics (Fig.
cording to Le Maitre (1989). In addition, the rhyolites plot in the field of alkali-feldspa
yolites on the Q-ANOR diagram (Fig. 4) of Streckeisen and Le Maitre (1979).

Based on the molar ratio of $Al_2O_3/CaO + Na_2O + K_2O$, the rhyolites may lassified as peraluminous. The peraluminous nature is best manifested in t ppearance of normative corundum in the norm. On the R1-R2 diagram (De La Roche *l.*, 1980), the rhyolites lie within the fields of anorogenic and post-orogenic grani which represent an extensional tectonic regime.

The rhyolites are characterised by a restricted range of silica alongwith high SiC K_2O, Rb and low Na_2O, characteristic of S type granitoid (Chappell and White,198 These rhyolites fall in the S type field in the K_2O - Na_2O diagram of Hine *et al.* (197 The rhyolite sample (MB 75) analysed for oxygen isotope composition gave a 18O value - 9.42 per mil. The high 18O values (+9 per mil) are genrally assumed to be produced nelting of metasediments (Taylor and Turi, 1976).

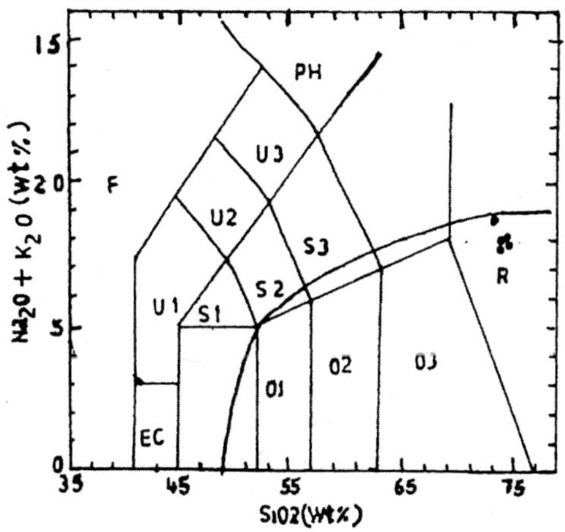

Fig. 2. TAS diagram (Le Bas *et al.* 1986) showing the position of Mandor rhyolites in alkal field of Miyashiro (1978)

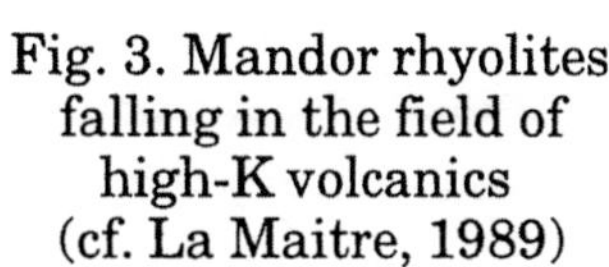
Fig. 3. Mandor rhyolites falling in the field of high-K volcanics (cf. La Maitre, 1989)

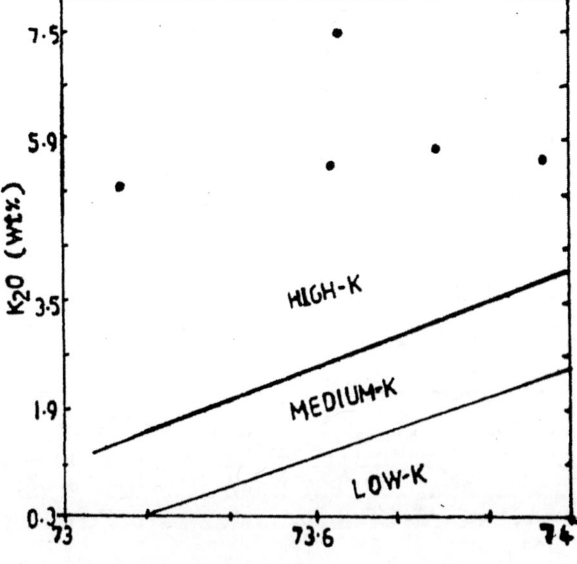

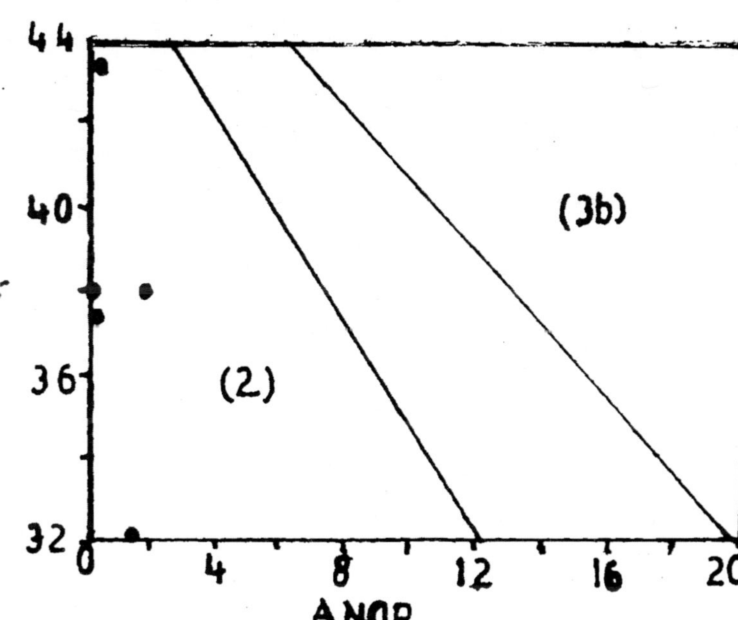

ɡ. 4. Q-ANOR diagram showing the position of Mandor rhyolites in the field of alkali-feldspar rhyolites (cf. Streckeisen and La Maitre, 1979)

Volcanic suite wit unusally large volum of rhyolite and a scarcit of associated basalts a peculiar of continent rift zones (Wilson, 198 The rhyolitic rocks Mandor area hav geochemical featur typical of A-typ granitoids. The chemic features also point their association with a extensional tecton regime. The region gravity anomaly map western Rajastha indicates th Mohorovicic discontinuity at shallow depth of 30 Kr The elevated mantle an thickening of crust h resulted due to extensic under the continent rift mechanis (Bhushan, 1989).

The generation of Mandor rhyolites by fractional crystallization of mantle derive agmas are not viable mechanism in the light of chemical composition, normative dat eponderance of acidic volcanism and the absence of intermediate and basic rocks. Th le of crust in the generation of Malani rhyolites has already been reported (Bhusha 89; Maheshwari *et.al.* 1996). The rhyolite of the investigated area analysed for oxyge otope composition also gave a high 18O value of +9.42 per mil, which is attributed e role of crust in the generation of Mandor rhyolites.

ACKNOWLEDGEMENTS

Vinod Agarwal is grateful to CSIR, New Delhi for providing necessary financi sistance. Anil Maheshwari is thankful to LABISE, Brasil for providing facilities rry out oxygen isotope studies.

REFERENCES

ig, M.M. 1994. Geochemistry and eruptive mechanism of the felsic volcanics around Mando Jodhpur area, Rajasthan. *Unpublished Ph.D. Thesis,* M.L.S. University, Udaipur, 102 p.

ıushan, S.K. 1989. Late Proterozoic granitoid plutonism in northwestern parts of Indian shiel(Int. Symp. "*Structure and Dynamics of the Indian Lithosphere*". Hyderabad. 53-55.

ıappell, B.W. and White, A.J.R. 1984. I and S types granite in the Lachlan fold bel southeastern Australia. In : Xu Kegin Tu Guangchi (Ed) *'Geology of granites and thei metallogenetic relations*. Science Press, Beijing, China. 87-101.

La Roche, H., Leterrier, J., Grand, C.P. and Marchal, M. 1980. A classification of volcanic an plutonic rocks using R1-R2 diagrams and major element analyses : its relationship wit current nomenclature. *Chem. Geol.*, 29 : 183-210.

ne, R., Williams, I.S., Chappell, B.W. and White, A.J.R. 1978. Contrasts between I and S typ granitoids of the Kosciusko Batholith. *Jour Geo. Soc. Aust.,* 25 : 219-234.

Touche, T.H.D. 1902. Geology of western Rajputana. *Mem. Geol. Surv. India,* 35(1) : 1-116.

Bas, M.J., Le Maitre, R.W., Streckeisen, A. and Zonettin, B. 1986. A chemical classification (volcanic rocks based on total alkali silica diagram. *Jour. Pet.,* 27 : 747 - 750.

Maitre, R.W. 1989. A classification of Igneous rocks and glossary of terms : Recommondatior of the IUGS sub-commission on the systematics of Igneous rocks, Blackwell, Oxford.

ıiselle, M.C. and Wones, D.R. 1979. Characteristics and origin of anorogenic granites. *Geol. So Am. Abst. Prog.,* 11 : 468.

aheshwari, A., Coltorti, M. and Sial, A.N. 1996. Crustal Influences in the petrogenesis of Malaı rhyolites, Southwestern Rajasthan : combined trace elements and oxygen isotop constraints. *Jour. Geol. Soc. India.*, 47 : 611 - 619.

iyashiro, A. 1978. Nature of alkalic volcanic rock series. *Contrib. Mineral. Petrol.*, 66, 91-104.

ıliwal, B.S. 1994. Tectonics of the Post-Aravalli Mountain building activity and its bearing o the accumulation of sediments along the western flank of the Aravalli Range, Rajasthaı India. In : Ahmed, R. and Sheikh, A.M. (Eds.) *Geology in South Asia - I.* January 19-2 1992. Hydrocarbon Development Institute of Pakistan, Islamabad, Pakistan. 52-60.

ıreek, H.S. 1981. Petrochemistry and petrogenesis of the Malani igneous suite, India, *Bull. Geo Soc. Am.* Part II. 92(2) : 206-273.

reckeisen, A. and Le Maitre, R.W. 1979. A chemical approximation to the modal QAP classification of Igenous rocks. *N. Jb. Mineral Abh.,* 136 : 169-206.

ıylor, H.P. Jr. and Turi, B. 1976. High 180 igneous rocks from the Tuscan magmatic provinc Italy. *Contrib. Mineral Petrol.*, 55 : 33.

halen, J.B., Currie, K.L. and Chappell, B.W., 1987. A type granites : Geochemical charcteristic discrimination and petrogenesis. *Contrib. Mineral. Petrol.*, 95 : 407-419.

ilson, M. 1989. *Igneous Petrogenesis*, Unwin Hyman, London, 373 p.

ieological Evolution of Northwestern India
!.S. Paliwal (Ed.)
cientific Publishers (India), Jodhpur, 1999, pp. 109 - 117.

BIVALVE BIOSTRATIGRAPHY OF THE JAISALMER FORMATION, WESTERN RAJASTHAN, INDIA

R.P. Kachhara and R.L. Jodhawat*

Department of Geology, Nagaland University, Kohima - 797 002
**Department of Geology, M.L. Sukhadia University, Udaipur - 313 001*

ABSTRACT

A sequence of marine sediments deposited during the Jurassic period is well exposed around Jaisalmer town, Rajasthan. These sediments are grouped into Jaisalmer, Baisakhi and Bhadesar Formations in the ascending order. Ammonoids - the precise time indicators occur but in case of the Jaisalmer Formation confined to the upper horizons only. On the other hand, many horizons throughout the Jaisalmer Formation reveal the presence of a rich and varied bivalve taxa. A few taxa have short vertical range thus they are of immense value in the biostratigraphic zonation. Altogether, seven bivalve zones are recognized. Basing on these it is inferred that the Jaisalmer Formation is equivalent of the Patcham and Chari series of Kutch and ranges from Lias to Oxfordian but Bajocian and Upper Callovian stages are not well represented.

{ey words : Biostratigraphy, Bivalve, Jaisalmer Formation, Western Rajasthan.

INTRODUCTION

The Jaisalmer Formation has an intertonguing relationship with the underlyi ,athi Formation and in turn overlain unconformably by the Baisakhi Formatio ccording to Pareek (1981) this Formation comprises a sequence of cream coloured, bu rown limestone, fossiliferous limestone, oolitic limestone and greyish brown sandston he sequence is well exposed especially in an area of about 400 sq. kms. arou aisalmer town. The sediments are dominantly buff and yellow in colour, forming eries of parallel escarpments, in the vicinity of Jaisalmer town. The beds trend NE-S vith very low dips (about 3° to 5°) towards NW. These strata entomb mega invertebrat ımely; bivalves, gestropods, brachiopods, corals, cephalopods, echinoids and crinoids ıe decreasing order of abundance. Among these, bivalves are quite common, w ·eserved and distributed throughout the sequence. Earlier workers have hardly pa ıy attention to bivalves in view of their long range. During the course of present stuc ıthors observed that a few bivalve taxa have short vertical range thus they play gnificant role in formulating the biostratigraphy of the Jaisalmer Formation vis-a-v ırrelation with the sequences of other areas.

SUCCESSIONS

The Jaisalmer limestone of Blanford (1877) is now known as the Jaisalm ›rmation. The Formation has been divided into the Joyan, Fort, Badabag, and Kuldh ıits. Of these Dasgupta (1975) has retained Joyan, Fort, Badabag and Kuldhar as t embers adding one more towards the base namely Hamira Member. Kachhara aı ›dhwat (1981) have favoured the division of the old Kuldhar Member into Kuldhar aı ıjiya Members restricting former from Kuldhar village to Mayajal village and latt om Mayajal village to Jajiya village.

The Hamira, Joyan, Fort and Badabag Members are extensive and more or le ntinuous from southwest to northeast whereas the Kuldhar and Jajiya Members a stricted to southwestern side. The lowest Member i.e. Hamira is bit irregular due tertonguing phenomenon. It is well exposed in a low scarp 3 km. north of Dabha ong Barmer-Jaisalmer road. The top horizons of the Joyan Member are uniform stributed but the lower ones are exposed only around Saran-ki-Dhani and in the ar ›tween Chundan and Joyan 'Kharin'. The Fort Member exhibits uniformity mo earty than any other member especially in the vicinity of Jaisalmer town. Its scarps a nspicuous but the beds pinch out towards southwest. Different lithounits of t ıdabag Member crop out mainly between Jaisalmer-Ramgarh road aı marsagar-Lodorva road and show pinching phenomenon laterally. The Kuldh ember is exposed around abandoned village of Kuldhar particularly along the Masur ream, whereas the Jajiya Member crops out mainly around Jajiya village. These la 70 Members are overlain by the strata of the Baisakhi Formation unconformabl arlier workers have envisaged several en-echelon step faults trending NNW-SSE NE-SSW in the region near Kuldhar and 1.5 km. east of Manpiya. T ;hostratigraphy of the Jaisalmer Formation along with important lithologies and the ıickness is given in Table 1.

ıble 1. Jaisalmer Formation

[ember	Gross lithology	Thickness in metre
	Baisakhi Formation	
	--- Unconformity ---	
jiya	Alternating sandstone and shelly limestone, golden oolite, and marly limestone	16
uldhar	Yellow limestone, greenish shale, and golden oolite.	7.5
ıdabag	Yellow and buff limestone, shelly limestone, yellow sandstone, greenish shale and conglomerate	20
›rt	Buff, Yellow to greenish yellow limestone, and soft, light coloured sandstone	27
›yan	Shelly limestone, marl, shale and sandstone	12
amira	Yellow arenaceous limestone, shale and marl.	13

BIOSTRATIGRAPHIC ZONES

All the lithounits were searched for fossil collection by taking traverses along th ike. About 3200 fossil individuals of bivalves were collected. Of these 1100 left ou :ause of poor preservation or fragmentary nature. The reamining 2100 have bee ntified which constitute 171 species assigned to 76 genera grouped into 41 families, lers and 5 subclasses. Majority of the forms are found to be long ranging considerin ıole western India and on global basis. Still a few have a restricted range suitable fo biostratigraphic zonation. Accordingly seven bivalve assemblage zones aı :ognized. An attempt is also made to equate them with the available ammonoid zone India.

NE 1. *Palaeonucula hausmanni, Eomiodon lunulatus* zone

This zone is the part of the Hamira Member. It is well exposed in a scarp. 3 kn rth of Dabhala village on Barmer-Jaisalmer road. The lithounits are (Fig. 1) :

Laminated limestone - 2m.

Dirty white shales - 3m.

Brown fossiliferous marly limestone – 1m.

Greenish yellow shales - 2m.

Yellow arenaceous limestone - 1m.

Of these, the horizon of brown fossiliferous limestone has yielded a handful of sma stropods, a few brachiopods, and bivalves - oysters, nuculids and others. The undersic the limestone is studded with nuculids. A check list of bivalve taxa is -

Restricted to zone : *Palaeonucula husmanni* (Roemer) and *Eomiodon lunulatu* okoyama)

Associated long ranging taxa : *Trigonucula* (?) *stoliszkai* (Cox), *Palaeonucul neiformis* (J. Sowerby), *Catinula sandalina* (Goldfuss), *Corbula lyrate* (J. Sowerby e first one extends from zone 1 to 5, next two range from zone 1 to 7 and the last or m zone 1 to 6.

The species *Palaeonucula hausmanni* is known to occur in the Upper Liassic beds rope (Cox, 1940) and *Eomiodon lunulatus* in the Liassic strata of Japan. This zone erefore, assigned to Upper Lias.

NE 2. *Rhaetavicula contorta, Myoconcha crassa* zone

This zone is a part of the Joyan Member. It crops out in the area betwee ran-ki-Dhani to Joyan 'Kharin'. The lowermost unit is exposed further 3 km rthwest of zone-1. The various litho units are (Fig. 2) -

Marl—3.5m.

Yellow sandstone studded with oysters—1.25m.

Light yellow shale — 4.25m.

Bivalve fossils other than oysters are comparatively rare. The most fossil yieldir ›rizon is marl, exposed between Chundan and Joyan 'Kharin'. All the recorded bival› xa confined to this zone only namely *Rheetavicula contorta* (Port lock), *Myoconch* *assa* (J. Sowerby), *Modiolus hillanus* (J. Sowerby), *Isognomon*, rikuzenicı 'okoyama), I. sp., and *Trigonia* sp. These species following various authors (Trechman ›23; Hayami, 1957) are restricted to the Lower Jurassic saving *Myoconcha crassa* whic nges from Liassic to Bajocian. The zone in question, therefore, may be assigned pper Lias. Below marl, the horizon of yellow sandstone near Saran-ki-Dhani is so muc udded with parallelly arranged oysters that it can be named as oyster reef. In Europ 'ster reefs and ostrea beds are restricted to Lower and Middle Jurassic (Hudson, 196 :oedsson, 1951). Thus this zone is rightly placed in Lower Jurassic.

)NE 3. *Eomiodon indicus, E. baroni* **zone**

This zone is part of the Fort Member. All the arenaceous lithounits from groun ırface to two-third height of the scarps in the vicinity of Jaisalmer town form this zon ıe lithounits are (Fig. 3) :

Pink calcareous sandstone — 3m.

Pale yellow to dirty white shale — 6m.

Pink sandstone — 4m.

Friable sandstone — 8m.

All the fossil individuals belong to genus *Eomiodon*. Main bulk has come from th ›rizon of pale yellow to dirty white shales. The various species identified are *Eomiodc* *dicus* Cox, *E. baroni* (Newton), *E.* sp. indet, and *E.* spp. All of these are restricted ıis zone only.

Globally also these forms are recorded only from the Bathonian strata of Kutc ttock, Melagasy, Tanganyika and Myanmar (Cox, 1935, 1936, 1952, 1965, Agrawal ar acker, 1980; Singh and Rai, 1980). Thus this zone is definitely Bathonian.

)NE 4. *Modiolus patchamensis, Modiolus sahai* **zone**

This zone is also a part of the Fort Member. Lithounits are calcareous in nature ar op out from the top of the scarps down to upper one-third height. The various units aı 'ig. 3).

Hard, buff fossiliferous limestone — 2m.

Yellow fossiliferous marl — 1.5m.

Greenish yellow unfossiliferous limestone — 2.5m.

This zone has yielded varied mega-invertebrates as bivalves, gastropod :achiopods and echinoids. Main bulk has come from yellow fossiliferous marl. The mo ›mmon bivalve genera are *Dacryomya*, *Modiolus*, *Corbula*, *Pholadomya* and *Homomy* he check list of bivalve taxa is as follows.

Restricted to zone : *Modiolus patchamensis* (Cox, M. Sahai, Singh and Kanjila

. *imbricatus* (J. Sowerby), *M. durnovarie* (Arkell), *Bteroperna shuklai* (Singl *ɪkevellia waltoni* (lyeett), *Trigonia distincta* (Kitchin), *Lucina lirata* (Phillip *actromya mehrotrai* (Agrawal), *Neocrassina (Pruvostiella) major* (J. Sowerb *ocyprine* cf. *Cyreniformis, Arctica antiopa* (d'Orbigny), *Corbula bassae*, (Singh and Ra *daghaniensis* (Cox), *Pholadomya* cf. *hemicardia* (Roemer), *P. (Bucardiomya)* sp. ju *otei* (Brongniart), *Homomya gibbosa* (J. Sowerby), *Ceratomya concentrica* (J. Sowerb *euromya alduini* (Brongnirt), *Thracia greppini* (Cossmann) and few other forms l ɪder open nomenclature.

Associated long ranging taxa besides those mentioned in zone-1 are : extending ne-5 *Nuculoma blakei* (Cox.) *Dacryomya lacryma* (J. Sowerby), *Placunopsis radia* hilips), *Protocardia dynosisea* (Buvignier); extending to zone-6 *Camptonectes indic* ox); extending to zone-7 are *Palaeonucula blanordi* (Cox), *camptonectes aurit* chlothiem), *chlamy curvivarians* (Dietrich), *Neocrassina rajnathi* (Agrawal), a *euromya uniformis* (J. Sowerby).

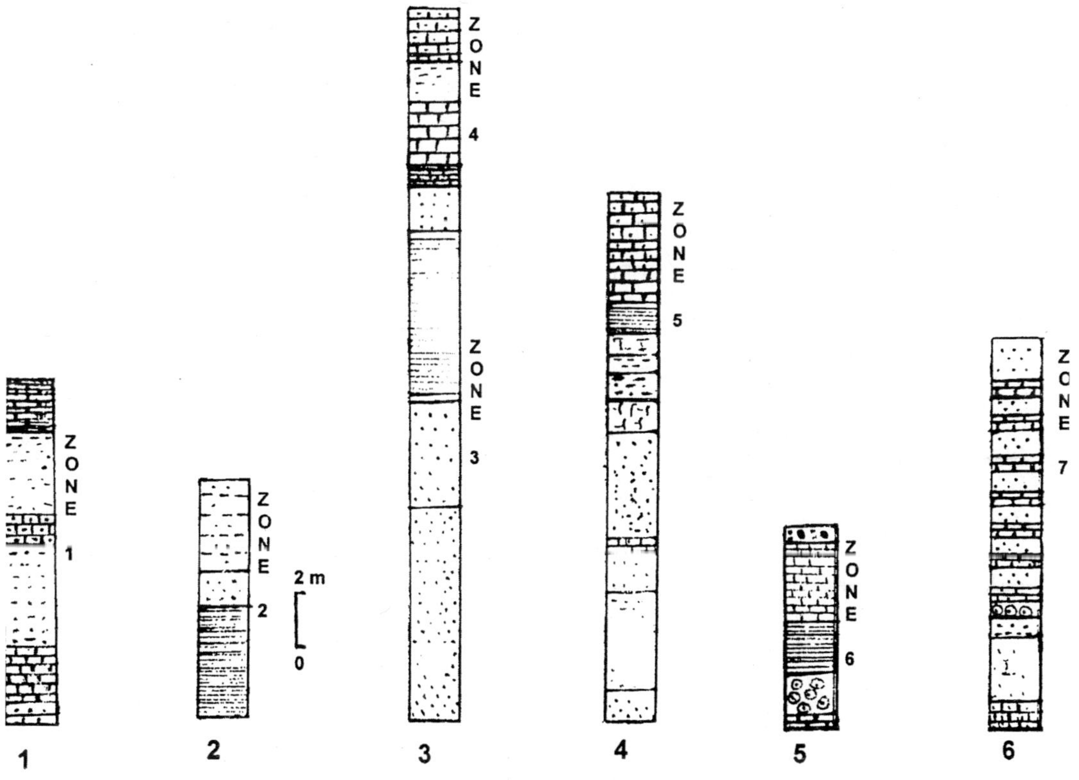

g. 1 to 6. Biostaticgraphic zonation of the Jaisalmer formation

Majority of the forms are Callovian and a few are Callovo- Oxfordian. Howeve odiolus patchamensis* and *M. sahai* are known to occur only from the Lower Callovi rata of Kutch alongwith *Macrocephalites macrocephalus*, (Cox, 1940; Singh a anjilal, 1974) thus correspond to macrocephalus zone of Spath (1933). Overa semblage also favours Lower Callovian age for this zone.

Sahni and Bhatnagar (1958) also recorded an echinoid species *Hemicidar isalmerensis* of Callovian age from these strata.

ONE 5. *Limatula jaini, Falcimytilus* **sp.,** *Coelastarte* **sp. zone**

The lowest unit exposed at about 3 kms. from Jaisalmer on Ramgarh road where her higher units successively continue for further 7 km. on the same road upto t akni stream. This zone directly overlies the zone-4 and forms a part of the Badaba ember. The various units are (Fig. 4) —

Arenaceous limestone — 3m.

Yellow limestone — 1.5m.

Greenish shale — 0.7m.

Shelly limestone — 1m.

Yellow shale — 0.7m.

Flat pebble conglomerate — 1.2m.

Scarp forming limestone — 1.5m.

Current bedded yellow sandstone — 4.2m.

Buff limestone — 0.5m.

Soft sandstone — 1.5m.

Yellow marl — 3m.

Dull yellow sandstone—1.2m.

Of these units yellow marl, greenish shale, yellow limestone and arenaceo nestone are fossil yielding. The check list of bivalve taxa from this zone is given here –

Restricted to zone : *Palaeonucula menkii* (Roemer) *Inoperna sowerbian* orbigny) *Falcimytilus* sp. *Camptonectes laminatus* (J. Sowerby), *Limatula jaini* (Singl *igonia brevicostata* Kitchin, *Neocrassina* (*coelastarte*) sp. besdies a few left under ope menclature. Other associated long ranging taxa in addition to those mentioned wit wer zones are *Inoperna glendayi* (Weir), *chlamys ambigu* (Muenster), *Mesoling rawali* (Singh) ranging upto zone-6 and *Catinula stolizkai* (Cox) upto zone-7.

The forms restricted to this zone are either callovian or Callova-Oxfordian b nus *Falcimytilus* and subgenus *Coelasterte* hitherto not known from the strata young an Middle Callovian in Kutch. Moreover, *Limatula jaini* is Middle Callovian on grawal and Kachhara, 1973). Therefore, this zone may be referred to Reineckeia Be Agrawal (1956) and rehmani/exceps zone of Spath (1933).

)NE 6. *Tellurimya regeri, Meleagrinella echinata Cercomya* **sp. zone.**

This zone is a part of the Kuldhar Member and related units are well expos ound abandoned village of Kuldhar along the Masurdi stream. This village is about n. from Jaisalmer on Khaba road. The various horizons entomb a good wealth of faun ıe ammonoids have made their first appearance. Besides bivalves, belemnites aı achiopods-terebratulids and rhynchonellids are quite common. The different horizo this zone are (Fig. 5).

Nodular band — 0.6m.

Yellow shelly limestone — 3m.

Greenish shale — 2m.

Golden oolite — 1.6m.

Yellowish white limestone — 0.3m.

As compared to other concerning bivalves lower two horizons are high ssiliferous. A check list of identified bivalve taxa leaving aside those left in op menclature is as follows -

Restricted to zone : *Brachidontes laitmairensis* (de Loriol), *Pinna orna* 'Orbigny), *Pecten rochi* Agrawal, *Eopecten tegulatus* (Morris and Lyceet), *Meleagrinel hinata* (W.Smith), *crenostreon probescideum* (J. Sowerby), *Nicaniella numm* auvage), *Tellurimya rogeri* (Agrawal and Kachhara), *Goniomya literata* (J. Sowerb *ercomya* sp., *Thracia depressa* (J. Sowerby), *T. triangularis* (d'Orbigny). Oth sociated long ranging taxa extending to zone-7 are : *Grammatodon jurianus* (Co *hlamya subtextoria* (Muenster), *Catinula alimena* (d'Orbigny), *Lopha* (*Actinostreo arshii* (J. Sowerby), *Trigonia elongata* (J. Sowerby), *Vaugonia* (*Orthotrigonia*) *exorti* itchin), *Pholadomya* (*Bucardiomya*) *lirate* (J. Sowerby), *Entolium corneolus* (You ıd Bird), *E. partitum* (J. Sowerby), *Plagiostoma aciculata* (Muenster) and *Pseudolim ıplicata* (J. Sowerby).

All the zonal species are Callovian and *Tellurimya rogeri* is reported from Midc allovian strata of Kutch (Agrawal and Kachhara, 1979) in association with ammonoi *anceps* zone of Spath (1933) or Reineckia Beds of Agrawal (1956). Remaining speci e either restricted to Callovian or range beyond it. Moreover, from the horizon olden Oolite the ammonoid genera *Reineckeia, Subkossmatia Idiocylocer inkeliniceras* and *Sivajicaras* are of Middle Callovian only.

)NE 7 : *Parallelodon keyserlingii, Coelopis buvignieri* **zone**

This zone is a part of the Jajiya Member. It is well exposed around Jajiya villa ıe lithounits are (Fig. 6).

Alternating sandstone and shelly limestone — 10m.

Golden oolite — 0.6m.

Fossileferous marl — 0.6m.

Current bedded sandstone — 2.5m.

Marly limestone — 1m.

The lowermost horizon entomb varied fauna as bivalves brachiopods, ammonoi stropods, and crinoids. Other fossiliferous horizons concerning bivalves a ssiliferous marl and golden oolite. The check list of bivalve taxa other than those left en nomenclature is : *Parallelodon Keyserlingi* (d'Orbigny), *Modiolus* sp. jiv. *bipartit* . Sowerby), *Gervillella siligua* (J.A. Eudes-Deslongchamps), *Pecten* aff. *Kachhen* ox), *Plagiotoma* aff. *blanfordi* (Cox), *Grypheae* (Bilobissa) *rathodi* (Singh), *Lop ctinostreon) gregarea* (J. Sowerby), *Trigonia haboensis* Agrawal, *Vaugon rthotrigonia) hispida* (Kitchin), *V. (O.) kutchensis* (Kitchin), *Sphaera rogeri* Agraw *haeriola pruvosti* (Agrawal), *Nicaniella extensa* (Phillips), *Neocrassina (Pruvostiel eneixae* (Agrawal), *Coelopis greppine* (de Loriol) *C. (Trigonopsis) buvignieri* (d'Orbign) *Carollina* (Damon), *Ceratomyopsis basochiana* (Defrance), *Eocallista* sp. ju *ncrediformis* (Blacke and Hudlestone), *Eopecten* sp. juv. aff. *abjectus* (Phillip *agiostoma* off. *onata* (Arkell), and *Anisocardia minima* (J. Sowerby). Other ta tending to this zone from the lower ones are mentioned while describing the vario nes. In this checklist majority of the forms are of Oxfordian and a few are Callov xfordian. Therefore, this zone is definitely Oxfordian. The ammonoid genera collect om this zone namely *Lissoceratoides*, *Brightia*, *Distichoceras*, *Dhosaites*, *Mayaites*, a *imayaites* also favour Oxfordian age for this zone as they are also recorded from Dho lite of Kutch. Zonal forms are confined to Oxfordia strata of Europe (Arkell, 1923-37).

In between Kuldhar and Jajiya one of the horizons i.e. yellow compact san nestone near abandoned village of Mayajal has yielded the bivalve taxa *comptonec ritus* (Schlothiem), *C. indicus* (Cox), *Chlamys (Radulopecten)* sp. *Catinula stolizk* ox), *C. sandalina* (Goldfuss), *Nanogyra nana* (J. Sowerby) *Lopha (Actinostrec arshii* (J. Sowerby) all with Callovo-Oxfordian range. From its stratigraphic position pears to be of upper Callovian equivalent of *athleta* zone of Spath (1933).

CONCLUSIONS

Based on bivalve assemblages 7 biozones are proposed for the Jaisalmer Formatic ixa of zone-2 and zone-3 are very distinct and do not occur in any other zone. Zone id zone-7 also entomb ammonoids. From bivalve point of view zone-7 is distinct as is majority of taxa restricted to Oxfordian. Zone-1, zone-4, zone-5 and zone-6 are mair signed to different ages based on overall assemblages. Lubimova *et al.* (1961) have al corded Callovian and Oxfordian ostracodes respectively from zone-5 and zone-7. T isalmer Formation based on bivalve taxa ranges from Upper Lias to Oxfordia owever, for Bajocian and Upper Callovian, authors have not come across ve aracteristic bivalve assemblages.

ACKNOWLEDGEMENT

The authors are thankful to the Head, Department of Geology, M.L. Sukhad niversity, Udaipur for providing the laboratory and library facilities. Thanks are al ie to Prof. B.S. Paliwal of J.N.V. University, Jodhpur for the help rendered duri eparation of the text.

REFERENCES

ɡrawal, S.K. 1956. Contribution a etude Stratigraphique et paleontologue du Jurassique Kutch (Inde). Ann. Centre et Document. *Paleont.* Paris, 19 : 188.

ɡrawal, S.K. and Kachhara, R.P. 1973. Habo Beds near Ler (Kutch); Pt. 3- Biostratigraphy of t beds on the east of Ler. *Proc. Indian Nat. Sci. Acad.* 45(A) : 129-146.

ɡrawal, S.K. and Kacker, A.K. 1980. Biostratigraphy of Bathonian- Callovian beds of Mouwa dome, Eastern Bela Island, District Kutch (Gujarat). *Recent Researchers in Geolo* Hindustan Publishers, New Delhi. 631- 643.

ːkell, W.J. 1923-37. A monograph of British Corallian Lamellibranchia, pt. I-X, *Mon. Palaeo Soc. Lond.*, xxxviii, 392 p.

.anford, W.T. 1877. Geological notes on the Great Indian Desert between Sind and Rajputaı *Rec. Geol. Surv. Ind.*, 10(1) : 10-13.

ɔx, L.R. 1935. Triassic, Jurassic and Cretaceous Gastropoda and Lamellibranchia of the Attc District. *Pal. Indica, Mem.* 20(5) : 27.

ɔx, L.R. 1936. Mollusca from Southern Persia (Iran) and Bahrein Island. *Pal. Indica*, Mem, 2 69.

ɔx, L.R. 1940. Jurassic Lamellibranch fauna of Kachh (Kutch) : *Pal. Indica,* 3(3) : 157.

ɔx, L.R. 1952. Jurassic Lamellibranch Fauna of Kachh (Kutch) *Pal. Indica,* 3(4) : 128.

ɔx, L.R. 1965. Jurassic Bivalvia and Gastropoda from Tanganyika and Kenya. *Bull. Brit. M Nat. Hist. Geol. Suppl.* 213 p.

as Gupta, S.K. 1975. A revision of the Mesozoic-Tertiary stratigraphy of the Jaisalmer bas Rajasthan. *Jour. Ind. Earth Sci.* 2 : 27-94.

ayami, I. 1957. Liassic *Gervillia* and *Isognomon* in Japan. *Japan Jour. Geol. Geograph.* 28 (1- 95-106.

udson, J.D. 1963. The recognization of salinity controlled Molluc assemblages in the Gr Esturine series (Middle Jurassic) of the Inner Hebrids. *Palaeont.* 6 : 318-326.

achhara, R.P. and Jodhawat, R.L. 1981. On the age of Jaisalmer Formation, Rajasthan, Ind *Proc. IX Indian Coll. Micropal, Strat* 235-247.

ıbimova, P.S., Guha, D.K. and Mohan, M. 1961.-Ostracoda of Jurassic and Tertiary deposits Kutch and Rajasthan (Jaisalmer), *India. Bull. Geol. Min. Met. Soc. India,* 22 : 60.

areek, H.S. 1981. Basin configuration and Sedimentary Stratigraphy of Western Rajasth *Jour. Geol. Soc. Ind.*, 12 : 517-527.

ahni, M.R. and Bhatnagar, N.C. 1958. New fossils from the Jurassic rocks of Jaisalm Rajasthan, India. *Rec. Geol. Surv. Ind.*, 87(2) : 418-437.

.ngh, C.S.P. and Kanjilal, S. 1974. Some fossil mussels from the Jurassic rocks of Habo Hill Kutch, Gujarat, Western India. *Ind. Jour. Earth. Sci.* 1(1) : 113-125.

.ngh, C.S.P. and Rai, J.N. 1980. Eathonian-Callovian Fauna of western Bela Island (Kut Pt.I-Bivalve families Cardiidae, Neomiodontidae and Corbulidae, *Jour. Palaeont. Soc. I* 23-24 : 71-80.

ɔath, L.F. 1933. Revision of the Jurassic Cephalopod Fauna of Cutch. *Pal. Indica* N.S., 4(6).

rechmann, C.T. 1923. The Jurassic rocks of New Zealand. Quert. *Jour. Geol. Soc. Lond.*, 7 246-312.

roedsson, G. 1951. On the Hoganes series of Sweden (Rhaeto-Lias) : *Lund Univ. Arsskriff, N.* Abl. 2, Bd. 47(1) : 25.

eological Evolution of Northwestern India
.S. Paliwal (Ed.)
cientific Publishers (India), Jodhpur, 1999, pp . 118 - 128.

MICROPALAEONTOLOGICAL STUDIES OF LIGNITE SAMPLES FROM WESTERN RAJASTHAN

R.P. Tripathi, M.S. Sisodia*, K.L. Srivastava* and B.D. Sharma

Department of Botany and Department of Geology,*
Jai Narain Vyas University, Jodhpur - 342 001 (India)

ABSTRACT

Description is given of plant microfossils isolated on maceration of lignite samples collected from two localities of Eocence age, i.e., Giral in Barmer and Barsinghsar in Bikaner districts of Rajasthan. The plant microfossils are represented by algal filaments and nets, fungal spores and hyphae (Thallophytes), bryophytic thalli and capsules (?), pteridophytic cuticles, sporangia and spores, cuticles of gymnosperms and various kinds of pollen grains; cuticles and peculiar fructifications (?) of angiosperms from both the localities. There are also present many kinds of plant microfossils of unknown affinities and wings of insects in the lignite. Palaeoecological conditions that existed during the Eocene period in Western Rajasthan are discussed.

ːey words : Micro-plant-Fossils, Lignite, Mesozoic, Tertiary Western Rajasthan.

INTRODUCTION

The Western Rajasthan has both Mesozoic and Tertiary (Cenozoic) rocks exposed everal places e.g. Akal, Habur, Kapurdi, Bikaner, Palana, Barsinghsar, Giral, Ba Kolayat etc. which preserve plant and animal fossils as impressions, incrustations a etrifactions. Several kinds of plant fossils have been described by a number of worke Bose 1949, Lakhanpal and Bose 1951, Rao and Vimal 1950, 1952, Lakhanpal 196 Deshmukh and Sharma 1978, Gularia 1984, 1990, Harsh and Sharma 1988, 1992, Har t al. 1992). Lignite is found at many places in the Western region of the State Rajasthan e.g. Palana, Barsinghsar, Giral etc. All lignite deposits in the state are Eocene age (Bhatia 1977, Das Gupta 1975, 1977, Verma 1982).

Harsh and Sharma (1992) studied the chemistry of a lignite sample from Palar
ìikaner) and identified both inorganic and organic constituents. Rao and Vimal (195
)52) described a number of pollen grains and spores from Palana lignite. Otherwise, n
uch work has been done on plant microfossils found in the lignite samples of Westeı
ajasthan. Investigations on plant microfossils as well as megafossils are going on in tl
alaeobotany Laboratory of the Department of Botany of J.N.Vyas University, Jodhpu
ı the present investigation maceration has been done of 50 samples of lignite collect
om Giral (Barmer) and Barsinghsar (Bikaner) localities of Western Rajasthan. Tl
icrofossils obtained during maceration of lignite from Giral include algal filaments ar
ıts, fungal hyphae and spores, *Riccia* like thalli and moss capsules, pteridophyt
ıticles, sporangia and spores, gymnospermous cuticles and some angiospermic (
icrofossils of unknown affinity. In addition to plant fossils a wing of insect is al
eserved in a lignite sample. Microfossils isolated from lignite samples of Barsinghs
ave representatives of all major groups of plant kingdom i.e., algae, fungi, bryophyte
eridophytes, gymnosperms and angiosperms. These include filaments, nets, hypha
orangia, spores, cuticles, pollen grains and peculiar kinds of small seeds ar
uctifications. On the basis of plant microfossils, the palaeoecological conditions at tl
me of deposition of Eocene rocks in Western Rajasthan have been discussed.

MATERIAL AND METHODS

The lignite samples for the present investigation have been collected from Gir
$6^{0}83'$ N - $71^{0}15'$E) in Barmer district and Barsinghsar (6 km N - W of the Palar
llage) Bikaner district of Rajasthan (Fig. 1). The lignite at Giral is mined in an op
posure (Fig. 2). Samples at Barsinghsar were collected by boring a well in the ar
ig. 3). The lignite samples were properly washed with alcohol (90%) and distilled wat
eliminate surface contamination. It was then crushed to small pieces and then treat
ith conc. HNO_3 for 12-24 hours. From the digested material, acid was removed l
peated decantation and washing with distilled water. Acid free material was treat
ith 5-10% KOH solution for 10 minutes. The alkali was removed by decantation ar
ashing with distilled water, and the material was sieved through 70-200 grades
eves. Plant microfossils were separated both by plucking up individuals as well as l
ntrifugation and were mounted in freshly prepared glycerine gelly.

GEOLOGY OF THE AREA

Lignite occurs in the sedimentary rocks of the Tertiary period at several places i
e districts of Barmer, Bikaner, Jaisalmer and Nagaur. At Giral the lignite occurs i
least seven seams intercalated with clay and gypsum deposits of Eocene age (Dut
)83). While at Barsinghsar a thick layer of lignite, varying from 15.10 metres to 31.
etres in thickness is found (Fig. 3). It is covered by top sand, sandstones and clay, ar
sts on clay and black shales. The basement is again made up of sandstones (Fig. 3
ie lignite has been taken out by boring wells in the area.

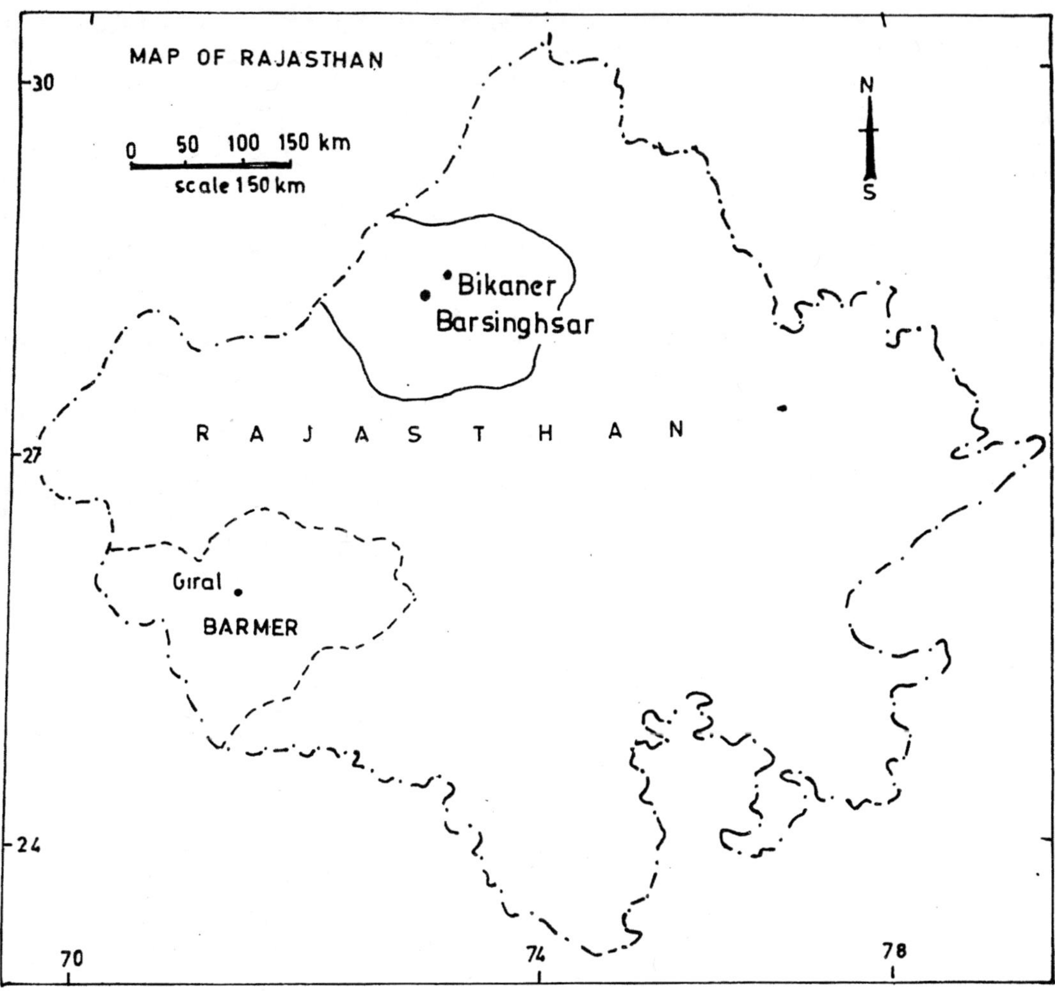

Fig. 1. Map of Rajasthan showing places of study

DESCRIPTION

lant Microfossils

Thallophytes - Algae : Algal filaments and nets occur frequently in ligni .mples of both the localities. The filaments are unbranched; faint markings ansverse septa are visible in some of the filaments (Pl. 1, Fig. A) but details of septa ıd cell organelles, i.e. chloroplasts, etc. are not preserved. Filaments vary in size i. ngth 400-3230 µm and thickness 45-140 µm. Regular nets (Pl. I, Figs. B, C) as well : regular nets occur frequently in the material. The former are 170 x 100 µm to 1120 :0 µm in size and made of regularly interconnected filaments surrounding sever ›ints. Encircling each point there are 5-7 inverted triangle like cavities. The thickne:

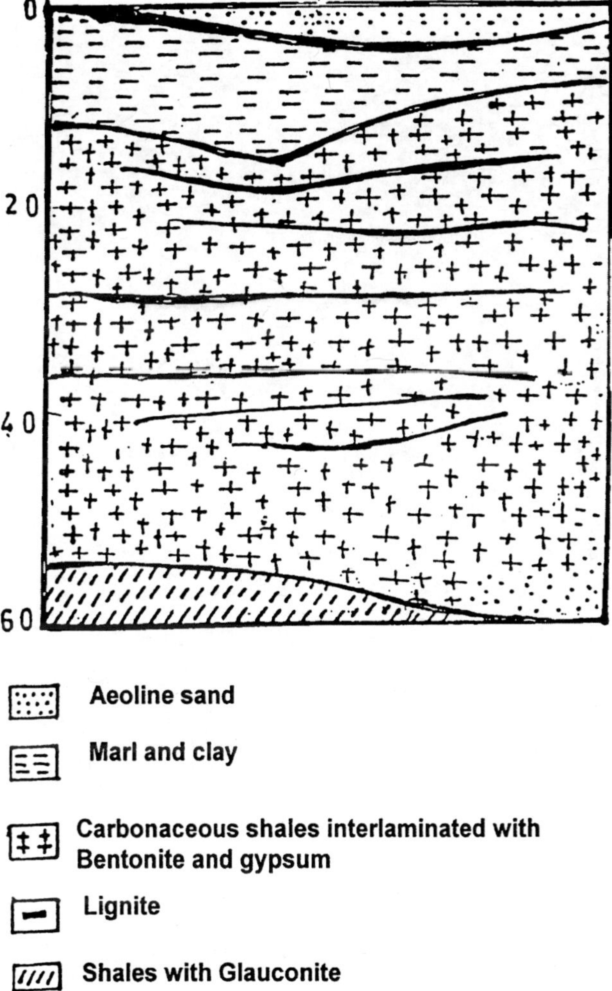

Fig. 2. Lithostratigraphy based on Borehole Data at Giral, Barmer.

filaments which make the nets is variable and also the size of the inverted triangul vities. In some of the nets the filaments are narrow and cavities are large (Pl 1, Fig hile in others filaments are thick and cavities become small (Pl. 1, Fig B). The irregul ts have filaments of variable thickness and the cavities formed are of variable shap d sizes. These nets may be compared with the freshwater extant green algal gen *ydrodictyon* (Bold *et al.,* 1987). Whether the regular nets have affinities wi *ydrodictyon* or not, is yet to be decided.

Fungi : A number of kinds of fungal hyphae and spores found in the lignite of Gir d Barsinghsar are assigned to the group fungi. Dark coloured cleistothecia like bodi e seen in some slides prepared from the samples of Giral. From the margin of whi ise septate and branched hyphae (Pl. 1, Fig. D). Some of the hyphae have hook li

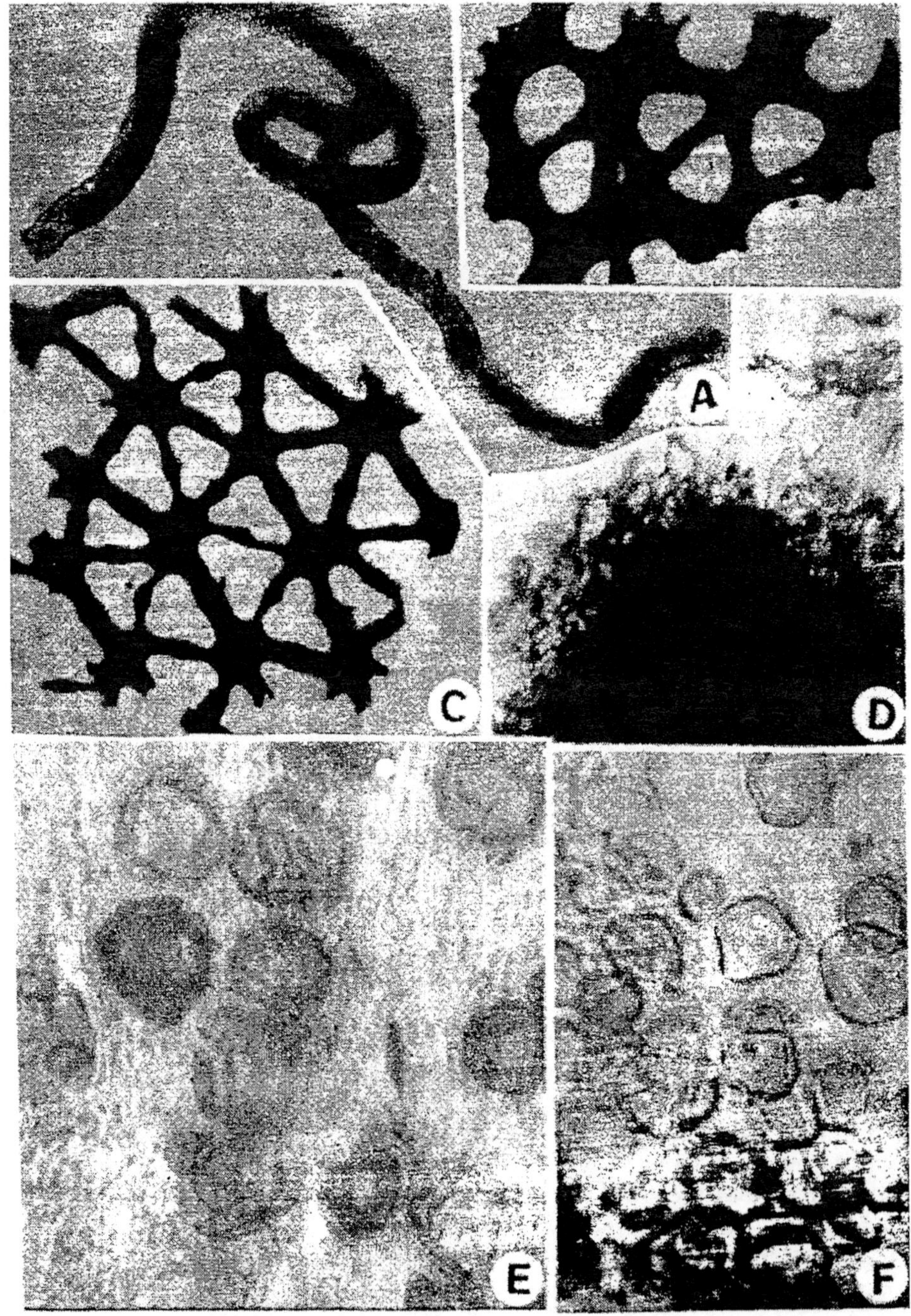

Plate 1.

rminal ends. Presence of septae in hyphae and cleistothecium like body sugg(lationship with Ascomycotina (Bold et al., 1987). However, neither asci nor ascospo uld be seen in any slide.

Fungal bodies and spores isolated as well as adhered to the host cuticle are found e lignite of Barsinghsar, The individual spores are either unicellular or multicellul ıe former are small, circular and double walled resembling uredospores of *Puccinia* me rust (Pl. 1, Fig. F). A few large multicellular spores resembling *Phragmothyri* lwards (Jain and Gupta 1970) are also present in slides. The spore is circular wi ccessive rings and radiating hyphae which give the appearance of eudoparenchymatous tissue. The central cells are small and dark while the outer ce e rectangular and thin walled. Ascospores could not be seen could not be seen.

Uniseriate multicellular spores are either spindle shaped or rod shaped. The form e *Alternaria* type but without verticle septae. The spore has a small distinct stalk a e spindle shaped body is made up of 8-12 cells. In each cell at the transverse sep ere are seen two valve like structures of unknown nature. The rod shaped spo semble *Helminthosporium*. However, further investigations are required for assigni act relationship. Some of the multicellular spores united irregularly, each consisting to 15 spores. Such spores occur in some of the members of Deuteromycoti orangia or acervuli are also found in the lignite. There is no sterile cellular w rrounding the sporangium suggesting their affinity with the Thallophyta but a th n-cellular membranous covering is present which gives circular/ globular shape to t orangium. The spores come out on bursting off this membranous wall of sporangiu ıe spores are similar, circular, 20-25 mm in diameter with double walls (Pl. 1, Fig. sembling uredospores of some rust (?).

A number of host cuticles with chlamydospores or zygospores are also present in t mite (Pl. 1, Fig. E). Each large double walled spore size 25-40 µm is accompanied wi smaller spore (?) size 8-12 µm or bulbous base of the spore without stalk (?) similar e zygospores of many VAM genera (Gerdemann 1965, Schenck and Perez 198 ırther investigations are going on in this regard.

Bryophyta : Bilobed and unilobed thalli full of black material with lateral scal ıd rhizoids (?) have been isolated from the lignite samples of Giral and Barsinghs ıese may be compared with the thalli of *Riccia* (Pl. 2, Fig. A). Cellular structure of t alli is not visible.

Elongated bodies identical to a typical Moss capsule are present in the samples th localities, which measure 119 x 225 µm to 2166 x 425 µm. The upper portion nical, like an operculum and lower portion resembles seta of a sporogonium. Neith llular details nor spores are visible in the capsule.

Pteridophyta : Sporangial bodies, trilete spores and leaf cuticles of pteridophy ants could be isolated from the lignite samples. Megasporangium and dispers egaspores of *Isoetes*, fern sporangium, spores of *Dandotiaspora* and other trilete spo ıd leaf cuticles have been collected from the lignite of Giral. A structure similar egasporangium of *Isoetes* is present in the material. This has transverse trabeculae a arkings of a large number of scattered megaspores (Pl. 2, Fig. B) in between t abeculae. The wall of megasporangium is made of isodiametric cells. Details egaspores inside sporangium are not preserved. However, a large number of dispers egaspores have been isolated from the same material which resemble the spores tant genus *Isoetes* (Pl. 2, Fig. D).

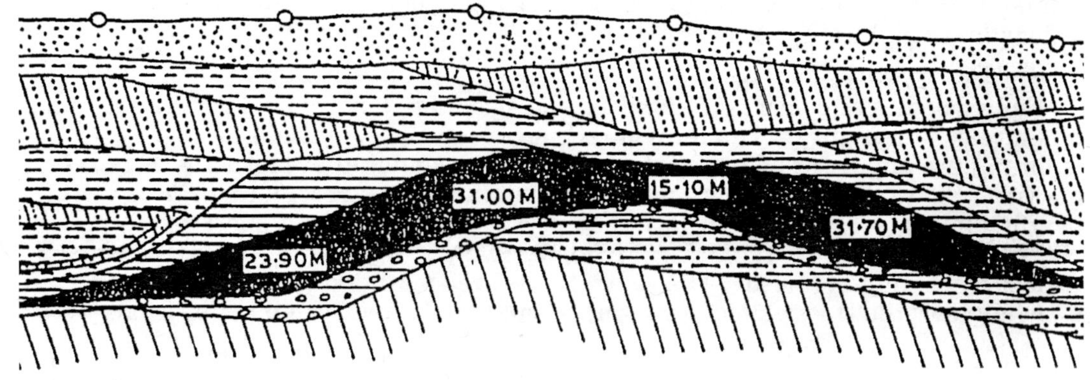

INDEX

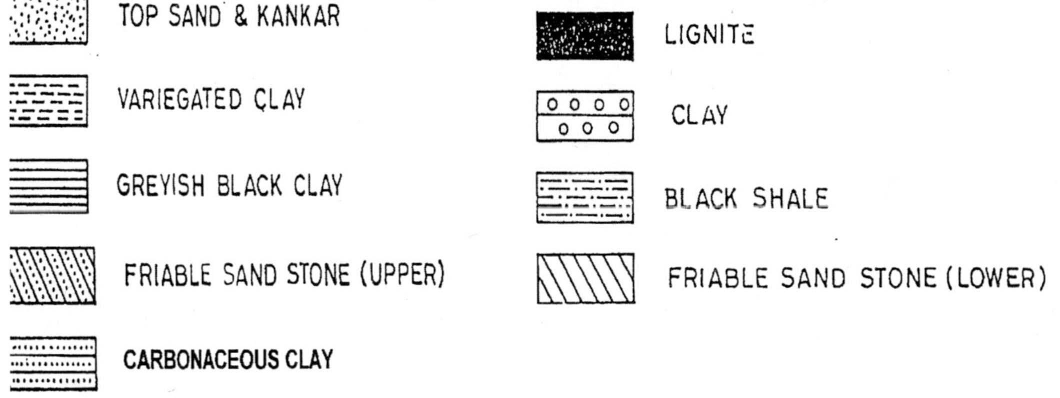

ig. 3. Lithological succession at Barsinghsar (Based on borehole data) Scale : Vertical 160 metr Horizontal 2250 metres.

A globose sporangium with a thin stalk is seen in a slide. The sporangium has ertical/oblique annulus and a few spores in it. The sporangium resembles a typic ptosporangiate sporangium of Gleicheniaceae or Hymenophyllaceae. The spor resent inside the sporangium have tri-radiate markings on them.

A number of thin cuticles with stomata are well preserved in the lignite sampl Pl. 2, Fig. E). The lignite collected from Giral has a number of spores of ferns; some hich resemble *Dandotiaspora* (Sah *et al.,* 1971) in morphology. This spore has be escribed by them from the Lower Eocene of Rajasthan. There are many mo teridophytic spores of unknown affinities.

Similarly, a number of leaf cuticles with various kinds of epidermal cells ar uperficial stomata are present in the material from Barsinghsar in addition to th rilete pteridophytic spores. One of the cuticles is multiveined having lateral veins ght angles forming more or less squarish aerioles (Pl.2, Fig. C). This resembles th iticle of the fern family Dipteridaceae (Bower 1926) and the fossil frond *Hausmann* unker (Bose and Sah 1967). However, details of stomata are not clearly preserved.

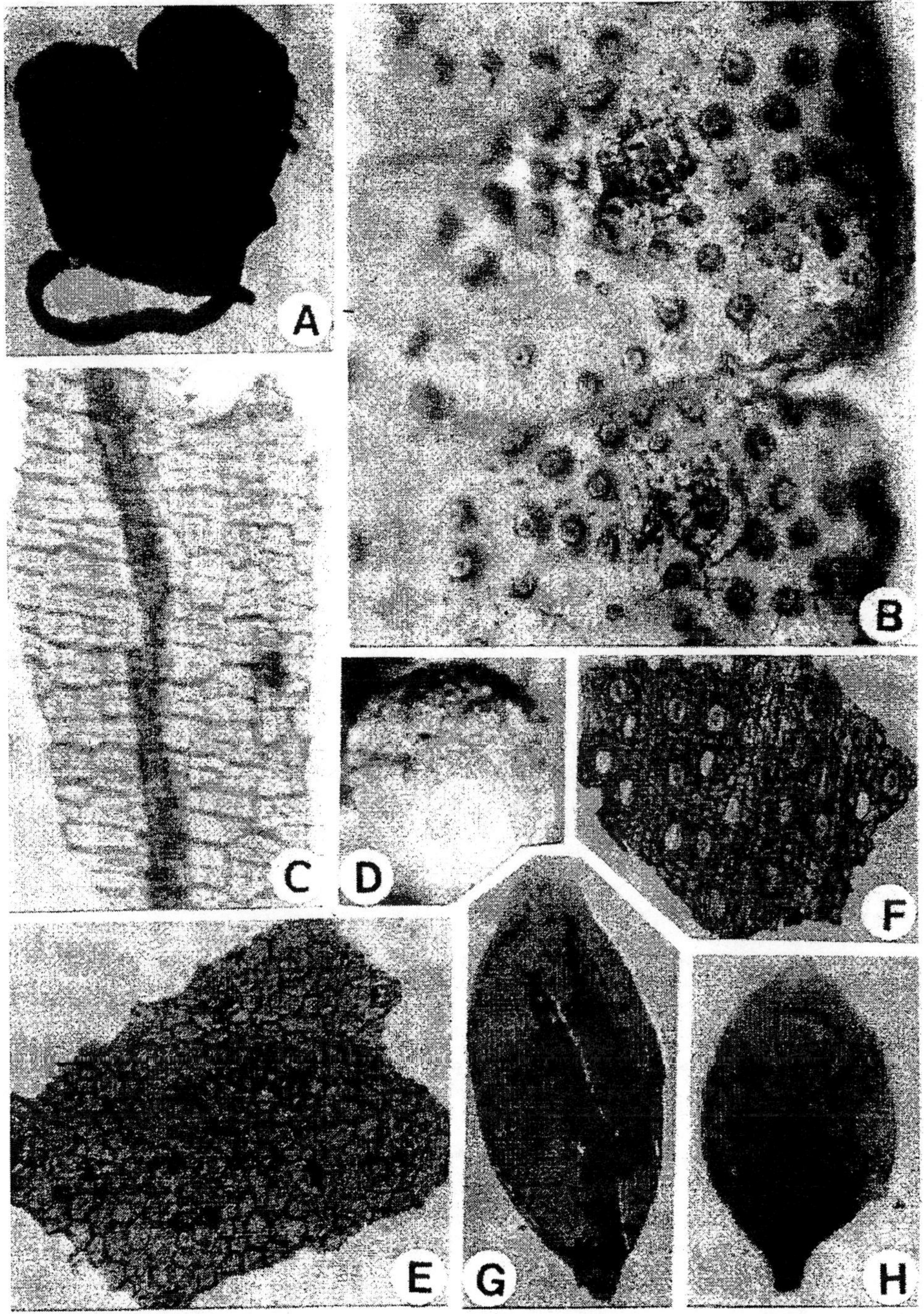

Plate 2.

A number of pteridophytic spores with distinct triradiate markings are present i e material. Some of them resemble *Dandotiaspora* sp. (Sah et al. 1971) in having pa ;e structures at the ends of tri-radiate markings. Quite a good number of megaspore ıve been collected which are under investigations.

Gymnosperms : A number of cuticles with haplocheilic stomata are preserved i e material from Giral and Barsinghsar. The distribution, orientation and number (bsidiary cells surrounding a stoma are variable (Pl. 2, fig. F). The epidermal cells hav mparatively thick walls identical to those of conifers and cycads (Greguss, 196 ockey Ko, 1986). Neither pollen grains nor seeds of gymnosperms could be collecte om the samples of lignite studied for the present purpose.

Angiosperms : Angiosperms were very common in the vegatation during th ocene period in Rajasthan. Cuticles, pollen grains and peculiar types of fructificatior e found frequently in the lignite of Giral and Barsinghsar. The cuticles have variou pes of epidermal cells and morphology of stomata. The cuticles have thin or thic alled epidermal cells. Stomata are scattered or regularly arranged. Subsidiary cells a stinct or indistinct. However, the systematic relationships of these cuticles need furthe vestigation. Various kinds of pollen grains have been isolated from the samples. Thes e circular with smooth walls or have different types of ornamentations on their exin ıe pollen grains are sulcate, porate tricolpate or hexacolpate types.

A number of peculiar types of fructifications (?) and seeds (?) occur frequently in th gnite samples of both the localities. The former is an oval stalked structure with outh at the opposite end and surrounded by a number of teeth like structure sembling the peristome of a moss plant (Pl. 2, Figs. G and H). Inside this either 1- rpel like structures are seen or it is filled with large number of spores and spo trads. But neither filaments nor anthers could be seen in any fruiting body. Oval seec) with markings of converging rays are also present in the material. These structure cur isolated from the fructifications (?) whether these fructifications and seeds are (ıgiospermic affinities or not, is yet to be decided.

DISCUSSION

The Tertiary rock formations in Western Rajasthan have plant fossils preserved i ndstones, fuller's earth and lignite. Fossil leaves are found preserved as impressior eshmukh and Sharma 1978) while the stems as over silicified decorticated wooc ıularia 1990, Harsh and Sharma 1995, Harsh et al., 1992). The microfossils in ligni om Palana were studied by Rao and Vimal (1952). They identified spores (eridophytes and pollen grains of angiosperms. In the present investigation existence (l major groups of plant kingdom have been reported which shows that the deposition (gnite took place either in muddy soil or in a lake. Presence of algal filaments and ne ıggests existence of aquatic habitat and favours fresh water deposits of Giral an ırsinghsar lignites, while the occurrence of *Riccia* like thalli with scales, moss capsule eridophytic cuticles of Dipteridaceae, sporangia and spores, and gymnospermou ıticles point towards the occurrence of mesophytic conditions during the deposition.

The concept of occurrence of marine conditions in the entire Thar desert (ajasthan during Eocene (Kaul 1957, Singh 1952, Lakhanpal 1964, Bhatia 1977, Da upta 1977) seems not very valid on the basis of investigations carried out on woo om Bikaner (Gularia 1990, Harsh and Sharma 1995, Harsh et al., 1992) and the pla icrofossils isolated from the Giral and Barsinghsar lignite samples (described in t esent paper). Algal genus *Hydrodictyon* is found in fresh water. Similarly, the thalli *iccia*, the spores and cuticles of pteridophytes, etc. favour the presence of non-mari nditions at the time of deposition of lignite found at Giral and Barsinghsar. Probabl th marine and non-marine conditions existed in Rajasthan during the Eocene period.

ACKNOWLEDGEMENTS

The financial support extended to BDS and RPT by the University Gran ommission and to MSS by Shell India Production Development B.V. is gratefu knowledged.

REFERENCES

natia, S.B. 1977. Palaeontology of Rajasthan : A Review. *Natural Resources of Rajasthan* 2 : 8 - 906.

old, H.C., Alexopoulos, C.J. and Delevoryas, T. 1987. *Morphology of plants and fungi*. Harper a Row, New York.

ose, M.N. 1949. Angiospermic remains from Barmer sandstones. *Rec. Geol. Surv. India* 18 : 24 247.

ose, M.N. and Sah, S.C.D. 1967. Some pteridophytic remains from the Rajmahal Hills, Bih *Palaeobotanist* 16 : 12 - 28.

ower, F.O. 1926. *The ferns (Filicales).* II. Cambridge, London.

as Gupta, S.K. 1975. A revision of the Mesozoic Tertiary stratigraphy of the Jaisalmer bas Rajasthan. *Indian Jour. of Earth Sci.* 2 : (1) : 77 - 99.

as Gupta, S.K. 1977, The stratigraphy of the West Rajasthan Shelf. *Proc. Collq. Indi Micropalaeontol. and Strat.* O.N.G.C. Dehradun : 219 - 233.

atta, A.K. 1983. Geological evaluation and hydrocarbon prospects of Rajasthan basin. *Petrole Asia Jour.* 1 : 93 - 100.

eshmukh, G.P. and Sharma, B.D. 1978. Fossil plants from the Eocene of Barmer, Rajasth (India). *Trans. Isdt. and Ucds.* 3 : (2) : 88 - 90.

erdemann, J.W. 1965. Vesicular-arbuscular mycorrhizae formed on maize and tulip tree *Endognne fasciculate. Mycologia.* 57 : 562 - 575.

reeguss, P. 1968. Xylotomy of the living Cycads with a description of their leaves and epiderm *Akademiai Kiado*. Budapest, Romania.

ularia, J.S. 1984. Occurrence of anacardiaceous woods in the Tertiary of Western Ind *Palaeobotanist* 32 : 35 - 43.

ularia, J.S. 1990. Fossil dicotyledorious woods from Bikaner, Rajasthan, India *Geophytology* 1 182-188.

arsh, R. 1990. Studies on the petrified woods from the Tertiary of Bikaner (Rajasthan) a palaeoecology of the area. *Unpublished. Ph. D. Thesis,* J.N. Vyas Univ., Jodhpur.

ɪrsh, R. and Sharma, B.D. 1988. *Araucarioxylon bikaneriense* sp. nov. from the Tertiary Bikaner, Rajasthan, India. *Phytomorphology*, 38 (2 - 3) : 111 - 115.

ɪrsh, R. and Sharma, B.D. 1992. Chemistry of an extinct wood from Palana lignite (Bikan Rajasthan. *Ind. Jour. of Earth Sci.* 19 : (1) : 50 - 52.

ɪrsh, R. and Sharma, B.D. 1995. Petrified Tertiary woods from Bikaner (Rajasthan). *Ind. Jo of Earth Sci.* 22 : 384 - 389.

ɪrsh, R. Sharma B.D. and Suthar, O.P. 1992. Anatomy of petrified woods of Lecythidaceae a Combrataceae from Bikaner (Rajasthan), India. *Phytomorphology.* 42 (1&2) : 87 - 102.

.in, K.P. and Gupta, R.C. 1970. Some fungal remains from the Tertiaries of Kerala Coa *Palaeobotanist* 18 (2) : 177 - 182.

ɪul, K.N. 1957. A palm fruit from Kapurdi (Jodhpur, Rajasthan Desert) *Cocos sahnii* sp. n *Curr. Sci.* 20 : 138.

ɪkhanpal, R.N. 1964. Specific identification of the Guttiferous leaves from the Tertiary Rajasthan. *Palaeobotanist* 12 : 265 - 266.

ɪkhanpal, R.N. and Bose, M.N. 1951. Some Tertiary leaves and fruits of the Guttiferae fr Rajasthan. *Jour. Indian Bot. Soc.* 30 : 132 - 136.

ɪo, A.R. and Vimal, K.P. 1950. Plant microfossils from Palana Lignite (Eocene). Bikaner. *Cu Sci.* 19 : 82- 84.

ɪo, A.R. and Vimal, K.P. 1952. Tertiary pollen from lignites from Palana (Eocene) Bikaner. *Pr Nat. Inst. Sci India* 18(6) : 595-601.

[illegible], S[illegible] Kar, R.K. and Singh, R.V. 1971. Stratigraphic range of *Dandotiaspora* gen. nov. in t Lowe[illegible] Eocene sediments of India. *Geophytology* 1(1) : 54-63.

henck, N.C. and Perez, V. 1987. *Manual for the identification of VA mycorrhizal fungi.* 2 Edition International Culture Collection of VA mycorrhizal Fungi (INVAM).

ngh, S.N. 1952. On the extension of the Kirthar Sea to Rajasthan. *Proc. Nat. Acad. Sci. In* 22(3) : 7-10.

ockey, R.A. and Ko, H. 1986. Cuticle micromorphology of *Araucaria* de Jussieu. *Bot. Gaz.* 14 508-548.

ərma, K.K. 1982. The fossil record and environment of desert covered areas of Westrn Ind *Geol. Surv. India Mis. Pub.* 49 : 141-152.

Geological Evolution of Northwestern India
B.S. Paliwal (Ed.)
Scientific Publishers (India), Jodhpur, 1999, pp. 129 - 147.

RIPPLE MARKS AND SYNTHESIS OF BEACH SEQUENCES : A STUDY OF EARLY PALAEOZOIC JODHPUR SANDSTONE, WESTERN RAJASTHAN

D.S. Chauhan and Bhanwara Ram

Department of Geology, J.N.V. University, Jodhpur - 342 005
**Department of Mines and Geology, Jodhpur - 342 001*

ABSTRACT

The lower Palaeozoic Jodhpur sandstone, representing the basal siliciclastic rocks of the Marwar Supergroup, is exposed in the vicinity of Jodhpur. It comprises a package of shales and sandstones which is divisible into three distinct lithofacies. The lower deltaic, the middle beach and the upper fluvial facies. The middle facies consisting of a fine to midium grained sandstone of uniform nature, is chosen for the present study. This sandstone facies shows prolific development of wave ripples, cross bedding, planar bedding etc. The ripple morphology is found to be a very important attribute. It has been observed that the ripples are developed in hierarchical order showing straight crest ripples at the base followed upward by undulatory ripples, sinuous ripples, interference ripples and finally adhesion ripples/adhesion warts at the top.

The development of ripples in sandstone of the middle facies is discussed in the light of modern concepts on wave generated structures. It is envisaged that the straight crest ripples have formed under relatively deep water conditions and undulatory ripples, sinuous ripples and interference ripples under gradually shallowing water conditions. Finally, the above hierarchy of ripples culminates into adhesion ripples/adhesion warts denoting subaerial condition. It is suggested that the sequence of ripples developed in a hierarchical order as noted in the present area represents a shallowing upward cycle. Accordingly, the sandstone of the middle facies is a product of beach processes characterized by a number of shallowing upward cycles.

Key words : Palaezoic, Jodhpur Sandstone, Marwar Supergroup, Beach Sequences, Ripple Marks, Rajasthan.

INTRODUCTION

The lower Palaeozoic Jodhpur sandstone exposed between Jodhpur city and Bales widely used as a building stone (Fig. 1). Historical monuments such as Mehranga rt (AD 1459) and Umaid Bhawan Palace (AD 1946) of Jodhpur city are good example is still being used as a principal building material as well as decorative stone whi autifies the townships in the major part of Western Rajasthan.

Geological examination of the outcrops reveals that it is only the middle horizon e sandstone that is being quarried for building stone. Obviously, this sandstone of t iddle horizon ought to be of specific nature. It has been observed that the sandstone ver horizon is composed of interbedded sequence of lenticular bodies of sandy shal aly sandstone and medium to coarse grained sandstone of variable dimensions; whe the sandstone of the upper horizon comprises mainly medium to coarse grain ndstone and very coarse grained sandstone with intercalations of pebbly nglomeratic sandstone. It is therefore implicit that both of these varieties of sandsto not make good quality of building stone.

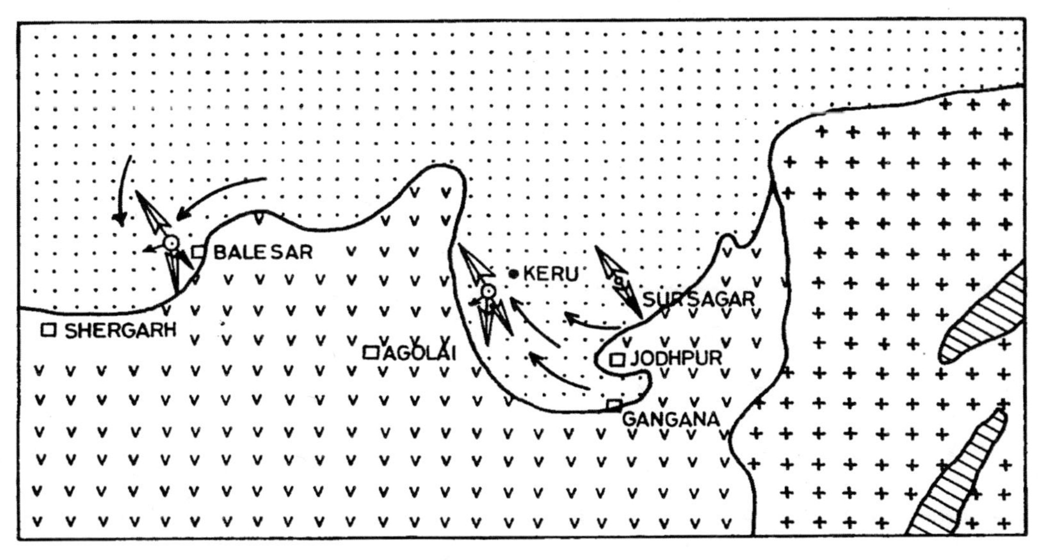

g. 1. Geological map of Jodhpur and Balesar areas showing principal palaeocurrent directions.

Unlike the above two varities, the sandstone of the middle horizon is medium ıe grained reqularly bedded and possesses fascinating light pinkish yellow to pinkis own colours. It easily splits into regular slabs and can be shaped into suitable blocl ıd bricks, the material that makes buildings of the region beautiful and attractive. It is sandstone of the middle horizon that has attracted the attention of the authors ke up its intensive sedimentological study (Figs. 2, 3 and 4).

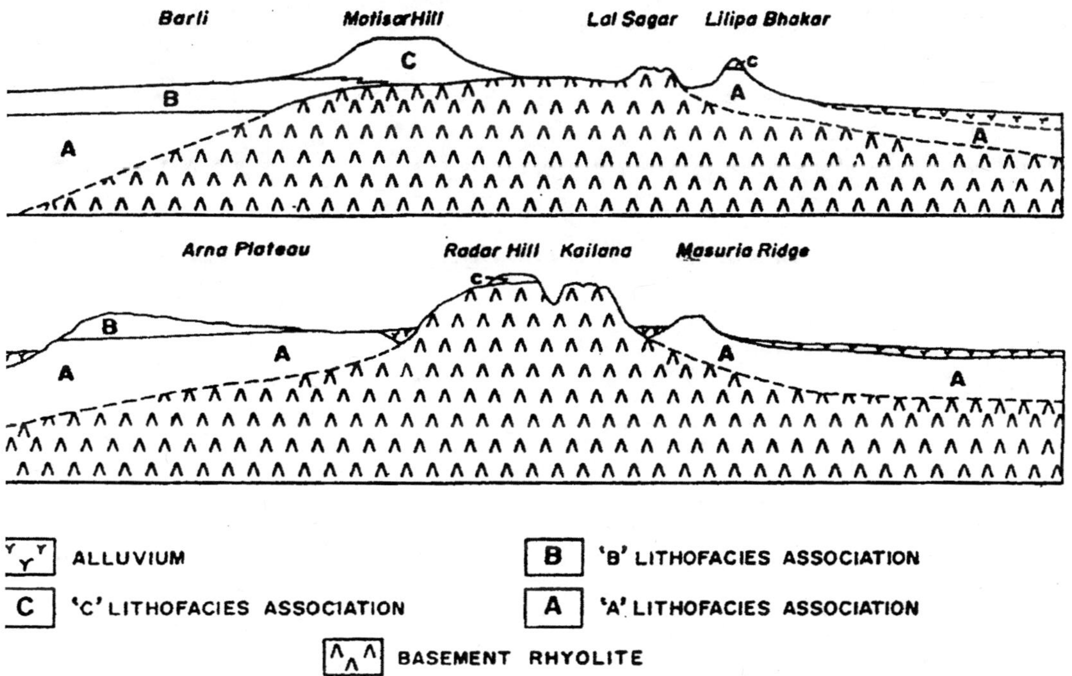

;. 2. Geological sections along A - A' and B - B' lines indicated in Fig. 3 showing relationshi between the basement and various lithofacies of the Jodhpur Group in the area.

In an earlier study (Chauhan *et al.*, 1991) the geological attributes of the sandston the lower, middle and upper horizons have been evaluated. It was concluded that thes rizons of sandstone represent three distinct sedimentary facies which are designate ccessively as A,B and C lithofacies and these facies are the products of deltaic, beac d fluvial processes respectively. The present study is confined to the synthesis of th olutionary history of the 'B' sandstone facies – a gift of ancient beaches to this regio ig. 4).

The sandstone exposed during day to day quarrying operations has served as luable source of geological information. Among the large variety of sedimentar uctures preserved in the sandstone, ripples have been found most useful for this work

GEOLOGICAL SETTING

Jodhpur sandstone rests unconformably over a highly rugged basement of Malar yolite (745 ± 10 Ma) and granite (approximately 600 ± 70 Ma) (Crowford and Compsto

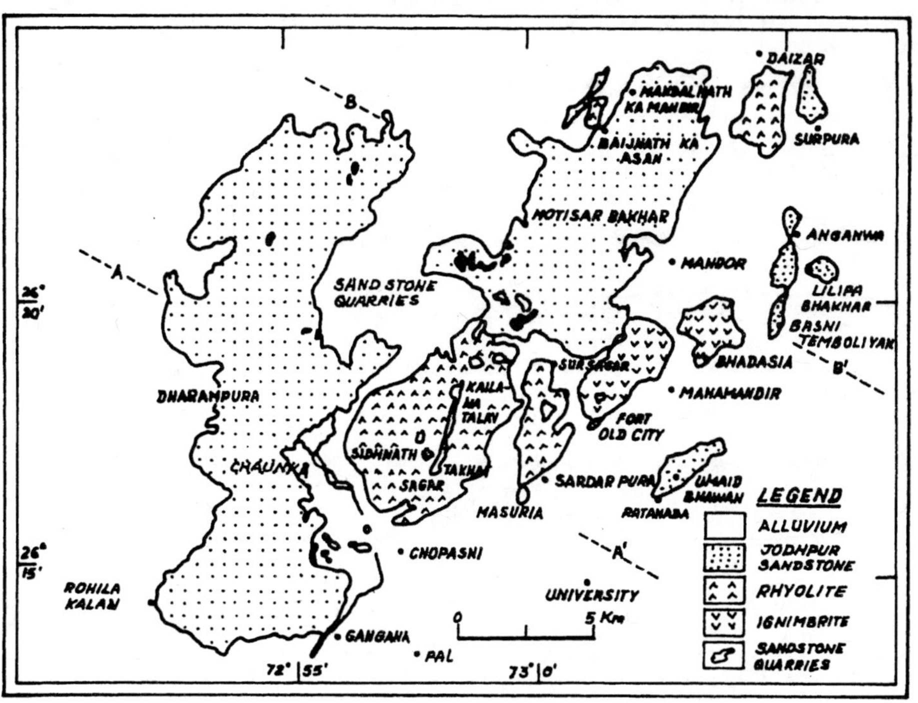

Fig. 3. Geological map of the Jodhpur city and adjoining area.

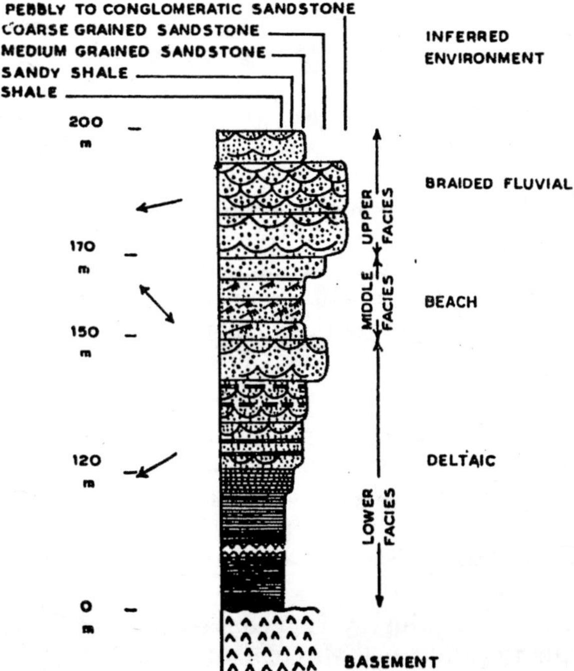

Fig. 4. Composite stratigraphic sections of Jodhpur Group of rocks in the vicinity Jodhpur town showing lithofacies associations, dominant plaeocurrent directions and inferred depositional environments.

971, Sharma *et al.*, 1975). It is considered to form the basal sequence of the Marwa upergroup, designated as Jodhpur Group, followed by sequences of carbonate lithofacie nd siliciclastic-evaporite facies named as Bilara Group and Nagaur Group respectivel his entire succession of siliciclastic, carbonate and evaporite facies which togethe epresent the Marwar Supergroup, is considered to be deposited in a NNW-SSE trendin ide and far flung basin called as Marwar basin (Pareek 1981, 1984). This basin start om the vicinity of Jodhpur city and continues, subsurfacially further north-westwar p to the Salt Range of Pakistan. Accordingly the rocks of the Marwar Supergroup ar quated with the Cambrian succession of Saline Series of the Salt Range. Jodhpu andstone resting unconformably over the rhyolite basement exposed along the Jodhpu okaran section represents the western margin of the Marwar basin (cf. Narayana 971, Datta 1983, Pareek, 1984).

GEOLOGICAL ATTRIBUTES

The middle horizon of the Jodhpur sandstone i.e. the 'B' lithofacies, is characterize y reddish brown, greyish brown, pinkish brown, creamy pink to pink coloured, fine t edium grained sandstone. These colour changes are systematic from base upward an epend upon the nature and amount of iron oxide present. The basal units of sandston re reddish brown, greyish brown, the middle pinkish brown and the upper creamy pin pink in colour. The grain size also increases, though slightly, from base upwards. Th aximum thickness of sandstone of 'B' lithofacies is approximately 20 m. Petrographi tudy suggests that the sandstone is mainly composed of rounded to well rounded grain f monocrystalline quartz. A small fraction of polycrystalline quartz is also presen uthigenic quartz developed as partial overgrowths constitutes the main cementin aterial. Tourmaline occurs as well rounded grains within the range of fine sand. Fro e overall predominance of quartzose sand (98%), the sandstone of 'B' lithofacies is ypical quartz arenite.

SEDIMENTARY STRUCTURES

The middle horizon of the Jodhpur sandstone shows excellent preservation edimentary structures through out its entire thickness. Surface structures includ ipple marks, adhesion warts, rain drop imprints, rill marks and scour marks. Th iternal structures are ripple bedding, parallel bedding and cross lamination/cros edding. Ripples are most widely developed. These are mostly small wave rippl ollowed less commonly by small current ripples and mega ripples (cf. Reineck and Sing 986). Ripples have been found useful for the present works, hence their detailed accour our concern.

WAVE RIPPLES

Most of the ripples show smooth regularly developed long crest line often wit ning-fork like bifurcations. These attributes are typical of wave ripples (cf. Reineck an ingh 1986). In vertical profile the ripples are commonly asymmetrical; howeve ometimes perfectly symmetrical ripples also occur. The nature of crest of wave ripple ows considerable variation in plan view. Accordingly there are straight crest rippl dulatory crest ripples and sinuous crest ripples (Figs. 5, 8 - 12). When one set of suc ples is intersected by another set of ripples interference type of ripples develop. (Fi).

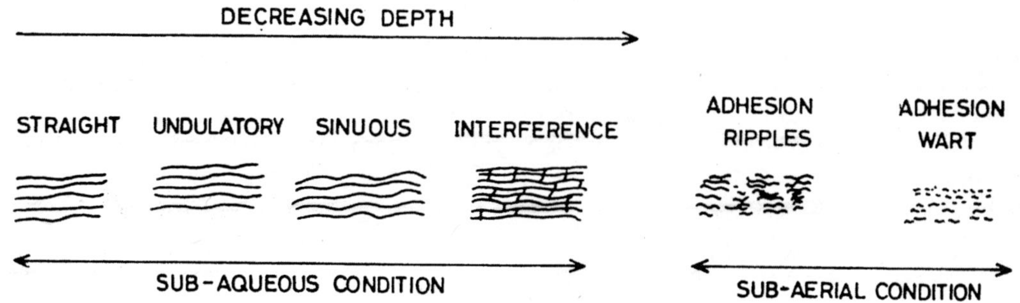

;. 5. Diagram showing different types of wave ripples of the area and the relative conditions their formation.

The wave ripples have variable ripple length of 28 mm to 53 mm and ripple heig 3 mm to 5 mm; their ripple index ranges from 5.6 to 10.6 and symmetrical index vari m 1.80 to 2.78 (Table 1).

ble 1. Ripple parameters of "B" Lithofacies

pes of Ripple	L	H	L_1	L_2	RI	SI
aight crest ripple	28	5	18	10	5.6	1.80
dulatory crest ripple	28	3	19	9.0	9.33	2.10
erference ripple	53	.5	39	14	10.6	2.78

ata of the dominant ripple pattern.

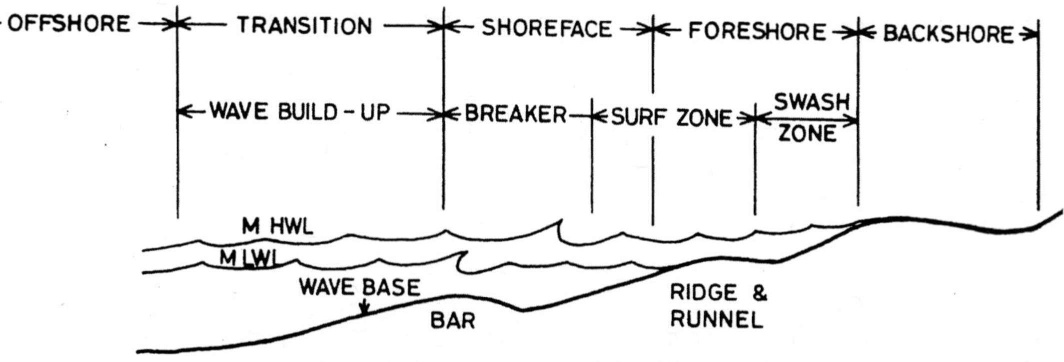

;. 6. Schematic representation of the various geomorphic and dynamic zones of a beach profi (Modified from Emery, 1960)

i) Straight Crest and Undulatory Crest Ripples

Perfectly straight crest ripples are uncommom. They usually show gent ındulations (Fig. 11). However, when the undulations are closer, sinuous ripples a ɔrmed. Undulatory ripples are widely developed. These ripples commonly show smoo utline and rounded crest. Sometimes they also display flat crests, and in other cas ointed troughs. These features indicate that the ripples have been modified by wa rosion (Fig. 8) on account of lowering of water level (cf. Reineck and Singh, 1986).

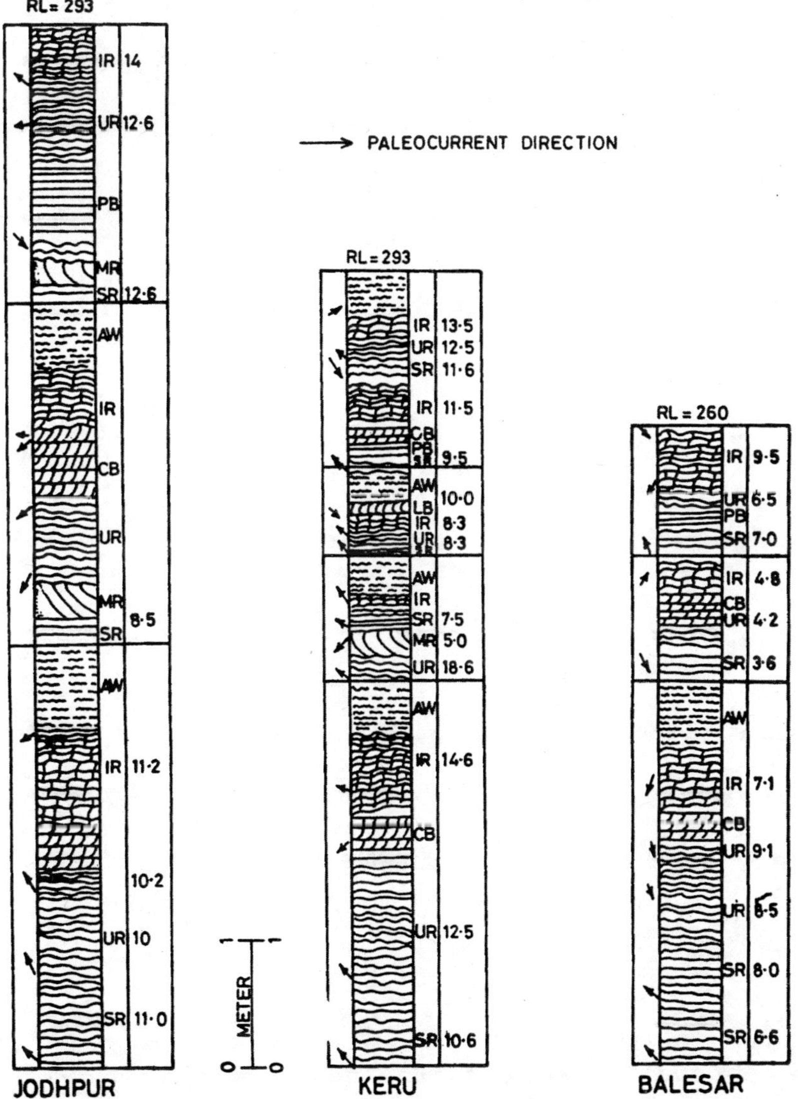

ig. 7. Schematic vertical sequences of ripple types and associated other beforms of bea lithofacies of Jodhpur Group.

g. 8. Almost symmetrical straight crest ripples. Note that their crest have become flat due subsequent erosion. Sursagar area Jodhpur.

g. 9. Small wave ripples shwoing undulating to sinuous crest line and distinct tongue li bifurcation of the crest. Note, the direction of wave induced flow appear from left to rig side, Balesar area.

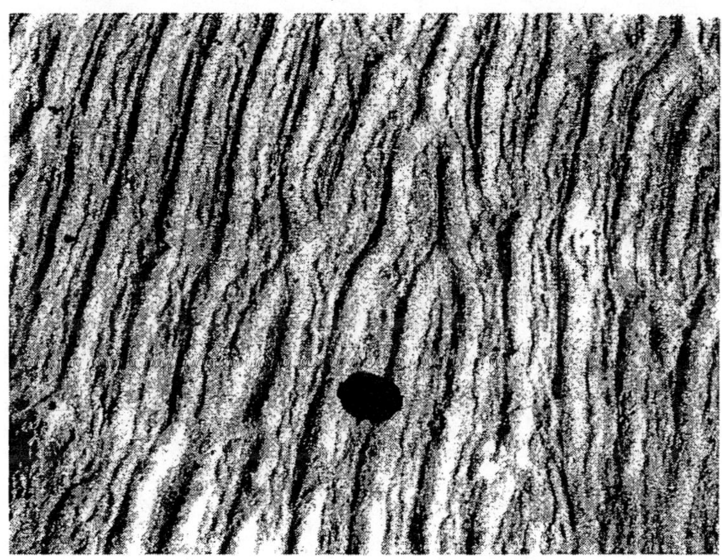

g. 10. Straight to undulatory wave ripples. Note tunning fork like bifurcation in the crest of tł ripples, the trough of these ripples are marked by a few low thin but parallel cres Sursagar, Jodhpur.

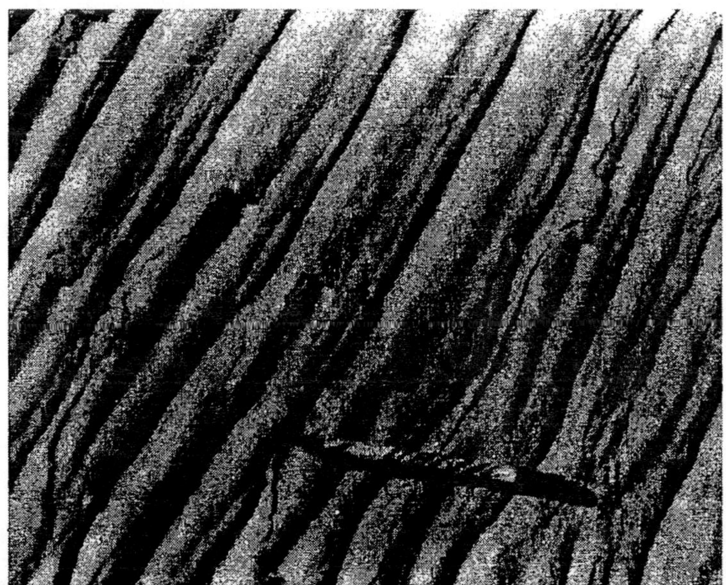

g. 11. Multiple crest ripples. Notes that the main crests are divided into two conspicous crest one of the smaller crest is further divided into still smaller crests. Such ripples denoı successive events of falling water level, Sursagar area, Jodhpur.

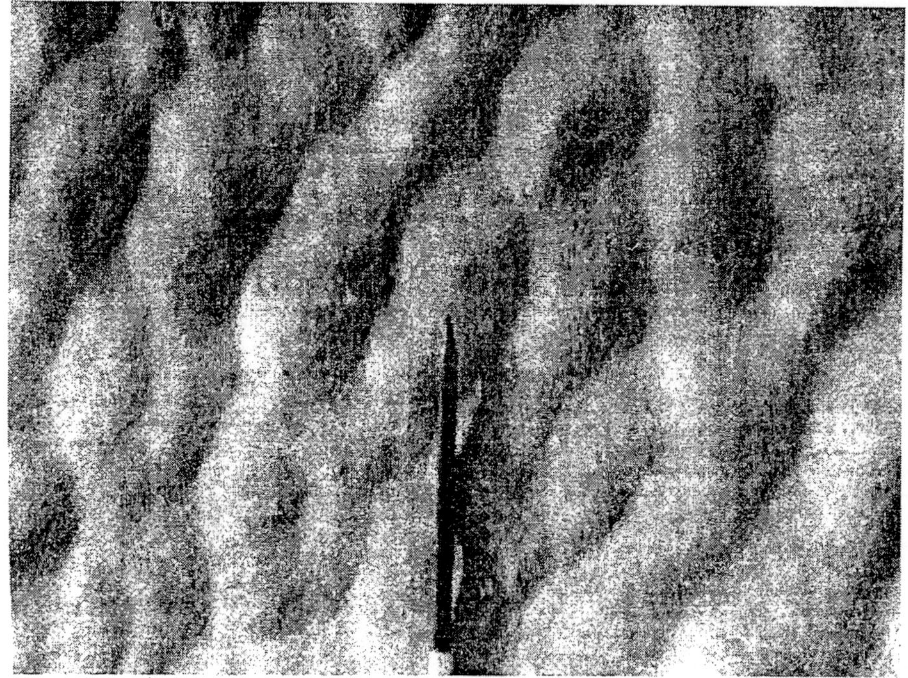

g. 12. Interference ripples showing lobes and saddles.

g. 13. Rain drop imprints developed in Jodhpur sandstone indicating subaerial condition deposition.

i) Interference Ripples

There is a large variety of ripples under this category. The undulatory ripples a nuous ripples have been found to grade into interference ripples. In such cases t pples become more and more sinuous and the ripple train becomes out of phase. ıch points the crests become discontinous and there appear knobs and saddles formi ıterference ripples (cf. Allen 1984). There are other types of ripples which roughly sh ctangular to rhombic pattern.

There are still other types of ripples which display a reticulate pattern. A clc oservation of these ripples suggests that they are composed of two intersecting sets pple trains, one superposed over the other. The ripple train of the first set of ripp ıay appear prominent. These must have formed by progressive waves. The second set ıore or less right angle to the first one and is more conspicuous. It clearly divides t est of the first set of ripples possibly by operation of a second set of waves more or le ormal to the first set. Hence there results a reticulate pattern of ripples. In some cas has been noted that, wave length of the first set of ripples is usually much larger th ıat of the second set. Ripples of the second are usually very small, and can be referr s mini ripples (Reineck and Singh 1986).

Interference ripples are considered to have been formed from two trains of sh ested waves propagating in directions 90° apart (Allen 1984). It has be xperimentally found that one of the superimposed waves might have developed from t olique reflection of a train of progressive wave (Silvaster 1972 in Allen 1984 p. 45 ig. 12).

ii) Ladder Back Ripples

These are small wave ripples the troughs of which are occupied by mini ripples. T end of the mini ripples is normal to the trend of main ripples. It has been observed th ı some cases the mini ripples traverse the adjacent crest of the main ripples. Su pples also show the reticulate pattern. They are supposed to have been formed reshore environment due to subsequent flow of returning shallow water in the runn ormal to the trend of initially developed ripples (Reineck and Singh 1986; Davis 1978)

v) Double/Multiple Crest Ripples

In some of the ripples the main crest line appears divided into two or more smal ests (Figs. 8 and 10).

) Adhesion Ripples and Adhesion Warts

There are some faintly developed ripples like structures which show irregul iscontinuous crestlines with intervening irregular flat troughs. Such ripples are referr s adhesion ripples. These ripples appear asymmetrical in cross-section having steep oss side than the lee side. Hence they are also called antiripplets (cf. Allen 19 eineck and Singh 1986).

'urrent Ripples

There are other rarely occurring ripples showing irregularly developed crestlin(hey may show small tongue like projections in the downstream direction. They hav ıng crests and are highly asymmetrical in profiles. These appear to be current ripples.

Iega Ripples

Mega ripples showing different morphology are observed. There are long creste ndulating mega ripples having 0.7 m to 2.3 m length and 0.04 m to 0.06 m height. The re usually overprinted with small wave ripples. Less commonly, there occur lunat ıega ripples having cresentic crests and conspicuous troughs. Occasionally there ar ıterference type of mega ripples which show large reticulate to irregular pattern.

'ross Bedding

Cross bedding is commonly found in sandstone of 'B' lithofacies. It is mostly tabula ype of cross bedding occuring as solitary set. Occasionally, cosets are also present. Th ıickness of the cross bedded units varies from 5 cm to 15 cm. The foreset laminae are ım to 5 mm in thickness showing a dip of 10° to 16°. Cross bedded sets showin i-directional flow trend are also present.

RIPPLE HIERARCHY

The various types of ripples and the associated sedimentary structures as describe ı the previous section have been examined in terms of time and space relationship. Th udy of sandstone quarry sections from base upward revealed that the various types c pples have developed in a hierarchical order. There are straight crest ripples at th ase followed successively by undualatory, sinuous and interference ripples and finall dhesion ripples/adhesion warts at the top (Fig. 5). Incidently, this hierarchy of ripple orresponds to the hierarchy of current ripples as suggested by Allen (1984). Hence i ught to be similarly useful to infer the wave energy and water depth condition esponsible for the development of sequence of lithounits of 'B' lithofacies. For thi urpose the important aspects of formation of various types of ripples due to interplay c 'aves is discussed and relative water depth conditions are inferred.

WAVE DYNAMICS AND RIPPLE MARKS

Ripples are produced by interaction of waves and currents with noncohesiv ediment. In nearshore-beach environment it is mainly the wave activity that produce ipples and other related structures such as cross laminations/cross bedding. It is sai ıat under shallow water conditions the behaviour of waves becomes similar to that c urrents and hence more or less similar sedimentary structures are produced.

According to Stockes theory, waves coming from open sea are oscillatory waves. Tł brations created by such waves dictate water particles to describe circular patł ithout any net displacement of water body. When waves reach shallow water, withi ιe limit of wave base, they start interacting with the substrate, as a result oscillatoı aves change into solitary waves. Under such wave conditions water particles follo liptical paths leading to over all water motion and as consequence motion of sedimeı ırticles in the direction of wave travel (Inman and Bagnold 1963). With decreasir ater depth, wave builds in height and eventually breaks. Concurrently the wave attaiı gher bottom orbital velocity which continues to increase even in the surf zone wheı ιe solitary nature of the waves is lost (Fig. 6) (Clifton *et al.*, 1971).

It is said that the advancing solitary waves within the nearshore-beac ıvironment have two components, called landward and seaward surges. It is normal ιe landward surge that dominates. It leads to formation of landward oriente ructures, such as asymmetrical ripples and/or cross bedding. In cases where tł ·award surge dominates, seaward oriented structures will be produced. That is wł polar sedimentary structures are considered to be characteristic of nearshore beac ıvironment (Clifton *et al*., 1971; Reineck and Singh 1986).

As per bottom configuration the incomming waves in a nearshore- beac ıvironment, may be refracted. As a result the waves become oriented parallel to tł ottom contours of the substratum. Such waves give rise to ripples aligned parallel to tł ıoreline. Further, decrease in depth and increase in roughness of substratum make tł ave profile more undulatiory. This possibly implies that the change from straight cre undulatory crest ripples and ultimately to sinuous crest ripples is indicative ıallowing water conditions (cf. Clifton *et al.* 1971). It has also been noted that und ery shallow water conditions the waves may be reflected from the substra regularities and coastal highs. This would generate secondary waves. The interactic etween primary and secondary waves may give rise to interference ripples. Suc ructures imply very shallow water conditions (Allen 1984; Komar 1972; Davis 197 eineck and Singh 1986).

In some cases where the waves strike the shore obliquely, longshore currents aı ·oduced. These currents disperse sediments parallel to the shore line and give rise pical shore directed current ripples, mega ripples and/or cross bedding. Finally, tł ·litary waves lose their identity after passing through the surf zone and transform in thin sheet of water in the swash zone. This sheet of water returns after strikir ıckshore as backwash flow. Therefore in the swash- backwash zone no ripples aı rmed, only the parallel bedding is produced.

ipple morphology versus water depth

From the above discussion of the various types of ripples, especially the ιorphology, and the environment of their formation, some broad generalizations can ł ıade about the relation of ripple morphology to the water depth conditions. Accordingl ·raight crest ripples must have formed in relatively deep water conditions. Wit ecreasing depth they are followed by undulatory, sinuous and interference ripples.]

ddition shallowing water conditions are indicated by associated flat crest ripple ıultiple crest ripples and ladder back ripples. Further, adhesion ripples/adhesion war nd rain drop imprints reported from the successive sandstone sequence of the arı ignify subaerial conditions (Fig. 7).

In light of above the following model showing ripple morphology/ripple type versı ecreasing water depth can be envisaged. This model of wave ripple morphology versı 'ater depth is roughly comparable to the one suggested by Allen (1969) for curreı .pples.

'yclic Sedimentary Sequences

A proper analysis of the sedimentary structures, specially ripple marks as describı ı an earlier section, suggests that the sandstone of 'B' lithofacies comprises distin ədimentary cyclic sequences. Each complete cycle is characterized by straight cre .pples at the base, followed successively by undulatory, sinuous, interference, aı nally adhesion ripples at the top. The cycle may be completely or partially developı 'ig. 7). Obviously each cycle reflects shallowing water conditions, culminating in ubaerial conditions. It thereby constitutes a typical shallowing upward sequence.

The analysis of ripple parameters such as wave length and height, and special .pple index seems to corroborate the above observation (Table-1). For example, thoug ss regularly, the value of ripple index increases from that of straight crest ripple 'hich form the base of a cycle, to the interference ripples forming the top. From the: ata it can broadly be deduced that higher the value of ripple index shallower are tl 'ater conditions. These data also corroborate that each cycle of sandstone of the ' thofacies represents a shallowing upward cycle.

CAUSES OF CYCLIC SEDIMENTATION

It has been noted that the compositional, textural and structural characteristic earshore and coastal sediments vary cyclically in many sedimentary sequences. The: hallow marine cyclic sequences have been attributed to alternate storm and fa 'eather conditions (Hobday and Reading 1972; Dott and Bourgeois 1979; Hunter aı lifton 1982). However, recognizing a sedimentary facies purely as a product of fa 'eather or storm deposit is normally difficult. Hobday and Reading (1982) whi escribing the late Precambrian shallow marine sandbar sequences of Finmark, reportı ıat the cross bedded sandbars and horizontally bedded siltstone and sandstone are tl roduct of fair weather environment; whereas the sandstone marked by erosion urfaces as well as showing symmetrical wave ripples are the product of storm processe eriodic operation of these processes has given rise to sandbar sequences. Similarl [unter and Clifton (1982) have reported that the upper Cretaceous rocks of the Caı ebastian area, Oregon is a cyclic sandstone deposit formed due to the periodic operatic f storm and nonstorm processes. They have recognized, mainly on the basis of variatic ı the sedimentary structures, as many as fourteen cycles. These cycles are separated l rosional surfaces. Each one of these complete cycles consists of three components, ›wer hummocky cross stratified sandstone, inferred to be a storm deposit, a midd lanar and ripple bedded sandstone with shale beds in its middle, to be a fair weathı eposit, and an upper bioturbated sandstone, inferred to be deposited during fa eather or during minor storm separated by long intervals of fair weather conditions.

Examination of various units of a cycle of 'B' lithofacies in terms of fair weath ersus storm deposits suggests that a major part of the cycle appears to be a fair weath eposit. The various ripple bedded and plane bedded units of a cycle usually sho radational contacts and a very little change in the grain size from base upwar oncurrently, these units are completely free from typical storm deposit features lil ummocky cross bedding. These points support the aforesaid contention. Howeve rosional features such as scour marks and other discontinuities do occur occasional ithin some of the lithounits of a cycle. Such erosional surfaces are more clearly seen ne area, in association with the cross bedded units of a cycle where erosional surfac onspicuously form lower bounding surfaces of the cross strata. These features seem to l ne result of storm effects. The incomplete cycles observed in Keru and Balesar area robably indicate that the erosional processes were more effective in causing eliminati f some of the units of the cycle (cf. Clifton *et al.* 1971). Hence, the missing part off a ncomplete cyclic sequence denotes destructional phase in cyclic sedimentation (Fig. 7).

In view of the above discussed climatic/seasional and atmospheric factors, the fa eather versus storm conditions cannot be considered as the main cause of cyclicity ne area. Scrutiny of the various lithounits of a cycle of the area from base upwar uggests that it is the water level that should have fluctuated periodically to give rise ne cyclic sequence of the area. The basal unit comprising straight crest ripples follow y undulatory and sinuous ripples suggests that there was a gradual fall in water lev n the course of deposition of these units.

able 2 : Fair weather versus storm deposits of the area

air weather constructional phase	Plane bedded sandstone, Cross-bedded sandstone, and Ripple bedded sandstone
torm Destructional phase)	Erosional surfaces

The occurrence of adhesion ripples and adhesion warts at the top of each cyc uggets that the sedimentation culminated into subaerial conditioins. The subaeri acies is again followed by the more or less similar lithounits of a second cycle, i. aving straight crest ripples at the base and adhesion warts at the top. These cycl must have formed by periodic rise of water level, followed by sedimentation creati hallower water conditions due to sediment aggradation. Thus, each cycle of the area is ypical retrograding sequence, comprising relatively deeper water lithounits deposited subaerial facies (cf. Curray 1964). However, it is to be made clear that the rise in wat evel due to transgression was of a minor proportion. Cyclic sedimentation of coast marine shelf facies caused by periodic transgressions of different proportion and tin uratioin on global scale has been explained by Allen (1984). Preservation of su etrograding shallowing upward sequences which are the product of transgression is sa o be uncommon. However consideration of the following work is useful in this regard.

Hails and Hoyt (1968) have pointed out that in the coastal plains of Georgia, tl

)osition of a Holocene and six Pleistocene shorelines suggests that the barrier islan(:annot have originated as offshore bars after transgression, but they might have result(lue to various marine transgressions during Pleistocene and Holocene periods. Similarl Gohren (1975) studied aggradation of longshore bars infront of the North Sea tidal flat)bserved that these bars migrate towards the coast when exposed to wave action.

In view of the above it is envisaged that there might have been intermittent rise i ;ea level, though of very small proportion, followed by sand accumulations giving rise lifferent lithofacies.

Cyclic Sequences of the 'B' lithofacies

The mechanism of sedimentation of the lithounits of 'B' facies may broadly l nferred from shallowing upward carbonate cycles, discussed by Wilson (1975 Accordingly, it can be conceived that each cycle has formed when the rate of sar iccumulation exceeded the rate of subsidence. This has permitted the sediment to bui .o the sea level or slightly higher as indicated by adhesion ripples/adhesion warts of ea(:ycle in the present area. Following such subaerial conditions (back shore environmen :here had been a relative rise in sea level, giving an effect of transgression. Formation ıew cycle starts with a fresh deposition of sand following a rapid rise in sea level. Thu :yclic sedimentation takes place due to sediment aggradation-progradation interrupt( )y rapid transgressive events.

Environment of Deposition

The 'B' lithofacies mainly comprises thinly bedded sandstone having a maximu :hickness of 20 m in Sursagar area. It shows, in order of falling abundance, sma ısymmetrical wave ripples, planar cross bedding, planar bedding and mega ripple These structures appear to have formed under moderate to low energy conditions (c Komar 1976; Reineck and Singh 1986). The grain size of the sequences ranges from 0. nm to 0.1 mm representing mostly fine to medium sand. This further suggests that onl noderate to low energy conditions were prevailing during the deposition of sandstor ;equence of the 'B' lithofacies.

A systematic analysis of the sedimentary structures specially wave ripples will als)e helpful to understand the wave energy conditions vis-a-vis the relative depth of wate n this respect the absolute depth of water in which the deposition of the sandstor ;equences of 'B' lithofacies took place is envisaged below.

Symmetrical wave ripples are reported from water as deep as 204 m from Orego :ontinental shelf (Komar 1972). However, these ripples are considered to have formed b 'ery high storm surges and represent exceptional cases. Hence if they can constitute th)art of a regular ancient sedimentary sequence is doubtful. Normally wave ripples a1 'eported to form in shallow water. Symmetrical ripples are reported to form in wat(lepth upto 4 m in Gulf Gaeta, Italey (Reineck and Singh 1986). On the contrar ısymmetrical wave ripples are considered to form in much shallower water. It is sai :hat if the beach slope is very gentle, the shoaling waves create asymmetrical wave vhich produce asymmetrical wave ripples (Reineck and Singh 1986). In this respect th ork of Clifton, *et al.* (1971) and Hunter and Clifton (1981), as mentioned earlier is vei ;eful. Clifton, *et al.* (1971), have reported asymmetrical wave ripples from offshoi gion and lunate mega ripples from wave build up zone and planar bedded facies fro ırf and swash zone of Oregon coast. According to them asymmetrical wave ripples hav rmed in water depth of more than 3 m and the rest of the planar bedded facies forme 1 m to 2 m deep water. These ripples mostly have developed beyond the breaker in th ner offshore region.

In view of the above discussion it is necessary to take a fresh stock of th dimentary structures preserved in a complete cycle of sandstone of 'B' lithofacies. A entioned earlier, asymmetrical wave ripples are most widely developed through out th ındstone sequence of 'B' lithofacies. This makes one believe that the major part of th quence has deposited in the inner offshore region (cf. Clifton, et al., 1971). However, imperative to mention here, that the asymmetrical wave ripples preserved in th hounits of the area very often show other specific features, like double cresting and/ ultiple cresting, flat cresting and are associated with ladder back ripples. Thes eatures suggest that the sequence was developed in very shallow water of the shorefa foreshore region (cf. Reineck and Singh 1986).

The planar cross bedded units of a cycle suggest that there was a slightly hig locity condition causing ripple migration, leading to formation of cross bedded unit ıe bipolarity sometimes shown by a single cross bedded set and reversals of directio ıown by a number of cosests in a cross bedded unit demonstrate the effect of tide ıring the deposition of these units (Fig. 1) (cf. Allen 1984; Reineck and Singh 1986). Th esence of mega ripples in the sandstone sequence suggests that sometimes there wei ightly high velocity wave induced currents giving rise to mega ripples as observed fro any foreshore regions of modern beaches (Komar 1976; Davis Jr. 1978; Reineck an ngh 1986).

As mentioned in the earlier section plane bedded units must have formed in swas ne where the maximum orbital velocity is higher (Clifton and Hunter 1971). Finall ıe occurrence of adhesion ripples and adhesion warts at the top of each cycle of 'l hofacies suggests that subaerial conditions were existing denoting backshore beac ıvironment (Allen 1984; Reineck and Singh 1986).

To epitomise, the environment of deposition of different lithounits of 'B' lithofacie ınged from inner shoreface to backshore. The inner shoreface is marked by purel raight crest to slightly undulatory asymmetrical wave ripples; outer foreshore enoted by undulatory to sinuous asymmetrical wave ripples, mega ripple and cros edded units; the inner foreshore is signified by interference ripples, plane bedde ındstone; and backshore environment is indicated by adhesion ripples, adhesion war ıd rain drop imprints (Fig. 13). The water depth might have not exceded a few tens ntimetres. As such it remained within the range of moderate to low energy beaches (c eineck and Singh 1986). From the above it is presumed that the entire sandstor quence of 'B' lithofacies got deposited within the confines of breaker zone and as such presents a typical low to moderate energy beach sequence (Table 3).

able 3. Sedimentary structures and inferred beach environment in 'B' lithofacies of tl area

ɜdimentary structure	Geomorphic unit	Dynamic zone
.dhesion ripples .dhesion warts lane bedded sandstone	Backshore	Swash backwash
lane bedded sandstone ıterfĕrence ripple	Inner foreshore	
Iega, ripples ross bedded units inuous ripples 'ndulatory ripples	Outer foreshore	Surf zone
lightly undulating pples .symmetrical straight :est ripples	Inner shoreface	

REFERENCES

llen, J.R.L. 1969. *Current Ripples*, Amsterdam North Holland Publ. Co., 433p.

llen, J.R.L. 1984. *Sedimentry Structurs*, Developments in Sedimentology, 30, Elsevi Amsterdam, 663 p.

hauhan, D.S., Dubey, J.C. and Ram, B. 1991. Geological analysis of part of Nagaur basin in t vicinity of Jodhpur City. In : Tandon, S.K., Pant, C.C., Casshyap, S.M., (Eds.) *Sedimenta Basins of India : Tectonic Context*, Gyanodaya Prakashan, Nainital.

lifton, H.E., Hunter, R.E., Phillips, R.L. 1971. Depositional structures and processes in t nonbarred high energy nearshore. *J. Sediment. Petrol*, **41** : 651-670.

lifton, H.E. 1976. Wave-formed sedimentary structues - A conceptual model. In Davis, Jr. R. Ethington, R.L., (Eds.), *Beach and Nearshore Sedimentation Soc. Econ.* palaeontologis Mineralogists Spec. Publ. 24, 126-146.

urray, J.R. 1964. Transgressions and regressions : In Miller, R.L., (ed.) *Papers in Marine Gelo* (Shepard Commemorative Valume), 175-203, McMillan : New York

atta, A.K. 1983. Geological evolution and hydrocarbon prospects of Rajasthan Basin. *Petroleu Asia Jour.*, **6** : 93-100.

avis Richard, A., Jr. 1978. Beach and nearshore zone. In Davis R.A. Jr. Ed., *Coastal Sedimenta Environments*, Springer- Verlag, New York, 237-280.

ohren, H. 1975b. Dynamics and Morphology of sand banks in the surf zone of outer tidal fla *Proc. 14th Int. Conf. on Coast. Eng.* 871-883.

obday, D.K., Reading, H.G. 1972. Fair weather versus storm processes in shallow marine sa bar sequences in Late Precambrian of Finnmark, North Norway. *Jour. Sediment. Petrol.* **4** 318-324.

oyt, J.H., Henry, V.J. Jr., Weimer, R.J. 1968. Age of Late Pleistocene shoreline deposits Coast Georgia. *"Means of Correlation of Quaternary Successions"*. *Cong. Int. Associati Quaternary Res.* **8** : 381-393.

unter, R.E. and Clifton, H.E. 1982. Cyclic deposits and hummocky cross stratificatioin probable storm origin in Upper Cretaceous rocks of the Cape Sebastian area, southweste Oregon. *Jour. Sediment. Petrol.* **52** : 127-143.

man, D.L. and Bagnold, R.A. 1963. Beach and nearshore processes In : Hill, M.N., (Ed.) *The se*
3, 529-553, Wiley and Sons, New York.

mar, P.D., Neudeck, R.M., Kulm, L.D. 1972. Observations and significance of deep-wat oscillatory ripple marks on the Oregon Continental shelf. In : Swift, D.J.P., Duane, D.B Pilkey, O.H., (Eds), *Shelf Sediment Transport Process and Pattern*, 601-619. Stroudsbour Pa. : Dowden Hutchinson and Ross.

mar, P.D. 1976. *Beach Processes and Sedimentation*. Prentice-Hall Inc. New Jersey, 429p.

rayanan, K. 1971. Problems of stratigraphy of the Rajasthan shelf. *Proc. Symp. Problems Arid Zone of India*, Jodhpur, 92-100.

reek, H.S. 1981. Basin configuratioin and sedimentary stratigraphy of Western Rajastha *Jour. Geol. Soc. India*, **22** : 517-527.

reek, H.S. 1984. Pre-Quaternary Geology and Mineral Resources of northwestern Rajastha *Geol. Surv. India Mem.* **115** : 1-95.

ineck, H.E. and Singh, I.B. 1986. *Depositional Sedimentary Environments*, Springer-Verla Berlin. 439p.

arma, K.K., Nagpal, K.K. and Jhingran, A.G. 1975. Fission track geochronology of ear Precambrian of Rajasthan, India. *Chayanica Geologica*, **1** : 196-211.

ilson, J.L. 1975. *Carbonate Facies in Geological History* Springer-Verlag, New York : 471p.

eological Evolution of Northwestern India
.S. Paliwal (Ed.)
cientific Publishers (India), Jodhpur, 1999, pp. 148 - 166.

THE CHANGING SCENARIO OF WESTERN RAJASTHAN THROUGH MINERAL DEVELOPMENT

D.S. Mehta

Department of Mines and Geology, Government of Rajasthan, Jodhpur - 342 001

ABSTRACT

Minerals are playing the key role in overall development of western Rajasthan through industrialization. Today, not only the number of leases and production of these minerals has increased many folds, but many new minerals of vital importance for industrial development have also been added to the inventory list of minerals. New marble and granite policies launched by the State Government have helped in attracting heavy investments in the mining and processing of these decorative stones, though a real break-through is still awaited in striking oil and natural gas deposits. Many other minerals are still the monopoly of Western Rajasthan. The time is not far when Western Rajasthan may be known not for its Thar desert but for its mineral resources and industries.

ey words : Minerals, Development, Industries, Western Rajasthan.

INTRODUCTION

The western Rajasthan, comprises districts of Sri Ganganagar, Hanumanga ikaner, Churu, Nagaur, Pali, Jodhpur, Sirohi, Jalore, Barmer and Jaisalmer. Althou is constitutes the 61% of the total area of the state, yet, out of about 2 lac sq. kms. la ea of Western Rajasthan, only about 52% of the land is considered suitable ricultural use. Furthermore factors such as scanty and erratic rainfall, scar getation, extreme climatic conditions, adverse ground water conditions etc. are addi the hardship, poverty and under development of the inhabitants of this part ajasthan.

However, nature has also been generous in compensating for these adversities stowing it with a variety of mineral resources (Pareek, 1981b; Joshi, 1993). The ineral resources have been playing a vital role in the economic and industrial growth estern Rajasthan (Govt. of Rajasthan, 1989). Minerals such as gypsum, clays, marble akrana, sandstone of Jodhpur and Nagaur have been known and worked since p

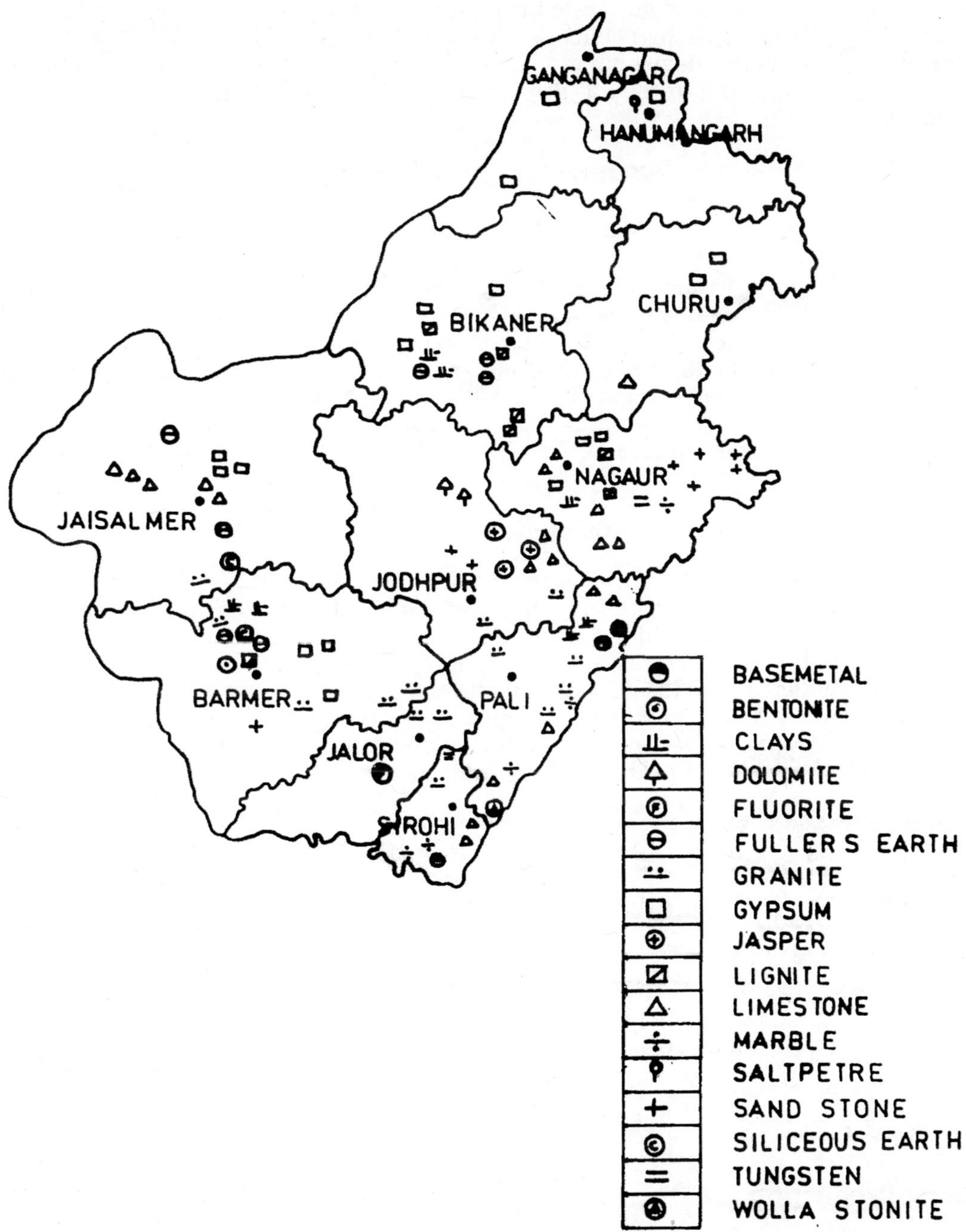

Fig. 1. Map showing important mineral occurrances in Western Rajasthan

dependance time but their exploitation and utilization had been limited. Not only t: oduction and utilization of these minerals has increased many folds sin dependance but many new minerals have also been added to the inventory of miner sources of Western Rajasthan (Pareek, 1981a and b; Paliwal 1993, 1997). An overvie the pace of mineral development (Govt. of Rajasthan, 1989) and its impact dustrial, economic and over all development of Western Rajasthan is present reunder :

MINERAL DEVELOPMENT

inerals of Rajasthan

Today, 69 types of minerals are being produced in Rajasthan (46 major and : inor) out of which Western Rajasthan has been contributing the production of : inerals (17 major minerals and 13 minor minerals) (Fig. 1). Furthermore, out of the minerals, 12 minerals (9 major and 3 minor) are the monopoly of Western Rajasth: one. These minerals are tungsten ore, ball clay, gypsum, selenite, jasper, siliceo rth, wollastonite, lignite and natural oil and gas (major minerals); bentonite, fuller rth and rhyolite (minor minerals). The list of minerals produced in Rajasthan is shov Table - 1.

ie Parameters of Mineral Development

Numer of mineral concessions, mineral production, sale value of minerals produce nployment in mineral sector and revenue from minerals are some of the key areas f sessing the pace of mineral development. 1972-73 has been taken as the base year f is study and data for the years 1976-77, 80-81, 86-87, 90-91 and 94-95 have be mpiled to study the trends of mineral development. Table 2 shows the trends ineral development at a glance and mineral wise number of leases, production, sa lue, employment, revenue from major and minor minerals in Western Rajasthan f e year 1994-95 are shown in Tables 3 to 6.

ineral Concessions

) *Major Minerals* - There were only 101 major mineral concessions in Weste Rajasthan in 1972-73 which increased to 223 in 1994-95. This increase comes to 2 times or 120.79%. The corresponding increase in entire Rajasthan has been 1. times or 48.4% only.

) *Minor Minerals* - In case of minor minerals in Western Rajasthan, the rise has be from 286 (72-73) to 1808 (94-95) that is 6.32 times or 532.17%. The corresponding ri for entire Rajasthan has been only 5.37 times or 436.65%.

Quarry Licences (QL) - The growth of Quarry Licences in Western Rajasthan has be 2.93 times or 192.90% from 3385 (72-73) to 9915 (94-95). The corresponding figur for the entire state comes to 2.09 times or 108.87%.

ineral Production

) *Major Minerals* - There has been a substantial rise in the production of maj minerals in Western Rajasthan during the past 12 years. The total production

ıble 1. Major and Minor minerals produced in Rajasthan

S.No.	Major Minerals	S.No.	Minor Minerals
01.	Copper Ore	01.	Aslet
02.	Iron Ore	02.	• Bentonite
03.	Lead, Zinc and Silver	03.	• Brick Earth
04.	Manganese Ore	04.	Chert
05.	• Tungsten Ore	05.	Chips and Chips Powder
on Metallic Minerals		06.	• Fuller's Earth
06.	Amethyst	07.	• Granite
07.	Aquamarine	08.	• Kankar-Bajri
08.	• Asbestos	09.	• Lime Kankar
09.	• Ball Clay	10.	• Limestone (Burning)
10.	Barytes	11.	Limestone (Building)
11.	• Calcite	12.	• Marble
12.	Chalk	13.	• Masonry Stone
13.	• China Clay and White Clay	14.	Mill Stone
14.	Corundum	15.	• Murram
15.	• Dolomite	16.	Patti-Katla
16.	Emerald	17.	Quartzite
17.	Epidote	18.	• Rhyolite
18.	Feldspar	19.	• Salt Petre
19.	• Fire Clay	20.	• Sand stone
20.	• Fluroide	21.	Serpentine
21.	Garnet	22.	Shale
22.	Graphite	23.	Slate Stone
23.	• Gypsum		
24.	• Jaspar		
25.	Kyanite		
26.	Laterite		
27.	• Limestone		
28.	Magnesite		
29.	Mica		
30.	Ochres		
31.	Pyrophyllite		
32.	• Quartz		
33.	Red Oxide		
34.	Rock Phosphate		
35.	• Selenite		
36.	Shale		
37.	Silica Sand		
38.	School Slate		
39.	• Siliceous Earth		
40.	Sillimanite		
41.	Soapstone		
42.	Sulphur		
43.	Vermiculite		
44.	• Wollastonite		
45.	• Lignite		
46.	• Oil and Gas		

Table 2. Development at a Glance western Rajasthan vs Rajasthan (1972-73 to 1994-95)

Year	MINERAL CONCESSIONS (NOs.)									PRODUCTION (In Lac Tonnes)					
	M.L. Major Minerals			M.L. Minor Minerals			Quarry License			M.L. Major Minerals			M.L. Minor Minerals		
	W.R.	Raj.	%	W.R.	Raj.	%	W.R.	Raj.	%	W.R.	Raj.	%	W.R.	Raj.	%
72-73	101	880	11.48	286	1809	15.81	3385	9221	36.71	10.80	46.68	23.14	32.56	56.82	57.30
76-77	167	1162	14.37	476	2833	16.80	4790	12262	39.06	7.09	83.05	8.54	17.66	73.69	23.97
80-81	140	1317	10.63	873	4035	21.64	6485	14992	43.26	9.40	59.28	15.86	69.16	196.43	35.21
86-87	NA	1825		NA	5731		NA	16963		27.15	121.48	22.35	99.89	234.18	42.66
90-91	200	1391	14.38	1385	9181	15.09	8957	17408	51.45	31.10	147.76	21.05	101.09	283.72	35.63
94-95	233	1306	17.08	1808	9708	18.62	9915	19260	51.48	37.22	184.59	20.16	145.38	322.56	45.07
GROWTH (94-95 OVER 72-73)															
Times	2.21	1.48		6.32	5.37		2.39	2.09		3.45	3.95		4.46	5.68	
%	120.79	48.4		532.17	436.65		192.91	108.87		244.63	295.44		346.50	467.69	

Year	SALE VALUE (In Crore Rs.)						EMPLOYMENT (Persons per day)						REVENUE (Lac Rs.)		
	M.L. Major Minerals			M.L. Minor Minerals			M.L. Major Minerals			M.L. Minor Minerals					
	W.R.	Raj.	%	W.R.	Raj.	%	W.R.	Raj.	%	W.R.	Raj.	%	W.R.	Raj.	%
72-73	1.29	9.81	13.15	3.92	6.49	60.40	2826	25285	11.18	18258	57893	31.54	54.72	233.91	23.39
76-77	15.01	41.35	36.30	4.93	13.14	37.52	3123	24554	12.72	17182	50670	33.91	96.27	501.79	19.19
80-81	NA	53.21		NA	54.27		NA	26737		NA	106921		235.62	905.84	26.01
86-87	9.46	143.93	6.57	37.67	137.29	27.44	4186	27465	15.24	55458	216025	25.67	700.05	2454.3	28.52
90-91	25.33	225.1	11.25	107.34	417.45	25.71	5831	27283	21.37	60479	310268	19.49	1557.40	5598.40	27.82
94-95	62.71	355.09	17.66	256.58	600.75	42.71	8936	47557	18.79	58384	308580	18.92	3378.26	15396.47	21.94
GROWTH (94-95 OVER 72-73)															
Times	48.61	36.20		65.45	92.57		3.16	1.88		3.20	5.33		61.74	65.82	
%	4761.24	3519.67		6445.41	9156.55		216.21	88.08		219.77	433.02		6073.72	6482.22	

Table 3. Major Minerals Numbers of Leases, Production Revenue, Sale Value and Employment 1994-95 in Western Rajasthan.

S.No.	Minerals	No. of Leases Raj.	No. of Leases W. Raj.	No. of Leases %	Production ('000 tonnes) Raj.	Production ('000 tonnes) W. Raj.	Production ('000 tonnes) %	Persons Engaged (per day)	Revenue ('000 Rs.)	Sale Value (Lac Rs.)	District of Occurrence
1	Ball clay	50	50	100.00	194.19	194.19	100.00	464	7169.37	149.63	Bikaner (L), Pali Jaisalmer
2	Gypsum	35	35	100.00	1709.59	1709.59	100.00	2201	37503.86	2049	H.Garh, G.Nagar, Nagaur Barmer, Jalore, Jaisalmer
3	Jasper	26	26	100.00	5.11	5.11	100.00	105	245.99	12.76	Jodhpur (L) Jaisalmer
4	Siliceous Earth	7	7	100.00	1.10	1.10	100.00	117	209.6	2.66	Barmer (L), Jaisalmer
5	Wollastonite	3	3	100.00	69.08	69.08	100.00	1641	5851.19	412	Sirohi
6	Selenite	3	3	100.00	11.98	11.96	100.00	413	702.12	57.86	Barmer (L), Bikaner
7	Lignite	2	2	100.00	Production commenced from May 95				64.99		Barmer, Bikaner, Nagaur
8	Tungsten	2	2	100.00	No Production during 1994-95						Nagaur, Sirohi
9	Fire Clay	26	15	57.69	110.54	107.16	96.94	220	90.89	80.37	Bikaner (L.)
10	China and White Clay	74	17	22.97	335.75	98.92	29.46	212	1360.25	74.19	Bikaner (II L), Pali
11	Calcite	26	7	26.92	66.72	60.20	90.23	715	3290	138.76	Sirohi (L) Pali
12	Fluorite	17	6	35.20	1.49	0.75	50.34	142	282.08	26.31	Jalore (L)
13	Asbestos	52	6	11.54	24.92	5.36	21.31	27	204.7	8.3	Pali
14	Limestone (Major)	42	14	33.33	19628.38	1447.01	7.37	2601	38036.8	2171.65	Sirohi, Jaisalmer, Nagaur and Jodhpur
15	Quartz	69	19	27.54	102.80	8.05	7.83	35	474.4	6.03	Pali (II L)
16	Oil and N. Gas	–	–	–	–	–			6786.9		Jaisalmer
17	Dolomite	10	1	10.00	8.55	0.12	1.40	33	2.91	0.1	Jaisalmer, Jodhpur
		444	213		22270.18	3718.55		8926	100826.61	5189.62	

Importance of Minerals in Western Rajasthan

No. of Leases
1. Ball Clay
2. Gypsum
3. Jasper
4. Quartz
5. China/White Clay
6. Limestone

Production
1. Gypsum
2. Limestone
3. Ball Clay
4. Fire Clay
5. China/White Clay
6. Wollastonite
7. Calcite

Revenue
1. Limestone
2. Gypsum
3. Ball Clay
4. Oil and N. Gas
5. Wollastonite
6. Calcite

Table 4. Minor Minerals Number of Leases, Mineral Production, Revenue, Sale Value and Employment 1994-95 in Western Rajasthan

S.No.	Minerals	No. of Leases			Production	Persons	Revnue	Sale	District of Occurrence
		Raj.	W. Raj.	%	('000 tonnes) Raj.	Engage (Per day)	('000 Rs.)	Value (Lac Rs.)	
1	Bentonite	16	16	100.00	56.49	58	1974.44	62.05	Barmer
2	Fuller's Earth	13	13	100.00	16.58	49	452:32	165.82	Barmer, Bikaner
3	Rhyolite	57	57	100.00	100.91	207	866.13	69.14	Jodhpur, Pali
4	Saltpetre	2	2	100.00	21.51	450	204.64	90.6	Hanumangarh
5	Sandstone	2076	100	4.82	932.52	16078	45571.62	2455.84	Jodhpur, Jaisalmer
6	Masonry Stone	574	1871	—	7620.84	19992	10549.73	2777.86	Jodhpur (L) Nagaur, Churu, Jaisalmer, Sirohi
7	Marble	5	215	6.69	689.41	11135	82605.78	14910.7	Nagaur (L) Sirohi, Churu, Jaisalmer
8	Kankar-Bajri	11	—		1017.04	1383	10046.88	191.7	Bikaner, Nagaur, Jaisalmer, Sirohi
9	Murram	1	1	100.00	183.98	534	263.07	51.95	Nagaur (L) Jodhpur, Sirohi, Jaisalmer
10	Limestone (Burn)	518	183	35.33	1943.92	3057	58763.49	1480.57	Nagaur (L), Jodhpur, Pali. Sirohi, Churu
11	Granite	728	361	49.59	27.68	1228	7625.96	329.13	Bikaner (L) Jalore, Sirohi, Pali
12	Lime kankar	14	5	35.71	22.92	69	566.65	9.24	Churu (L) Nagaur, Bikaner
13	Brick Earth	421	278	66.03	1913.75	4319	10663.03	3006.9	G. Nagaur, Bikaner, Jodhpur, Jaisalmer, Nagaur, Churu
		8944	1805		14550.35	58559.00	230153.74	25601.39	

IMPORTANCE OF MINERALS IN WESTERN RAJASTHAN

No. Leases	Production	Revenue
1. Masonry Stone	1. Masonry Stone	1. Marble
2. Granite	2. Limestone (Burn)	2. Limestone (Burn)
3. Brick Earth	3. Brick Earth	3. Sandstone
4. Marble	4. Kankar Bajri	4. Brick Earth
5. Limestone (Burn)	5. Sandstone	5. Masonary Stone
6. Sandstone	6. Marble	6. Kankar Bajri

ıble 5. Major Minerals Production and Employment Status 1972-73 versus 1994-95 Western Rajasthan

No.	Minerals	Production ('000 Tonnes)		Persons engaged (Per day)	
		72-733	94-95	72-73	94-95
	Ball Clay	–	194.19	–	464
	Gypsum	984.93	1709.59	1143	2201
	Jasper	0.41	5.11	30	105
	Siliceous Earth	0.4	1.10	5	117
	Wollastonite	2.29	69.08	55	1641
	Selenite	10.19	11.96	34	413
	Lignite	–	–	–	–
	Tungsten	0.02	–	414	–
	Fire Clay	15.25	107.16	322	220
).	China and White clay	44.5	98.92	411	212
..	Calcite	7.28	60.20	180	715
:.	Fluorite	–	0.75	–	142
;.	Asbestos	1.47	5.36	90	27
:.	Limestone	7.16	1447.01	13	2601
;.	Quartz	1.21	8.05	68	35
;.	Oil and N. Ga	–	–	–	–
'.	Dolomite	2.61	0.12	24	33
;.	Barytes	0.33	–	26	–
		1078.05	3718.6	2815	8926

ıble 6. Minor Minerals Production and Employment Status 1972-73 versus 1994-95 Western Rajasthan

No.	Minerals	Production ('000 Tonnes)		Persons engaged (Per day)	
		72-73	94-95	72-73	94-95
	Bentonite	4.01	56.49	77	58
	Fuller's Earth	8.50	16.58	45	49
	Rhyolite	–	100.91	—	207
	Saltpetre	1.02	21.51	16	450
	Sandstone	–	935.32	—	16078
	Masonry stone	1363.50	7620.84	12619	19992
	Marble	66.44	689.41	3253	11135
	Kankar-Bazri	–	1017.04	—	1383
	Murram	–	183.98	–	534
)	Limestone (Burn)	290.30	1943.92	791	3057
l	Granite	–	27.68	–	1228
!	Lime Kankar	–	22.92	–	69
}	Brick Earth	1523.20	1913.75	1457	4319
		3256.97	14550.35	18258	58559

major minerals in Western Rajasthan in 1972-73 was only 10.80 lac tonnes but 1994-95 this increased to 37.22 lac tonnes. This rise has been 3.44 times or 244.62 The comparative status of production of more important major minerals duri 1972-73 versus 1994-95 is shown in Annexure-5. Important minerals contributing the rise in production of major minerals in 1994-95 are limestone (20109%), calci (726.9%), wollastonite (2916.6%), gypsum (73.57%), ball clay (194.1%), fire cl (602.68%), china clay/white clay (122%) etc.

b) *Minor Minerals* - In the case of minor minerals, growth in production has been fro 32.56 lac tonnes (72-73) to 145.38 lac tonnes (94-95) registering an increase by 4. times or 346.49%. While there was no production of rhyolite, kankar-bajri, murra granite and lime kankar in 72-73, there has been appreciable increase in producti of bentonite, salt petre, sandstone (masonry stone), marble and limestone in 94-9 Annexure-6 shows the comparative status of important minor mineral producti and employment in 1994-95 over 1972-73.

ale Value of Minerals

Although increase in sale value of minerals produced is not a direct measurement nineral development, but it has an indirect bearing on the overall mineral developme n Western Rajasthan the sale value of major and minor minerals from 1972-73 to 94- as shown an increase of 48.61 times (4761%) and 65.45 times (6445%) respectively.

mployment

Mining of these minerals has generated employment in the state. It is all the mo nportant as mineral sector provides employment to the labour class and weak ections and their families. During the year 72-73 about 2826 persons per day we mployed in the mining of major minerals in Western Rajasthan. This number increas about 8936 in 94-95 which is an increase by 3.16 times (216%). In the case of min ninerals the employment increased from 18258 (72-73) to 58384 (94-95) registering rowth of 3.19 times (291%).

evenue

Revenue from minerals is the direct revenue earned by the state exchequer in t rm of dead rent, excess royalty, application fee, lease money etc. The progress in t eld of revenue has also been excellent. It has recorded an increase of 6482.21% (52 imes) from a meagre amount of Rs 54.72 lacs in 1972-73 to Rs. 3378.26 lacs in the ye 994-95.

THE IMPORTANT MINERALS OF WESTERN RAJASTHAN AND THEIR RESOURCE POSITON

ignite

The discovery of lignite in Rajasthan (Fig. 2) dates back to 1886 when Pala ignite was discovered near Bikaner and the mining operations started there in 1898. mall thermal power station was also set up but the power station and mining w topped due to the problem of frequent fires inside the mines.

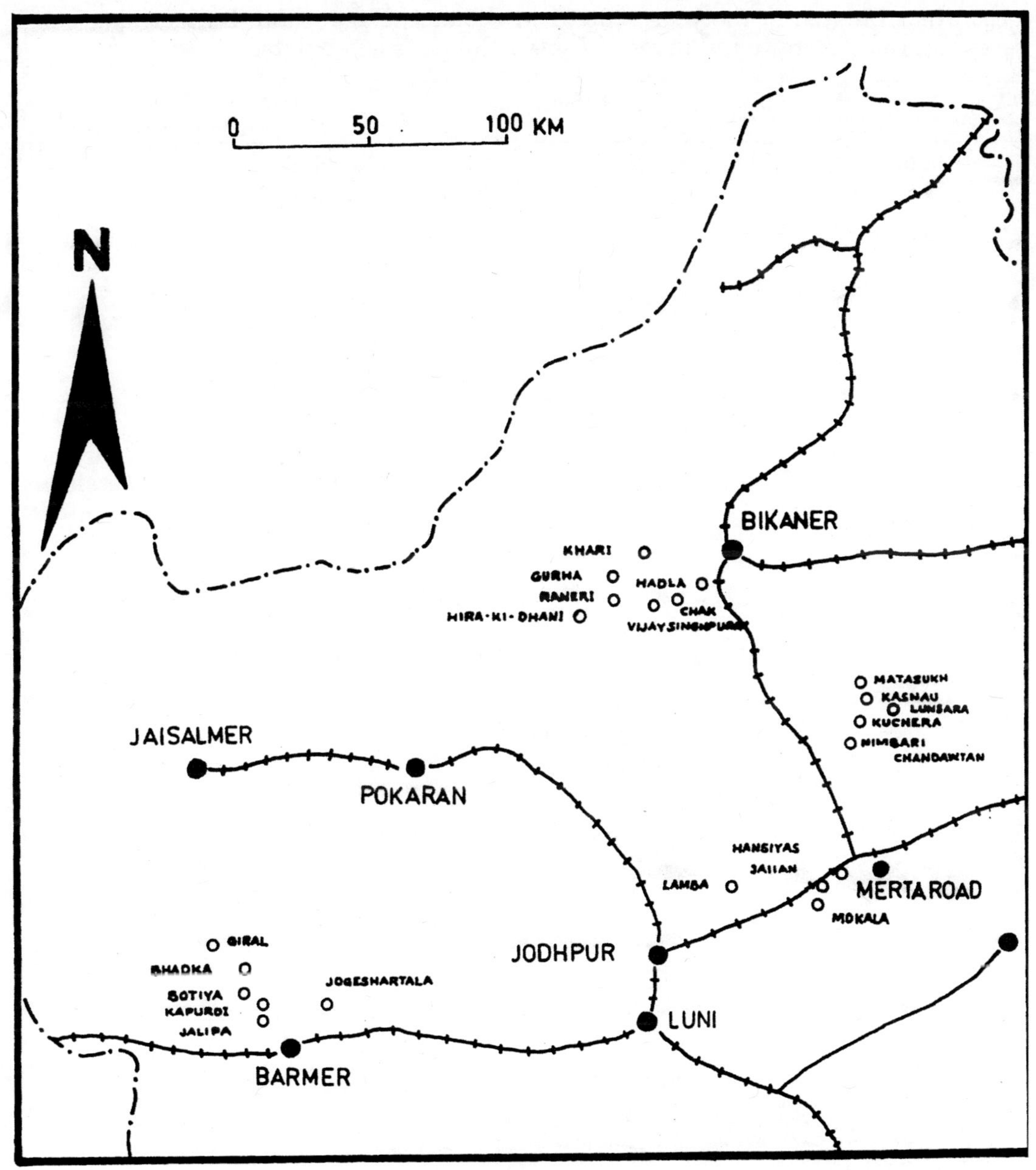

ig. 2. Location map of north-western Rajasthan showing occurrences of Lignite in the state

Based on the information of lignite cores in the tube wells of Ground Wat epartment systematic exploration by drilling was again started by the Sta epartment of Mines and Geology, Geological Survey of India and Mineral Explorati orporation Limited in the year 1980-81 in Bikaner, Nagaur and Barmer districts. T resent status of Lignite resources of Table 7 the state is tabulated below -

No.	District	Area/Block	Total Geol. Reserves (in million tonnes)	
	Barmer	Kapurdi	150.40	
		Jalipa	316.28	
		Bothia	151.67	
		Giral	101.90	
		Jogeshwar Tala	34.50	
		Sonari	43.50	
		Sachha Sauda	28.70	
		Bharka	9.45	
		Bothia - Bhadka	10.40	
			846.89	846.89
	Jaisalmer	Khuri	13.80	13.80
	Bikaner	Palana	23.60	
		Barsinghsar	77.80	
		Gurha (East & West)	79.76	
		Bhoblasar	3.90	
		Bithnok Extn.	20.00	
		Gadiyala & Bithnok West Extn.	40.00	
		Raneri	31.84	
		Mandal Charana	15.00	
		Hadla	5.00	
			296.90	296.90
	Nagaur	Merta Road & Meera Nagar	83.20	
		Mokala	36.90	
		Kasnau - Igyar	64.90	
		Kuchera - Indawar	12.00	
		Matasukh	7.69	
		Lunsara	7.34	
		Nimri-Chandawatan	6.10	
			217.79	217.79
		Grand total		1375.38

›il and Natural Gas

Exploration for oil and gas in Western Rajasthan was started by Oil and Natur ˧as Corporation (ONGC) in 1954 and by Oil India Ltd. (OIL) in 1983. Oil and Natur ˧as Corporation has drilled 59 wells (86600 Mts.) and OIL has drilled 23 wells (524 ıts.) in Jaisalmer, Bikaner, Barmer and Sri Ganganagar districts. Gas reserves prov ɔ far in Western Rajasthan are tabulated below (Table 8)

'able 8.

Sl. no.	Agency	Gas Fields	Reserves in MCM	Million Cu. Mts. Recoverable
1	ONGC	Manhera Tibba	863	486
2	ONGC	Ghotaru	973	660
3	ONGC	Bhakri Tibba	55	31
4	ONGC	Kharotal	205	115
5	ONGC	Bankia	57	32
6	OIL	Tanot	1400	
7	OIL	East Tanot	550	
8	OIL	Ramgarh	100	
9	OIL	Dandewala	6166	6150
10	OIL	Jalalwala	225	
11	OIL	Bagitibba Tibba	300	
12	OIL	Gamnewala	47	
			10941 (10.94 BCM)	7474 (7.47 BCM)

The natural gas fields of Rajasthan are in a position to produce upto 14 lakh cul ıetres per day (CMPD) of gas. However at present about 6.00 lakh CMPD of natural g ; being produced in Rajasthan (5.5 lac CMPD by OIL and 0.5 lac CMPD by ONGC).

In addition to natural gas, heavy oil has also been reported in Rajasthan for t! rst time in Bagewala, Kabrewala and Tavriwala blocks of Bikaner-Nagaur Basin. T! stimated reserves of heavy oil stand at 250 million tonnes.

MS and Cement Grade Limestone of Jaisalmer District

In Jaisalmer district, about 650 sq. kms. area around Sanu-Habur-Khuiala-Hen ;i Dhani-Tulsi Ram Ki Dhani-Khinya Khinsar-Mandha etc. is occupied by S.M.S. aı ɜment grade limestone deposits of Khuiala Formation of Lower Eocene age. The SN rade limestone is forming the upper most layer of the Khuiala Limestone Formation. ; bouldary, fine grained and hard in nature and varies in colour from whitish to lig ellowish to buff and pinkish. CaO content varies from 52% to 54% and SiO_2 from 1.0ɛ ɔ less than 1%. The cement grade limestone occurs below the SMS grade limesto: orizon. It is fine grained chalky, white to yellowish in colour. Its thickness varies frc 2 mts. to 4 mts. CaO content varies from 42% to 51% and SiO_2 from 3% to 12%. T ˧eological Survey of India carried out regional survey and sampling but the departme f Mines and Geology of Rajasthan (DMGR) started its detailed and systema xploration by drilling on 400x400 mts. grid in 1987. Since than, the DMGR has carri ut 21517 mts. of drilling over 1092 bore holes covering an area of about 182 sq. kms. t Iarch, 1996. As a result of this exploration, 5182 million tonnes of reserves of all grad f limestone have been proved (920 MT SMS Grade + 629 MT chemical grade + 3627 N ɜment grade).

Two mining leases of 10 sq. km. area each have been granted to M/s Rajasth :ate Mineral Development Corporation and Rajasthan State Mines and Minei mited to the north of Sanu for mining of SMS grade limestone for supply to vario eel plants of Steel Authority of India Limited. About 3.4 MT of SMS grade limestoi ılued at Rs. 64 crores has been mined till March 1996, earning a revenue of Rs. 12. ores to the state exchequer. About 2500 to 3000 persons are engaged in the mining MS grade limestone per day.

As a result of exploration by DMGR, 6 blocks of cement grade limestone have be otified for setting up of major cement plants of 1.5 MT per annum capacity as und 'able 9).

ıble 9.

Miyon ki Dhani	Block	7.08 Sq. Km.
Tulsiram Ki Dhani	Block	7.20 Sq. Km.
Khinya-Khinsar	Block 1	7.00 Sq. Km.
Khinya-Khinsar	Block 2	7.00 Sq. Km.
Khinya-Khinsar	Block 3	7.00 Sq. Km.
Tulsiram Ki Dhani	Block B	5.00 Sq. Km.

ther Limestone Reserves

Chemical and cement grade limestone deposits are also available in the districts agaur, Pali, Sirohi and Jodhpur. The reserves and production of limestone in the stricts during 1994-95 is tabulated below (Table 10)

ıble 10.

istrict with localities	Reserves (MT) Chemi.	Cement	Prod. (lac tonnes)	Rev. (lac Rs.)	Labour
AGAUR	50	500	16	331	2200
otan, Gangawana, Mundawa, eh, Tantwas, Madpura, Gujron Ki hani, Tadas, Tankla, Mankupur, c.					
ALI	10	500	3.35	84.3	400
igrana, Sinla, Ras, Gudia, Devli, ulan, etc.					
ROHI	—	400	7.9	106	400
ndwara, Watera, Banas Amli, etc.					
DHPUR	10	50	6	185	600
alauna-Rampuria, Ransigaon - ariyadhana, Borunda, etc.					

ypsum

Gypsum is the monopoly mineral of Western Rajasthan. About 90% of countr oduction of gypsum is from Nagaur, Bikaner, Sri-Ganganagar, Hanumangarh, Barm ıd Jaisalmer districts of Western Rajasthan. District wise reserves and position oduction revenue etc. of gypsum for 1994-95 is tabulated below (Table 11).

ıble 14.

istrict with localities	Reserves in MT	Prod. in Lac. T.	Rev. Rs. Lacs	Employment (person/day)
[KANER	5	0.94	23.34	40
.msar, Dhirera, Kaoni, Bharru, ɔonkaransar, etc.				
ʀI GANGANAGAR	4	13.31	278.00	600
aghunathpur, Karnisar, Dhadra, Pallu :c.				
'AGAUR	950	1.24	28.18	970
hadwasi, Got-Manglod, Jeevanbera etc.				
ARMER	3	0.66	22.19	275
awas, Uttarlai, Phalsund, Chitar Ka Par, .kharda, Bhimarlai etc.				
·ISALMER	12.4	0.34	6.91	260
[ohangarh, Lakhesar, Hamirwali Nadi 'achna, Chandan, Nokh, Lakha, etc.				

So far gypsum was reserved for public sector only and Fertilizer Corporation c ıdia, RSMM and RSMDC have been the only producer of gypsum. But recently th tate Government has issued notification whereby gypsum mining has been opened fo :ivate sector also for captive use in cement plants, gypsum based industries etc.

lays

A variety of clays are found in Western Rajasthan such as ball clay, fire clay, whit ay, China clay etc. out of which Western Rajasthan is the sole producer of ball clay an re clay. Production of ball clay/fire clay has been reported from Bikaner, Pali an ıisalmer districts but Bikaner district is the largest and main producer and accounts fo ɔout 44 million tonnes of estimated reserves of ball clay/fire clay. White clay occur i ikaner, Barmer and Jaisalmer districts.

entonite

Bentonite is another monopoly mineral of Western Rajasthan and Barmer district i ıe sole producer of bentonite. About 110 million tonnes of reserves of bentonite hav ɜen assessed in Barmer and Jaisalmer districts. Bentonite of Rajasthan is largel ɔn-swelling type and is used as bleaching agent in oil and sugar industries etc.

uller's Earth

In Rajasthan, entire production of fuller's earth is coming from Barmer and Bikane ıstricts. Bikaner district alone accounts for 111.35 million tonnes of fuller's eart ɛserves. Barmer and Jaisalmer districts account for 0.84 million tonnes of reserve :inging the total resource positin of fuller's earth in Rajasthan to 112.19 million tonne ; against 133.45 million tonnes of total reserves of the country. Fuller's earth is als ɜed for refining of oils, fats and grease.

iliceous Earth

Deposits of siliceous earth are confined to Barmer and Jaisalmer districts. Tot eserves are of the order of 10 million tonnes. It is used in filter aids and refractory sill oducts.

'olframite (Tungsten Ore)

Wolframite is another important mineral of Western Rajasthan. Its deposits a nfined to the Rewat hill area near Degana, district Nagaur. A total reserve of 334 nnes of the ore (65% WO_3) is estimated. It is used in the production of alloy steels an gh speed cutting tools. This is the only deposit of the country.

ranite

Sirohi, Jalore and Barmer districts of Western Rajasthan have the best quali anite deposits of Rajasthan. Jodhpur and Pali districts are also rich in commerci anite deposits. Number of leases and production of granite during 1990-91 and 1994-9 as as under (Table 12).

able 12

Year	Lease	Production	Remarks
1990-91	172	7539 tons	Prior to launching of Granite policy.
1994-95	361	27680 tons	After launching of gGranite policy

ase Metals

Sirohi and Pali district have a large number of small base metal deposits. Already umber of small deposits of base metals have been prospected and proved at De asantgarh, Ajari, Pipela and Goria in Sirohi district and Nayakhera, Kunej iranthiya etc. in Pali district.

ther Minerals

Wollastonite, jasper and salt petre are also the monopoly minerals of Weste ajasthan having good industrial applications and demand. Apart from these minera aporite minerals like polyhalite, sylvite and halite have also been proved in parts huru, Hanumangarh, Bikaner and Sri-Ganganagar districts of Western Rajasthan. Th eserves of halite are more than six trillion tonnes, while those of potash minerals are 6 illion tonnes. However as these deposits are very deep seated (at a depth of 350 to 9 etres) their economical exploitation is a remote possibility for the present.

MINERAL DEVELOPMENT AND FUTURE SCENARIO

Discovery of new mineral deposits and their systematic exploration have paved t ay for rapid industrial and economic development in Western Rajasthan. The mo nportant areas of development are as under -

. Power Generation

Power is the backbone of development. Without sufficient and assured pow ıpply, no industrial development can take place. Rajasthan has always been facing a :ute power shortage. It always had to depend on the external resources for its pow ɜneration/power supply, whether it was hydel power, thermal power or atomic powe ow with the discovery of lignite deposits and natural gas reserves in Westeı ajasthan, the stage has been set for generation of about 1860 M.W. power by the turn ıis century - and this power generation will be solely based on internal resources a nder -

ı) Gas Based Power Generation

3 MW and 35.5 MW power plants have already been commissioned by Rajasthan Sta Electricity Board at Ramgarh in Jaisalmer district recently, based on the supply gas by ONGC and OIL respectively.

. Rajasthan State Electricity Board is contemplating to set up more plants to genera additional power of about 124.5 MW based on gas reserves proved by OIL.

›) Lignite Based Power Generation

M/S Hindustan Vidhyut Corporation has signed the contract to set up a pow generation plant of 2 x 250 MW capacity on the lignite deposits of Palan Barsinghsar, Gurha and Bithnok in Bikaner district.

. The Power Purchase Agreement in favour of M/S West Power is under process f setting up of 4 x 300 MW capacity power plants based on Jalipa and Kapurdi Ligni deposits of Barmer district.

Setting up of these power projects will involve an investment of more than Rs. 40(:ores.

. Cement Industries

In 1972-73 no cement production was coming from Western Rajasthan; but toda ot only the production has started, in years to come Western Rajasthan alone will l roducing about 15 to 18 million tonnes of cement annually. The present status ar ıture scenario of cement production in Western Rajasthan is as under (Table 13)

ı) *Cement plants under operation*

able 13.

Iame of the Plant	Location	Capacity
. Laxmi Cement	Banas, Dist. Sirohi	10.80 Lac TPA
. J.K. White Cement	Gotan, Dist. Nagaur	0.80 Lac TPA
. Birla White Cement	Khariya, Dist. Jodhpur	1.50 Lac TPA

›) *Cement plants under construction* - One cement plant - Binani Cement is und construction stage near Pindwara, Dist. Sirohi. This plant will be of 15 lacs TP capacity.

) *Mining leases sanctioned for cement plants* - Two mining leases have been sanctione in favour of D.C.M. and Shree Cement near Amli, in Sirohi district and near Ras, i Pali district respectively. Cement plants on these leases are likely to come up in nea future.

) *Cement plants to be sanctioned in near future* - Six blocks of limestone have so fa been notified in Jaisalmer district for the setting up of major cement plants of 15.0 lac TPA capacity. Applications received for these blocks have already bee scrutinised and recommendations for the sanctions have been forwarded to th Central Government. Final sanctions are expected soon.

) *Areas under prospecting/notification stage* - Blocks of cement grade limestone nea Tankla and Dehru, district Nagaur are under prospecting by the Department Mines and Geology. Exploration work in Tankla block has almost been completed an the area will be notified soon for inviting the applications. Explorations at Dehru ai still under way.

Setting up of cement plants as above will involve an investment of about Rs. 6000 t 00 crores. Limestone mining and the cement plants will provide avenues fo nployment, particularly to backward class and labour class families.

In addition to the major cement plants, a number of mini and tiny cement plant e also running in districts of Jodhpur, Pali, Nagaur and Churu.

Other Limestone Based Industries

The mining of SMS grade limestone in Jaisalmer district by RSMM and RSMDC major industry. Since 1987 about 34 lac tonnes of SMS grade limestone has alread en mined from Jaisalmer district, valuing about Rs. 64 crores. The State Governmer s earned a revenue of Rs. 1277 crores from the SMS grade limestone of Jaisalmer an out 2500 to 3000 persons per day are employed in the mining of SMS grade limeston Jaisalmer.

A number of lime kilns, hydrated lime plants and other chemical industries ai erative in Jodhpur, Pali and Nagaur districts based on the high grade limeston posits of Sojat - Bilara - Gotan belt. A number of new industries of this nature ai ely to come up in near future also.

Lignite Mining For Commercial Use

Lignite mining for commercial and industrial use has also started in Wester ajasthan recently. RSMDC is mining lignite in Giral area in Barmer district. So fa ey have produced 30000 tonnes of lignite since May 1995. This lignite is being used i ick kilns in Rajasthan and Hariyana. The lignite deposits of Kasnau and Mata Sukl strict Nagaur are being under taken by to RSMM for mining of lignite. Small deposit lignite will be given to cement plants coming up in Western Rajasthan for their captiv e.

Gas Based Other Projects

The State Government has prepared a shelf of seven projects based on natural ga posits of Western Rajasthan for joint sector in various industries. These project visage an investment of about Rs. 51,600 crores.

Exploration Activities Under Joint Venture

The liberalization policy of Government of India for exploration and exploitation (ineral wealth has invoked a lot of public interest in Western Rajasthan.

) *Oil and Gas* - 12 blocks have been identified in Jaisalmer, Bikaner, Sri Ganganaga: Churu, Nagaur and Barmer districts and MOU's are to be signed for this with She India, Essar Oil, Phoenix Overesea etc. Shell India has already started geophysic: exploration in Sanchore-Barmer Basin at an estimated cost of about Rs. 1500 Crore: The preliminary results are said to be very encouraging.

) *Base Metals* - M/S Metmin Finance and Holdings Ltd. is interested in signing a MO for aerial survey in Sirohi district for base metals and noble metals.

Gypsum Based Industries

Gypsum is another important industrial mineral which is supporting man dustries of Plaster of Paris. Gypsum is also an essential raw material for cemer dustries and with the setting up of many cement plants in Western Rajasthan, it wi : an added advantage to these plants as gypsum is readily available in the vicinity (ment plant sites. Earlier, mining of gypsym was restricted and leases were grante ily to the public sector under takings like F.C.I., RSMM and RSMDC, but recently th ;ate Government has relaxed these conditions and now mining leases have bee lowed in the private sector also for captive use. Thus, cement plants can have their ow ining lease for gypsum also for use in their industry. M/S Indian Rayon ar ntemplating to set up a Gypsum Fibre Board plant to produce about 1000-1500 sq. mt: 'GFB per day with an investment of about Rs. 220 crores in joint collaboration.

Granite Industries

Granite is an important decorative stone and provides a good scope for setting up ('anite processing units in Western Rajasthan. To encourage granite mining and settin) processing units, the State Government has further revised its granite polic creasing the plot size from 100x100 metres to 150x150 metres (250 Hect.) to facilitat echanised mining of larger blocks for the export. In Western Rajasthan 1041 plots ('anite have been delineated so far in Jalore, Sirohi, Barmer and Jodhpur districts.

Ceramic Industries

Western Rajasthan has all the basic raw material like clays, quartz, feldspa ollastonite, calcite etc. for setting up of ceramic industries.

). Other Mineral Based Industries

The bentonite and siliceous earth deposits of Barmer district also have a very goo :ope for setting up of processing plants for these minerals in Barmer district itself.

Thus, the future of Western Rajasthan in terms of industrial and econom: evelopment is very bright and the day is not far when Western Rajasthan will be know r it's mineral based industries rather than for its desertic conditions.

REFERENCES

ɔvt. of Rajasthan, 1989. *Minerals in Rajasthan* Department of Mines and Geology. Governme of Rajasthan, Udaipur, *Spel. Publ.* 1-121.

ıshi, K.N. 1993. Mineral Resources in Rajasthan. In : Chauhan, T.S. (Ed.) *Natural and Hum Resources of Rajasthan.* Scientific Publishers (India), Jodhpur, 147-166.

ıliwal, B.S. 1993. Geology of Rajasthan – a summarium. In : Chauhan, T.S. (Ed.) *Natural a Human Resources of Rajasthan.* Scientific Publishers (India), Jodhpur, 1-22.

ıliwal, B.S., Pareek, U.S. and Vyas, A. 1997. Structural framework of the Makrana mar deposit and its bearing on the Precambrian Stratigraphy of Rajasthan. In : Wijayanand N.P. Cooray, P.G. and Mosley, P. (Eds.), *Geology in South Asia-II,* Geological Survey a Mines Bureau, Sri Lanka, Professional Paper - 7 : 123-136.

ıreek, H.S. 1981a. Basin Configuration and Sedimentary Stratigraphy of Western Rajastha *Jour. Geol. Soc. India.* 22(11) : 517-527.

ıreek, H.S. 1981b. Mineral Resources of Western Rajasthan. *Indian Minerals,* 34(2) : 41-58.

eological Evolution of Northwestern India
:.S. Paliwal (Ed.)
cientific Publishers (India), Jodhpur, 1999, pp. 167 - 182.

EXPLORATION CHALLENGES FOR HYDROCARBONS IN WESTERN RAJASTHAN

T.K. Banerjee, Reba Devi and P.K. Mitra
Oil India Limited, Rajasthan Project, Jodhpur – 342 001

ABSTRACT

Oil India's venture for hydrocarbon in western Rajasthan started in 1983 and within a span of 13 years it has established 8.8 billion standard cubic metres of gas in Jaisalmer Basin and about 15 million tonnes of heavy oil (excluding bitumen) in Bikaner- Nagaur Basin by drilling 37 wells. In Jaisalmer Basin Mesozoics and Tertiaries are gas producers; whereas, in Bikaner-Nagaur Basin Paleozoics are oil bearing. The gas composition in Jaisalmer Basin vary significantly both in lateral and vertical directions indicating long migration path possibly due to tectonic activity. The gas encountered in deeper horizons are having more carbondioxide and nitrogen as compared to shallower horizons supporting vertical migration. In contrast, heavy oil of Bikaner-Nagaur Basin is found in insitu condition. The above two basins though adjacent to each other (separated by Pokaran-Nachna high trend) are having specific characteristics and its associated problems which need integrated approach to resolve. In Jaisalmer Basin, the Pariwar sands are of low resistivity due to its mineralogical composition and shallow marine condition of deposition as evident from electrologs and cores. The identification of hydrocarbon bearing zones thus becomes extremely difficult. Similarly, the Upper Goru sands do not show typical density neutron crossover for gas due to the presence of high density minerals. In Sanu and Khuiala formations of Tertiary age, the gas is richer in hydrocarbon but are difficult to be explored as they are locally developed sand bodies unlike the extensively developed Pariwar and Goru sands. In Bikaner-Nagaur Basin, so far heavy oil has been discovered within Jodhpur sandstone only in Baghewala structure. In the neighbouring structures viz; Tavriwala and Charanwala, Jodhpur sandstone has been been encountered and appears to have pinched out against the basement. A synergistic approach has resulted in solving some of the above problems and to achieve a greater success in the exploration of hydrocarbon.

ℹey Words : Hydrocarbons, Exploration Challenges, Jaisalmer Basin, Bikaner-Naga Basin, Rajasthan.

INTRODUCTION

Oil India Limited (OIL) started its activities for exploration of hydrocarbons i ʋestern Rajasthan in 1983. So far it has acquired 11643 Glks of 2D seismic data and 3(q. kms of 3D seismic data and has drilled 37 wells in its PEL area of 28600 sq. kms : 'aisalmer and Bikaner-Nagaur Basins (Fig. 1). Within a span of 13 years OIL ha stablished 8.8 billion standard cubic metres of hydrocarbon gas in Jaisalmer Basin ar .5 million tonnes of heavy oil (excluding bitumen) in Bikaner - Nagaur Basin.] 'aisalmer Basin, hydrocarbon gas was first discovered in Tanot structure in tl ⁄lesozoics. Subsequently several other gas bearing structures viz; Dandewala, Jalalwal łamgarh and Bagitibba were discovered (Fig. 2). The Tertiaries have also been proved ›e hydrocarbon gas bearing in some of the structures. The gas composition vari‹ ignificantly both laterally and in vertical geological coloumn. The gas encountered : leeper horizons has more carbon dioxide and less hydrocarbon percentage as compare‹ o that of shallower horizons (Figs. 13 and 14). In Bikaner - Nagaur Basin, heavy oil ha ›een discovered in Baghewala structure. A thick coloumn of bituminous oil has also bee ncountered in the upper ıorizon of Baghewala tructure which is present n the neighbouring tructures viz; Tavriwala, ζalrewala, Charanwala ınd Deviwali (Fig. 3). The lifferent hydrocarbon ›earing zones encountered n the two basins have pecific geological haractertics and the 'arious associated ›roblems which are hallenging for the ›xploration of ıydrocarbons. The integ-ated multidisciplinary tudies have provided olutions for some of the ›roblems to achieve a ;reater success in the ›xploration of ıydrocarbons so far.

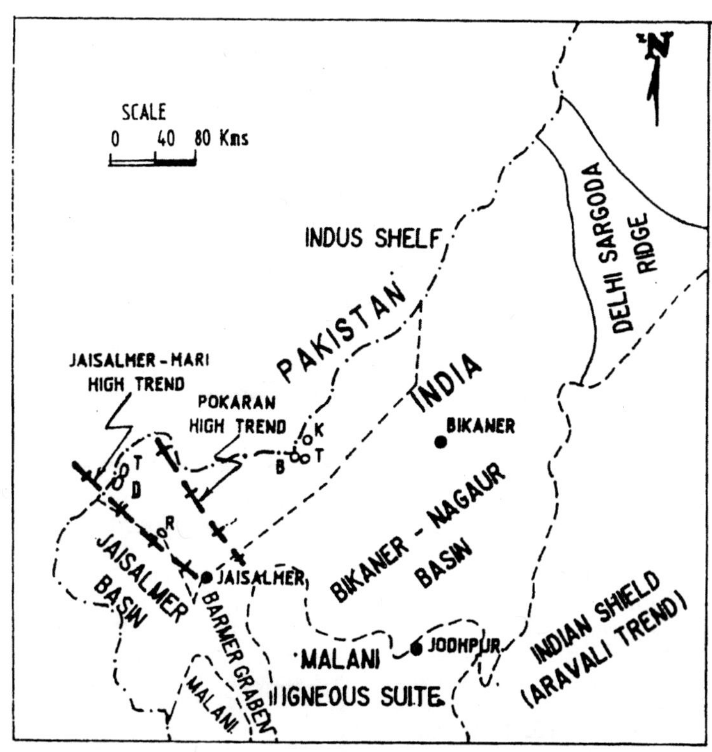

Fig. 1. Map showing sedimentary basins of Western Rajasthan

JAISALMER BASIN

The commerically exploitable quantities of hydrocarbons are confined to fo rmations namely Pariwar, Goru, Sanu and Khuiala in Jaisalmer Basin. However, th ajor portion of OIL's established gas reserves belong to Goru Formation.

In Pariwar Formation several sand units separated by thin shale breaks have bee entified which are characterized by low resistivity (lower than the shales) (Fig. 4 ough the exact reason for the low resitivity of these sands is not known but it ferred that this could be due to the presence of some conductive minerals such a derite and pyrite (Figs. 7 and 8). This is further supported by the density - neutron lo sponse as in Pariwar Formation the density of abundant siderite and pyrite which a high density minerals a compared to silicates and oth carbonates such as calcite an ankerite. The Neutron porosi also reads higher than in Go Formation (Fig. 4). This is al probably due to the presence siderite which gives hig neutron porosity respons Because of the above reaso the quick look identification hydrocarbon bearing zon using logs often becom deceptive.

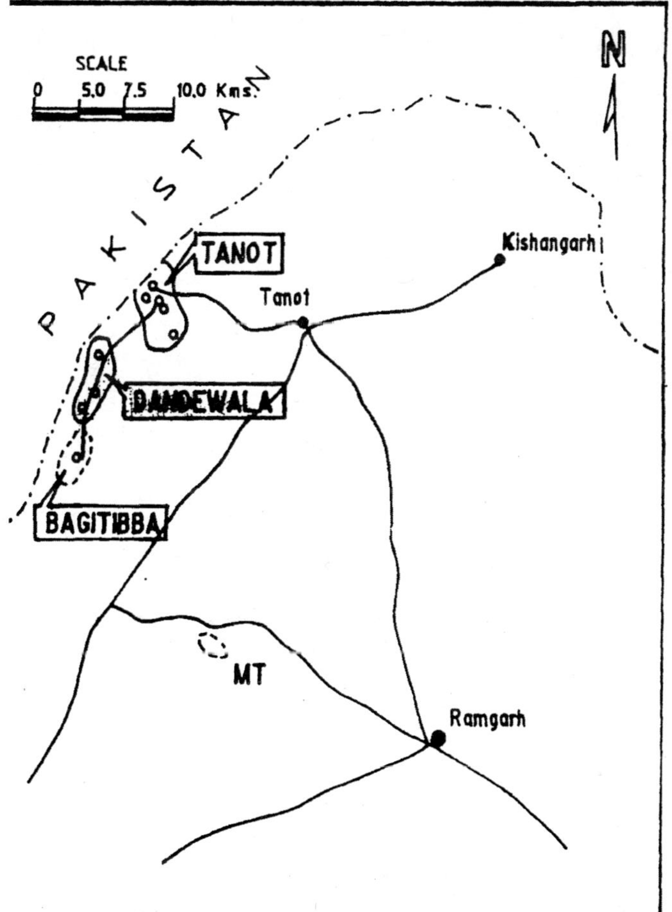

'ig. 2. Major gas fields of Jaisalmer Basin under OIL's PEL area.

The lower part of Pariwa Formation consists of massiv coarse to medium graine sandstone with interbedde claystones and were probab deposited in non-marine marginal marine environmen The upper part of Pariwa Formation consists of fi grained sandstone ofte glauconitic and has high detrit clay and mica content. They a strongly bioturbated sands an the effect of bioturbation sometimes very stron destroying the primary beddi as evident from the core (Fi 10). The biogenic burrows acro the bedding have also bee

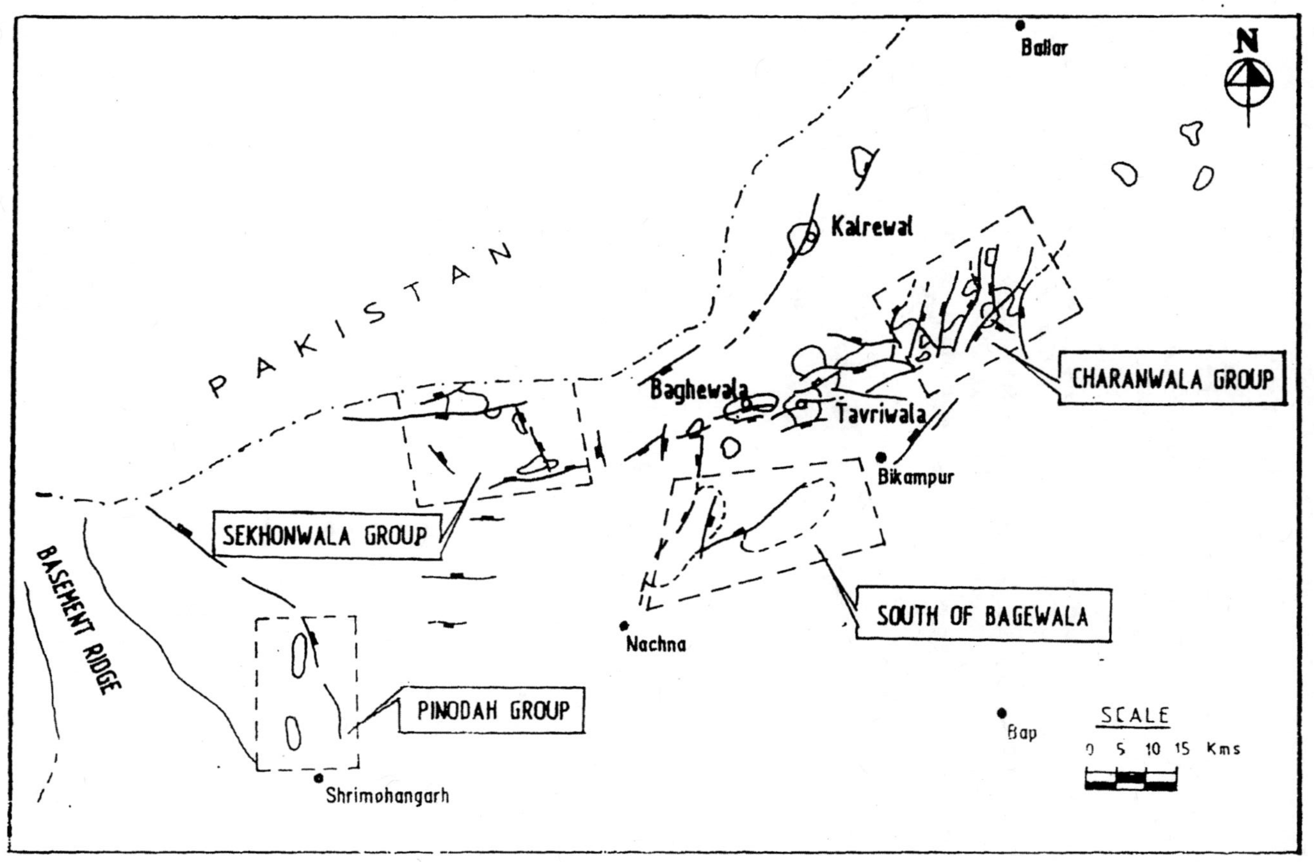

Fig. 3. Identified structures of Bikaner-Nagaur Basin under OIL's PEL area.

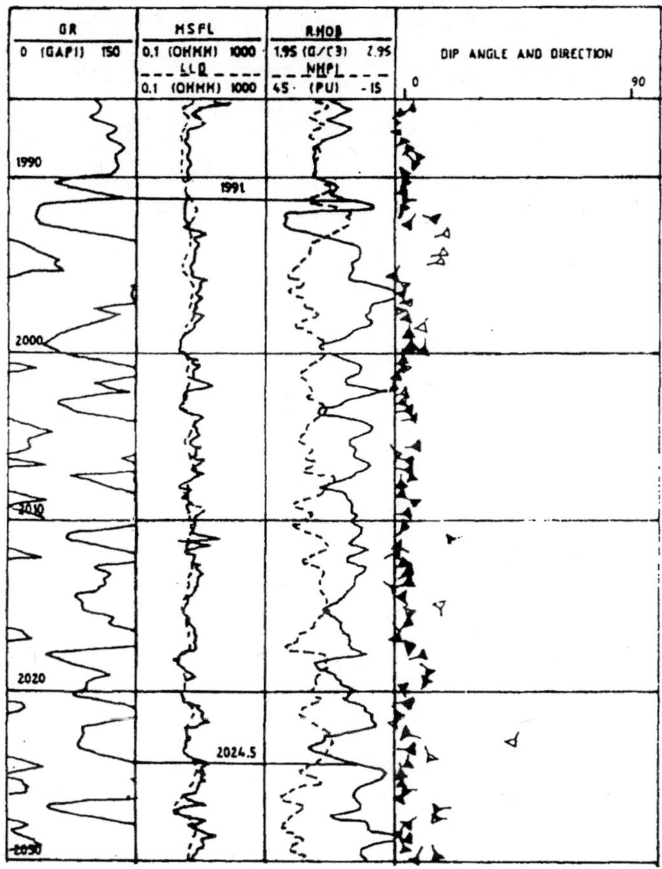

Fig. 4. Electrolog characteristics of Pariwar sand units.

›served in the cores (Fig. 11). The above evidences indicate that these sands we: ›posited in a delta front/prodelta environment. Presence of abundant siderite furth‹ ıpports shallow, swampy and marshy type of depositional environment. These san‹ ıve poor reservoir quality because of their argillaceous and micaceous nature which h: e net effect of clogging the pore throats and hence reducing the permeability.

The Goru formation has been divided in two units - Upper Goru and Lower Goru.] ›wer Goru, two prospective sand units, separated by a zone with very high GR readir ıve been identified (Fig. 5). This zone with high GR reading serves the purpose arker for correlation. The sand unit lying below the high GR marker is well develope ıd is the producing sand in Bagitibba area. The other sand unit lying above the hig R marker is well developed and proved to be gas bearing in Dandewala area. Tl pmeter data shows high random dips or absence of dips in the top part of the sar hich indicates that the sand was deposited in high energy environment, characterist distributary mouth bar. The coarsening upward sequence and the effect of bioturbatic

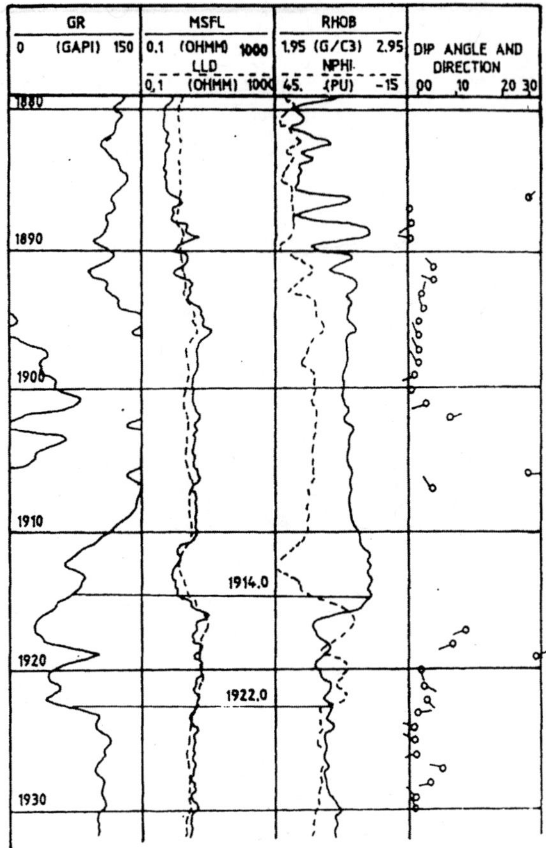

Fig. 5. Electrolog charactristics of Lower Goru sand unit

observed in the part conventional cores is supportive of the above fact (Fig. 1ʕ Biostratigraphic and sedimentological studies suggest that the Lower Goru sand is lower deltaic origin.

In Upper Goru, three prospective sand units have been identified (Fig. 6) and a found to be present in all the wells drilled so far, which indicates a widesprea development of these sand units over the entire area. The total thickness of the thr sand units do not vary much and lies between 25-30 m. They also maintain unifor lithological features all across the area. The dipmeter log show that the formation dip very low (2-3 degrees). The dip data is mostly scattered and not much of regular patte can be seen. Biostratigraphic studies show the presence of abundant and diversifi planktonic formainifera which suggests a bathymetry of 75-100 m with transgress shore line at the time of deposition. The above observations indicate that the Upper Go sands are shallow marine beach barrier sands. Granulometric studies of the lithologic samples is also suggestive of a mature beach environment for the Upper Goru sand These sands are medium to coarse grained clean sands as compared to the Lower Go and Pariwar sands and hence are expected to have better reservoir quality.

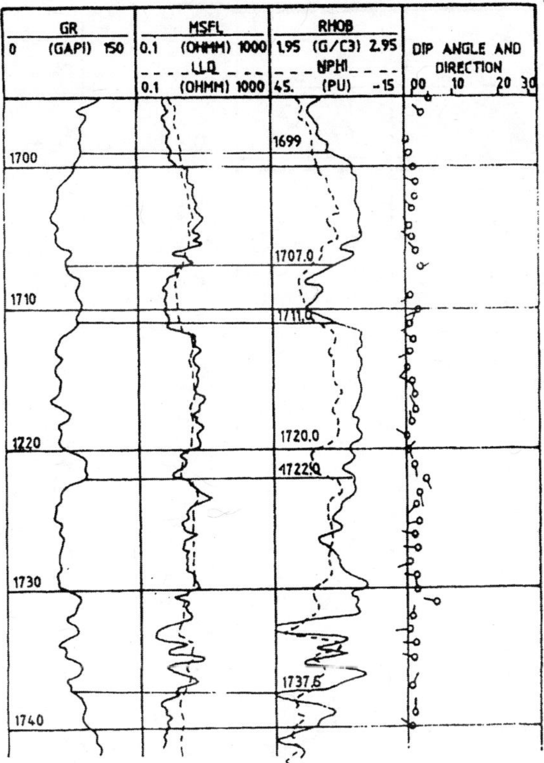

Fig. 6. Electrolog characteristics of Upper Goru Sand units.

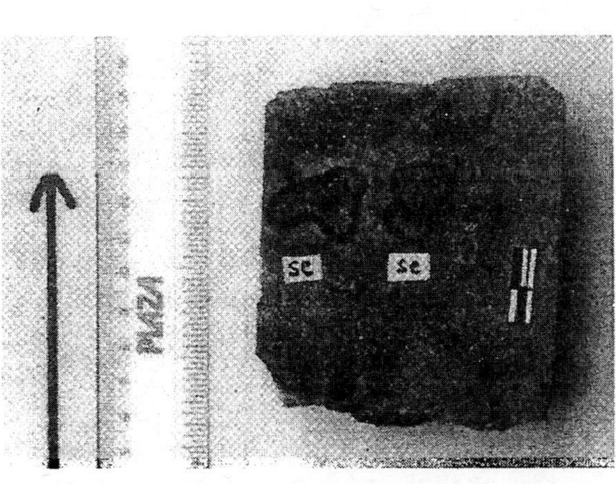

Fig. 7. Presence of siderite concentration (SC) in Pariwar

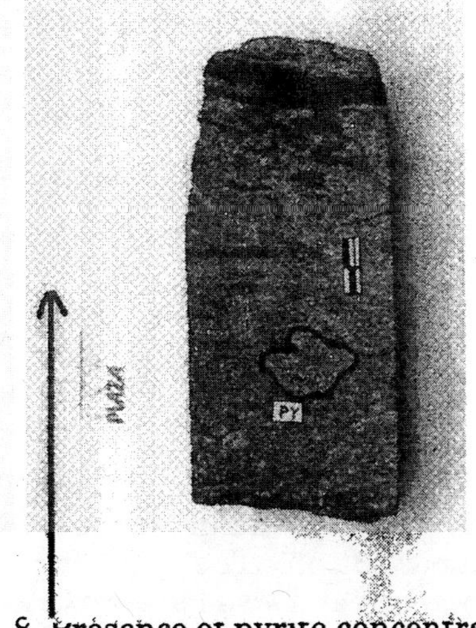

Fig. 8. Presence of pyrite concentratio (PC) in Pariwar

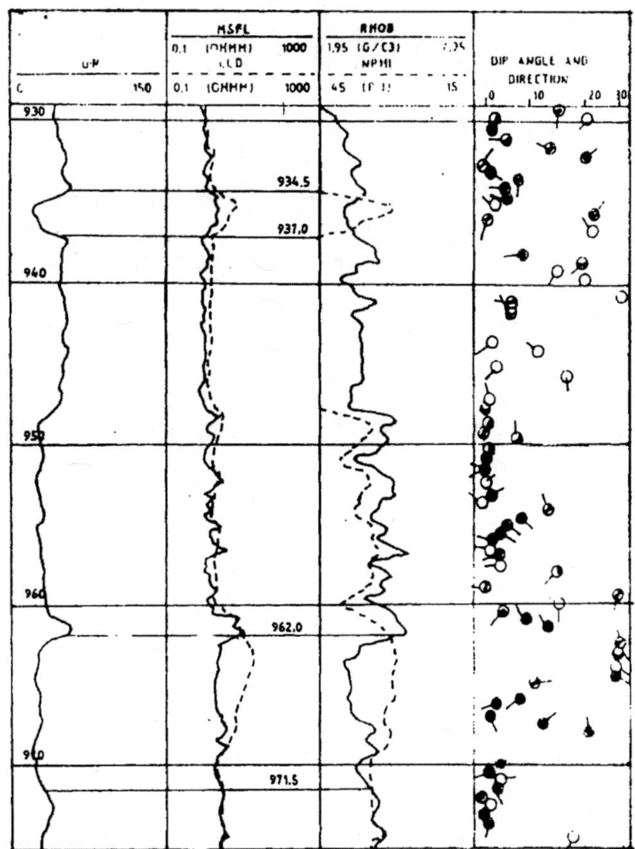

Fig. 9. Electrolog characteristics of Sanu sands

The density - neutron log does not show any crossover tendency particularly in t p two sand units but on testing, both these sand units have been found to 'drocarbon gas bearing. Unlike in a hydrocarbon gas bearing zone, the bulk densi ads very high in these sand units probably due to the lithology. The neutron porosi wever reads conventionally low. Therefore, the conventional quick look interpretati density-neutron log in these zones often becomes deceptive. The density-neutron l wever, show some cross over tendency against the third bottom most sand unit.

In view of above, it is likely that all the three sand units of Upper Goru a 'drocarbon gas bearing unlike what was inferred earlier.

The Sanu and Khuiala Formations of Tertiary age occurring within a depth range 00m, are the main shallow gas bearing horizons in OIL's operational areas isalmer Basin. The gas encountered in these shallow horizons are very rich 'drocarbon and also have very rich carbon-dioxide as compared to deeper horizons.

Fig. 10. Bioturbated sands of Lower Goru

Fig. 11. Presence of vertical burrows in Lower Goru.

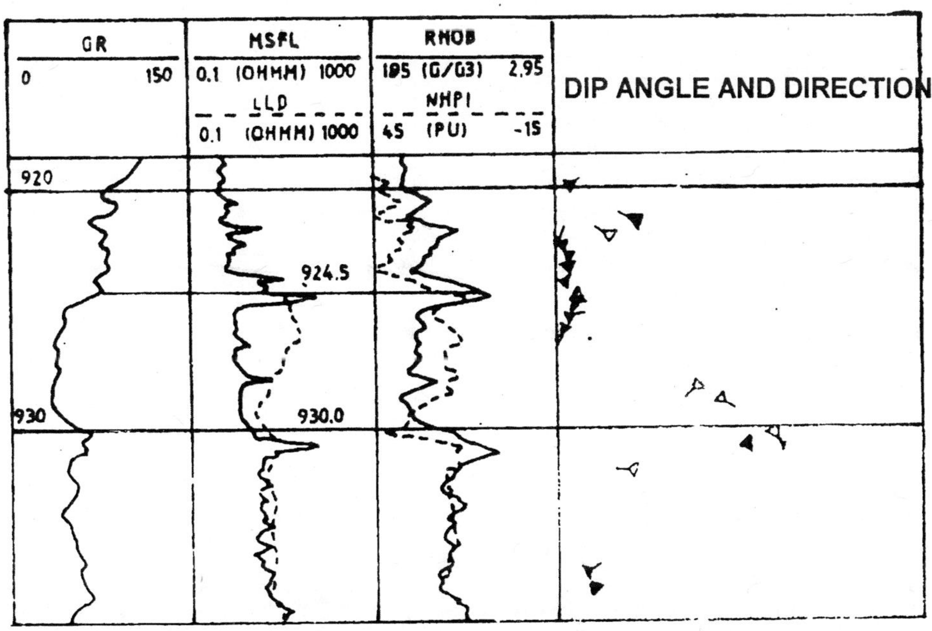

Fig. 12. Electrolog characteristics of Khuiala sand

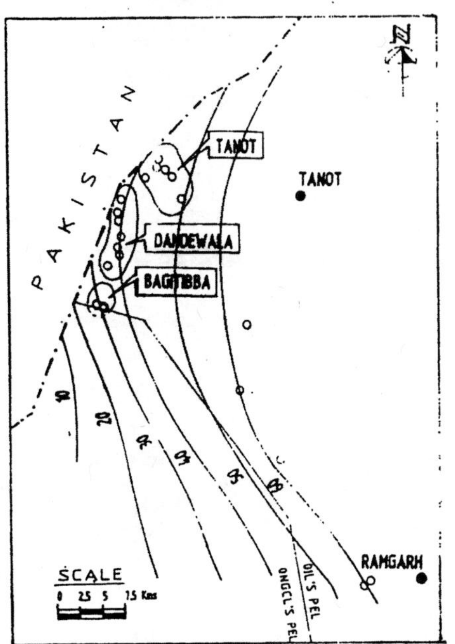

Fig. 13. Map showing hydrocarbon content of gas from Pariwar/Goru Formations

is not found to be commercially viable ɔm reserves and reservoir pressure ıint of view. Unlike Dandewala area, .e Khuiala sand unit above the nestone section is not developed in ınot area. The Khuiala limestone ction underlying the sand unit is also ıssibly gas bearing (as per log ridence) but on testing, no encouraging sults were obtained possibly because its very tight nature.

It can be inferred from the above scussions, that both Pariwar and the ɔwer Goru sands are distributory outh bar sands deposited in delta ɔnt environment (Fig. 15). ıbsequently, the shoreline was ansgressive, when the waves were fective in redistributing the sediments sulting in development of ·ach-barrier sands of Upper Goru ɔrmation along the exposed stretches coast line.

The Sanu Formation of Paleoce age represents alternating sequence fine grained clastics and carbonat deposited under an unstable sea lev condition, wherein the environment w shallow marine for the carbonates aı fresh water for the sandstones. T Sanu sands are gas bearing in Bagitib area (Fig. 9).

The overlying Khuiala Formati of Lower Eocene age consists of shall marine shales and fossilifero limestone deposited still under slight oscillating sea level condition. T localised development of the sandsto unit on top of limestone section like th in Dandewala area is the consequence regressive fluctuations in transgressi shore line supporting the oscillating s level condition. The Khuiala sand uı (lying just above the limestone unit) gas bearing in Dandewala area (Fig. 1: In Gamnewala area, though the Khuia sand unit is proved to be gas bearing b

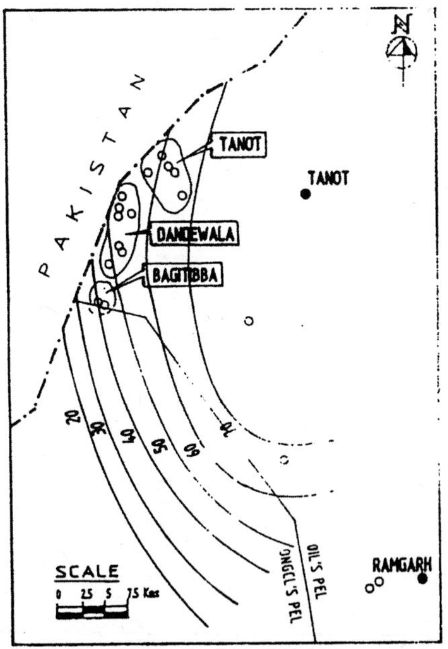

Fig. 14. Map showing hydrocarbon content of gas from Sanu/Khuiala Formations

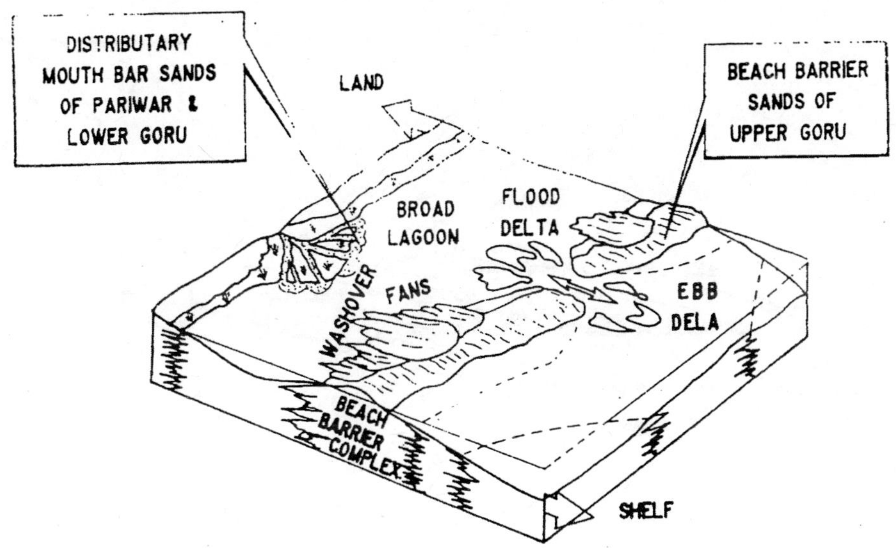

Fig. 15. Depositional environment of Pariwar and Goru sand bodies.

BIKANER - NAGAUR BASIN

In Bikaner-Nagaur Basin numerous ıoderate to high amplitude structures are resent but the main prospective structures re lying along the Bikaner arch. Till date 0 wells have been drilled in this basin on ifferent structures viz., Baghewala, vriwala, Kalrewala, Deviwali and haranwala (Fig. 3). Out of these ;ructures, Baghewala is found to be the ıost promising one.

For the first time in India presence of eavy crude oil was established in alaeozoics in Baghewala structure. The ıain prospective horizons encountered in ·aghewala structure are as follows :

. Upper Bilara Dolostone-Bitumen/ Extra heavy oil.

. Middle Bilara Limestone-Heavy oil.

. Lower Bilara Limestone/Dolostone - Heavy oil.

. Jodhpur Sandstone - Heavy oil.

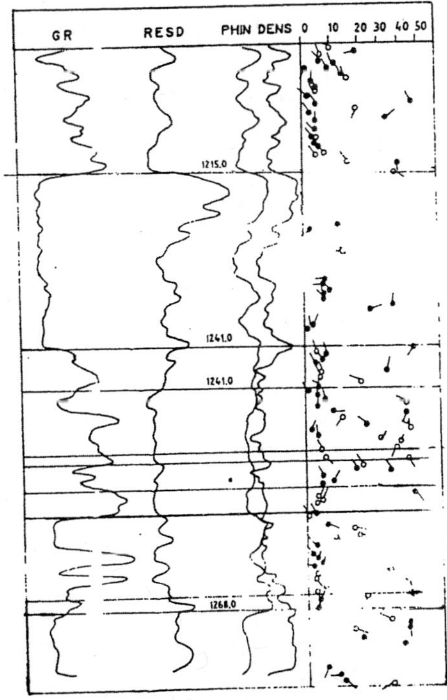

Fig. 16. Electrolog characteristics of Jodhp Sandstone and Lower Bilara Carbonates

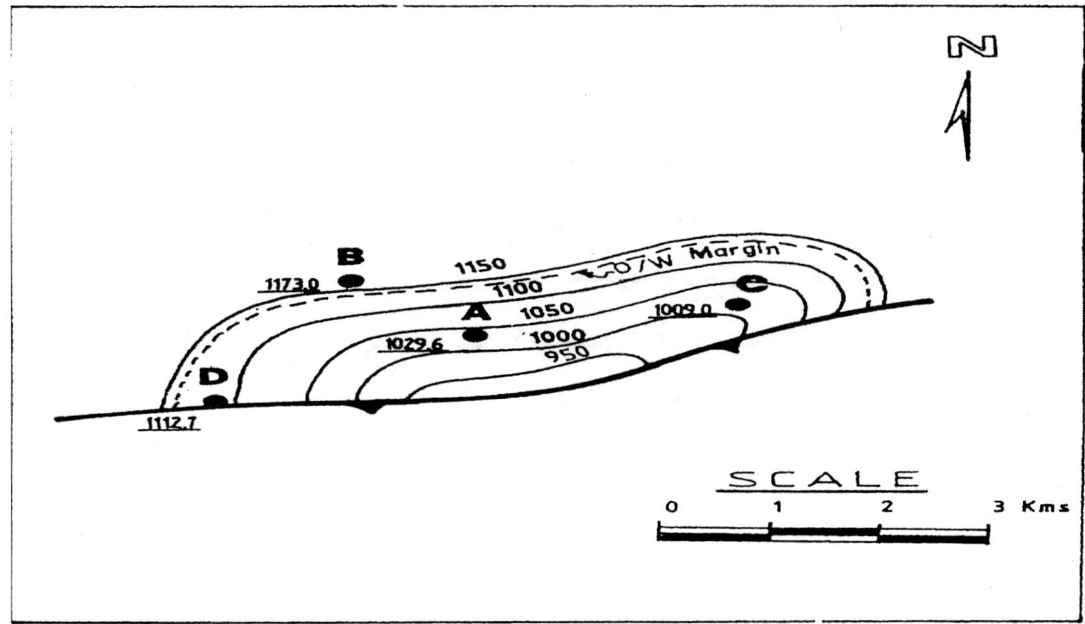

Fig. 17. Structure map on top of Jodhpur Sandstone on Baghewala structure

Jodhpur Sandstone is the lowermost unit lying over the basement and is the mai eavy oil producing zone (Fig. 16). From seismic studies it is observed that this unit iirly widespread. The basement relief has played a role in the deposition of this unit a vident from the well drilled on Tavriwala structure. In Baghewala structure, we rilled on structurally higher part shows a less thickness of Jodhpur Sandston ndicating the role of basement high on the deposition of Jodhpur Sandstone.

Jodhpur Sandstone unit has been subdivided into eleven subunits separated b istinct shale bands (Fig. 18). Bottom part of this unit is coarse grained and more porou nd hence have better reservoir characteristics. It seems to have been deposited in fluvi nvironment. The upper part of the unit is shaly and hence less porous. It shows marin nfluence as supported by the presence of biogenic burrows and the siliciclastic arbonate mineralogy. It is observed that the structurally higher part is more porou han the flanks probably indicating pinching out of the upper part against the basemen igh in the higher part of the structures. The average pay thickness of Jodhpu andstone unit is 18 metres.

The preliminary analysis of the produced crude oil from Jodhpur Sandstone show hat the pristane/phytane ratio of the saturated fraction is about 0.57 indicating anox nvironment of deposition. The very high resin and asphaltic content (42%) is suggestiv f immature to marginally mature oil that has undergone a small amount iodegradation.

A gradation of crude oil characteristics has been observed in different wells drille n the same structure. The API gravity varries from 9 to 19 degree possibly due to :

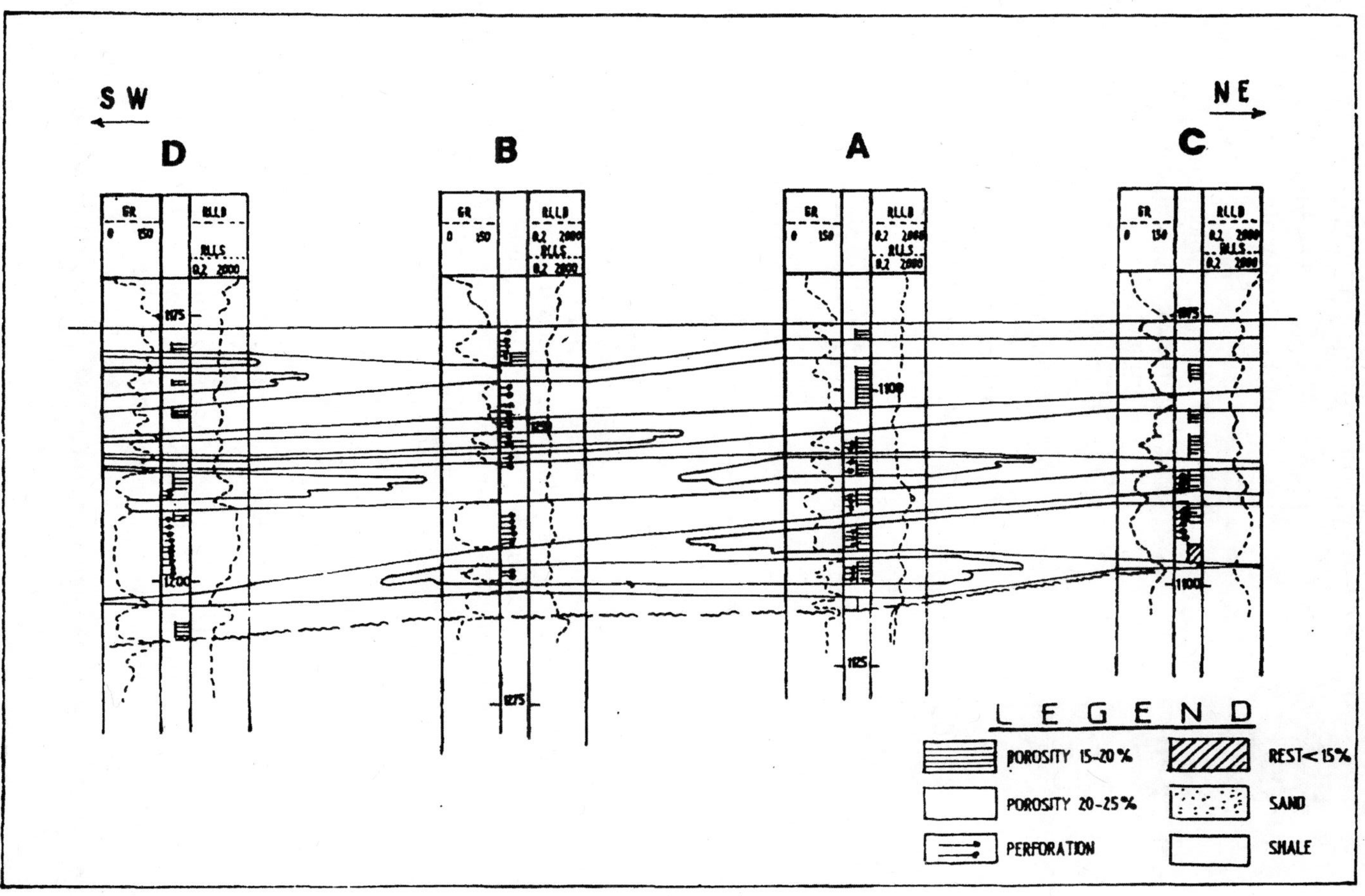

Fig. 18. Correlation of different sand units of Jodhpur Sandstone between wells A, B, C & D.

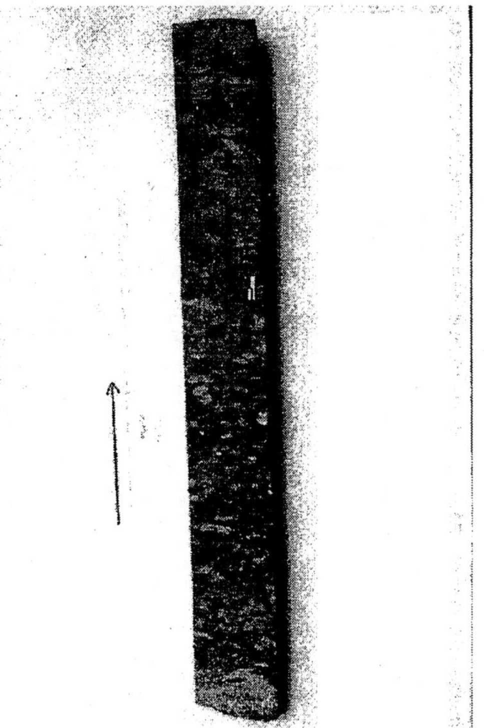

Fig. 19. Coarsening upward sequence within Lower Goru

(a) Gravity segregation crude oil as per structur position (Fig. 17).

(b) Oxidation due proximity of water zone.

Lower Bila Limestone/Dolostone li immediately above Jodhp Sandstone (Fig. 16). The low part of Lower Bilara is poro dolomitic limestone while t upper part is non poro dolostone as evident from t cores. However, the presence crude oil has been established this zone from the core and l evidences. The crude characteristics of this zone a Jodhpur Sandstone is same. T average pay thickness of th reservoir is 23 m and is present all the wells drilled on Baghewa structure.

The Middle Bilara sand u is about 9 m thick and contai relatively lighter oil as evident s oozing out nature in the conventional cores (Fig. 20). Patches of extra heavy oil ha so been observed in the conventional cores. However, on testing, this unit has main oduced water possibly due to hard and compact nature of the sandstone resulting fro lomitic cement and the presence of anhydrite.

Upper Bilara carbonate unit is bitumen bearing and has been encountered in all t ells drilled so far in Bikaner-Nagaur Basin. Seismic data shows that Upper Bila lostone is limited in extent and is pinching out both in north east and south we rections. Total pay thickness of this unit is about 75 m in Baghewala structure. Vario pes of dolostones such as hard and compact dolostone, massive dolostone, laminat lostone, etc. have been identified in Upper Bilara carbonate unit. Different kinds rosities such as matrix porosity, fracture porosity, vuggy and moldic porosity have al en identified in this reservoir. The vugs present, seems to have been formed due ssolution of an evaporite mineral viz; anhydrite, gypsum, halite, etc. The upper part lostone is hard and compact and contains bitumen both in vugs and fractures as w in matrix porosity (Fig. 21). The lower part of dolostone however, contains relative hter oil. The Upper Bilara reservoir has produced water on testing possibly due to t mobile nature of the crude at formation temperature while the bottom water channe rough fractures.

Fig. 20. Oil soaked Middle Bilara Sandstone

Fig. 21. Semisolid oil along fractures and pores in Upper Bilara Dolostone

ACKNOWLEDGEMENT

The authors are grateful to the management of Oil India Limited for granti ermission to publish this paper. Thanks are also due to the colleagues for the ssistance.

REFERENCES

Aitra, P., Mukherjee, M.K., Mathur, B.K., Bhandari, S.K., Qureshi, S.M. and Bahukhandi, G. 1993. Exploration and hydrocarbon prospect in Jaisalmer Basin, Rajasthan. *Proc. Seco Sem. on Petroliferous Basins of India*, 2 : 235-284.

)atta, A.K. 1983. Geological evolution and hydrocarbon prospects of Rajasthan Basins. *Petr Asia Jour.*, 6(4) : 93-100.

nterim Biostratigraphic and sedimentological report of Goru formation, Rajasthan - Phase I & (*ONGC Report - Unpublished*).

Chandra, A., Saxena, S.M., Singh, S.K., Chaterjee, M. and Malasi, S.K. 1986. Proposal for drilli locations in Kishangarh sub basin (*OIL Internal Report-Unpublished*).

)asgupta, R., Mitra, P.K. and Jonathan, R. 1996. Regional geological studies of Jaisalmer basin Rajasthan with special reference to Goru formation (*OIL Internal Report-Unpublished*).

Chitale, J. and Devi, R. 1992. Occurrence of heavy oil in Baghewala (*OIL Internal Repo Unpublished*).

Banerjee, T.K., Devi, R. and Sharma, M. 1995. Reservoir characterisation of Baghewala heavy field. In *Annual Convention of AEG*, Bangalore.

EDS, 1995. A sequence stratigraphy study of the Lower Goru-Sember formations and equivalen of the Lower and Middle Indus basins of Pakistan and Rajasthan (India), IEDS (*Integrat Exploration and Development Services Ltd.) Report* (1995), Vol. 1.

eological Evolution of Northwestern India
.S. Paliwal (Ed.)
:ientific Publishers (India), Jodhpur, 1999, pp. 183 - 190.

GEOLOGY OF THE MIDDLE PROTEROZOIC RAS MARBLE)EPOSIT - A POTENTIAL RESERVE FOR THE CEMENT INDUSTR

B.S. Paliwal and Arun Vyas*

Department of Geology, Jai Narain Vyas University, Jodhpur - 342 005
**Department of Geology, Government Bangur College, Didwana - 341 303*

ABSTRACT

The Ras marble belt is one amongst the fifteen important limestone belts that are potential reserves for the cement industry associated with different geological formations in Rajasthan. This belt extends from Ras to Kundal and Tunkara on one side to Asarlai in Pali district on the other. H re the good reserves of crystalline limestone have a high CaO and low Fe_2O_3 and MgO content, putting these deposits into the cement grade.

.ey words : Ras Marble, Raialo Formation, Delhi Supergroup, Cement Indust Rajasthan.

INTRODUCTION

The study area around Ras (74°11' 39" : 26°18' 45") lies in the Pali district 'estern Rajasthan (Fig. 1). The calcareous rocks of Ras area form a part of t recambrian metasediments (Pascoe, 1950) deposited along the north-western margir ıe Indian peninsu.ar shield. This calcareous sequence was assigned a separ: :ratigraphic identity as the "Raialo series" by Heron (1953) sandwiched between t nderlying Aravalli system and the overlying Delhi system of rocks. However, la 'orkers observed that this is the part and parcel of the Delhi Supergroup. (Gupta et 980; Paliwal, 1993; Paliwal et al. 1997).

Workers of the Geological Survey of India (G.S.I., 1977) included this unit into t 'elhi Supergroup (as the lower most Raialo Group). As per Gupta *et al.* (1980), it h

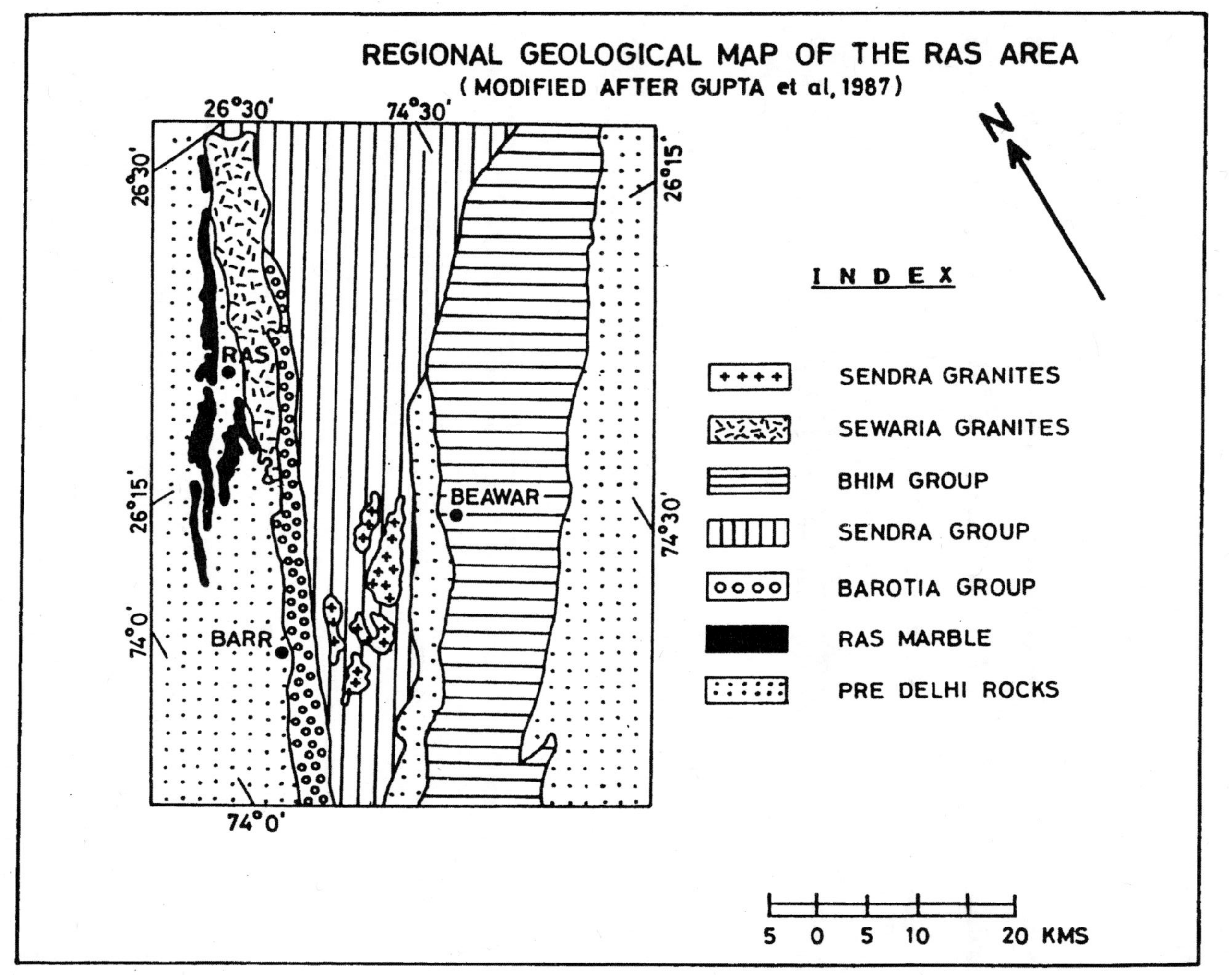
REGIONAL GEOLOGICAL MAP OF THE RAS AREA
(MODIFIED AFTER GUPTA et al, 1987)
N
26°30'
74°30'
26°30'
26°15'
26°15'
74°30'
74°0'
74°0'
RAS
BEAWAR
BARR
INDEX
SENDRA GRANITES
SEWARIA GRANITES
BHIM GROUP
SENDRA GROUP
BAROTIA GROUP
RAS MARBLE
PRE DELHI ROCKS
5 0 5 10 20 KMS

›en stratigraphically assigned to Kumbhalgarh Group of Delhi Supergroup. Sinha-R(*al.*, (1993) considered the Ras marble to be the part of the Middle Delhi Group as tl ıs Formation lying, above the Barr Conglomerate. However, the status of tl .lcareous sequence, as part of the Raialo Group is still debated. Anyway, the pre-Del ıaracter of the migmatites enveloping Ras marble has been confirmed by Chowdhary . (1993) on the basis of structural pattern.

GEOLOGICAL SETTING OF THE RAS AREA

The bed rock geology of the area comprises crystalline limestone, marbl .lc-gneisses, schist and pegmatites, (Table - 1) exposed in a terrain of wind blown sar ıd recent alluvium (Fig. 2). The calcareous unit is predominantly composed ystalline white, sacharoidal, coarse grain marble, interbedded with bands of fine le ystalline blue grey limestone with diopside and white mica. Dark calc-gneisses occı ›oradically. The grey foliated, porphyritic Pre-Aravalli gneiss crops out in valleys :tween the limestone ridges. The enveloping gneisses are intruded by pegmatites of tl rinpura age.

ıble 1. Geological Succession of the Area

Pleistocene to Recent	Recent alluvium and desert sand
Post Delhi Intrusive	— Pegmatites
Raialo Formation	— Calc granulite and schist — Crystalline limestone

DEFORMATION PATTERN

Two elongatged NE-SW trending overturned isoclines with axial planes dipping at .gh angle represent the outliers of the Raialo series of Heron (1953), around Ras. The ıtliers are situated on the north western side of the great Delhi synclinorium (Hero)53).

The Ras marble belt extends for about 80 km along the strike in the form of straig ıd almost continuous ridges. These NE-SW striking ridges show variable amount of d 'ig. 3). Generally, dip varies from 35° to 55° easterly but at few places it is recorded ; w as 25° and as high as 70°. Interestingly, at a few places near old Ras fo 'awangarh) westerly dips were also recorded. Here the angle of dip varies from 35° ›° westerly.

Detailed structural mapping carried out in the study area reveals a clear cut effe ' four generations of superposed foldings (Fig. 4). The earliest fold seems to be reclin oclinal in geometry with fold axis plunging 38° towards 113° N. The second fold is tig . nature, pulnging moderately (12°) towards 352° N. Axial planes of the first generati(

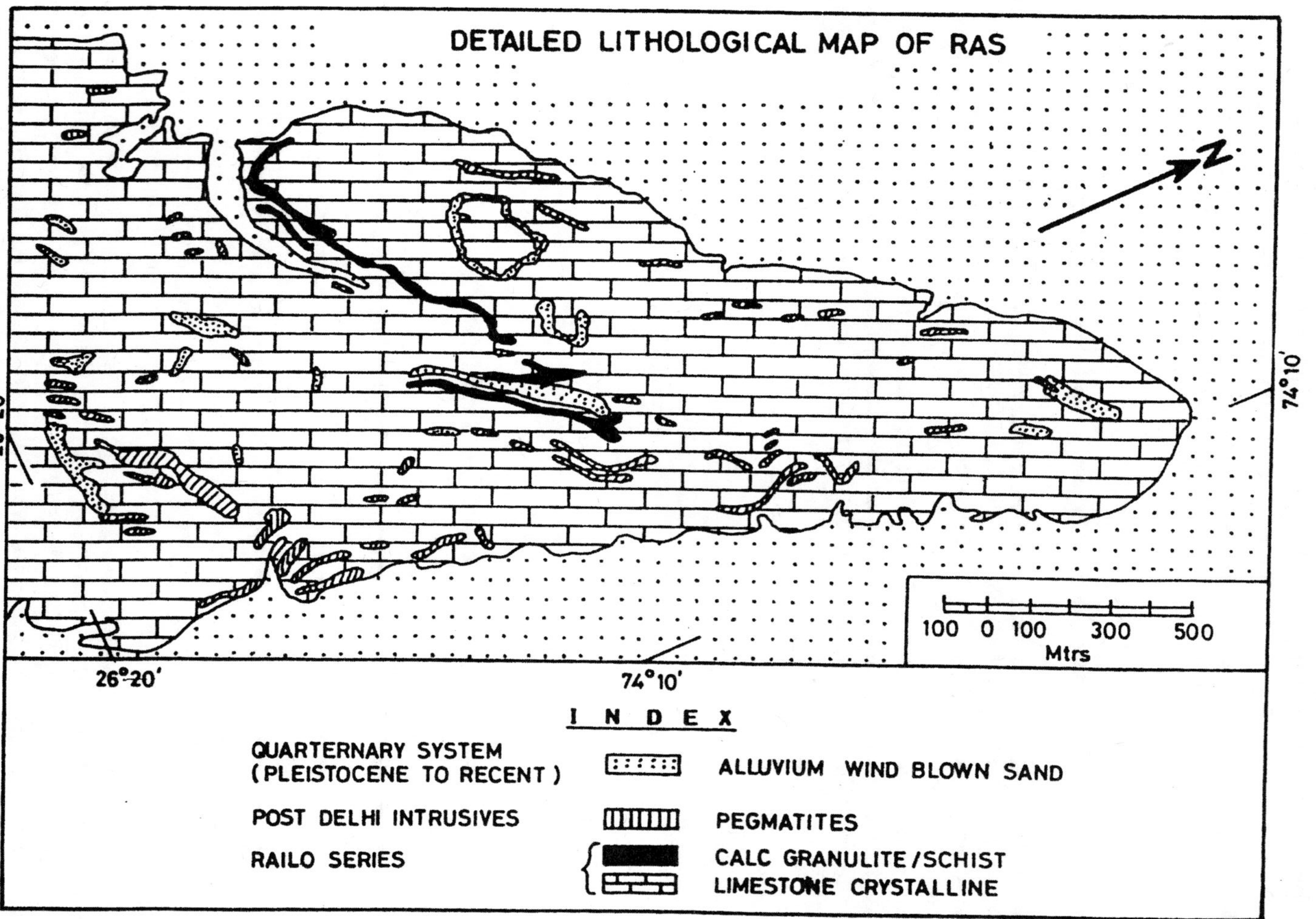
DETAILED LITHOLOGICAL MAP OF RAS
N
26°20'
74°10'
100 0 100 300 500
Mtrs
26°20'
74°10'
INDEX
QUARTERNARY SYSTEM
(PLEISTOCENE TO RECENT)
ALLUVIUM WIND BLOWN SAND
POST DELHI INTRUSIVES
PEGMATITES
RAILO SERIES
CALC GRANULITE/SCHIST
LIMESTONE CRYSTALLINE

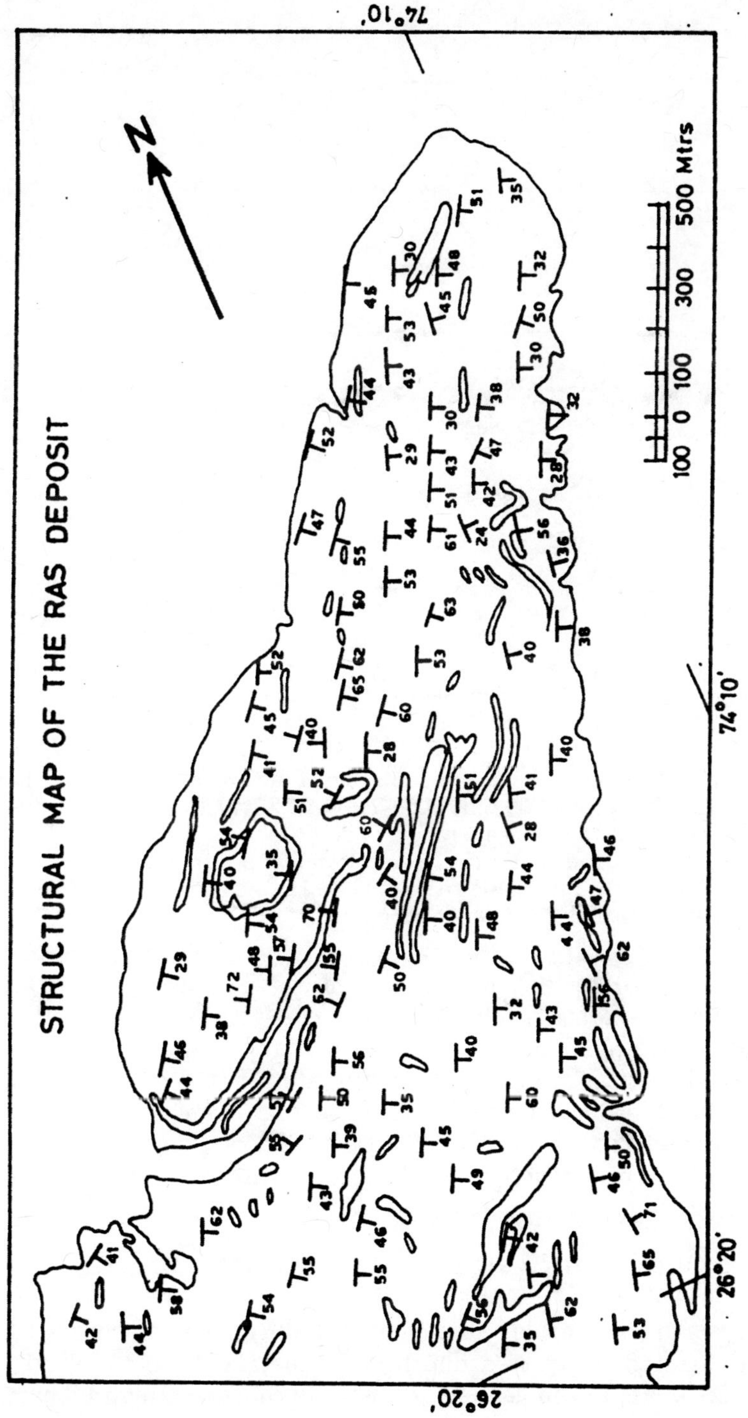

Fig. 3. Structural Map of the Ras area

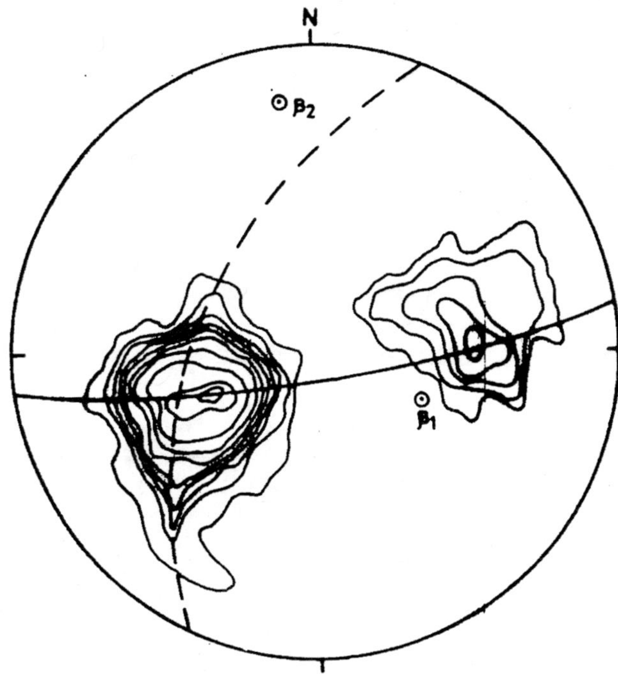

Fig. 4. S-pole diagram of the Ras area showing orientation of different generation of folds.
Contours : 0.80-1.60-2.40-3.20-4.00-6.50-9.80-21.30-27.80-40.10%
Dotted girdle corresponds to First generation fold.
Continuous girdle corresponds to Second generation fold :
β-1 - Represents attitude of First generation fold.
β-2 - Represents attitude of Second generation fold.

folds are dipping about 30° to 40° towards southeast. A strong down dip lineat associated with the first generation folding is very conspicous in the study area. Ax planes of the second generation folds and pervasive crenulation cleavage developed ax planar to them, are subvertical in nature, trending almost in the north-south directi Surprisingly the Ras marble shows four phases of the folding episodes as compared to five folding episodes reported from the pre-Delhi rocks. However, the envelop migmatites certainly by show a structural pattern comparable with the pre-Delhi rocks

ECONOMIC SIGNIFICANCE OF THE RAS MARBLE

Chemical analysis of the marble samples obtained by diamond core drilling - (us inclined drilling calyx method) up to a depth of 50 metres, from the study area sh excellent results as far as its use in cement industry is concern. Mean laboratory valu of the samples collected, are as next page.

able 1. Mean laboratory values of the Ras Marble

.No.	Contens	Value in percentage
	LOI	38.04
	SiO_2	10.10 (as I.R.)
	CaO	47.28
	MgO	00.63
	Fe_2O_3	00.73
	Al_2O_3	01.05
	$CaCO_3$	84.37
.	$MgCO_3$	01.32
.	Total Carbonate	85.69

Chemical analysis results are quite encouraging and indicate the econom gnificance of the Ras marble. With high CaO and low Fe_2O_3, MgO, and the tot ırbonate (T.C.) values more than 85%, Ras marble is suitable enough to be used in tl ement manufacturing industry.

CONCLUSION

Cement industry plays very significiant role in country's industrial growth. Ind roduced 69.3 million tonnes of cement in the year 1995-96 from about 125 big and 3(ıini cement plants.

Rajasthan has approximately 3000 million tonnes of limestone reserves of all grad ; accounts for more than 16.5% of the country's total limestone reserves. Each tonne ement produced requires about 1.6 tonnes of limestone and on an average about 75% mestone produced is used for cement manufacturing. That way, Rajasthan state alo as limestone reserves large enough to produce approximately more than 1405 milli onnes of cement.

Ras marble belt of the Delhi Supergroup extends from Ras to Kundal and Tunka n one flank and upto Asarlai on the other flank, for about 80 km stretch, hosts go eserves of cement grade limestone. Geological investigation carried out in the ar ıggests that the Ras marble deposits of Pali district are best in quality and sufficient uantity to be used in the cement industry.

ACKNOWLEDGEMENT

Arun Vyas is grateful to Dr. V.B. Khilnani for providing facilities for field aı ıboratory work.

REFERENCES

howdhury, S., Chowdhury, I. and Gupta, P. 1993. Structural geometry of Ras marble and relationship with the pre-Delhi rocks. Rec. Geol. Surv. India. 31 : 217-230.

.S.I. 1977. Geology and mineral resources of the States of India. Part XII : Rajasthan - *Ge Surv. India Misc.* Publ. No. 30 : 1-75.

upta, S.N., Arora, Y.K. Mathur, R.K., Iqbaluddin, Balmiki, Prasad, Sahai, T.N. and Sharn S.B. 1980. *Lithostratigraphic map of Aravalli region southern* Rajasthan and NE Gujarε G.S.I. Pub. Hyderabad.

eron, A.M. 1953. The geology of the central Rajputana. *Geol. Surv. India Mem.,* 79(1) : 389.

aliwal 1993. Geology of Rajasthan – a summarium. In : Chauhan, T.S. (Ed.) *Natural and Hum Resources of Rajasthan.* Scientific Publishers, Jodhpur, 1-22.

aliwal, B.S., Pareek, U.S., Vyas, A. 1997. Structural frame work of the Makrana Marble depo and its bearing on the Precambrian Stratigraphy of Rajasthan, India : In : Vijayanan N.P., Coorey, P.H., and Mosely, P. (Eds.) *Geology in South Asia - II* . Proceedings , Seco south Asia Geological Congress Colombo, Sri Lanka Jan. 19-24, 1995. Geological Survey a Mines Bureau Sri Lanka Professioal Papers, 7 : 123-136.

ascoe, E.H. 1950. *Manual of Geology of India and Burma,* Geological Survey of India, Calcut third edition, I : 349 - 350.

nha - Roy, S., Mohanty, M. Malhotra, G., Sharma, V.P. and Joshi, D.W. 1993. Conglomer horizons in south-central Rajasthan and their significance on Proterozoic stratigraphy a tectonics of the Aravalli and Delhi fold belts. *Jour. Geol. Soc. India,* 41(4) : 331-350.

ɔlogical Evolution of Northwestern India
S. Paliwal (Ed.)
ientific Publishers (India), Jodhpur, 1999, pp. 191 - 205.

PROTEROZOIC AND PHANEROZOIC PHOSPHORITE OF RAJASTHAN : PRESENT STATUS

S.C. Mathur, S.K. Trivedi and D.S. Chauhan
Department of Geology, J.N.V. University, Jodhpur

ABSTRACT

Among the phosphorites discovered in Rajasthan since 1966, only Proterozoic phosphorites are economically viable as compared to Phanerozoic phosphorites. The total reserves in the state are of the order of 90 million tonnes of which 80% is of low grade (20-28% P_2O_5) and require benifeciation before their use. The present paper provides an elaborate note on the present geological status of phosphorite of Rajasthan covering basic information on geology, mode of occurrence, depositional environment and phosphogenesis.

ey words : Phosphorite Deposits, Proterozoic, Phanerozoic, Present Status, Rajasthaı

INTRODUCTION

Phosphatic rocks, mainly of sedimentary origin, are widely distributed ajasthan, both geographically and geologically (Fig. 1). Several phosphogenic episod ıve been identified in the state, of which the best developed are associated with tl ·oterozoic period and the rest of the low grade are confined to the Phanerozc diments. A lot of work on phosphorites of Rajasthan has been carried out in the pa ree decades especially on Proterozoic phosphorite deposits (Sheldon, 1966; Muktinat)67, 1974; Adhia, 1969; Ram Babu and Majumdar, 1969; Srikantan *et al.* 196 ıauhan, 1970, 1973, 1979; Banerjee, 1971, 1975, 1986; Paliwal, 1980; Roy and Paliwa)81; Pareek, 1984; Hussain and Banerjee, 1986; Chauhan and Mathur, 1987, 199

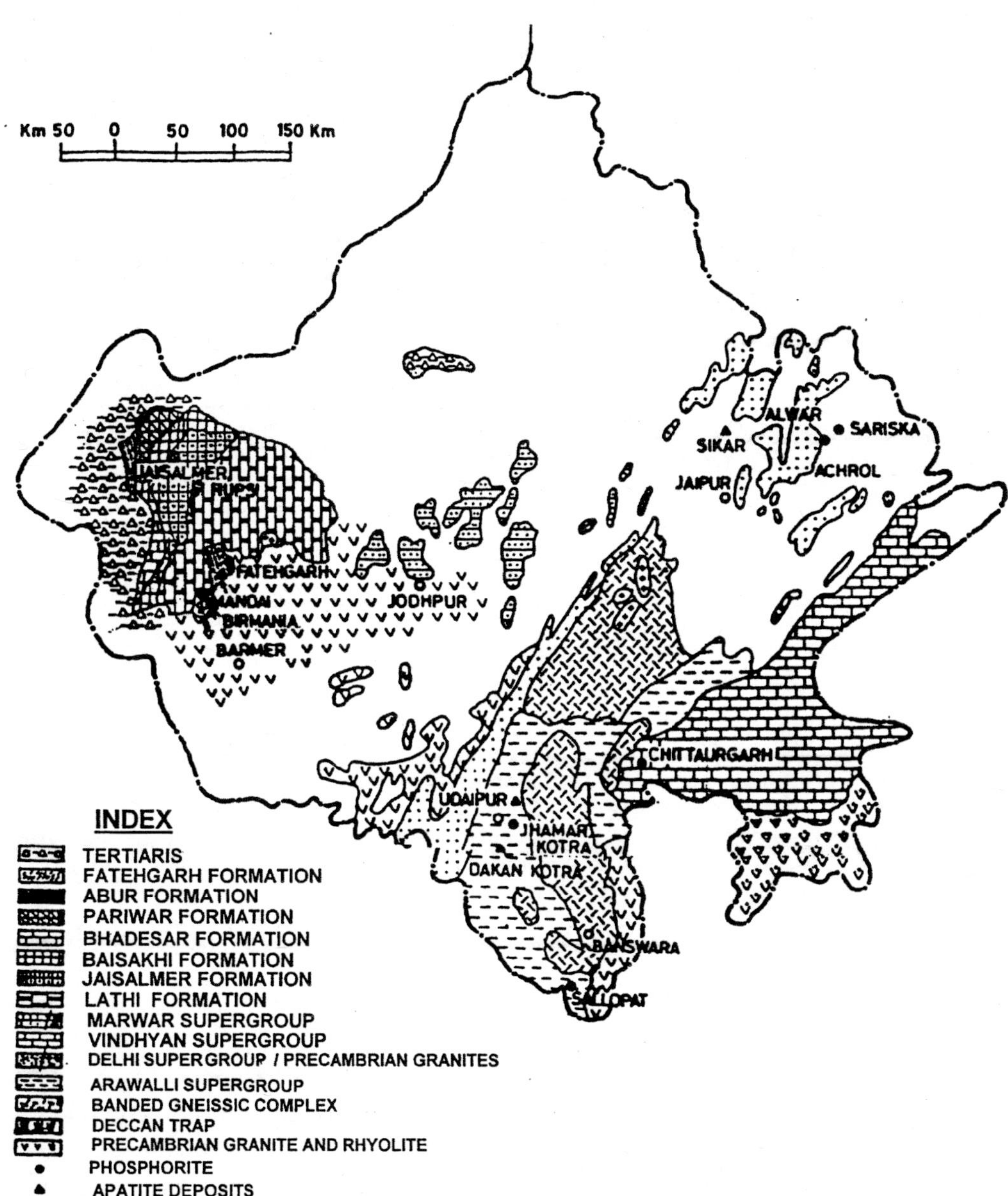

Fig. 1. Geological map of Rajasthan showing phosphorite occurrences.

athur, 1987, 1997; Banerjee and Sood, 1989; Chauhan and Sisodia, 1989; Sisodia ar ıauhan, 1990; Trivedi, 1992; Mathur and Chauhan, 1995). At present the knov serves are inadequate to fulfil our growing demand. Rajasthan contributes about 70 the total production of India from its seven workable deposits of Udaipur ar ınswara districts. The production of phosphorite in India is only 31% of its tot mand. As a consequence, India is importing about 1.8 million tonnes of rock phospha ery year and the demand upto 2000 AD is likely to be of the order of 2.4 million tonn MYB, 1991). In these circumstances the low grade deposits have become mo gnificant and are hope for the future. The paper embodies detailed basic geologic formation about Proterozoic and Phanerozoic phosphorite deposits of Rajasthan (Tab and 2) which may faciliate the beneficiation of low grade deposits and enhance ances of their workability. The distribution pattern of phosphorites in Rajasthan is di ssed under two regional headings : Proterozoic and Phanerozoic phosphorite deposits.

ıble 1. Characteristics of Proterozoic phosphorites of Rajasthan

/pe of deposit	Jhamar Kotra	Others (Udaipur)	Sallopat & others	Achrol
ıosphorite types	Stromatolitic, oncolitic, brecciated, fragmental & botryoidal	Stromatolitic, brecciated & fragmental	Stromatolitic, bedded, brecciated & bended	Encrustations, bended & stromatiolitic
ıickness (metres)	1-30m	1-30m	1-15m	6 cm - 3.5m
eserves (million tonnes)	65.86	17.00	1.5	—
ajor minerals	Carbonate fluorapatite, calcite & dolomite	Carbonate fluorapatite, calcite & dolomite	Francolite dolomite, quartz, felspar & mica	Francolite, quartz, clay & muscovite
ost Rocks	Dolomitic limestone	Dolomite	Dolomite limestone	Limestone, shal & quartzite
ıemical composition (%)				
O_5	32.50	5.25	15-30	3.96
O_2	6.65	7.8	11.80	-
ıO	55.25	54.85	42.72	-
$_2O_3$	0.60	0.60	0.34	-
gO	5.85	2.80	3.50	-
terfering oxides	SiO_2, MgO & F	SiO_2, MgO & F	SiO_2, MgO & F	SiO_2, F & Fe_2O
ıvironment of deposition	Subtidal	Intertidal to supratidal	Subtidal to intertidal	Subtidal to intertidal

ıble 2. Characteristics of Phanerozoic phosphorites of Rajasthan

ype of deposit	Birmania	Fatehgarh	Mandai	Rupsi	Chittor & Bundi
ıosphorite pes	Stromatolitic, bedded & grainstone	Skeletal, grainstone, mudstone & casts	Grainstone & mudstone	Nodular	Laminated & blotches
ıickness (metres)	1-9m	1-5m	1.3 m	—	0.5-1m
eserves million tonnes)	4.89	—	—	—	—
lajor minerals	Fluorapatite, calcite & quartz	Collophane, quartz, feldspar & iron oxide	Collophane, quartz, & iron oxide	Callophane, quartz calcite & iron oxide	Collophane quartz & calcite
ost Rocks	Calcareous sandstone	Ferruginous sandstone	Ferruginous sandstone	Shale and sandstone	Limestone & sandstone
hemical composition (%)					
$_2O_5$	5-20	5-20	5-10	—	—
O_2	9-30	40-45.00	42-46.6	—	—
aO	41-53	9.5	7.4	—	—
$_2O_3$	1-2	11.22	9.4	—	—
lgO	1-4	1–2	—	—	—
ıterfering xides	SiO_2 & Fe_2O_3	SiO_2 & Fe_2O_3	SiO2 & Fe_2O_3	—	—
nvironment of eposition	Subtidal-intertidal	Subtidal-intertidal	Subtidal-intertidal	Shallow-subtidal	Shallow-subtidal

PROTEROZOIC PHOSPHORITE DEPOSITS

daipur deposits

In the Udaipur area phosphorite occurs in a peripheral pattern coveri ıamarkotra, Matoon, Nimach Mata, Bargaon, Dakan Kotra and Sisarma (Roy a aliwal, 1981). The phosphorite is associated with dolomitic limestone and dolomite ıe Aravalli Supergroup. The phosphorite is primarily associated with *algal stromatoli* 'ig. 2 and 3). Other types are massive, fragmental, brecciated, discontinuo ıicrosphorite layers, all derived from the stromatolitic phosphorite and seconda ıosphate veins. (Chauhan, 1970, 1979; Banerjee, 1971; Paliwal, 1980; Roy and Paliw 981.

The dominant phosphatic constituent of Udaipur phosphorite is microcrystalli ırbonate fluorapatite, with minor amount of calcite, dolomite and quartz. Replaceme apatite by secondary calcite and opaques such as iron oxides and iron carbonate occ ıore commonly. (Chauhan, 1979; Banerjee, 1986).

The stromatolites at Jhamarkotra, Kanpur and Bargaon are mainly columnar ıape and are predominantly composed of phosphate, whereas stromatolites akankotra, Nımach Mata and Sisarma are dominantly composed of carbonate a

Fig. 2. Columnar stromatolites showing phosphate and carbonate laminations at Jhamarkotra.

contain a very littl phosphate. Th stromatolites in th former area are mainl columnar structure showing uniform shape size and maximur growth density per uni area. Here th intercolumnar materia is usually free fror stromatolitic an siliciclastic materia There is a complet absence of erosiona structures lik desiccation cracks c fenestral fabri suggesting that the

ere deposited under oderately deep subtidal nditions with moderate wave tion (Chauhan, 1979; hauhan and Sisodia, 1984). In ntrast, at Dakankotra and imach Mata stromatolites ow considerable variation in eir morphology and hosphate content. Their tercolumnar material is full stromatolitic intraclasts. nere occurs intraformational nglomcrate of stromatolitic erivation (Fig. 4) and ratifera showing desiccation acks and saucer shaped ructures. These biogenic ructures suggest that the owth of stromatolites and eposition of associated

Fig. 3. Folded stratiform stromatolites at Dakan Kotra

hosphorite took place under intertidal to supratidal setting. (Chauhan, 1979; Chauha nd Sisodia 1984; Sisodia 1986). The reworked, brecciated, fragmented or slumpe hosphorite, asbsence of dolomitic matrix and bedding structure indicate a high energ nditions prevailed during their formation. The secondary calcareous apatite occurs a he network of veins controlled by fracture pattern of the host rock. The colloform an

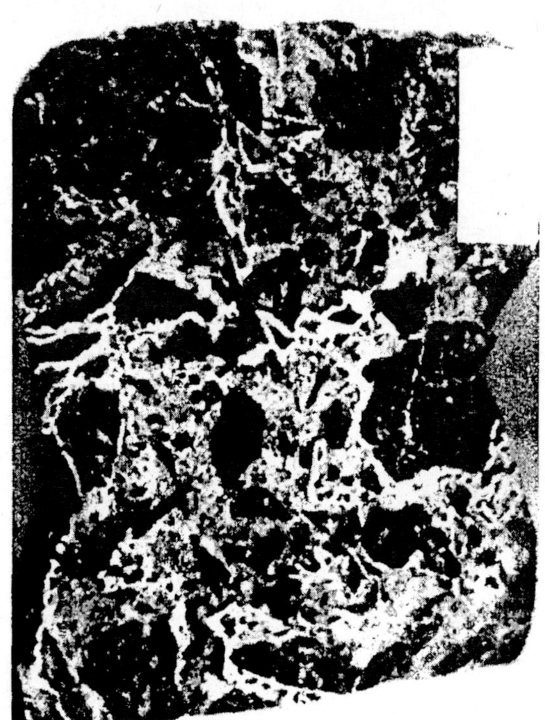

Fig. 4. Intraformational conglomerate with pebbles of columnar stromatolites Nimach Mata) (Size of the sampel 16 x 10 cm)

globular accreting bands, a Jhamarkotra give a deceptiv appearance of bedded agate. The oolitic pisolitic, reniform and botryoidal form indicate subsequent secondar phosphatization during later stage (Chauhan and Sisodia, 1984; Banerjee 1986; Choudhary and Roy, 1986 Choudhary 1990; Sisodia, 1990).

Banswara deposits

The phosphorite at Sallopat Ram-Ka-Munna, Monadunger, Zhardo and Jhermoti areas in Banswar district is hosted by dolomiti limestone. This phosphorite i primarily stromatolitic in nature. Othe rocks are orthoqurtzite, phyllite, arkos and cherty limestone. These rock belong to Lunawada Group of th Aravalli Supergroup. Due t deformation and surficial weatherin the initial nature of phosphorite ha changed. As such, the phosphortie o Sallopat and near by areas seemingl occurs in stromatolitic, bedded brecciated and clayey forms. Th osphorite horizon runs for a strike length of 2.0 kms in the form of lenses. There hav

Fig. 5. Fold developed in quartzose layer in dolomitic limestone. Cleavages are parallel to axial plane of the fold (Sollopat)

Fig. 6. Columnar stromatalites showing surface orientation and development of lateral bridges. (Sallopat)

been three distinct phases deformation and the effects deformation have been ve pervasive on the rocks of the ar including the phosphorite horiz (Fig. 5). This would be explicit fro the subsequent description.

The stromatolitic phosphori is characterized by column stromatolites. These column structures are 1 cm. to 30 cm. length and 2 to 5 cm. wide, and a enclosed in a matrix of dolomit limestone (Fig. 6). The stromatoli columns are composed of thi irregularly disposed conv phosphate laminae with th smaller intervening carbona laminae. The phosphate compone of stromatolite is composed microcrystalline carbona fluorapatite and a little quantity calcite and quartz. The bounda between phosphate and calcite sharp. The matrix compris coarsely crystalli calcite/dolomite, quartz ar phosphate intraclasts.

The columns of stromatolit columns normally gro rpendicular to the bedding and appear as circular to slightly elliptical in transver ction. But due to deformation they have been laid down parallel to the bedding wi e development of bedding cleavage and updip convexity (Fig. 7). Often the columns a und flattened, along the cleavage planes, (Paliwal, 1980; Chauhan and Sisodia, 198 some cases the stromatolites aquire sheet like configuration. The elongated colum e observed in the limbs of the folds whereas sheet like structures are seen close to the nge areas. This suggests that there was a dominant stress induced flowage of materi rmal to the fold hinges. Secondly, the flowage seems to be more or less of equal exte direction parallel and perpendicular to the fold hinges (Trivedi, 1992; Banerjee ar od, 1989).

The composition of bedded and brecciated phosphorite is similar to that romatolitic phosphorite. The clayey phosphorite occurs in highly deformed and sectio nere weathering is intensive. It appear earthy, friable and often powdery in nature. Tl trographic and geochemical attributes of phosphorite of Banswara and Udaip dicate that geologically they are similar in nature and have common mode of genesis.

The formation of Aravalli phosphorite (Udaipur and Banswara) seems to be ful ntrolled by microbial communities involved in the growth of phosphate beari romatolites. The absence of phosphate in intercolumnar spaces suggests that initial

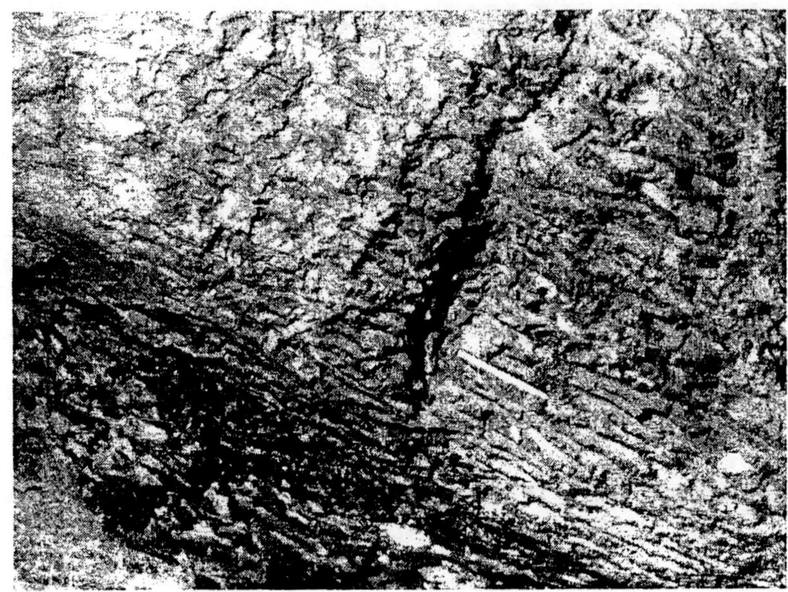

Fig. 7. Stromatolites appear laid down parallel to the cleavage and indicate updip convexity (Sallopat).

romatolites grew due to direct in situ precipitation of calcium carbonate influenced b rtain forms of cyanobacterial communities, under subtidal to intertidal condition imultaneously, other variety of filamentous cyanobacterial community changed th iemical environment of the ambient water and initiated apatite formation b ntemporaneous replacement of initially accreting carbonate structures (Chauhan 197 989; Chauhan and Sisodia, 1984 and 1990; Trivedi, 1992).

irouli and Nathdwara deposits

The uraniferous phosphorite is reported to occur in a belt running about 40 km om Nathdwara to Kirouli-Madar in Udaipur district. The phosphatic limestone ssociated with marble and dolomitic limestone of the Aravalli Supergroup. Ten maj nses of uraniferous phosphorite have been delineated in the limeston etrographically, the phosphorite consists of microcrystalline phosphatic materi mprising predominantly of carbonate fluorapatite and dahlite together with calcit iartz goethite and specks of pyrite. The phosphorite varies texturally from massiv minated, bedded, granular to pelletal and nodular and can be put in the category ickstone to wackstone phosphorite (cf. Cook and Shergold, 1986). The studies indica iat Kirouli phosphorite shows characteristic near shore euxenic conditions of depositic Kant, *et al.*, 1989).

aipur and Alwar deposits

The type area of Delhi phosphorite is around Achrol-Nawabas (Jaipur district) an duka-Andwari-Dalalpur (Alwar district). The phosphorite is associated with (1) impur mestone, mudstone and shale (Kushalgarh Formation), (2) breccia, stromatolit ırtzite, limestone and shale (Sariska Formation) and (3) stromatolitic calcareoı yllite (Thanaganj Formation) belonging to the Ajabgarh Group of Delhi Supergrou e phosphorite occurs in form of lenses and bands spreading over a strike length of 60 tres. Petrographically, phosphorite occurs as (1) thin microsphorite bands in fir ıined sandstone, (2) columnar to pseudocolumnar stromatolites in quartzite an yllite and (3) as encrustations in hornstone breccia. The phosphorite here is compose ımorphous to cryptocrystalline carbonate fluorapatite, quartz, clay (Kaoline and Illit d iron oxide alongwith muscovite, tourmaline and pyrite. The crystalline phase ıtite appears to be metamorphosed product of collophane (Dayal *et al.* 1984). Th eliminary geological studies indicate that phosphorite of these areas is formed due genic-diagenetic and associated orthogenic processes operating in subtidal ertidal environment of unstable shallow shelves of Delhi basin.

ıittore and Bundi deposits

Five minor phosphorite occurrences have been reported from limestone an ıdstone sequence of Vindhyan Supergroup near Jhiri, Savaria, Sirbu, Balwas, Javd ora, Jhar and Bari in the Bundi and Chittore districts. The thickness of phosphat rizon is 10 cm to 1 metre and the P_2O_5 contents is low (3-10%). The phosphorite ittore and Bundi occurs in form of thin laminations of microsphorite alongwit osphate clasts and are rarely stromatolitic. The phosphate is associated wit bonate and clay minerals. The presence of cross bedding, ripple marks an

ig. 8. An outcrop showing quartz arenite in form of stone debris in fore ground. In the back ground contact of phosphorite and micritic dolostone is visible (Birmania).

Fig. 9. The bedded to laminated nature of phosphorite is distinctly visible due to differential weathering (Birmania).

Fig. 10. Outcrop showing circular to elleptical bodies of phosphorite embeded in calcareous sandstone mastrix. Note. these circular bodies actually represent transver section of columnar to pseudocolumnar stromatolites (Birmania).

romatolites indicate very shallow depositional environment during the formation osphorite (PPCL Report, 1978; IBM Report, 1984).

PHANEROZOIC PHOSPHORITE DEPOSITS

rmania Deposits

The phosphorite occurrences around Kohra, Birmania and Ladu Singh-ki-Dha long to Birmania Formation (Cambrian ?). A few centimetres to 3 metres thic osphorite horizon stretches for a strike length of 6.8 km. The phosphorite of Birman sin occurs in a mixed assemblage of siliciclastic (Fig. 8) and carbonate lithofacies ar ows bedded and stromatolitic nature (Figs. 9 and 10). The stromatolites are column pseudocolumnar structures and are mainly composed of thin alternate layers osphate and carbonate minerals, often studded with silt sized quartz grains. Th tercolumnar spaces are filled with quartz and phosphate clasts cemented wi rbonate minerals.

The bedded phosphorite is composed of three types of laminations. The fir rimary) type of phosphorite is found as 1 to 3 mm thick microsphorite lamination mposed of cryptocrystalline fluorapatite. They are occasionally studded with silt size artz grains. The top surfaces of these laminations are highly irregular and are marke numerous erosional features such as scour channels, polygonal cracks ar icrofractures etc. (Chauhan and Mathur 1987). The second type of laminations are 3 mm thick, mainly composed of phosphate grains, pellets and quartz grains cemente sparry calcite. This type of phosphorite can be termed as grainstone phosphorite (c ggs 1979; Cook and Shergold, 1986; Chauhan and Mathur 1994; 1995; Mathur, 199 e third type of laminations are 4 to 10 mm. thick, mainly composed of phospha ains, pellets and quartz grains cemented by phosphate. This phosphate ceme croaches the incorporated quartz and phosphate clasts and replaces them to varyir tent. As a result it leads to the formation of a homogeneous secondary phospha minations. It can be termed as packstone to boundstone phosphorite (cf. Cook ar ergold, 1986; Mathur, 1987; Chauhan and Mathur 1994).

The phosphorites of Birmania seems to have formed due to interplay of biogeni agenetic and associated allogenic processes operating in the subtidal to intertid vironment of Birmania basin.

upsi deposits

Small nodular phosphorite occurrences have been reported from Rupsi ar isakhi villages about 15 km. north of Jaisalmer. The rounded to well rounded nodul cur within the repetitive sequence of shale, ferrugineous sandstone, calcareo ndstone and concretionary sandy limestone of Baisakhi Formation of Jurassic A areek, 1981, 1984, Dave and Chaterjee, 1996). The phosphatic nodules are composed mogeneous concentric layers of phosphate and carbonate developed around nuclei artz and shell fragments. Sometimes, carbonate inclusions occur within the phosphat yers which, together with the vaguely defined patches of micrite constitute grumulo ructures (Bathurist, 1975, Southgate, 1986).

ıtehgarh deposits

The exposures of low grade phosphorite occurrences of Fatehgarh extend froı ›ora-Sandha to Bhuyar over a distance of 40 kms. The type area are around Bariyar: .ndha, Lordi, Jogmaya temple and Dharvi Kalan in Barmer and Jaisalmer district. ıese phosphorite occurrences belong to Fatehgarh Formation (Cretaceous to Eocene? ıe phosphorite of Fatehgarh basin is associated with repetitive sequences ı nglomerate, ferruginous sandstone, calcareous sandstone, quartz arenite and siltstor Fatehgarh Formation. It overlies the Lathi Formation (Jurassic) and underlain b ıpurdi Formation (Upper Paleocene to Lower Eocene). The individual lithofacies show stinct variation in lithology, thickness and sedimentary structures. The planar, wav ıdulating and trough cross laminations, ripple marks, mud cracks, skeletal remains ı stropods have been observed in siliciclastic and phosplorite rocks of the area indicatin allow water conditions.

The phosphorite beds are interbedded with ferruginous and calcareous sandstone. yellow to brown in colour. There are some spread beds containing well preserved fossi gastropods (Lymnaea, Galba, Acteon, Pleurotomaria, Diplozone and Moculites) an lecypods (Mytilus). The thickness of phosphorite bed ranges from a few centimetres t maximum of 1 metre. There is sharp change in composition of the phosphorite horizo ›m calcareous at the base to more ferruginous at the top. This change is also reflecte colour (yellow to creamish brown). The phosphorite displays a wide spectrum ı ıosphorite types viz. mudstone, grainstone-wackestone and packstone-bandstor minations of discontinuous, disrupted and irregular nature. The phosphorite comprise ›tched fragments, phosphate grains, pellets, oolites, ovules, coprolites and bor ıgments along with the phosphatized skeletal remains of gastropods. Some ostracod lecypod shells, crustacian casts and imprints of appendages and fragments of bone ıd spicules of fishes are also seen. These phosphate fragments and other phosphat tritus occur in association with subangular to rounded grains of quartz, orthoclas rcon and flakes of biotite. They are cemented together by phosphat nonite/hematite/siderite and or carbonate cement. The ironoxide cement, during th alm of late diagenesis, pervaded through cracks and fractures of most of th rrigeneous minerals causing ferruginization of the phosphate. The main phosphat btrance in the rock is isomorphic collophane which appears amorphous in thi ctions. In some thin sections collophane is seen replacing the carbonate minerals. Th udstone laminae and composite laminae of grainstone-wackestone show numerou ırrow structures. Quartz and calcite are sometimes rimmed by ferruginous cement. Th esence of phosphatized fossils, shells, coprolites, ovules and feacal pellets indica ıosphatization of carbonate faunal debris.

The presence of angular to rounded phosphate clasts shows reworking processe ıe phosphorite of Fatehgarh seems to be formed due to biogenic-diagenetic ar sociated reworking processes alongwith replacement in the restricted shallow wate ıvironment of Barmer basin.

andai deposits

About 15 kms of Fatehgarh a persistent horizon of phosphatic sandstone runs for rike length of about 2 kms. It seems to be south-western extensions of Fatehgar ıosphorite. It is associated with 80-100 metre thick repetitive sequence of conglomerat rruginous sandstone, calcareous sandstone and fossiliferous sandstone of Mand ›rmation/Fatehgarh Formation. Planar wavy, lenticular undulating lamination ɔugh cross bedding, mud cracks, casts and moulds of lamellibranchs and gastropod ıve been observed in ferruginous and calcareous sandstone of Mandai. These feature note shallow water conditions of deposition. The phosphorite comprises phosphatize ells of lamellibranchs and gastropods alongwith phosphate blotches, discontinuou minations and phosphatic clasts. The phosphate is highly ferruginous and is compose cryptocrystalline mass of collophane. Associated ferruginous minerals present ar ıartz, orthoclase, biotite and zircons. These are cemented with limonite/siderit condary phosphate and minor amount of carbonate minerals. Similar mineralogic mposition and mode of occurrence of Mandai and Fatehgarh phosphorite suggest th ey have a common mode of origin.

DISCUSSIONS

All known phosphorite deposits of Rajasthan are of marine origin, occurring a ecific type of sedimentary deposits, formed under typical phosphogenic conditions. Th ıosphogenesis is a complex multi stage process which might have operated under fo odels : (a) Direct precipitation model (b) $CaCO_3$ replacement model (c) Primary-biogen odel (d) Biogenic - Diagenetic model.

The phosphorites of Rajasthan are the product of a single or combination of two c ore above mentioned models. It is interesting to note that the Proterozoic phosphorite Rajasthan are characterized by a marked deficiency in terrigenous component. I ntrast, Phanerozoic phosphorite is rich in terrigenous material indicating highl verse environmental conditions of deposition. These phosphorite occurrences ajasthan, though concide with the major global phosphogenic episodes, the sizab ıosphorite deposits are confined only to the Aravalli sediments of the Precambrian ag owever, in course of formation of Phanerozoic phosphorites of Rajasthan there wa nsiderable influx of terrigenous detritus as such fixation of phosphate remaine stricted and only poor grade phosphorite could form. Thus, it is sedimentary malie at controlled phosphorite formation in Rajasthan through ages since Early Proterozo ne.

The phosphatic stromatolites of Udaipur and Banswara indicate a deep oderatly stable shallow shelf condition of deposition with reasonably suitab ıvironment for phosphorus fixation.

ACKNOWLEDGEMENTS

Authors are deeply indebted to Prof. V.K. Vaya, Department of Geology fo oviding working facilities. The authros are also greatful to Prof. B.S. Paliwal and D .S. Sisodia for their help and for critically going through the manuscript. The financi sistance to SCM from UGC (FAT) is gratefully acknowedged.

REFERENCES

hia, J.D. 1969. - Indigenous rock phosphate for superphosphate manufacture : *Fertilizer New* 14(11) : P. 13-19.

nerjee, D.M. 1971. Precambrian stromatolitic phosphorites of Udaipur, Rajasthan, India. *Ge Soc. Am. Bull.* 82 : 2319-30.

nerjee, D.M. 1975. Environment of formation of some older and younger phosphorites of Indi *Proceedings of the IXth International Sedimentological Congress*, Nice (France), Theme - 9 1-7.

nerjee, D.M. 1986. Proterozoic and Cambrian phosphorites regional review : India subcontinent. In : Cook, P.J. and Shergold, J.H. (Eds.). *Phosphate Deposits of the Worl* Cambridge Univ. Press, London, 1 : 131-148.

nerjee, S.K. and Sood, N.K. 1989. - Geology of Sallopat phosphorite deposit. *Geol. Soc. Ind Mem.* 13 : 39- 45.

turin, G.N. 1982. *Phosphorite on the sea floor; origin, composition and distribution*, Elsevei New York, 343 p.

audhary, R. 1990. Two decades of phosphorite investigations in India. In : Nothholt. A.J.G. ar Jarvis, I (Eds.), *Phosphorite Research and Development, Geol. Soc. India Spel. Publ.* 52 305-311.

audhary, R. and Roy, A.B. 1986. The stomatolite bearing Precambrian phosphorite deposits Jhamarkotra, Rajasthan, India In : Cook, P.J. and Shergold, J.H. (Eds.) *Phosphate Deposi of the world, Proterozoic and Cambrian Phosphorites,* Cambridge Univ. Press, Cambridge, 202-219.

auhan, D.S. 1970. Phosphorite bearing stromatolites from Aravalli of Rajasthan (Abs.) 5 Himalayan Geology Seminar, Chandigarh, 10.

auhan, D.S. 1979. Phosphorite bearing stromatolites of the Precambrian Aravalli phosphori deposits of Udaipur region, their environmental significance and genesis of phosphorit *Precambrian Research*, 8 : 95-126.

auhan, D.S., 1989. Microbial activities and genesis of Aravalli Phosphorite, Udaipu Rajasthan. *Him. Geol.* 13 : 39- 51.

auhan, D.S. and Sisodia, M.S. 1984. Nature of Udaipur Phosphorite and its genet implications. *Geol. Surv. India. Spel. Publ.*, 17 : 79-104.

auhan, D.S. and Mathur, S.C. 1987. Geological nature of Birmania phosphorite. *Jour. Indic Assoc. Sedimentologist*, 7 : 71-81.

auhan, D.S. and Mathur, S.C. 1994. Lithofacies evaluation and environmental interpretation Birmania basin, Jaisalmer district, W. Rajasthan, India. In : Ahmed, A. and Sheikh, A.l (Eds.). *Geology in South Asia I* : Proceedings of GEOSAS-I, February 23-27, 199 Hydrocarbon Development Institute of Pakistan, Islamabad, Pakistan. 61-68.

auhan, D.S. and Sisodia, M.S. 1989. Phosphorite of Rajasthan. *Geol. Surv. India, Mem.* 13 9-23.

ok, P.J. and Shergold, J.H. 1986. Proterozoic and Cambrian phosphorite : An Introduction. In Cook, P.J. and Shergold, J.H. (Eds.). *Phosphorite Deposits of the World.* Cambridge Uni Press, London, 1 : 1-8.

yal, B. 1983. Some aspects of phosphogenic provinces of India. *IGCP Publ. China*, 1 : 35-48.

ussain, V. and Banerjee, D.M. 1986. Birmania Phosphorite : Petrominerological characters an Palaeogeological implications. *Jour. Geol. Soc. India.* 27 : 450-455.

dian Mineral Year Book, 1991. IBM pub. Nagpur, 534 p.

athur, S.C. 1987. Geological Investigation of Birmania phosphorite deposit, district Jaisalme Rajasthan. *Ph.D. Thesis (Unpublished),* University of Jodhpur, Jodhpur. 120 p.

athur, S.C. and Chauhan, D.S. 1995. Analysis of geological data using computer application f the interpretation of geological nature of Birmania basin, *Jour. Sedi. Lithofacies an Paleogeography*, Chengdu, China, 15(2) : 28-39.

athur, S.C. 1997. *Genesis of grainstone phosphorite.* In : Wijayananda, N.P., Cooray, P.G. ar Mosley, P. (Eds.) *Geology in South Asia-II. Proceedings of GEOSAS-II Geological Survey an Mines Bureau Sri Lanka Professional Paper.* 7 : 349-356.

uktinath, 1967. Phosphorite deposits in Rajasthan. *Indian Minerals*, 21(2) : 83-101.

uktinath, 1974. Phosphorites-its present status and guides for further search in India : *India Science Congress*, 61st Ses., 91-114.

aliwal, B.S. 1980. Systematics of the Precambrian stromatolites of Rajasthan - a review in tl light of tectonic deformation. *Geol. Surv. Indian Misc. Publ.*, 44 : 68-74.

areek, H.S. 1981. Basin configuration and sedimentary stratigraphy of Western Rajasthan. *Jou Geol. Soc. India.* 22 : 517-527.

areek, H.S. 1984. Pre-Quarternary geology and mineral resources of Rajasthan. *Mem. Geol. Sur India.* 115 : 1-95.

eport of the Sub Group-I (Chemical and Fertiliser minerals) of the working group in non metall (Industrial) Minerals constituted by planning commission, IBM publication, 1984, p. 142.

eport of the Sub Group-II on fertilizer minerals. PPCL Publication 1978, p. 39.

iggs, S.R. 1979. Petrology of the Tertiary phosphorite system of the Florida. *Econ. Geol.*, 7 :195-220.

oy, A.B. and Paliwal, B.S. 1981. Evolution of the Lower Proterozoic epicontinental deposits stromatolite bearing Aravalli rocks of Udaipur, Rajasthan India. *Precambrian Research*, 14 49-74.

heldon, R.P. 1966. Phosphate deposits of India. The discovery, geology and potential ar proposed development. *Tech. Letter, Phosphate.* April 5, *USGS, Agency for International Development.*

isodia, M.S. 1986. Studies of Precambrian Aravalli Stromatolitic phosphorite of Bargao Nimachmata and Dakankotra of Udaipur region *Ph.D. Thesis* (Unpublished) University Jodhpur, India. 124 p.

isodia, M.S. and Chauhan, D.S. 1990. The influence of magnesium ions during the formation stromatolitic phosphorite of Udaipur, Rajasthan, India. In : A.J.G. Nothholt and I. Jarv (Eds.). *Phosphorite Research and Development.* 52 : 313- 320.

outhgate, P.N. 1986. Middle Cambrian phosphorite hardgrounds, phoscrete, prolite stromatolites and their implication for phosphogenesis. In : P.J. Cook and J.H. Shergo (Eds.). *Phosphate Deposits of the World*, Cambridge, Univ. Press, London, 1 : 327-368.

rivedi, S.K. 1992. Geological Investigation of stromatolitic phosphorite deposit of Sallop Banswara district Rajasthan, *Ph.D. Thesis* (Unpublished). University of Jodhpur, Indi 125p.

ological Evolution of Northwestern India
S. Paliwal (Ed.)
ientific Publishers (India), Jodhpur, 1999, pp. 206 - 209.

SILICEOUS EARTH FOR SOLAR THERMAL ENERGY STORAGE

N.M. Nahar, M.S. Sisodia* and M.M. Purohit

Central Arid Zone Research Institute, Jodhpur 342 003, India
**Department of Geology, JNV University, Jodhpur - 342 005. India*

ABSTRACT

Siliceous earth is identified as a suitable adsorbent material for solar energy storage. This material naturally occurs in Western Rajasthan, India. It is of very low cost and is available in plenty. The Siliceous earth can store 307.54 KJ, 345.16 KJ, 424.07 KJ and 361.0 KJ of energy per kg at temperatures of 100°C, 150°C. 200°C and 250°C respectively. The energy stored by Siliceous earth increases as temperature increases upto 200°C and than decreases. There may be minerological changes in the siliceous earth at more than 200°C. Therefore, the Siliceous earth can be used to store energy upto 200°C that is mostly the required temperature for space heating and cooling, solar refrigeration, solar cooking and process heat.

ey words : Siliceous Earth, Solar Energy Storage, Western Rajasthan.

INTRODUCTION

Different types of solar energy appliances have been developed and field test ıring the last hundred years all over the world. These devices are used for wat ating, space heating and cooling, cooking etc. These devices work very well during t ear sun, however, their performance is greatly affected due to intermittent nature lar radiation. The performance of solar cooker, particularly is most affected whenev ere are intermittent clouds, as a result, food remains half cooked and hence ge asted as it can not be recooked even by using conventional fuels. Thus, there is gent need for cheaper storage materials for solar cookers.

Various storage materials have been suggested (Telkes, 1974; Lorsch *et al.* 197 ose and Dunkle, 1977; Shegeishi *et al.* 1979; Scarmozzino *et al.*, 1980; Gopal *et a* 82; Garg *et al.*, 1985; Restuccia *et al.*, 1988; Marcos, 1990; Chauvet *et al.*, 1994) f ifferent applications for space heating and cooling to power generation. Very litt fforts have been made to find out a cheap suitable energy storage material for sol ookers. Considering this, a study was carried out to find the suitability of siliceous ear s an energy storage material. Siliceous earth is of very low cost and is easily availab 'he results discussed in this paper show that it can be used as an adsorbent material f nergy storage.

SILICEOUS EARTH

Siliceous earth can be of interest in solar energy because of its special structure, as a network of pores and cavities usually filled with water which can be removed wh eated and readsorbed when cooled.

In India a workable deposit of siliceous earth occurs in Barmer district of Rajasth etween the latitudes 26°20' to 26°30' and the longitudes 71°0' to 71°15'. This siliceo arth, yellowish white to white in colour, has low specific gravity : 0.8 to 2.1 and has ve igh 30% to 40% porosity. It fuses readily and shows inlumiscence and cation exchan roperty between Na and Ca. The Directorate of Mines and Geology, Rajasthan state h ivided the siliceous earth deposits of Barmer district into 14 blocks (DMG, 1984). T hemical composition and consequently the quality of the siliceous earth is not the sar 1 all the blocks. The SiO_2 percentage is found to vary from 71.62% to 80.68%, $Al_2($ aries from 4.34% to 9.89%, Fe_2O_3 0.63% to 3.42% CaO, 0.69% to 2.71% FeO 0.50% .22% and loss on ignition varies from 8.84% to 11.48%. Average chemical percentage 3azit' and 'Bariyada' blocks on which the present study is made as given below :

lock	Average analysis in per cent					LOI	Others
	SiO_2	Al_2O_3	Fe_2O_3	CaO	MgO		
azit	80.68	6.01	0.63	1.40	1.41	9.10	
ariyada	76.28	0.43	2.13	1.96	0.40	8.84	10.00

EXPERIMENTAL WORK

Ten samples of siliceous earth about 10 gm each were placed in a 25 ml beaker a vere weighed accurately by a top loading electronic balance with an accuracy of ± .01 n 'hese samples were dried in an infrared oven at 100°C, 150°C, 200°C and 250°C for ours so that water molecules present in the cavities were evaporated. The weight ried samples was determined to find out loss of water. These samples were put for ours in a pot having water to get them saturated again with water, gain in water amples was determined. These measurements were taken several times to find out t epeatability of the cycle. The average loss and gain of water was computed.

The total energy stored, Es, consisting of sensible heat or anhydrous material (ensible heat of water q_w and heat of adsorption qa is calculated for differe emperatures using following relation :

$$Es = q_s + q_w + q_a$$

$$= M_1 C_P T + M_2 C_P T + M_2 L \qquad \ldots (1)$$

here

Cp = specific heat of siliceous earth KJ/Kg.

Es = energy stored by siliceous earth KJ.

L = latent heat of vaporisation of water KJ/Kg.

M_1 = weight of dried sample Kg.

M_2 = weight of water Kg.

qa = heat of adsorption of siliceous earth KJ.

qs = sensible heat of siliceous earth KJ.

qw = sensible heat of water KJ.

RESULTS AND DISCUSSION

The average loss and gain in water after drying in an infrared oven and saturatin water atmosphere came out to be 7.70% and 9.06%, 7.70% and 8.66%, 8.21% an 79% and 4.76% and 4.97% at temperatures 100°C, 150°C, 200°C and 250°C respectivel

The material is to be used in the solar cookers where the operating temperature i 0°C. If an ambient temperature is 25°C then the energy stored (E_s) by 1.00 kg (iceous earth is calculated by the relation (1) :

$$E_{s100} = 0.9094 \times 1.09 \times 75 + 0.0906 \times 4.186 \times 75 + 0.0906 \times 2260$$
$$= 307.54 \text{ KJ}$$

The energy stored at 150°C (E_{s150}) is :

$$E_{s150} = 0.9143 \times 1.09 \times 125 + 0.0857 \times 4.186 \times 75 + 0.0857 \times 2260$$
$$= 346.16 \text{ KJ}$$

The energy stored at 200°C (E_{s200}) is :

$$E_{s200} = 0.9021 \times 1.09 \times 175 + 0.0979 \times 4.186 \times 75 + 0.0979 \times 2260$$
$$= 424.07 \text{ KJ}$$

The energy stored at 250°C (E_{s250}) is :

$$E_{s250} = 0.9503 \times 1.09 \times 225 + 0.0497 \times 4.186 \times 75 + 0.0497 \times 2260$$
$$= 361.0 \text{ KJ}$$

1.0 kg of siliceous earth can store 307.54 KJ of energy as compared to 193.00 KJ b olite (Scarmozzino *et al.* 1980). The specific gravity of the material is 0.8 to 2.: ierefore, this material can successfully be used in the solar cookers and in sola frigeration as an energy storage material in place of zeolite which is much mor pensive as compared to siliceous earth.

The siliceous earth can store 345.16 KJ, 424.07 KJ and 361.0 KJ of energy per kg a mperature of 150°C, 200°C and 250°C respectively. The energy stored by siliceous eart creases as temperature increases upto 200°C. However, at still higher temperature i creases which could be due to mineralogical changes.

CONCLUSION

The siliceous earth can be used as a suitable adsorbent material for solar energ orage. It can be used in place of zeolite which is very much expensive. It is found tha liceous earth can store energy at the temperature of 100°C, 150°C, 200°C and 250°(spectively. Therefore, the siliceous earth can be used to store energy upto 200°C that i ostly the required temperature for space heating and cooling, solar refrigeration, sola oking and process heat.

ACKNOWLEDGEMENTS

The authors are grateful to the Director, CAZRI, Jodhpur, the Head, Division V ıd Department of Geology, JNV University, for providing necessary facilities an constant encouragement for the present study. Help rendered by Mr. Harsh Vardha Ianager, RSMDC) and financial support to MSS by Shell Indian Productio evelopment BV is gratefully acknowledged.

REFERENCES

ıauvet, L.P., Probert, S.D. and Nevrala, D.J. 1994. Thermal Energy storage for supplyin domestic hot water and space heating, *Applied Energy*, 48 : 163-190.

lose, D.J. and Dunkle, R.V. 1977. Use of adsorbent beds for energy storage in drying and heatin systems. *Solar Energy*, 19 : 233.

epartment of Mines and Geology, Government of Rajasthan, 1984. D.M.G. *Technical Repo* (*Unpublished*) Udaipur, India.

arg, H.P., Mullick, S.C. and Bhargava, A.K. 1985. *Solar Thermal Energy Storage*, D. Reide Publishing company, Dordrecht.

opal, R., Hollebone, B.R., Langford, C.H. and Shigeishi, R.A. 1982. The rates of solar energ storage and retrieval in a zeolite water system. *Solar Energy*, 28 : 421-424.

ɔrsch, H.G., Kauffman, K.W. and Denton, J.C. 1975. Thermal energy storage for solar heatin and offpeak airconditioning. *Energy Conversion*, 15 : 1-8.

arcos, V.H. 1990. Investigation of a latent heat thermal energy storage systems. *Solar and Win Technology* 7 : 197-202.

estuccia, G., Cacciola, G. and Quagliata, R. 1988. Identification of zeolites for heat transforme chemical heat pump and cooling system. *International J. Energy Research* 12 : 101-111.

armozzino, R., Aiello, R. and Santucci, A. 1980. Chabazitic tuff for thermal storage. *Sola Energy* 24 : 415-416.

ıigeishi, R.A., Langford, C.H. and Hollebone, B.R. 1979. Solar energy storage using chemica potential changes associated with drying of zeolites. *Solar Energy*, 23 : 489-495.

elkes, M. 1974. Storage of solar heating/cooling. *ASHRAE Journal*, 16(9) : 34.

əological Evolution of Northwestern India
S. Paliwal (Ed.)
ientific Publishers (India), Jodhpur, 1999, pp. 210 - 215.

INDUSTRIAL GRADE LIMESTONE DEPOSITS OF TERTIARY PERIOD IN WESTERN RAJASTHAN

P.C. Bafna and B.S. Dhaka
Department of Mines and Geology, Jodhpur - 342 001

ABSTRACT

The industrial grade limestone of the Tertiary age, suitable for cement and chemical industry, is restricted to Khuiala Formation, exposed around Sam, Khuiala, Ramgarh, Hingola, Mohammad ki Dhani, Sanu, Tulsi Ram ki Dhani, Khinya, Khivsar and Mandha villages of Jaisalmer district. The State Department of Mines and Geology, Rajasthan, has carried out detailed exploration work for SMS grade, chemical grade and cement grade limestones in the entire Khuiala Formation. A total of about 20,000 m. of drilling done in this formation covering an area of 200 sq km indicated reserves of about 1000 million tonnes of SMS grade, 950 million tonnes of chemical grade and 2700 million tonnes of cement grade limestone.

ey words : Tertiary, Industrial Grade Limestone, Khuiala Formation, Jailsalme Rajasthan.

INTRODUCTION

During the Tertiary period different geological formations were deposited in th orthwestern part of Rajasthan (Pareek, 1981, 1984, 1977; Paliwal, 1993). The majo ck formations include shales (gypsious, carbonaceous, fossiliferous and silty shales nd stones, bentonite, fuller's earth, clays (ball clay, fire clay and china clay), lignite an mestone. The Tertiary basin situated in the west and north west of the Aravalli rang tends upto main Indus valley. The basin is a store house of several economic minerals.

GEOLOGY OF THE AREA

The limestones of the Tertiary age are restricted to the Sanu, Khuiala and Bandh rmations exposed in parts of Jaisalmer district (Lukose, 1977). The Sanu Formatic ainly comprises of sandstone and ocherous beds, best exposed near Mohammad-k ani. Sub surface limestone occurrences have been encountered at different horizor ring deep drilling for oil and described as Kharatar Marl Member. The Khuia rmation and Bandha Formation (Laul, 1980) are best developed in the Jaisalm strict around Sam, Khuiala, Bandha, Sanu and extend upto Khinya, Khinsar ar andha through Tulsi Ram-ki-Dhani in a semiarcuate manner (Fig. 1). Khuiala Fo ation includes sub bentonitic clays at the base, overlain by the foraminiferal limestor whitish and yellowish colour which is magnesian at places like Unrai and adjace eas. Overlain by the chalky limestone with Khaki shale intercalations and fossil shell ese beds are further overlain by the compact fossiliferous and/or non-fossiliferou alky limestone (Lukose, 1977) which is of cement and chemical grade. This limestone erlain by the creamish white to buff coloured hard compact, microcrystalline limeston ich is partly bouldery and partly bedded in nature. The foraminiferal limestor sting over this limestone, further west near Bandha village belongs to the Bandh rmation of Middle Eocene age. The pseudoconglomerate belonging to the Shum rmation of Pleistocene age is exposed 4 kms north of the Khuiala village (Table 1).

ble 1. Tertiary Stratigraphy of the Jaisalmer Basin. (Laul, 1980)

cent	—		Desert sand and sandstones.
b recent to eistocene	—	Shumar Formation	Mainly gravels with ferruginous nodules cemented, places limestone cobbles forming pseudoconglomerat occasionally coarse grained ferruginous sandstones.
			-Unconformity-
iddle Eocene	—	Bandha Formation	*Bakhri Tibba limestone member* Greyish white hard foraminiferal limestones. *Batrewala limestone member* greyish white to pinkish whi hard, compact, foraminiferal limestone becomir argillaceous at base.
			--Para Unconformity--
			Khinsar shale member Gypseous, ocherous, yellow shales with whitish yello argillaceous, thin limestone beds, calcareous silts ar foraminiferal test beds.
			Tetakkar limestone member
rly Eocene	—	Khuiala Formation	Creamish white to buff, hard compact, microcrystalli limestone in the upper part and alternations of yello shales and argillaceous limestone/calcareous claystone the lower part.
			--Unconformity--
			Mohammad ki Dhani sandstone member
leocene	—	Sanu Formation	Quartzose yellow to maroon, ferruginous, friable cro bedded sandstone with minor purplish white clays
			--Unconformity--
wer Cretaceous	—	Habur Formation.	

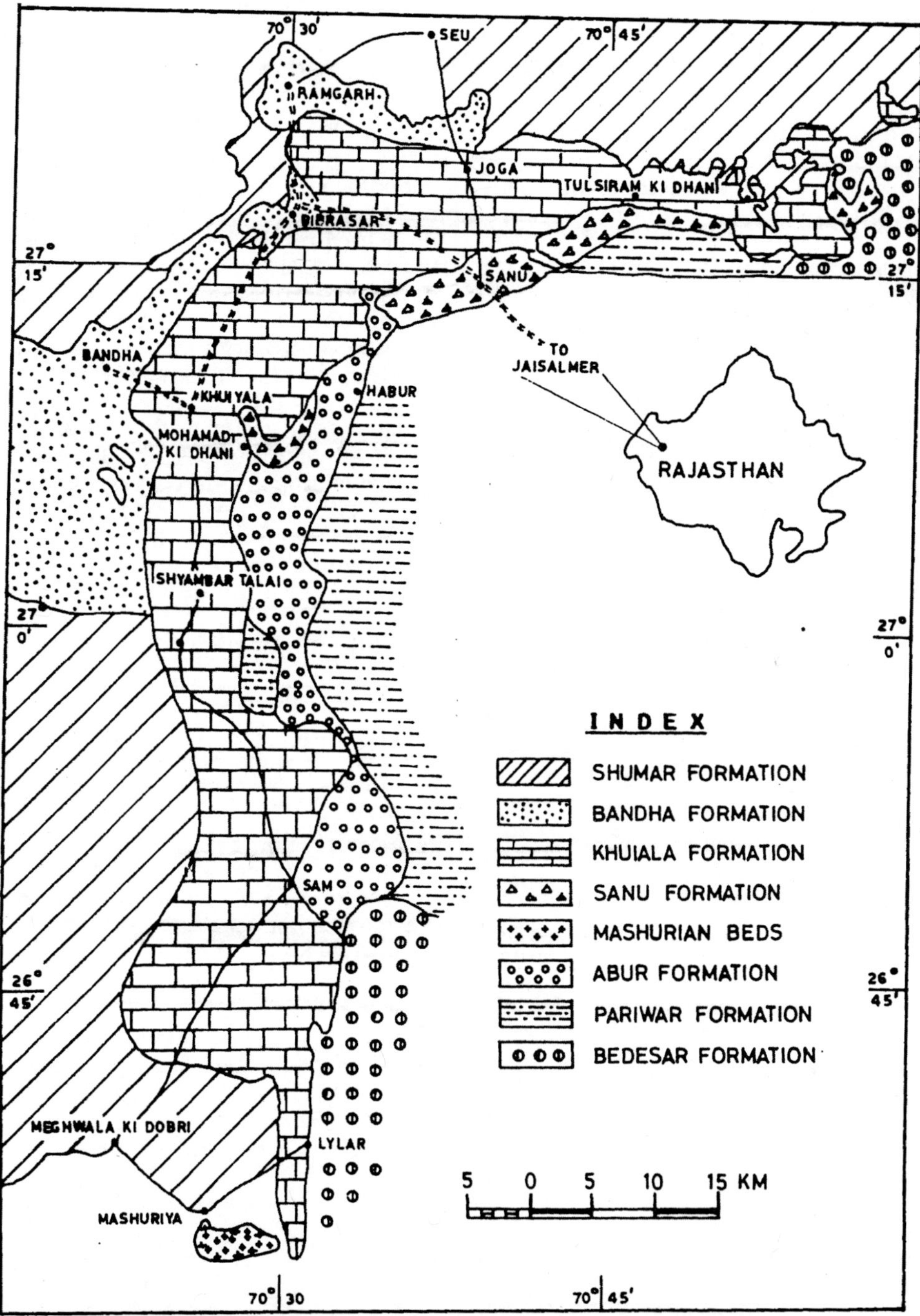

Fig. 1. Geological map of Sam-Khuiala-Ramgarh area of Jaisalmer district, Western Rajasthan

PROSPECTING FOR LIMESTONE IN THE AREA

The workable industrial grade limestone deposits are located in the Khuia ırmation of the Tertiary period was initially located by the Geological Survey of Ind aul, 1980). Department of Mines and Geology, Rajasthan, initiated prospecting wo: chemical grade limestone in parts of the Jaisalmer district during 1980-81 ar rried out mineral survey by scanning the area, mapping and finally by carrying o illing at 1 km grid interval during 1983-86.

As per the results of prospecting the attention of the State and Central agencies g verted to the limestone deposits of Jaisalmer because the country was facing scarcity v silica limestone after sudden closure of the Dehradun-Mussoorie limestone belt l e orders of the Supreme Court of India. In the year 1986 a team of experts visite nu-Khuiala area of Jaisalmer district and as per the discussion of the committ ospecting for exploration of S.M.S. grade limestone was augmented in the Sanu are orking permission in an area of 10 sq km was also granted to R.S.M.D.C. Ltd. to explc e S.M.S. grade limestone in Sanu-Ramgarh area and supply it to the SAIL through tl vernment of Rajasthan under Major Mineral Concession Rules.

EXTENT OF THE KHUIALA LIMESTONE

The Khuiala Formation of the Lower Eocene age (Laul, 1980) extends from about 3 n south of Sam i.e. south east of Mashuriyan to Khuiala, Unrai, Hingole and Biprasa a strike length of 75 km (Fig. 1). It takes a turn to east and extends from Bipasar ga, Tulsi Ram-ki-Dhani, Khinya, Khinsar, Rata, Mandha and further east for a stril ngth of 45 km. The width of this formation ranges from 2.5 km to 17 km, averaging) km and covers an area of about 800 sq km. Out of which an area of about 400 to 50 km contains S.M.S. grade limestone which is associated with the cement grade ar emical grade limestone. The remaining area comprises mainly chalky limestone ment and chemical grade. General trend of the formation between Sam Dhani ar prasar is N-S which swings to E-W between Biprasar and Mandha. The beds a nerally horizontal or have gentle dips of 4° to 5° due north. The thickness of tl nestone varies from 4 m to more than 50 m. Out of which S.M.S. grade limestone has ickness of 4 m to 16 m and the remaining part is either of cement grade or chemic ade. Thus a large deposit of the industrial grade limestone is available in the Khuia rmation.

RESERVES OF THE LIMESTONE

The State Department of Mines and Geology, Rajasthan took up further exploratic rk for the S.M.S., chemical and cement grade limestone since 1987 in a systemat y which is still under progress. A total of about 20,000 m of drilling has already bee rried out in the Khuiala Formation covering an area of about 200 sq km. Total 10(llion tonnes of SMS grade limestone, 950 million tonnes of chemical grade limesto d 2700 million tonnes of cement grade limestone have been estimated in the area.

Though potential areas have been prospected but a large tract is still le nexplored. 10 sq km area near village Sanu, district Jaisalmer has been leased out t ıe RSMM for exploitation of the S.M.S. grade limestone. A 7.2 sq km area has bee ·ased out to M/s. Mehar Cement, near Miyonki-Dhani, south of Ramgarh to install ıajor cement plant of 1.5 million tonnes cement per annum capacity on reserves of moı ıan 360 million tonnes.

LIMESTONE QUALITY

Quality wise the hard, compact, massive, bouldery and/or bedded microcrystallir mestone is generally termed as the S.M.S. grade limestone. The hard compact an ıassive S.M.S. grade limestone generally has SiO_2 less than 1.80% and CaO more tha 2%. It shows variation in SiO_2 from 0.50% to 5.5% and CaO 49.5% to 55%, oth onstituents are less than 1%. Boulders of S.M.S. grade limestone have CaO from 32% t 8%, Silica 15% to 22% and other ingradients in the permissible limits for its use i ement industry.

The chalky limestone is fossiliferous whitish, yellowish and creamish in colo aving CaO 42% to 53% (averaging 47%) and SiO_2 2% to 14% and more.

At some of the horizons, it becomes foraminiferal limestone. The yellow coloure raminiferal limestone has a higher magnesia and can only be utilized after blending.

able 2. Prospects of Mineral based industries on Tertiary limestone :

. Hard compact S.M.S. grade limestone (SiO_2 less than 1.8% and CaO more than 52%)	—	Already being supplied to Steel Plants in required size.
. Chemical grade limestone (SiO_2 2% to 4%, CaO more than 50%)	—	Huge fines generated during sizing for S.M.S. grad limestone can be utilized for producing precipitated CaCC and hydrated lime, quick lime, (upto certain size only i modern kilns).
	—	Other chemical grade limestone i.e. hard compact or chalk limestone can be utilized in producing precipiated CaCO soda ash, bleaching powder, hydrated lime, quick lime etc.
. Cement grade limestone (SiO_2 upto 13% and CaO about 45%)		Huge fines and intraburden produced during mining an sizing of S.M.S. grade limestone can directly be utilized fc cement manufacturing by proper blending. Besides the har compact cement grade limestone, chalky limestone, can b utilized for cement manufacturing. One major cement plaı is under construction near Ramgarh and five other areas aı readily available for leasing.

LIMESTONE MINING

The mining of S.M.S. grade limestone is continuing since 1986 and a huge quantit f fines (limestone below 25 mm size) have generated and accumulated on the mine site uge dumps of intra-burden with CaO 32% to 40% have also been accumulated, whi arrant immediate set-up of the cement plants. These Tertiary limestone reserv ıould be utilized immediately by preparing value added products which can be sc .digeneously and in the foreign market too. Huge deposits of S.M.S. grade limesto ·quire further research and development work for its utilization in other industries.

CONCLUSION

Tertiary formations of Jaisalmer host, a huge quantity of industrial grade limesto hich includes S.M.S. grade, cement grade and chemical grade limestones. Though the ertiary limestone deposits are located in a remote corner of the country but n .frastructural facilities are available in form of broadgauge railway line upto Jaisalm ıd water from the I.G.N.P. Of course power constraint is still there which can be solv ırough captive power plants based on lignite, gas, solar energy or wind energy.

ACKNOWLEDGMENT

Authors are thankful to Shri D.S. Mehta, Additional Director, Department of Mir ıd Geology, Jodhpur, for the help and to Prof. B.S. Paliwal of Jai Narayan Vy niversity, Jodhpur for examining the manuscript critically and offering valual ıggestions to improve it.

REFERENCES

aul, V.P. 1980. Prospects of industrial grade limestone in Khuiala Formation (Lower Eocen Jaisalmer district, Rajasthan. *Ind. Mins.* 33(4).

ıkose, N.G. 1977. Palynological evidence on climatic changes in Jaisalmer basin, Rajasth *Desertification and it's control. ICAR Chapter,* **5** : 31-41.

aliwal, B.S. 1993. Geology of Rajasthan - a summarium. In : Chauhan, T.S. (Ed.). *Natural a Human Resources of Rajasthan*, Scientific Publishers, Jodhpur, India. 1-22.

areek, H.S. 1977. The Permian - Mesozoic - Palaeogene Palaeogeography of the Rajasthan a Gujrat Shelf and correlation with that of the Indus basin. *IV International Gondwa Symposium*, Hindustan Publishing Corporation, New Delhi, India.

areek, H.S. 1981. Basin configuration and sedimentary stratigraphy of western Rajasthan. *Jo Geol. Soc. India* **22(11)** : 517-527.

areek, H.S. 1984. Pre-Quaternary Geology and Mineral Resources of Northwestern Rajasth *Mem. Geol. Surv. India,* 115 : 1-95.

ological Evolution of Northwestern India
S. Paliwal (Ed.)
ientific Publishers (India), Jodhpur, 1999, pp. 216 - 224.

A NOTE ON URANIUM MINERALISATION IN THE PERMIAN SANDSTONE OF BAP-BADHAURA AREA, JODHPUR DISTRICT, RAJASTHAN

G. S. Yadav, T. S. Miglani, T. Benarji Babu and B. M. Swarnkar*

Atomic Minerals Division, Uniara Garden, J.L. N. Marg, Jaipur-302004
**PO. Assam Rifles, Lalchand Basti, Shillong - 793011*

ABSTRACT

Anomalous concentration of uranium has been recorded for the first time in the Permian sandstone exposed at Badhaura, Bap and Bhimji Ka Gaon, areas of the Jodhpur district in Western Rajasthan. Badhaura sediments comprise yellowish brown, ferruginous sandstone, siltstone, claystone and shale with mollusca shells and plant fossils indicating marginal marine conditions of deposition. Uranium mineralization is confined to the hard and compact, dark brown, ferruginious, calcareous and phosphatic layers (10-70 cm thick) interbedded with fine grained, friable sandstone in the upper part of the formation. Dimensions of the radioactive zones vary from 10m x 5m to 500m x 100m. Grab samples have radiometrically assayed 0.012 to 0.06% U308, 0.016 to 0.077% U308 and <0.005% ThO_2. Low density alpha tracks on cellulose nitrate (CN) film correspond to ferruginous and phosphatic material. Uranium mineralization in the Badhaura sandtone is attributable to gradual build up of higher uranium concentration from the Malani Igneous Suite (MIS) of rocks. The Badhaura sandstone rests on Bap Boulder Bed (BBB) which has derived its unconsolidated and unsorted clasts mainly from MIS with minor contributions from rocks of Aravalli, Delhi and Marwar Supergroups. MIS is known to contain anomalously high uranium content. Four samples of granite occurring in Bap Boulder Bed have been analysed and found to contain from 13 to 39 ppm uranium. Uranium was brought into the host rock either as detritus or in solution and got fixed with ferruginous and phosphatic materials in the adsorbed state.

ey words : Uranium Mineralization, Bap-Badhaura, Permian Sandstone, Jodhp District, Rajasthan.

INTRODUCTION

The Permian non-marine sandstones are known to host mineable uraniu currences in France, ltaly, Yugoslavia and Hungary (Barthel, 1974). In western part e Nagaur basin outcrops of marine sediments were first reported by Misra *et al.,* 196 ;ar Badhaura containing the Permian fauna and flora.

Radiometric survey of the Bap Boulder Bed and Badhaura Sandstone was taken u ' the Atomic Minerals Division during the field session 1993-94. As a result of th ırvey, significant uranium anomalies were located (Fig. 1) for the first time around Ba 7^{0} 22'10"N; 72^{0} 20'23"E), Badhaura (27^{0} 20' 16 "N; 72^{0} 18'39"E) and Bhimji Ka Gao 7^{0}22'37"N; 72^{0} 16'33"E) in upper part of the Badhaura Sandstone in Jodhpur distric ajasthan (Babu and Yadav, 1994).

The aim of the present paper is to provide a brief account of these uraniun currences i. e. nature of uranium mineralization, provenance and its implications o rther uranium exploration in the Permian sediments of this area.

GEOLOGY OF THE AREA

The study area is mainly a peneplain country with scanty outcrops lying in th uthern part. It falls in the semiarid zone with scanty rainfall and internal drainag ie stratigraphic sequence (Table 1) is given below of the area.

ıblc 1. Stratigraphic Succession of the Bap-Badhaura area (After Pareek, 1984)

Lower Jurassic	:	Lathi Sandstone	
Permian	:	Badhaura Sandstone	
Upper Carboniferous	:	Bap Boulder Bed	
Cambrian	:	Marwar Supergroup	Nagaur Group Bilara Group Jodhpur Group
		Unconformity	
Precambrian	:	Malani Igneous Suite of rocks, Delhi Supergroup and pre-Delhi rocks. (Not exposed in the study area)	

Malani Igneous Suite of rocks occupy an area of 51000 sq. km. (Pareek, 1981) thes fusive and intrusive rocks alongwith pre-Delhi and Delhi metamorphics form th ısement for the Marwar Supergroup. The sedimentary sequence of the Marwa ıpergroup is divided into arenaceous Jodhpur Group, calcareous Bilara Group an enaceous-argillaceous Nagaur Group. Stromatolite bearing cherty dolomite of th lara Group is the oldest unit exposed in the study area. It forms a marker horizon an cupies the hill in the southern part. This unit probably contributed phosphate to th ıdhaura sandstone.

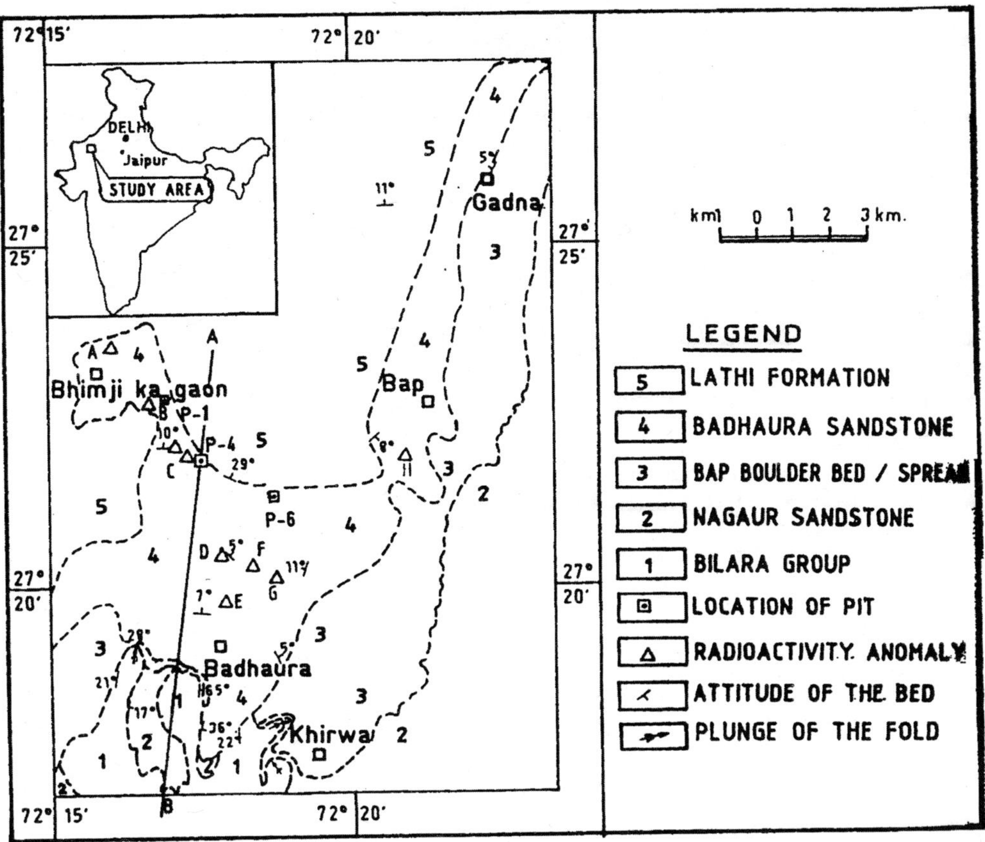

Fig. 1. Geological Map of Bap-Badhaura area showing uranium anomalies, District Jodhpur, Rajasthan.

Marwar Supergroup is overlain by Bap Boulder Bed of upper the Carboniferous ag 'he boulders extend from Khirwa as a linear belt for a length of 50 km on either sides [ational Highway No. 15. Another patch comprises cobbles, pebbles and boulders ıffs, rhyolite, granite, basalt, dolerite, quartzite, phyllite, slate, cherty dolomi egmatite etc, These clasts are embedded in the clayey matrix. Weathering and leachi f clay has given rise to boulder spread. It is overlain by the Badhaura Sandstone (Fig.2

Badhaura Sandstone is exposed as discontinuous outcrops around Badhau: himji Ka Gaon, Bap and Gadna covering an area of about 55 sq. km. It includ ıedium to coarse grained yellowish brown ferruginous sandstone, clay, shale, a: iltstone containing shells of lamellibranchs, brachiopods, cephalopods and gastropo long with plant fossils. It has a computed thickness of 30m (Pareek, 1984). The low art of the sandstone immediately overlying boulder bed is coarse grained gritty a: ɔntains cobbles and pebbles of granite, metamorphic rocks and cherty dolomite of Bila froup. Contact between the boulder bed and the sandstone is generally concealed a: ad been observed only at a few places like south of Badhaura, east of Magrasar tal nd north of Gadna village. Sandstone shows low basinward dips of 5 to 10° at all the ɔcations.

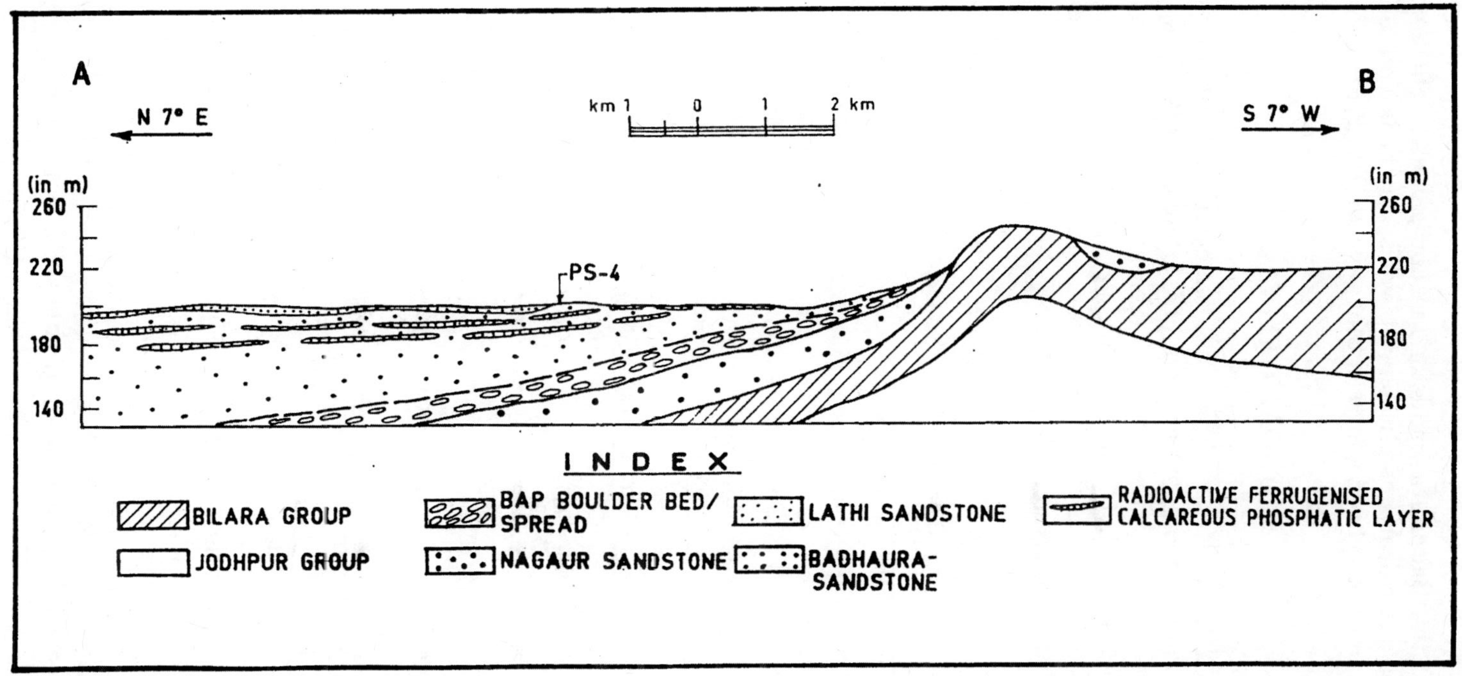

Fig. 2. Generalized Geological section through line AB, shown in Fig. 1, Bap-Badhaura area, District Jodhpur, Rajasthan.

The upper part of the Badhaura Sandstone is fine grained yellow coloured wi ıterbedded ferruginous and calcareous phosphatic layers, clay beds containing mollus nd plant fossils. One sample has been analysed that gave 18.85% P_2O_5.

RADIOACTIVITY

Radioactivity due to the presence of uranium has been located for the first time ıe upper part of the Badhaura sandstone north of Badhaura, west of Bap and arou hhimji Ka Gaon. 14 Grab samples collected from different locations have radiometrica ssayed 0.012 to 0.060%, U_3O_8, 0.007 to 0.077% U_3O_8 and 0.005% ThO_2. The chemic alues vary from 0.014 to 0.065% U_3O_8 (T). Results of samples alongwith dimensions ıe uraniferous occurrences are given in Table-2.

'able 2. Radiometric and Chemical Result of Samples of Badhaura Sandstone

ładioactivity Anomaly	Dimension (m x m)	Radiometric % eU_3O_8	Assay %U_3O_8	Results %ThO_2	Chemical Ass results %U_3O
A	10 x 5	0.014	0.016	< 0.005	—
B	20 x 10	0.060	0.077	< 0.005	—
C	500 x 100	0.019	0.024	< 0.005	0.022
		0.012	0.016	< 0.005	0.014
		0.013	0.007	< 0.005	0.015
		0.023	0.042	< 0.005	0.026
		0.052	0.077	< 0.005	0.065
		0.036	0.055	< 0.005	0.041
		0.032	0.049	< 0.005	0.033
D	50 x 10	0.031	0.038	< 0.005	—
E	50 x 10	0.028	0.041	< 0.005	—
F	300 x 12	0.022	0.029	< 0.005	—
G	50 x 20	0.016	0.019	< 0.005	—
H	Well dump	0.049	0.075	< 0.005	—

Uranium mineralization is confined to top 10 to 35 cm thickness of the sandstone vell as in the calcareous phosphatic lenses (upto 70 cm. thick) where ferruginisation ery prominent. Few pits were dug to know the behaviour of uranium mineralization epth. Channel samples collected from the pits P 1 (1.10 m depth) P- 4 (6. 05m dept nd P-6 2.0m depth) on analysis have given 86 ppm, 72 ppm and 80 ppm uraniu espectively. A section of the pit made at pumping station No. 4 (P5-4) of Jodhpur L 'anal is shown in Fig. 3. In the pit section 5. 25 m. thick yellow colourd fine grain riable sandstone is sandwiched between two clay bands. This sandstone contains tı erruginous and calcarious phosphatic layers with concentric laminations and pla ossils. The major oxides weight percentage of friable sandstone and phosphatic layers iven in Table 3.

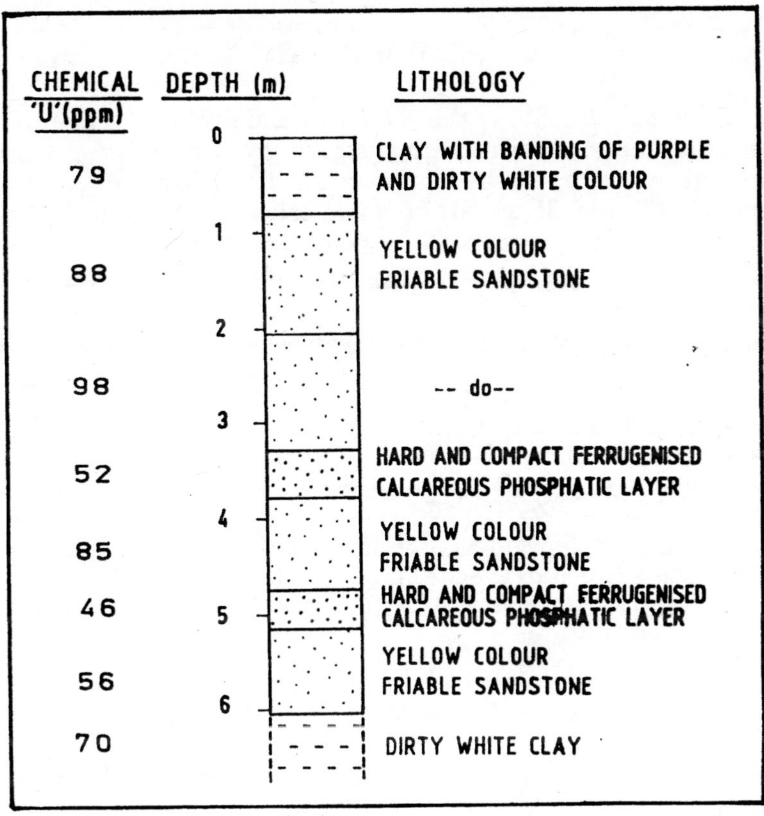

Fig. 3. Litholog of PIT-4 showing uranium values in the Bap-Badhaura area, District Jodhpur, Rajasthan.

Table 3. Major Oxides in samples of Badhaura Sandstone

Oxides	Friable sandstone	Friable sandstone	Phosphatic layers
SiO_2	44.18	47.06	9.07
Al_2O_3	14.32	12.28	4.57
CaO	6.75	6.94	34.51
MgO	1.09	1.01	0.6
Na_2O	1.68	1.68	1.01
K_2O	1.81	1.66	0.6
FeO	0.22	0.11	0.08
Fe_2O_3	14.96	13.71	13.53
MnO	0.17	0.25	0.23
TiO_2	1.61	1.98	0.54
P_2O_5	1.27	0.68	18.85
LOI	11.58	11.67	16.27

PETROGRAPHY

The framework constituents of the Badhaura Sandstone are mainly fine fragmen ıd tend to inequigranular texture with medium grain size. They are sub-angular ıb-rounded in nature and are set in a carbonate cement. Grain to grain contacts are n 'esent. Smearing of iron-oxide imparts reddish colour to the sandstone. The sandstoı rather immature. The framework constituents are quartz which is monocrystallin lspars mainly plagioclase with a few grains of K-felspar and biotite is the main mic eavy minerals include garnet, zircon and opaques.

In radioactive samples, grain size of the framework minerals tend to be equant. Tl 'ains are sub- angular to sub-rounded, loosely packed and some of them appear to])ating in the cement. It is cemented by large grains of calcite. The mineral constituen 'e quartz, felspar, apatite and biotite. Iron oxide smears the rock and it is found to] :h in some laminations. Ferruginised fossil remains are partially replaced by calcite. .e radioactive sample, low density alpha tracks on CN - film correspond to ferruginis ating and uranium appears to be in adsorbed state.

No discrete uranium mineral has been identified by X-ray diffraction technique. Tl her minerals identified include calcite, fluorapatite, goethite, quartz, pyrite, ilmenit ıtile, albite and microcline. Uranium is possibly associated with goethite aı ıorapatite in adsorbed state.

DISCUSSIONS AND CONCLUSIONS

Bap boulder bed is overlain by Badhaura Sandstone and has derived i ıconsolidated and unsorted clasts mainly from Malani Igneous Suite with min ntribution from rocks of Aravalli, Delhi and Marwar Supergroups. Malani Igneoı ıite comprises acidic rocks like granite, rhyolite ignimbrite, tuffs, welded tuffs, aı ;glomerates. All these units contain anomalous uranium content with good leachabili 'able 4).

ıble 4. Uranium content in the rocks of Malani Igneous Suite

No.	Area	Rock type	U (ppm)		Average %
			Average	Range	Leachability
	Malani Igneous Province	Granite (n = 73)	9.2	3-20	74.84 (n=13)
		Rhyolite (n = 31)	6.7	1-17	53.91 (n=7)
	Esrana Hills	Aplitic Venis (n = 13)	42	7-121	—
		Granite (n = 12)	29	5-61	—
		Rhyolite (n = 14)	27	4-61	—
		Tuff	19	4-49	—
	Siwana Fort hill section	Cong./Grit/Shale (n = 9)	51	20-93	—

The fact is also supported by the presence of anomalous uranium content of 52 pı ı an average (n 1493) in the water samples drawn from vast terrain of Malani Igneoı ıite in parts of Jodhpur, Barmer, Jalore, Pali and Sirohi districts of Western Rajasthε ıorikhan, 1988). 4 Samples of granite from the Bap Boulder Bed have shown 13°39 pp ·anium on analysis.

The Permian period was marked by marine transgression and arenaceous faci ·evailed in northwestern Rajasthan leading to formation of Badhaura Sandstoı 'areek, 1984). The loosely held igneous rich detritus of boulder bed in close proximity .e basin has supplied uranium in solution as well as with the detritus. This uraniu ight have fixed syngenetically in goethite-fluorapatite rich layers in adsorbed state aı so possibly in the lattice of fluorapatite as replacement of calcium. Presence of pyri ight have also played a role in reducing uranium. Further enrichment of uranium l fferential weathering and leaching processes has produced higher concentration in tl p layers of sandstone.

The Permian sediments of fluviatile origin have subsurface thickness of 255 m ıumrawali Talai (27°11' N; 70°17'E) borehole. The subsurface extension of the: ·diments in northern Rajasthan is not ruled out. There is a possibility of getting bett ineralization in fluviatile sediments. Thus the Premian fluviatile sediments may pro be a potential horizon and a future target for the uranium exploration in this area.

ACKNOWLEDGEMENTS

The authors express their gratitude to Director Atomic Minerals Division f ·anting permission to present this paper and also to Shri Rajendra Singh, Associa irector for guidance and encouragement. Thanks are also due to our laborato: ·lleagues for providing analytical support.

REFERENCES

ıbu T.B. and Yadav, G.S. 1994. Annual Report for the Field Session. 1993-94. *Unpublish Report, Atomic Minerals Division, Jaipur*, India. 1-17.

ırthel, F.H. 1974. Review of Uranium in Permian sedimentaries in Europe with special referen to Uranium mineralization in Permian sandstone, Procedings of Symposium on "*Formati of Uranium Deposits*" held at Athenes, May, 1974. 277-89.

onald L. Everhart 1983. *Uranium in Phosphates*", in Shanks Wyne C. (Ed.) *Cameron Volume Unconventional mineral deposits*, 61-67

orikhan, R. A. 1988. Report on the Status of Investigations in parts of Western Rajasthan f locating calcrete type of Uranium mineralisation, *Unpublished Report*, *Atomic Miner Division,* New Delhi, India, 1–12.

ıin, R.B. Hingorany, S.P. and Singh, J. 1995. Uranium and thorium mineralization in Lat sandstone of Western Rajasthan and its genetic relationship with Malani Igneous Sui India. Under publication in the EARFAM, *Atomic Mineral Division*, Hyderabad, 1-10

isra, J.S., Shrivastava, B.P. and Jain, S.K. 1962. Discovery of marine Permo-Carboniferous Western Rajasthan, *Curr. Sci* 30 : 261-262.

areek, H.S. 1981. Petrochemistry and petrogenesis of the Malani Igneous Suit,India,*Geol. S*
Amr. Bull., 92 (2) : 206-273.

areek, H. S. 1984. Pre-Quarternary Geology and Mineral Resources of Northwestern Rajastha
Mem. Geol. Surv. India. 1-99.

adav, G.S. and Jain R.B. 1995. Annual Report for the Field Session. 1994- 95, *Unpublish*
Report. Atomic Mineral Division, Jaipur India. 1-17.

eological Evolution of Northwestern India
.S. Paliwal (Ed.)
cientific Publishers (India), Jodhpur, 1999, pp. 225 - 231.

SOME MAJOR QUATERNARY EVENTS IN JAISALMER AREA, WESTERN RAJASTHAN

V.P. Laul and Virendra Kumar
Geological Survey of India, Western Region, Jaipur – 302 004

ABSTRACT

Jaisalmer area of western Rajasthan witnessed some major Quaternary events which include iron stone-pseudo-conglomerate, sandstone clay sequence of Shumar Formation of Pleistocene age which was chronologically followed by glaciation as evidenced by erratic boulders, faceted and striated pebbles of wide ranging rocks from Precambrian to Lower Tertiary in age. The pebble spread of heterogeneous composition includes quartz of various colours, quartzite, ironstone, sandstone, dolomite, rhyolite, granite, fossil wood, basic rock, phyllite etc. which mainly occur over Mesozoic-Tertiary rock formations. Their occurrence over Shumar Formation suggests a post Shumar age. The succeeding period witnessed humid climate as indicated by the presence of glacio-fluvial and fluvial deposits and palaeochannels. During this period, the area was drained by mighty river system of Saraswati. Evidences indicate its westerly flow through Kharariwala, Kanod, Khara and Mitha Rann and then taking a southward turn in Mitha Rann near Rupsi. Later on the Vedic Saraswati river dried up. This is attributed to tectonic adjustment and upheaval in Himalayas and development of subsurface ridges like those of Delhi-Haridwar and Delhi Sargoda. This event was followed by desertification due to wide spread aeolian activity and rise in aridity in the region.

ey words : Quaternary Events, Shumar Formation, Jaisalmer, Western Rajasthan.

INTRODUCTION

This paper is aimed to describe some of the major Quaternary events of Jaisalm rea, Western Rajasthan (Fig. 1). Quaternary formations represent the last chapter eological history. This time period is very important. It helps us to study the differe rocesses operating and changing faces of lands, formation of desert, drainage syste

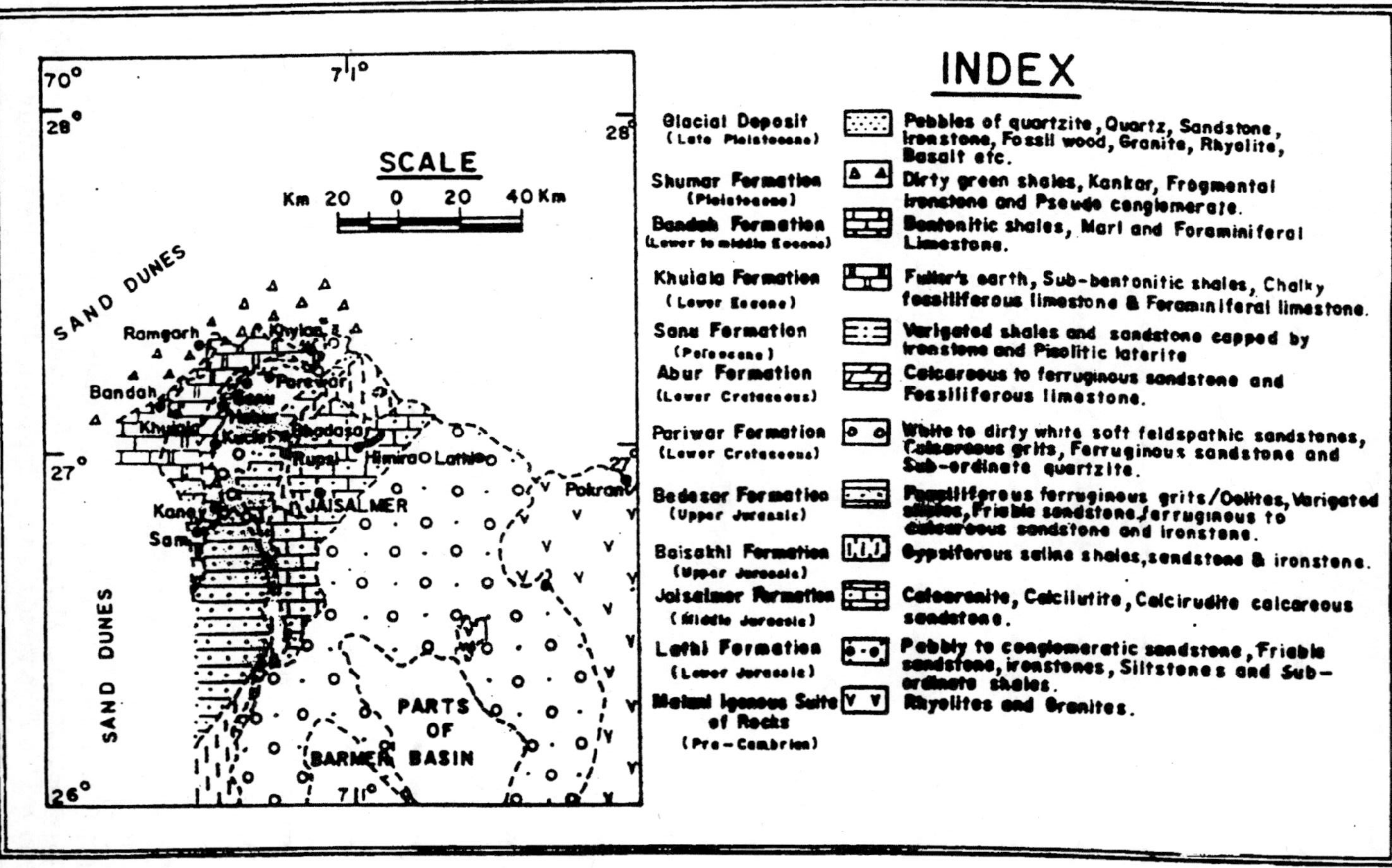

Fig. 1. Geological sketch map of Jaisalmer Basin, Rajastahn

nd their disappearance, human culture and civilisation growth centres, shifti opulations, nature of occupations and many other aspects. Studies also help to loc: ıineral and water resources. Such studies by experts of various disciplines are nmense value for planning and developmental schemes.

Some of the earlier workers who have made contributions in their respective fie ıclude Oldham (1893), Ghosh (1952), Singh (1971), Laul and Kumar (1980), Allchin a Ŧoudie (1974), Dasgupta (1975), Mukhopadhyay and Ghosh (1976), Lukose (197 .grawal *et al.* (1980) Roy (1982), Verma and Ray (1982), Mathur et al. (1982), Pare 1984), Kar (1987-92), Bakliwal and Grover (1988), Wadhawan (1988, 1994), a :ajaguru *et al* (1996), Kar and Ghosh (1988).

QUATERNARY SEQUENCE

The Quaternary beds and deposits in Jaisalmer area of Western Rajasthan maiı ıclude Shumar Formation, glacial spread, fluvial and aeolian deposits. They overı Ialani Igneous Suite of rocks and various Mesozoic Tertiary formations.

The generalised Quaternary sequence of the area is as follows :

. Ironstone pseudo-conglomerates, sandstone-siltstone-clay sequence of Shum Formation.

. Glacial pebble spread.

. Kankar, gypsite, older fluvial deposits/calcrete, abandoned and burried river chann (including Saraswati) deposits, fluvial pebble spread.

. Blown sand (dunes), sand sheet, newer fluvial deposits.

;humar Formation

Shumar Formation forms the oldest Quaternary formation in Jaisalmer area anc ; represented by an association of ironstone-sandstone-siltstone-clay sequence. Pseu onglomerates and ironstone unit is well developed in northern part of the Jaisalm rea. This formation in general overlies and overlaps Eocene limestone.

In southwestern part of the area a thick sequence of sandstone- siltstone-cl ccurring below the ironstone, pseudo-conglomerates forms the lower unit of Shum 'ormation and has been designated as "Dabri Beds". The maximum thickness of Shum 'ormation in this area varies from 20 to 60 m as intersected in dug wells and tubewell:

;lacial Pebble Spread

The succeeding event in Jaisalmer is marked by glaciation as evidenced by glac pread. Presence of glacial morains testifies the glacial activity in this area. Otl upporting evidences include the characteristic features of pebbles like heterogenec omposition, good faceting, grooves and polished surfaces. They consist of quaı uartzite, sandstone, ironstone, jasper, chert, rhyolites, basic rocks, granite, fossilwo hyllite and siliceous dolomite and have been recorded around village Bhojasar-Jasura nd Hamira in eastern part of Jaisalmer district. Such pebble spread has also be bserved on hill slopes between Jaisalmer and Damodara (Kumar and Laul, 1981).

The erratic boulders mainly of quartzite are also seen near Kanod village and o :-Tekkar-Sanu plateau.

Rounded to subrounded pebble spread has been recorded in Khiny- Danwar are: ıe glacial spread as sporadic occurrences extends from Pokaran to Kanoi to Saı curring over Mesozoic-Tertiary sequence while Pokaran to Sankara to Dangri are ıove Malani Igneous Suite.

The occurrence of spread over the Shumar Formation (Pleistocene) is suggestive (ıst Shumar age. Similar pebble-cobble/boulder spreads have also been recorded froı her parts of north-western Rajasthan (Mukhopadhyay and Ghosh, 1976).

.uvial and associated deposits

The succeeding period witnessed warm humid climate as indicated by fluvio-glaci: :posits such as rounded to sub-rounded pebble spread, calcrete and fluvial deposits i ılaeochannels. During this period, the area was drained by mighty river system (ıraswati. The reference on the account of palaeochannels of Saraswati and othe reams have been documented from time to time by various workers. One of the majc ıraswati channel along with its tributaries drained the eastern part of the area passin rough Ramdeora-Pokaran, Kharariwala Rann, Kanod Rann, Khara Rann and Mith ınn. In Mitha Rann it took turn southward and streams have been observed i ıdharva-Manpiya-Kuldhar depression.

The eastern side of this depression from Manpiya southward is faulted and decide e eastern limit of palaeochannels at places. Ephemeral streams still flow in thi :pression.

Fluvial deposits like rounded to subrounded pebble spread, alluvial soil, ephemer: reams, depressions and valleys, ranns, ground water seepage, wells, Berris, Kunds an ıod tree growth in Khara Rann together with supporting evidences like religious place ıch as Ramdeora, Baisakhi, Ludharva temple (old capital of Bhatti rulers of Jaisalmeı ıd number of present and past villages like Kanod, Rupsi, Manpiya, Kuldhar (now i ıins). Artisan condition of groundwater around Rupsi as revealed by tubewell discharg so support that ground water was highly recharged by Saraswati drainage system.

However, potential ground water resources (sweet water) in sandstone aquifers (ıisalmer Formation and Lathi Formation at places may be attributed to recharging (ound water by Saraswati river system.

eolian and associated deposits

The fluvial event was followed by desertification. A major part of western an ırthern part of Jaisalmer district is covered with thick pile of sand and high san ınes. The Jaisalmer central plateau tectonically known as Mari-Jaisalmer arch ha :ted as barrier. All types of sand dunes : longitudinal, transverse, barchans, paraboli c. can be seen. They cover extensive areas in the Shahgarh and Kishangarh sub-basin.

The longitudinal sand dunes and barchans are more commonly distributed in th ea west and south-west of Shahgarh and Sam. Sand hills are common in Shahgar gion and east of it. Sand hills around Kishangarh are mostly linear and of high relie

›uthwest of Kishangarh transverse sand dunes are commonly observed. Khuri-Miajla ırab area exhibits widely spaced sand dunes in flat terrain (Babu and Rao, 1974).

It is interesting to note that tectonic structures appear to have controlled tl cumulation and density distribution of sand and sand dunes before the active role ınd action. The Mari-Jaisalmer arch forming major part of central Jaisalmer plate ted as barrier while in Khuri-Maijlar-Sam-Dhanana quadrangle longitudin nd-dunes and sand hills are widely spaced west of Kanoi fault with increasing densi wards west. Longitudinal sand dunes and barchans are more common towards west ım-Samdhani fault (at 73°30' longitude).

In Ramgarh quadrangle, sand hills, longitudinal and transverse sand dunes a mmon around Kishangarh while west of Ramgarh the sand dunes are widely space ›uth-west of Kishangarh active sand dunes and shifting dune area are observed. ıthi-Derikot quadrangle the sand-dunes and sand sheets are sparsely and irregular stributed. Similarly, in Mohangarh quadrangle the sand dunes and sand sheets a ıarsely and irregularly distributed. In these interior places the sand accumulation ıe to transportation from western and northern parts-the main and major sand aı nd-dunes fields. The northeastward blowing winds transported the sand through wiı ıps and open areas to interior places.

Calcretes and gypsite deposits occur in the desert area of Rajasthan. Calcrete ca observed in aeolian plain and river valley sections while gypsite sandwiched betwe nd layers can be observed near Mohangarh, Lakha and Hamirwali nadi. Calcrete h en observed by other workers over various surfaces. The presence of calcrete aı psite deposits indicate the wet conditions followed by arid conditions in the area. It teresting to note that all along the palaeochannels of the Saraswati drainage systeı e deposits of calcrete/gypsite were formed under suitable geological conditions position.

DISCUSSIONS, CORRELATION AND CHRONOLOGY OF EVENTS

Authors favour glacial origin for heterogeneous composition, striated, facete ›lished pebbles occurring over various formations and occupying topographic hi aces. However, Rajaguru *et al.* (1996) and some other workers do not support this vie ıey argued that possibility of existence of glaciation at lower latitudes are not nformity with other evidences. They have suggest that gravels are not an anomalo ature in the landscape. They are rather 'lags' derived from local bed rocks. Authors fiı llowing observations in support of the glacial origin of the pebbles. Rajaguru et. servations are mainly confined to Lathi terrain in south-eastern part of Jaisalm strict and over very small area of Jaisalmer Formation. However, such pebble spre ıs also been observed over Baisakhi (Upper Jurassic), Khuiala (Lower Eocene) aı ıumar Formation (Pleistocene). Except Lathi Formation Rajaguru et al. failed serve conglomerate beds from any other Mesozoic-Tertiary formations, neither fro ırface nor subsurface. Drill cores, well dump materials and borehole sections do n ıpport the view that they have been derived from local beds. Moreover, the occurren these pebbles on the hill slopes and small hillocks also goes against their view.

In view of characteristic composition and features of the pebble spread, the omorphic position, also favours the glacial origin. This period of glaciation is

›nformity with the last Himalayan glaciation which occurred during 30,000-15,000 B. ɜriod.

Subsequently, warming up with the increase of temperature, provided water vers of western Himalayan region including Saraswati which flowed through the ar‹ nder consideration. The large scale melting of ice during post glaciation period releas‹ ırge amount of water and flooded the rivers of north-western India which in turn rais‹ ıe sea level. This rise was responsible for marine transgression in coastal areas ujarat and some adjoining parts of Rajasthan. Gupta (1972) has shown that in the La uaternary times there occurred three marine tansgression on Saurashtra Coast at (2,000 BP, 30,000 BP and 5000 BP.

It appears that rivers draining the northwestern part of India were at their pe: ıring 5000 BP which is in conformity with marine transgression of Saurashtra whi‹ ccurred around 5000 BP. The fluvial regime lasted upto 4000 BP when possibly Ved araswati dried up. This is attributed to tectonic adjustment and upheaval imalayas and subsurface ridges like Delhi-Hardwar and Delhi-Sargoda. After marii ansgression, the receding sea around this period left sands in low lying Tertiary are: Rajasthan and Gujarat. Subsequently aridity and powerful winds played importa ›le and sands were transported to different areas. Northeastward blowing winds align‹ ıe longitudinal sand dunes in NE-SW direction. The desert in all probability may not l der than 4000-5000 BP.

Evidences in favour of eustatic changes of sea level along Saurashtra coast durii uaternary period have been collected by many workers (Sastri and Puri, 1959, Murtl *al* 1969, Gupta 1972), Mathur et al. (1982) discussed Quaterna ıstatic/glacio-eustatic changes of sea level in relation to the origin of Rajasthan sɛ kes and desert formation.

ACKNOWLEDGEMENTS

Authors are grateful to Dr. S. Sinha Roy, Dy. Director General, G.S.I., W.R., Jaipı ›r granting permission to contribute paper in the Seminar. They are also thankful hri Y.K. Arora and P.C. Bakliwal, Directors, GSI, for stimulating working environme nd necessary encouragement for participation in the Seminar. Authors are al ıankful to Shri A.K. Madan and Bhagwan Sahai for computer work and preparation lates respectively.

REFERENCES

grawal, D.P., Datta, P.S., Hussain, Z., Krishnamurthy, R.V. Mishra, V.N., Rajaguru, S.N. aı Thomas, P.K. 1980. Palaeoclimate, stratigraphy and prehistory in north and wє Rajasthan. *Proc. Ind. Acad. Sci.* **89(1)** : 51-66.

grawal, D.P. 1982. Palaeoclimate and geochronology of the Arid Regions of India. *Geol. Suı India Misc. Publ.* 49 : 105-112.

llchin, B. and Goudie, A. 1974. Pushkar-prehistory and climatic change in W. India. *Wor Archaeology* **5(3)** : 358-368.

akliwal, P.C. and Grover, A.K. 1988. Signatures and migration of Saraswati river in Thar dese western India. *Rec. Geol. Surv. India.* **116(3-8)**.

sgupta, S.K. 1975. A revision of Mesozoic Tertiary stratigraphy of the Jaisalmer basi: Rajasthan, *Ind. Jour. Earth Sci.* : 77-94.

.osh, B. 1952. The Rajasthan desert-its Archaeological aspect. *Bull. Nat. Inst. Sci.* 1 : 37-42.

.pta, S.K. 1972. Chronology of the raised beaches and coral reefs of the Saurashtra coast *Jou Geol.* 84.

.r, A. 1987. Origin and transformation of longitudinal sand dunes in the Indian deseı *Geomorph.* **31** : 311-337.

.r, A. and Gosh, B. 1988. Evidence for neotectonism from the Indian desert. In : Singh, S. ar Tiwari, R.C. (Eds.) *Geomorphology and Environment Geog. Soc.* Allahabad, India., 300-310.

.r, A. 1989. Terrain characteristics of Jaisalmer district. *Geog. Rev. India* **51** : 48-59.

.r, A. 1990. Megabarchanaoids of the Thar-their environment, morphology and relationship wi longitudinal dunes. *Geog. Jour.*, **156** : 51-61.

.r, A. 1992. Geomorphology of the Thar desert in Rajasthan. In : Sharma H.S. and Sharma, M.] (Eds.) : *Geogrphical Facets of Rajasthan.* Kuldeep Publications, Ajmer (in Press).

ımar, V. and Laul, V.P. 1980. Records of Glaciation in Jaisalmer basin, District Jaisalme Rajasthan. *3rd Convention Indian Assoc. of Sedimentologists.* Varanasi (Abst.) 31.

ul, V.P. and Kumar, V. 1980. Marine transgressions, regressions, glaciation and desertificatic events of Jaisalmer basin. *3rd Convention. Indian. Asso. of Sedimentologists (Varanas (Abst.)* 28.

kose, N.G. 1977. Palynological evidence on climatic changes in Jaisalmer basin, Rajasthan. Ir Jaiswal, P.L. (Ed.) *Desertification and its Control.* ICAR, New Delhi, 31-41.

ıkhopadhyay, A.K. and Ghosh, R.N. 1976. The problem of the stratigraphic position of the Bε Boulders, Rajasthan. *India Jour. Earth Sci.* **3(2)** : 220 - 228.

ıthur, U.B., Verma, K.K. and Mehra S. 1982. Quaternary glacio- Eustasy in relation to tł origin of Rajasthan Salt lakes and desert. *Geol. Surv. India Misc. Publ.* 49 : 59-67.

dham, C.F. 1893. The Saraswati and the lost river of the Indian Desert. *J. Royal Asiatic Sc (N.S.).* **34** : 49-76.

reek, H.S. 1984. Pre-Quaternary geology and mineral resources of northewestern Rajastha *Geol. Surv. India Mem.* 15 : 1-95.

jaguru, S.N., Mishra Sheila and Ghate Savita. 1996. A re- interpretation of gravel spreads the Jaisalmer region, Thar Desert, India. *Jour. of Arid Environments*, **32** : 53-58.

y, A.K. 1982. Aspects of geology, landforms and hydrogeology of a part of Lathi area, Jaisalm district, Rajasthan. *Geol. Surv. India, Misc. Publ.* 49 : 37-45.

ıgh, G. 1971. The Indus valley culture seen in the context of post-glacial climatic and ecologic studies in northwest India. Archaeol. *Phys. Anthropol. Oceania, 6 : 177-189.*

rma, R.P. and Roy, P.K. 1982. Palaeodrainage studies and its utility in flood problems. A ca study from Luni basin. In : Sharma H.S. (Ed.) *Perspectives in Geomorphology*, Concept, Nε Delhi, 4 : 137-143.

ıdhawan, S.K. 1988. Evolution of Quaternary aeolian deposits of Jodhpur and Barmer districl Rajasthan, India, In : Patel, M.P. and Desai, N. (Eds.) *Proc. National Seminar Rece Quaternary Studies in India*, M.S. Univ., Baroda, 64-78.

ıdhawan, S.K. 1994. Dune dynamics and evoluation of aeolian landforms in parts of Jaisalm district, Rajasthan, India *Jour. Indian Soc. Remote Sensing.* **22(2)** : 65-77.

eological Evolution of Northwestern India
.S. Paliwal (Ed.)
:ientific Publishers (India), Jodhpur, 1999, pp. 232 - 240.

COMPUTER APPLICATION IN ORE BODY MODELLING

A.S. Sheoran and D. Sharma

Department of Mining Engineering, J.N.V. University, Jodhpur 342 005

ABSTRACT

Acurate estimation of reserves and the ore body modelling prior to actual mining is essential for judicious and sustainable exploitation of mineral resources. The paper enumerates the steps involved in modelling processes and different geological modelling methods available. The process of selecting a particular type of ore body modelling technique is also dealt with. Finally, the paper explains the working of a software developed in the 2D image ore body modelling.

:ey words : Ore Body, Modelling, Computer Application, Estimation, Mining Process

INTRODUCTION

With the passage of time optimal utilization of the earth's natural resources h ecome the need of the day. Acurate estimation of deposits and selection of suitab iining methods are essential for cost effective mine planning. Over the past twen ears there has been a vertical upsurge in the development of softwares dealing wi lmost all the facets of mining.

Ore body modelling is defined as a device to produce 2 and 3 dimensional visu nages of an ore body with respect to its size, shape, grade and other physic arameters. These models provide a mathematical and visual representation of tl eposit which is of direct use in the design and planning of the mine.

Computers are used for storage, retrieval and analysis of geologic data. Early effoı 'ere concentrated on automating manual methods to reduce error and increa fficiency. As the user's acceptance of the automated methods has increased mo ɪphisticated techniques for the complex ore bodies, have moved computer users to a ne ontier.

Now a days, computers are increasingly used for designing, monitoring aı mulation process. 3 D modelling and designing, geostatistics and compl athematical approximations of ore reserves, which otherwise practically impossible, a ɔw being done with the help of computers.

STAGES OF ORE BODY MODELLING

The various stages of ore body modelling (Fig. 1) are as follows :

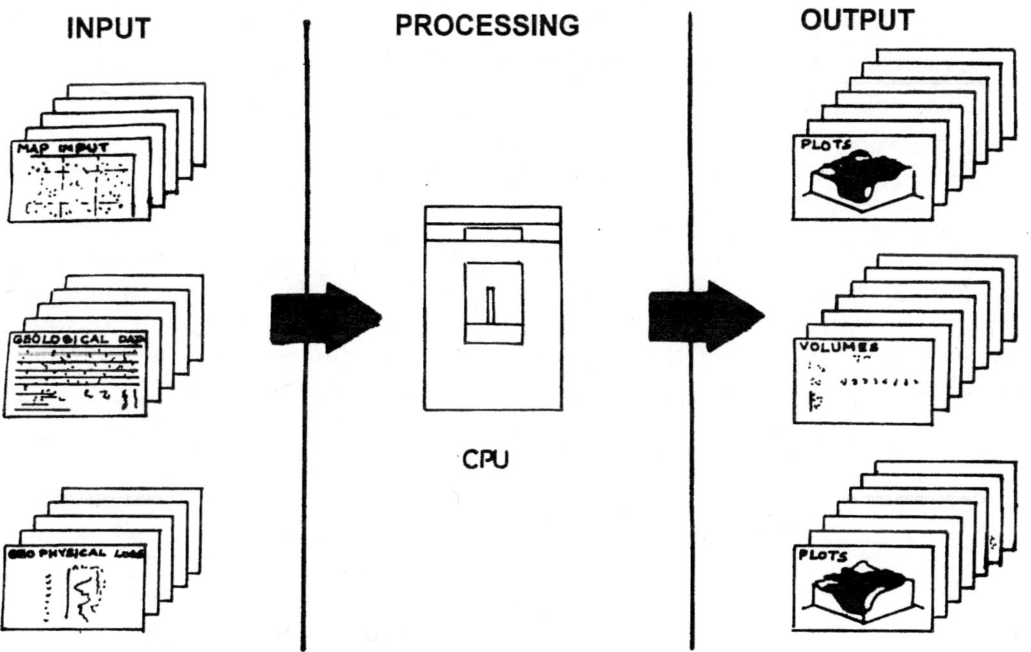

Fig. 1. Stages of computer processing

Collection of Data

Physical characteristics of the mineral deposits are never completely known, but a ferred from sample data. Noble (1993). have given some sample data types which cɛ used in modelling process are :

) Samples are taken by drilling, trenching, test pitting and channel sampling.

) The quantity of minerals in the samples is estimated through assaying or oth procedures.

) Direct observations such as geologic mapping and drill core logging. Geologic mɛ includes topography, cross sectional data, resistivity and density.

. Data Input

The data/information about deposit can be fed in the compuer by two ways :

.) Spread sheet based program

) Menu based interactive program

Both these types of programs can be used in data input, which create a data ba: his data base is subsequently used by main modelling program to generate ore bo odel. A data base should contain following information as a minimum requirement :

(a) Bore hole identification number

(b) Bore hole collor elevation and x,y coordinates

(c) Bore hole survey i.e. inclination, bearing, deviation etc.

(d) Seam/ore body top and bottom intercepts

(e) Grade/assay

(f) Lithology/rock type.

. Selection of Modelling Methods

Various models are used for specific conditions and specific mining operatioı [odelling choice is extensive and depends on the type of deposit, the commodity bei odelled, the value of interest, the complexity of the geologic deposit and the use :ceptance and familiarity with various modelling techniques. The suitablility ifferent modelling techniques for a deposit is discussed in later sections.

. Geological Modelling Methods

Various methods which are generally used for geological modelling include Gridd odel, Block model, Octree encoding model and Wire frame model.

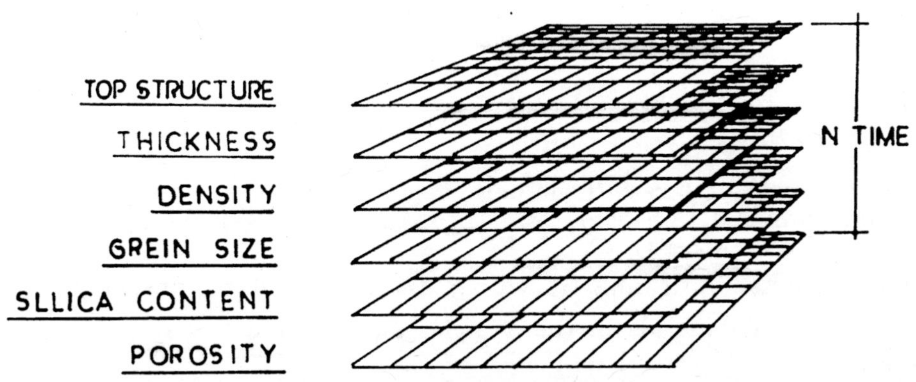

Fig. 2. Gridded model

) Gridded Model

The gridded model is a set of 2 dimensional matrices, (Fig. 2). Each representing ılue or a surface. These surfaces or values are the result of interpolation from a set regularly spaced data to a fixed grid or matrix. Structures, thickness and other valu acked on top of each other can easily be added, substracted, multiplied, divided gically composed to arrive at another set of data. The gridded model is normally us r bedded deposits such as coal, phosphate, sulphur, limestone, oil shale and tar sa c.

) Block Model

In this method, the mine area is divided into a series of 3-D blocks, typically 15-20 ı all sides, with each block assumed to be a uniform identity, either as waste or as e. Fig. 3 shows an ore body coded in a block model. Each block is identified by x, y and -ordinates of the block centre.

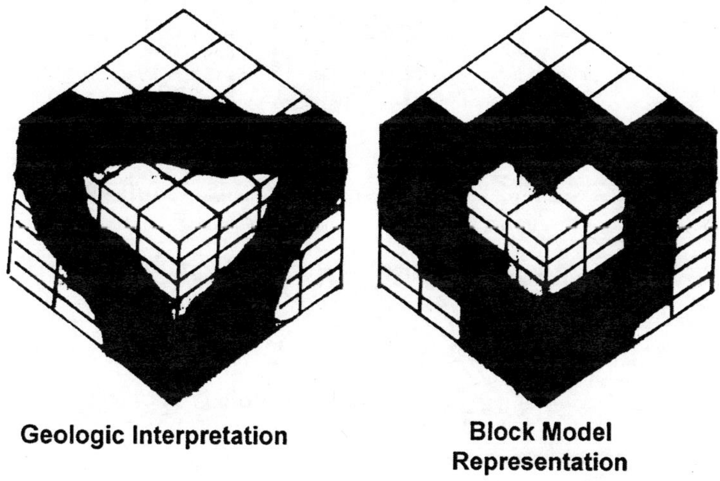

Fig. 3. Example of a Geologic Unit Coded into a Block Model

For open pit mines, block modelling technique is based on the division of the o earing region into a set of horizontal benches with a given height. The size of ea ench and the size of each block is governed by the mining equipments and the mini ethod used. The data along each diamond drill hole are then composted by weigh eraging along the length of the diamond drill hole and within each bench. T eighted average value so obtained is assigned to the centre of each segment of t amond drill hole intersecting the bench.

Following assignment of the diamond drill hole assay data to each bench, terpolation or extrapolation technique such as inverse distance or krigging is used timate the grade or any other desired variable value for each block. This estimati ocess continues for all blocks. These blocks can then be viewed in 2 or 3 dimensions anipulated to user's requirements to achieve desired results. The block model is usual ed for disseminated ore deposits such as copper, uranium, gold, lead, zinc etc.

Octree Volume Encoding Modelling Technique

In this method entire mine or ore body is enclosed in a large cuboid calle niverse". The cuboid universe is sub-divided into eight sub-cuboids of equal size. The: b-cuboids are known as voxels. Each voxel has its own colour depending on whether s inside (black), outside (white) or at the boundary (gray) of the ore body (Fig. 4). Th b-division continues recursively only for gray voxels and terminates when either ay voxel remains or when a preset resolution is reached. The smallest elements aft e last partition, are called resolution voxels.

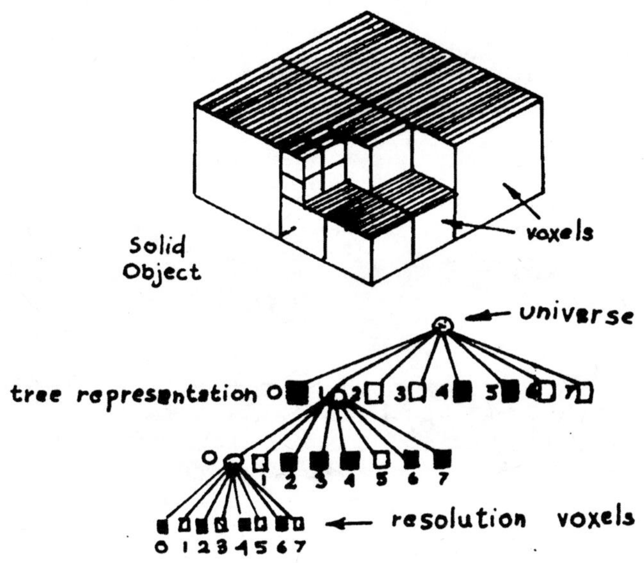

Fig. 4. Octree encoding of a solid object

These voxels are then treated in same manner as blocks in block modelli chnique but the resolution in case of voxels is very high and good and the operation aste and ore is achieved which is not possible in block modelling. This technique is us represent irregular geosolids and thin irregular ore bodies like that of gold deposits.

) Wire Frame Model

In this modelling method the boundaries are modelled rather then the thr mensional ore body. Known points on the surface of the ore body which is to odelled are joined in an irregular lattice pattern consisting usually of polygon sually triangular or rectangular) patches (Fig. 5) which represent faces on a polyhedr proximation of the desired 3-dimensional boundary.

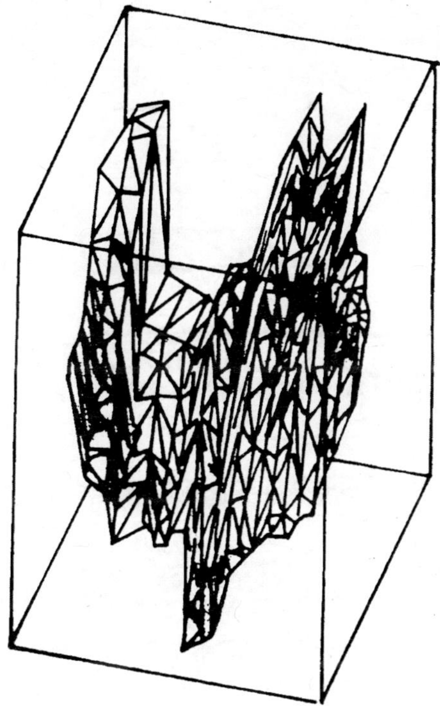

Fig. 5. Ore body wire frame model

To produce wire frame model geologist specifies the geometry, slice by slice, eith(section, or in plan and then these are joined up in 3-dimensions by linking parametei adjacent sections to produce wire frame models.

COMPUTER MODEL FORMULATION

A computer program OBMODEL is developed to illustrate basic principles of oı dy modelling. For this purpose a hypothetical copper ore body is assumed which vided into 50 m x 50m x 20m block size. Each block is identified by the x,y and z c dinates of its centre and is assigned a grade value (% of Copper content).

Block centre co-ordinates are stored in a ASCII file (COORD) and block grades aı red in another ASCII file (GRADE). These files can be modified using NON documeı e editor. Simplifed Flow Chart of OBMODEL is shown in Fig. 6.

By using the OBMODEL cross-section of ore body can be obtained in x,y and anes OR other words South, East and Reduced level. The output of OBMODEL aı own in Figs. 7 and 8. These figures show plan of ore body at 210 MRL at 0.2 and 0.5 t off grade.

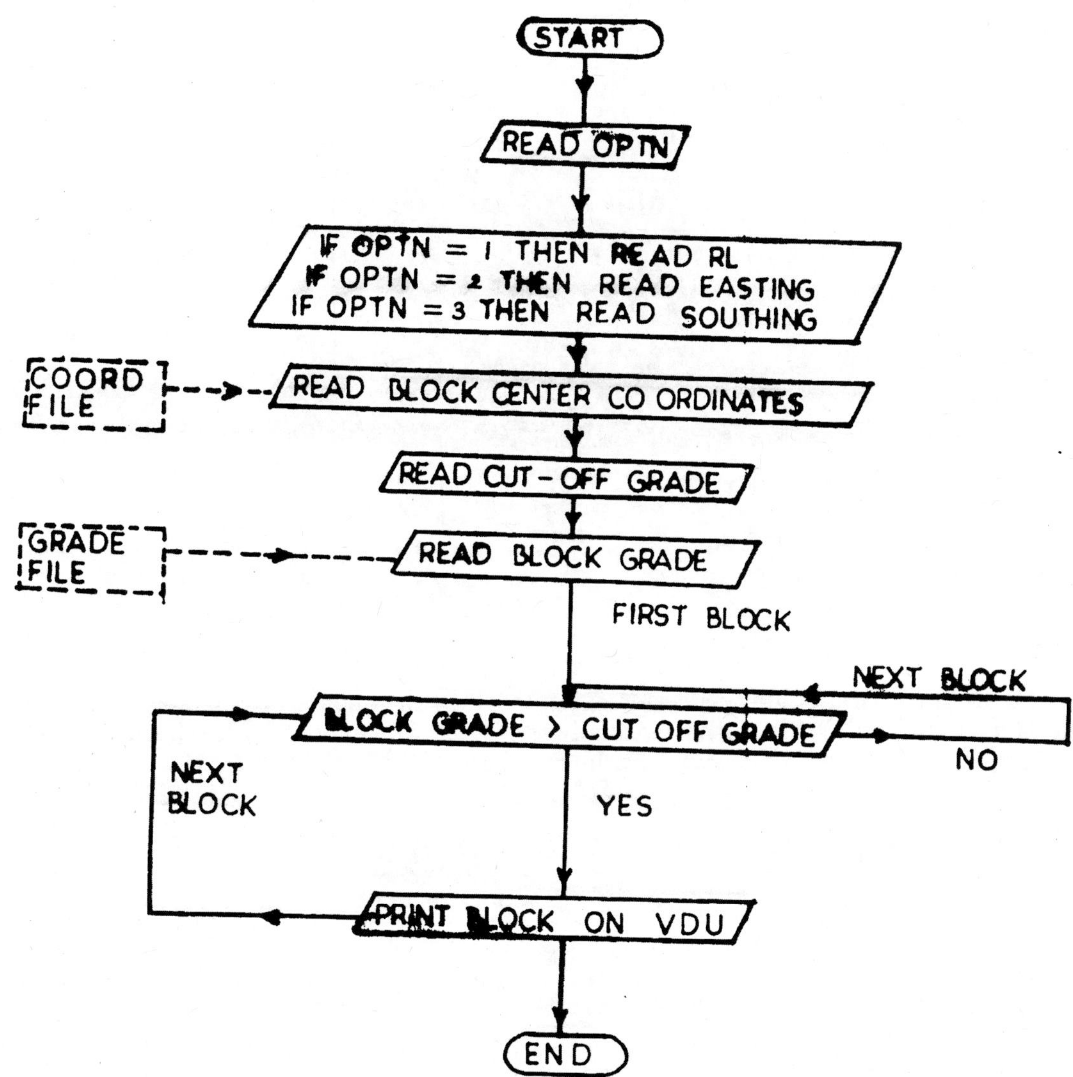

Fig. 6. Flow chart of OBMODEL

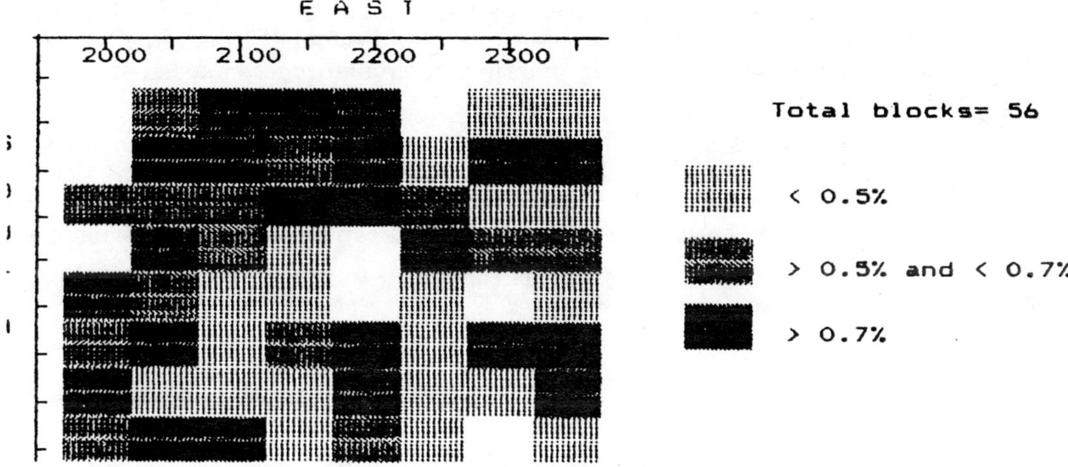

Fig. 7. Ore body cross section at 210 MT R.L. (0.2% cut off grade)

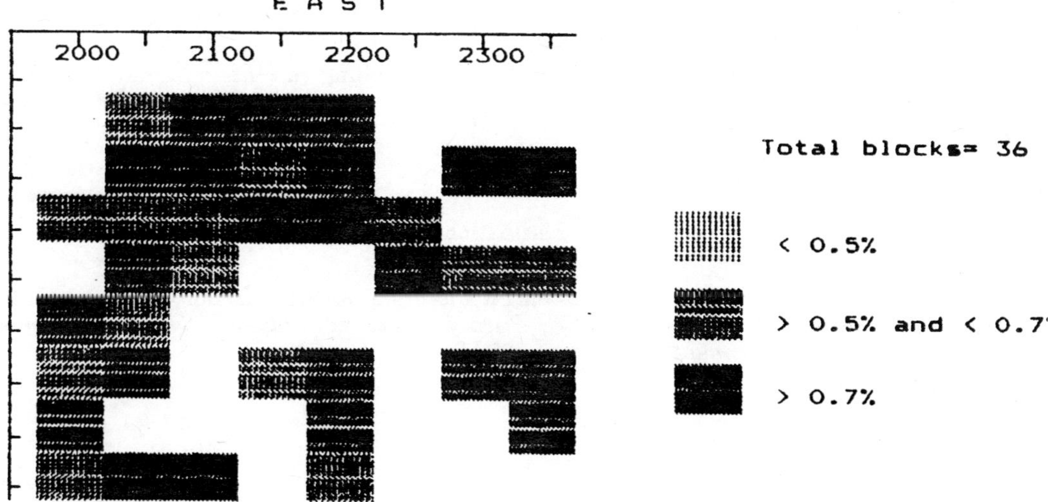

Fig. 8. Ore body cross section at 210 MT R.L. (0.5% cut off grade)

CONCLUSIONS

Computer based techniques are increasingly used in a number of industries and t enefits of such systems in mine planning are now becoming apparent. The use omputers to assist the ore body modelling, brings the advantages of speed and accur f calculations.

There are many softwares available in the market e.g. DATAMINE, MINER IINESTR, SURPAC, GEOLOG, DATAMINE-GUIDE, PC-MINE, which have got v ood ore body modelling facilities. An appropriate software may be selected according he user's requirements or the software like OBMODEL can be developed in house.

The cost of computerization is small as compared to other mining capital outl nd this can be used to save millions of rupees.

ACKNOWLEDGEMENTS

Authors are highly thankful to Mrs. V. Sheoran for her help in preparing the t nd figures.

REFERENCES

adiozamani, K. 1993. Computer Method. In : Hartman, H.L. (Ed.). *SME Mining Engineer Hand Book.* California, U.S.A. 599 - 613.

[aldar, S.K. and Choudhary, M. 1992. "Computer Aided Ore Body Modelling : An Integra Approach for Mine Planning at HZL", *Proc. of National Seminar on Computer Application Mining and Processing Industries*, Udaipur, 28-35.

[enley, S. and Wheeler, A.J. 1987. "The use of Solid Modelling in Underground Mine Desig *Proc. of 1st Canadian Conf. on Application of Computers in Mineral Industry*, Queb 359-364.

[oulding, S. and Eng. P. 1987. 3D Computer Modelling of Geology and Mine Geometry. *Min Magazne.* 156 : 226-231.

[oble, A.C. 1993. Ore Reserve/Resource Estimation. In : Hortman, H.L. (Ed.) *SME Min Engineering Hand Book.* California, U.S.A. 344-359.

akishev, B.R. and Sadykov, M.B. 1993. Open Pit Planning on the Basis of Computer Graph and Local Modelling of Ore Quality", *Proc. of 2nd International Symposium on M Mechanization and Automation*, Sweden, 625-631.

eological Evolution of Northwestern India
.S. Paliwal (Ed.)
cientific Publishers (India), Jodhpur, 1999, pp. 241 - 245.

DISPOAL OF TAILING FROM ROCK PHOSPHATE BENEFICIATION IN RAJASTHAN

K.L. Menaria, V.S. Palria*, and Deepshikha Sharma**

Department of Chemistry, J.N.V. University, Jodhpur - 342 005
**Department of Mining, Engineering J.N.V. University, Jodhpur - 342 011*
***Savitri Girls College, Ajmer - 305 001*

ABSTRACT

The mineral industry in many countries often accounts for a major percentage of their gross national products. This industry is also contributing significantly to environmental pollution at present and a potential hazard for future.

:ey words : Rock Phosphate, Beneficiation, Tailing Disposal, Jhamarkotra - Matoı Udaipur, Rajasthan.

INTRODUCTION

The mining and processing of rock phosphate generates a wide range of particles (ifferent size fractions as waste. These include coarse mine waste, fine clays, floatio ıiling, chemical precipitates and slimes. These wide range of size fractions requiı efitting design and construction of tailing dams and waste impoundment areas. Th esign should take into consideration the physical, chemical and mineralogical aspects.

TAILINGS AND SLIMES GENERATED IN ROCK PHSOPHATE BENEFICIATION

In the processing of rock phosphate for phosphoric acid or manufacture of fertilize ıe wastes generally consist of clay slime slurry that is only about 5% solids (Trichou ınd May, 1973). The actual volume of tailing is 1.5 times greater than the volur ›ccupied by the rock before mining. Tailings from such operations are disposed by :

(i) Slurry discharge into settling ponds or into ocean.

(ii) Dry stacking

(iii) By wet stacking using the upstream method

However, the method of disposal is dictated by the process, the climate and t ıydrogeological conditions as well as the economic and environmental constraints.

COMPOSITION OF PHOSPHATE SLIMES

Phosphate slimes vary in composition, but normally consist of –

a) *Clay minerals*

(i) Montmorillonite

(ii) Attapulgite

(iii) Kaolinite

(iv) Illite

b) *Phosphate minerals* : These minerals mostly consist of fine grained carbona fluorapatite, with small amounts of aluminium phosphates such as wavellite.

c) *Quartz* - Sand or silt sized particles.

d) *Other minerals* - Feldspar, heavy minerals and hydrated iron oxides.

GENERAL SCHEME FOR WASTE DISPOSAL

Many plants separate the slimes from the sands, dewater the sands for wat recycle and send the slimes to the tailing ponds. Considerble research (Bromwell aı Oxford, 1977) has been done to increase the rate of dewatering of the phosphte slim and tailings. The methods so evolved include :

a) Feeding slimes and sands separately to the same pond to achieve a sandwich effect.

b) Mixing of two fractions by pumping.

c) The use of flocculants.

However, the problems differ from plant to plant and therefore separate schem have to be designed for each operation and each plant.

CURRENT PRACTICE OF WASTE DISPOSAL IN INDIA

The rock phosphate industry in India is not a big industry as comapred to the woı standards because it contributes only 1% of the total world production.

Jhamarkotra and Maton mines are the only important mining centres located in t strict Udaipur. The production of Jhamarkotra is managed by M/s Rajasthan Stat ines and Minerals Ltd. Udaipur and Maton is managed by M/s Hindustan Zinc Limite daipur.

TAILING DISPOSAL IN JHAMARKOTRA

At Jhamarkotra currently there is no tailing disposal method. The piled up tailin increasing in bulk awaiting development of befitting technology. However disposed o iling containing above 10% P_2O_5 is being picked by farmers and small enterpreneu elding small revenue to the industry (Choudhary and Deb, 1994).

TAILING DISPOSAL IN MATON

Maton beneficiation plant was working till a few years ago producing calciu perphosphate and phosphoric acid but quite recently, owing to the problems linked t e high cost of production and tailing disposal the plant is closed and the run of mine i ing marketed. Adequate research support is desirable for meaningful utilization (aste. (Sisodia, 1995; Agrawal and Satynarayan, 1989; Chauhan and Sisodia, 1989).

SUGGESTIONS FOR PHOSPHATE TAILING DISPOSAL

In the light of above discussions and present investigations some specific remedia easures are being suggested which include.

) Improvement of floatation technology

Development of more efficient floatation process targeted to low silica contents i sirable. The current method using very costly phosphoric acid is to be discontinued an placed by the anionic and cationic guar gum derivatives which are recommended a tential reagents for the future in light of researches carried out in our laboratory.

) Blending of phosphorite and non-ferrous plant tailings

Sisodia (1995) has reported the use of sulphur rich mine waste from Zawar lea nc beneficiation plant to treat and solubilise the low grade phosphorite ore fro amarkotra. The mill discarded water from Zawar shows low pH (3 - 5), high Eh an ntains sulphuric acid. The powdered phosphorite ore dipped in this water for severa ys shows solubility due to formation of mono and dicalcium phosphate :

$$2\ Ca_5(PO_4)_3X + 7\ H_2SO_4 \rightarrow 3\ Ca(H_2PO_4)_2 + 7\ CaSO_4 + 2\ HX$$

$$2\ Ca_5(PO_4)_3X + 4\ H_2SO_4 \rightarrow 3\ Ca_2(HPO_4)_2 + 4\ CaSO_4 + 2\ HX$$

(Where X = F, OH, Cl)

This digested material was recommended for direct use as fertilizer

The work of Sisodia (1995) was reviewed during present investigation. This attem ɔpears to present an over simplified optimistic view based upon short ter ɛperimentation. The tailings from *lead-zinc* multi-metal beneficiation plants a kaline at the time of discharge but on standing for long time they generate ac ırough bio-chemical oxidation of pyrites and turn into a harmful acid discharge with p the order of 2-5. The typical analysis of a sample of water from *lead-zinc* beneficiati ant entering tailing dam is shown in Table -1.

In addition to the above mentioned constituents the tailing water also contai ırmful constituents viz a viz. cyanide, fluoride, mercury, cadmium, and chromate. Aft long residence period the sulphides are oxidised to sulphuric acid but this acid shall ntaminated with the harmful constituents which will built up the concentration aı ıis cumulative toxicity is required to be evaluated before this sulphuric acid is mix ith low grade rock phosphate of western Rajasthan to obtain soluble phosphates to a ; phosphatic fertilizer.

Partial acidification of tailings

In the case of Maton tailings rich in silica, partial acidification has been suggest / Lahari and Chulet (1993) in Choudhari and Deb (1994) and Marwaha (199 ccording to these workers the Maton rock phosphate deposit with citrate solubili 5%, lower organic matter and low carbonate content, is considered to be quite suitab r partial acidification with dilute sulphuric acid. The process liberates soluble $P_2($ hich is easily assimilable by plant.

able 1. Typical analysis of water sample from lead-zinc beneficiation plant enterii tailing dam

Colour	light brown
pH	8.0
TDS	885 ppm
COD	2 - 3 ppm
Chlorides (Cl^-)	200 ppm
Sulphates (SO_4^{--})	424 ppm
Sodium	130 ppm
Calcium	128 ppm
Magnesium	61 ppm
Lead	0.07 ppm
Zinc	0.10 ppm
Copper	0.01 ppm
Iron	0.87 ppm

Development of appropriate Biotechnology

Quite recently Purshottaman and Tamilvandan (1996) from Tamilnadu Agricultu niversity Coimbatore, have suggested the use of certain phosphorus solubilisi ıtrients, which can convert insoluble rock phosphate into a form readily assimilable ants. According to these workers soluble and viable soil phosphorus sometimes ge ked as insoluble phosphates. These insolubles are made soluble by phosphobacteri ɔfertilizers. The bacteria, contained in these biofertilizers solubilise phosphates ar ake them available for crops. Azospirillum enriches the soil with nitrogen. Howeve ey cannot solubilise phosphates. The two biofertilizers producing organisms a oduced separately but can be used together at the time of application. Further, it wa und that rock phosphate enrichment followed by metagenesis of the wild strain zospirillum brasilense yielded genetically altered strains of Azospirillum which exhibi osphorus solubilising character.

Phosphorus solubilising ability of mutant strains of Azospirillum are as high a osphobacteria.

ACKNOWLEDGEMENTS

Authors are thankful to the Hindustan Zinc Limited (HZL) Udaipur and Rajastha ate Mines and Minerals Limited (RSMM), Udaipur for providing facilities for th esent work.

REFERENCES

garwal, L.G. and Satyanarayan, K. 1989. Beneficiation of Indian Phosphorite ores. *Mem. Geo Soc. India* 13 : 117- 123.

omwell, L.G. and Oxford, T.P. 1977. *Waste clay dewatering and disposal, Geotechnical Practic for Disposal of solid waste materials*, ACSE, New York, 541-558.

hauhan, D.S. and Sisodia, M.S. 1989. Phosphorities of Rajasthan. *Mem. Geol. Soc. India* 13:9-2

houdhari, R. and Deb K. Pradeep, 1994. "The Beneficiality of Rock Phosphate RSMMI Experience and its Relevance in the Asian Context." *International Seminar on Minerals an Mineral based Industries in ESCAP Region*, Soaltee Holiday Inn Crowne Plaze, Kathman (Nepal) 6-9 Dec.

arwaha, B.C. 1992. Partially Acidificated Phosphate Rock (PAPR) A promising Fertiliz material - A review *Fertilizer News* December, 1992.

urshottaman and Tamilvandan, 1996. The Hindu Thursday, April 11, 24.

sodia M.S. 1995. 'Potential use of sulphur-rich mine waste from Zawar to utilize low gra phosphorite of Udaipur for Ecorestoration', In Ranawat P.S., Avadich P.C. and Agarwal (Eds.) : *Environmental Perspective of Small Scale Mineral Resource Utilization,* Agarw Publishers, Udaipur.

ichous, O. and May, A. 1973. *Electrophorisis and Coagulation Studies of some Flori Phosphate Slimes*'. U.S.B.M. Report RI 7816, U.S. Bureau of Mines .

Geological Evolution of Northwestern India
B.S. Paliwal (Ed.)
Scientific Publishers (India), Jodhpur, 1999, pp. 246 - 259.

CLIMATIC AND NEOTECTONIC CONTROLS ON DRAINAGE EVOLUTION IN THE THAR DESERT

Amal Kar

Central Arid Zone Research Institute, Jodhpur 342 003

ABSTRACT

Integrated stream networks are rare in the Thar desert, except in the south-eastern part where the Luni and its tributaries flow. In the north the dry valley of the Ghaggar is a classic example of drainage disappearance. The existence of the Saraswati-Drishadvati river system here, was essentially due to the Sutlej's perennial flow into the system. Early southward courses of the system appeared to be partly controlled by lineaments, while stratigraphic records in the dry valley of the Ghaggar (Saraswati) indicated tectonism at the end of the Harappan civilization. Yet, the influence of climatic change appears to be strong. Ground water samples from one of the Saraswati palaeochannels in Cholistan desert suggest that the latest event of recharge occurred during the last inter-glacial period. Three generations of superimposed palaeochannels can be recognised in the uplands-plains sequence near Merta. Gravels recycled from the Precambrian conglomerates constitute the upland palaeochannels, representing large spasmodic flood discharge with a hint of relief inversion and exhumation. Karst and pseudokarstic features in parts of the desert also point to former wetter climatic phases. In the Luni basin, paired terraces, alternate stream incision and braiding, abnormal shifts in channel courses, sudden thickening of weathered zone suggest recent tectonic movements along lineaments. The structural style and tectonic movements provided controls to the climate-driven fluvial processes. In other words, drainage evolution in the Thar desert may have taken place under a complex process involving both climatic and neotectonic elements.

Key words : Neotectonics, Climate, Drainage Evolution, Thar Desert India.

INTRODUCTION

Studies on the present and palaeochannel networks in the arid regions of the worl ve provided many significant clues to the hydrological process efficiencies in the dese: gions, as well as to landform evolution through fluvial processes and climatic chang er the time. In the Thar desert of Rajasthan a stream network exists mainly in th uth-eastern part, where the Luni and its major tributaries like Jojri, Guhiya, Band ıkri, Jawai, Sagi, etc. form a somewhat loosely integrated drainage network, called th ıni basin. Most of the streams originate in the Aravalli hill ranges with an averag ınual rainfall of more than 500 mm) and flow in direct response to rainfall during th mmer monsoon (June to September). In other words, the streams are ephemeral. rge part of the catchment area of the Luni and its tributaries lies over a moderatel ep and sandy older alluvial plain, with isolated hills and pediments. There are som ınefields in the western part of the basin. The combined flow drains into the Gre ınn of Kutch in the south (with an average annual rainfall of 300 mm).

The rest of the area in the Thar desert has very few streams, confined mostly to th cky uplands and isolated hills. In these areas the trunk streams hardly receive an ntribution once these are out of the rocky terrain and hence, loose discharge very fa rough infiltration in the sand and alluvium along their courses. The length of strea ıw is, therefore, largely a function of the stream network in the rocky upland area ovided rainfall variation within the catchment is not large enough and sedime aracteristics along the channel segments beyond the upland are not very dissimilar.

Apart from the above stream networks, the desert is endowed with almost ı rface drainage lines. Yet, there are remote sensing signatures of many former strean ıich continue beyond the present channels. While there have been some studies on th entification of these former courses, very little work has so far been undertaken ıderstand the external force which led to the creation and subsequent abandonment e channels, the process-form interactions along the courses over time, th perposition of the past channel systems, their impact on the terrain evolution, and th ıssible future use of the systems. We would like to discuss in the present paper th oad spatial pattern of the palaeochannels and the possible external forces, ıderstood from geomorphic evidence and researches in the allied fields.

PALAEOCHANNEL NETWORKS AND FACTORS RESPONSIBLE : EARLY RESEARCH

Systematic palaeochannel research in the Thar desert started in the early sixties e twentieth century when scientists at the Central Arid Zone Research Institu AZRI), Jodhpur, investigated the drainage pattern in the Luni basin area, using aeri ıotographs. It was noticed that many tributaries of the Luni river had disjoint urses and some were unable to meet the trunk streams. Many courses were complete sorganized and obliterated after a certain distance. Such features were especial arked in the Luni-Jawai plains. The scar marks of the abandoned courses and th ıooth, dark tone of the buried channel courses on the aerial photographs were used ap the palaeochannel pattern in the area (Ghose, 1964, 1965). The courses assum ore significance as some of these yielded potable ground water at shallow depth. Rise ·ound water level was noticed in wells along the courses whenever there was rainfall ıe upper catchment areas. It was suggested that the streams were active during rmer wetter climate and that these were disorganized after the present arid phase s .. Bandi, Sukri and Mitri rivers and the streams of Goliya, Bhailan and Kund amania tract suffered the most.

Researches since then at CAZRI, Geological Survey of India (GSI), Spa pplication Centre (SAC) of Indian Space Research Organization (ISRO) and oth ·ganizations, using aerial photographs and satellite imagery, helped to map a numb palaeochannel networks elsewhere in the Thar desert. Useful summaries are provid Singh (1977), Kar (1993a) and Rajawat and Narain (1996).

SARASWATI-DRISHADVATI SYSTEM

The most celebrated palaeochannel system of the Thar desert, t araswati-Drishadvati system in the northern fringe, was known to the mode searchers for more than a century, thanks to the pioneering work of Cunningha 871), Oldham (1886, 1893) and other scholars. These streams originated in t imalayas and used to flow through the present Punjab-Haryana alluvial plains. T araswati was identified by many scholars during the last one hundred years as t esent dry bed of the Ghaggar along the northern margin of the Thar desert in Harya ıd Rajasthan. The Raini, the Wahinda and the Nara in Pakistan have also be entified as some of the shifted courses of this river. A dry stream bed between His ohar and Bhadra was identified as the course of the Drishadvati, which is mentioned ıcient literature as a major tributary of the Saraswati river.

In the mid-nineteen seventies satellite image interpretation provided clues to son itherto unknown former, northeast to southwest-flowing stream courses through t har desert. The courses appeared to be the predecessors of the Saraswati- Drishadv ver system (Ghose *et al.*, 1979; Kar and Ghose, 1984; Kar, 1993a). A preliminary sear borewell lithologs and Quaternary stratigraphy along the marked courses suggest th ıe courses are buried within sand dunes and sandy plains of the Thar desert and th ıost of the courses could be older than 40,000 years. The early courses of the rivers we entified roughly through (1) the vicinity of Rajgarh, Hadyal, Ratangarh and the prese is-fit valley of the Jojri; (2) Nohar, Surjansar and Samrau; and (3) Sirsa, Lunkarans ıd east of Bikaner. A thick deposit of alluvium and Quaternary calcrete occur along t ›urses. The Luni, originating from the Aravalli range, and its major tributaries perha ·ained into the Himalayan system. The spatial pattern of the remote sensing signatur ıggested that the Saraswati- Drishadvati river system experienced several shifts duri ıe Late Quaternary period. The direction of shift was roughly from east to west, so th ıe above-mentioned northeast-southwest oriented courses could be the gradually shift ›urses of the Saraswati system over a period of time (Fig. 1). Subsequently, t araswati shifted away from the Luni and began to flow through another course, rough ırough Nohar, Anupgarh, Sakhi (in India), Khangarh (in Pakistan), Ghantial, west hahgarh (in the extreme western part of Jaisalmer district in India) and then the low ›urses of the Nara (in Pakistan). Further shifts took the Saraswati through the Rai ıd the Wahinda and the Hakra-Nara segment in Pakistan. Finally the river ceased

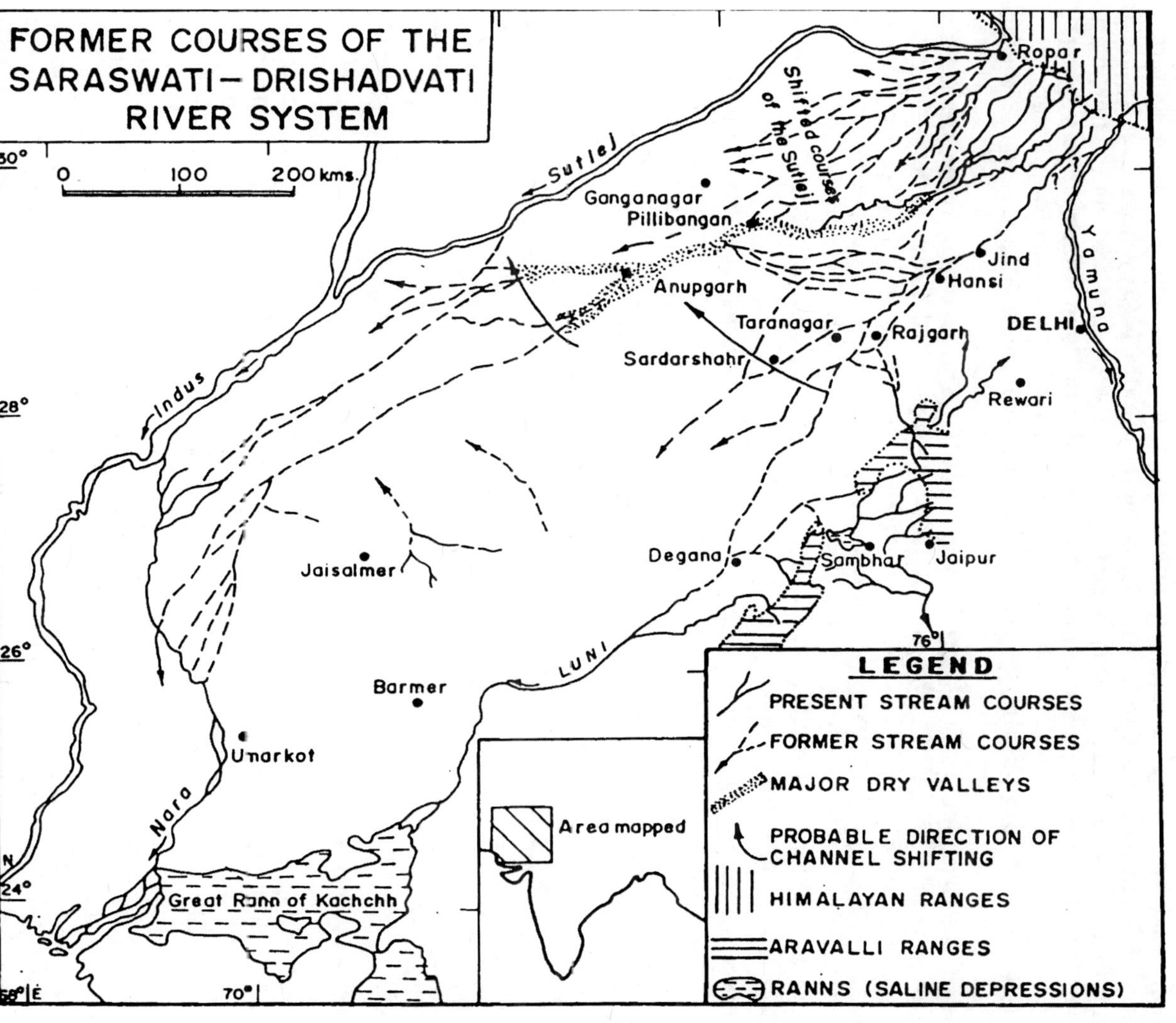

FORMER COURSES OF THE SARASWATI – DRISHADVATI RIVER SYSTEM
30°
28°
26°
24°
N
E
70°
76°
0
100
200 kms.
Ropar
Shifted courses of the Sutlej
Sutlej
Ganganagar
Pillibangan
Anupgarh
Jind
Hansi
Yamuna
DELHI
Taranagar
Rajgarh
Sardarshahr
Rewari
Indus
Jaisalmer
Degana
Sambhar
Jaipur
LUNI
Barmer
Umarkot
Nara
Great Rann of Kachchh
Area mapped
LEGEND
PRESENT STREAM COURSES
FORMER STREAM COURSES
MAJOR DRY VALLEYS
PROBABLE DIRECTION OF CHANNEL SHIFTING
HIMALAYAN RANGES
ARAVALLI RANGES
RANNS (SALINE DEPRESSIONS)

ow even through that course and met the Sutlej (Satadru of the early Indian literatur ia Anupgarh to the west of Ahmadpur East (in Pakistan). The Drishadvati al: radually shifted northwestward, ultimately occupying the Narnaul-Hansi-Hisa hadra-Nohar course to meet the Saraswati near Rangmahal. This last-mentioned tra as many east-west oriented palaeochannels, which are in various phases of obliteratic nd are bordered by sand dunes (Fig. 1). Some of the dry valleys have been canalize he findings provoked other researchers to come out with identical south- flowing ar est-flowing stream courses and almost similar shifts in them from remote sensing, (c ajawat and Narain, 1996). These findings replicated the philosophy inherent in ou ork and tended to compliment the basic structure of former drainage network that wa roposed, but did not step out to explore systematically any part in detail. The on xception was a collaborative effort between CAZRI and Ground Water Departmen overnment of Rajasthan, through which detailed geophysical and geological exploratic f the palaeochannels in the western part of Jaisalmer was undertaken for ground wat xploitation. The palaeochannel was identified in the difficult dune country throug eophysical depth soundings and a potable aquifer was found (Kar and Shukla, 199 ingh *et al.*, 1994), but problems of moving the heavy equipments into the high dun overed areas became insurmountable.

Remote sensing signatures suggest that the Sutlej used to nourish many of th arly courses of the Saraswati, and that the river shifted its course several times in th ub-Himalayan plains and ultimately shifted away from the Saraswati. The easternmo nd perhaps one of the earliest courses of the Sutlej that fed the Saraswati, was nea akhal, while the last major south-flowing channel of the river that fed the Saraswati i he Indian territory, is known as the Naiwal which used to meet the Saraswati nea nupgarh. Presently, the Sutlej meets the Chenab to the west of Bahawalpur i akistan and the combined flow of these streams meets the Indus further west. Th winging behaviour of the Sutlej is partly due to the loss of balance between the loa ischarge and grade, as the stream comes out from the steep mountainous terrain of th imalayas to the plains, but the channel changes could also be partly related eotectonic activities along the sub-Himalayan plains and the adjoining Siwalik ranges.

Neotectonic activities are also known from the Rann of Kutch where the Saraswa ver system used to drain. The last major earthquake in the region was in 1819 when major east-west lineament through the great Rann of Kutch and the Indus- Saraswa elta was activated, creating an earthen mound across the dwindling flow of the Dho uran and other distributaries of the Nara river in Pakistan, which once carried th ater of the Saraswati. The earthquake resulted in tilted upthrown blocks wit outhward dip, uplifted the Rann bed, beheaded the streams and partly sunk the Ko reek which was the mouth of the erstwhile Saraswati (Kar, 1993a, b). With suc vidence of neotectonics from the upper and the lower reaches of the river, it is qui kely that neotectonism played a significant role in the evolution of this stream networ ut the manifestations are mostly buried under a thick blanket of sand. Satellite imag nterpretation suggests the existence of some major NNE-SSW lineaments throug ajgarh-Churu, Bhadra- Sardarshahr and Sirsa-Lunkaransar, along which th alaeochannels of Saraswati system were marked (Fig. 2). It appears that th roto-Saraswati palaeochannel configurations are somewhat controlled by the tecton tyles. Some evidences of neotectonism have been found along the present dry bed of th haggar at the Harappan site of Kalibangan, where archaeological excavations ha evealed stratigraphic breaks in the alluvium.

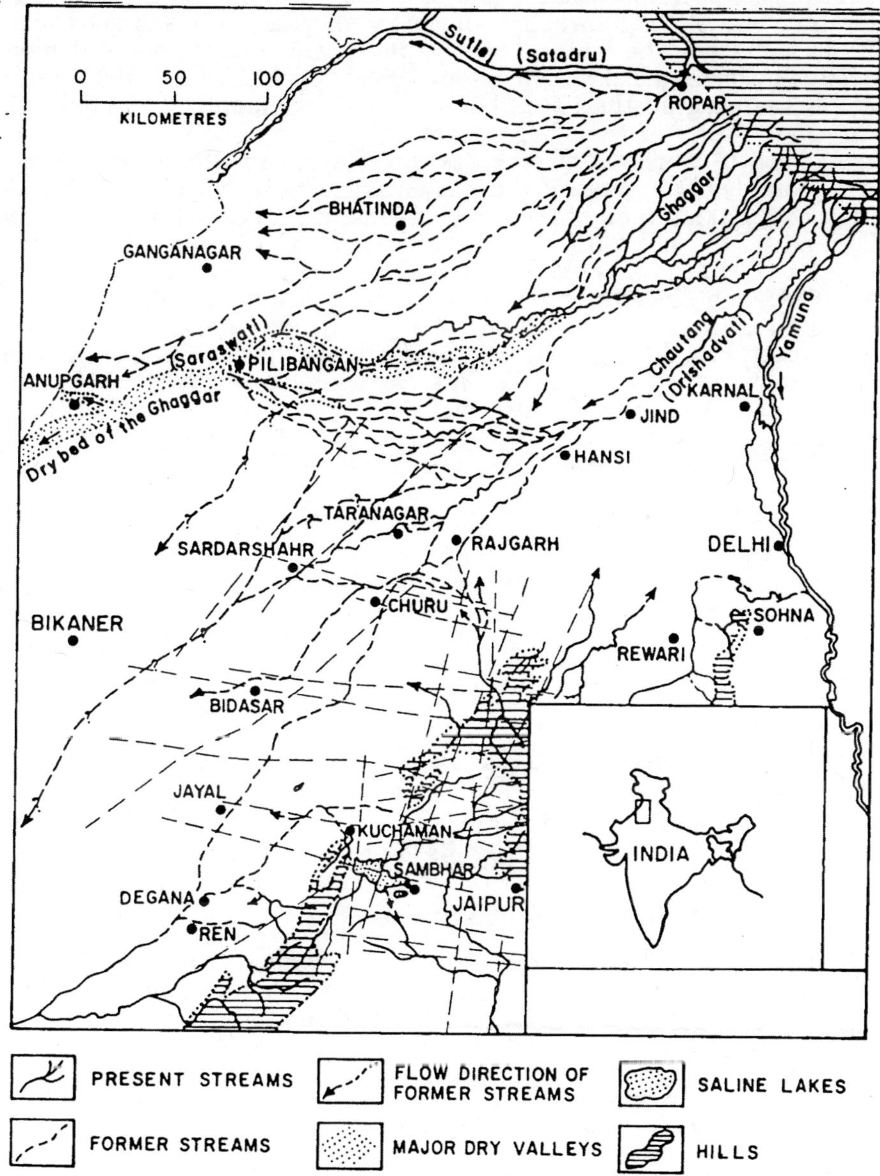

Fig. 2. Map showing the palaeochannels of the Saraswati system and distribution of lineaments in northwestern part of India

Although the evidence of neotectonism are difficult to trace, the evidence ıctuation in climate during the Late Quaternary period are numerous, especially ı ıvio-aeolian sequences in the alluvial and other sandy plains, as well as in the salir kes of the desert. Dating of pollens and sediments from across the Thar desert (Singh ., 1974; Chawla *et al.*, 1991; Singhvi and Kar, 1992) and correlation with informatic ɔm other areas of South Asia and Australia suggest a prolonged and vigorous summ

ıonsoon (wet phase) over the whole region between 140 and 125 kyr, as well as 60 aı
) kyr before present (BP), followed by somewhat shorter periods of good monsoon wi
ıtervening arid phases. The last major wet phase was approximately between 10 k
ıd 6 kyr BP (Bowler *et al.*, 1995; Wasson, 1995, Velichko *et al.*, 1997). Signatures
ımilar wet phases prior to the dated horizons are also found in the deep Quaterna
equences in many parts of the Thar desert. In the deep aeolian sequences these w
hases are marked by surfaces of stabilization. Geyh and Ploethner (1995) dated tl
round water samples from one of the pre-Ghaggar Saraswati palaeochannels in tl
holistan desert of Pakistan and found that the last recharge of fresh ground water to
lace in 12.9 and 4.7 kyr BP, with a break between 8 and 7 kyr BP. A similar age for tl
ıst ground water recharge was obtained from the Middle East and beyond (Geyh et a
985; Geyh, 1994), and since the recharge periods agreed well with the period of high
ainfall over the region, it was concluded that the recharge of ground water along tl
alaeochannel took place during early Holocene period. The cessation of the recharge w
lmost coterminous with the decline of the Harappan civilization. A recent study l
ingh *et al.* (1996) on the ground water along another palaeochannel of the Saraswa
lentified earlier in the extreme southwestern part of Jaisalmer district (Ghose et a
979; Kar, 1986), confirms that the recharge of the aquifer did not take place from tl
cal rainfall. While the periods of vigorous summer monsoon sustained fluvial activitie
ıe intervening arid phases favoured aeolian activities and disorganized the strea
etworks.

The sum total of evidences suggests that the climatic change, neotectonic activiti
ıd the load-discharge relationship along the Sutlej at the Himalayan foothills mig
ave played significant roles in the gradual disappearance of the Saraswati draina
etwork.

MAJOR FORMER STREAMS FROM THE ARAVALLI RANGE AND ELSEWHERE IN THE DESERT

A number of streams from the Aravalli range used to meet the south-flowi
roto-Saraswati system in the desert plains. The northernmost was the Kantli whi
sed to meet the system near Rajgarh or Dudwa Khara (Kar and Ghose, 198
resently, the Kantli hardly flows beyond Berasar to the south of Rajgarh. Upstream
ajgarh a number of former courses of the Kantli are controlled by east-west lineamen
'ig. 2). Ground water quality and discharge improves considerably along tho
neaments. Raghav and Grover (1991) mapped several other lineaments in the upstrea
eaches of the Kantli and suggested their controls on the channel shifts in the are
mong the other tributaries of the Proto-Saraswati from the Aravalli hills, one was fro
ıe hills to the north of Parbatsar passing through the present ephemeral saline lakes
argot and Kuchaman, while the others originated from the range to the south
arbatsar, flowed westward through Degana-Merta alluvial plain, and joined the maj
uthwest-flowing course through Ren, Kheduli and Pundlu to the west of Merta. Tl
ibutary which passed through Sargot and Kuchaman is now blocked by sand dunes
rm salt lakes. Nearby, the Sambhar lake at the confluence of the Medha and tl
upangarh streams had an outlet to the east, through which the combined flow used
rain into the Banas. That outlet is also blocked by sand dunes which were form
uring the period between 10-16 kyr before present due to climate change. Yet, sha
ngular bends in the courses of the surviving streams and their association with satelli
nage-derived lineaments suggest neotectonic activities as a complementary process
rainage disruption (Kar and Ghose, 1984; Kar, 1993 a; Paliwal 1993, 1994). Bagru aı
her places nearby Sambhar have registered mild earthquakes in the recent pa
)assarma, 1986).

An upland formed dominantly of the Bilara Limestone and Jodhpur Sandstone li) the northwest of Ren- Kheduli-Pundlu valley segment. It has a number of giant d alleys which appear to be the remnants of some earlier fluvial phases. The valleys er bruptly at the margin of the surrounding sandy alluvial plains, but create enormo amage to life and properties in those plains during rare spasmodic floods (i.e. that 996). This higher surface and a number of other low level planes have loosely cement ravel spreads. The gravels might have been detached from conglomerate bec ssociated with the Precambrian formations, and rolled significantly during the La ertiary- Early Quaternary fluvial regimes. The linear arrangement of the grav preads, a fining upward sequence in many of the sections and remote sensing signatur uggest at least three generations of superimposed palaeochannels, sustained possibly l urge, spasmodic flood discharge. Cementation of these channel gravels and relative uster erosion in the surrounding alluvial plains might have led to an inversion of reli uch that the earlier channels were left at a higher level than the alluvial surface bu y it (Kar, 1995). It is possible that the lower alluvial surface was then explored by tl roto-Saraswati system along the valley axis of Ren-Kheduli. Presently an under hannel occupies this valley from Pundlu downstream and is sustained by seasonal flc hrough several smaller dry valleys from the Aravalli hills. Similar raised chann ystems have also been reported from the Oman mountains, with implications f alaeohydrology (Maizels, 1990).

There are other evidences of such inversion of relief due to lithification of chann ed material in the Thar desert. For example, in the sandy plains between Jodhpu atkuriya and Osian the channel gravels of former small ephemeral streams have be emented into hard lithic calcretes through ground water mineralization, and are nc ft as higher surfaces due to selective erosion. Aeolian sediments occur both above ar elow many of these calcrete channel deposits. The findings suggest shifts in climate as najor force for drainage evolution. Karst and pseudokarstic features like sink holes ar oljes, as well as terra rosa in parts of the limestone hamada of Jaisalmer- Ramgarh ar n the Borunda-Rian-Merta area provide additional morphological evidence for tl eriods of higher rainfall in the past, but the variation in rainfall might not be too large

Most of the channels originating from the hills and rocky uplands within the dese sed to flow for shorter distances. For example, in the presently less than 200 m verage annual rainfall zone, the Lik river which was a major right bank tributary of tl uni, had its origin in the dominantly rhyolite hill-pediment complex of Pokaran ar haniyana. The stream built up its own narrow flood plain and flowed southward neet the Luni near Tilwara (Kar, 1988 a). Another major stream in the same dry be riginated in the uplands of Phalodi-Bap, flowed northward through the saline rann ap, the vicinity of Manchitiya, Naukha, Borana and Madasar, and ended in its own fa eposits further north (Kar, 1986). Wide dry valleys, often cut across by thick sa heets or high sand dunes, still mark the courses and carry water and sediments durin xtreme flood events like those of 1990 and 1996. The rocky tract of Jaisalmer has umber of dry valleys whose extent within the dune fields are difficult to trace throu onventional satellite remote sensing (using standard False Colour Composites, CCs). New radar remote sensing techniques, using Synthetic Aperture Radars, may seful in such terrain (Rajawat and Narain, 1996).

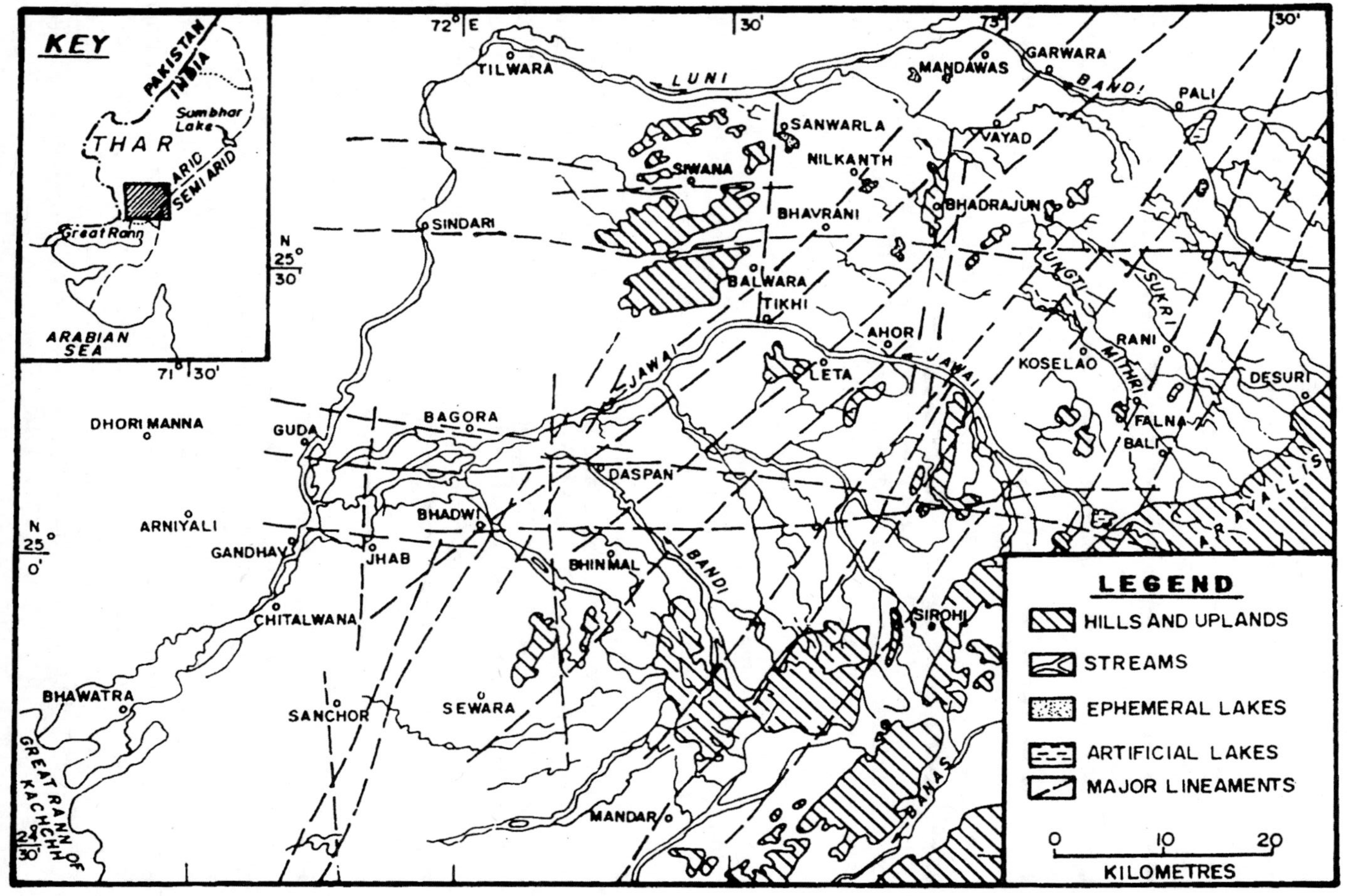

Fig. 3. Lineament map of south western Rajasthan showing major rivers, ephemeral lakes, artificial lakes and major lineaments in the region.

DRAINAGE NETWORK OF THE LUNI BASIN

The Luni basin in the south-eastern part of the Thar desert contains a bett ıtegrated drainage system, but it also has numerous examples of draina; isorganization. The disorganization was earlier attributed to recent aridity (Ghos 965). Interpretation of the satellite images of the Luni-Jawai plains, however, indicat ıe presence of a number of lineaments across the plains. Striking relationship w oticed between many of the NNE-SSW lineaments and drainage features like sudd ranching and widening of channel bed (e.g. that of Somesar, Sukri, Jawai and Sag isappearance and reappearance of channels (e.g. Mithri), and angular bend in tl ;ream courses (e.g. Sukri, Ungti, Bandi, Jawai, etc.). In addition to this, sudden incisi f stream beds and formation of tributary gullies, or terracing along many other strear .g. Somesar, Chhali, Sukri, Jawai, Bandi and Sagi), especially on the upstream side ıajor lineaments, are clear signatures of neotectonic movements. At least two terrac re noticed along the streams. The fall from the upper terrace to the lower terrace sually between 1.5 and 2 m, while that from the lower terrace to the stream bed etween 1.0 and 1.5 m. The maximum terrace relief of 10 to 12 m was noticed betwe aila and Daman along the Jawai river (Kar, 1984, 1988 b). Downstream of each maj neament the streams become braided and show signs of shifting till the next lineame cross them favours downcutting and terracing (e.g. Bandi, Sukri, Somesar, Jawai). If onsider the areas separated by the major NNE-SSW lineaments as individual fau locks, recent tectonic movements uplifted some of them, produced local base lev ıanges along the streams crossing them and led to downcutting and terracing. Tl maller tributaries might not have been able to keep pace with the downcutting of tl unk streams and, hence, responded by the formation of gully networks along them. Tl locks adjacent might have experienced relative down faulting, so that the streams ıem responded by braiding of courses and wider shifting of channel. Streams whi ould not maintain their courses over these differently moving blocks were obliterated arts and provided a segmented look (e.g. Ungti, Mithri). Since the amplitude of terra elief and also the depth of stream dissection has a tendency to increase from NNE SW, it may not be inappropriate to suggest that the lineaments represent faults whi lose to the northeast. The pattern of fluvial landform assemblages also suggest that tl ıovements produced a horst and graben sequence (cf. Paliwal, 1986) across the Lur awai plains, and possibly step faulting towards the west, the deepest segment lyi specially between the lower reaches of Jawai and Luni rivers in the west (rough etween Bhadwi, Jhab and Guda), and its linear extension further northeast throu alwara, Raithal, Motisara and Sanwarla (Kar, 1988 b, Fig. 3).

The satellite images taken after the flood of 1990 in the Luni basin, provi upportive evidence for neotectonically- controlled current channel processes. One of t est evidences was found in the Balwara-Debawas-Sanwarla sector where the ephemer treams flowing eastward from the Siwana hills in the west (e.g. the Luniwala and oth maller streams) and those flowing westward from the Aravalli hills (e.g. Mithri and i istributary networks), were suddenly lost in their own microfans along a maj neament, in spite of the fact that the streams were in spate (Kar, 1994). All the maj treams near the confluence of Luni with Jawai behaved abnormally while crossing ıajor NNE-SSW lineament. For example, the flood water of Sagi was divided through

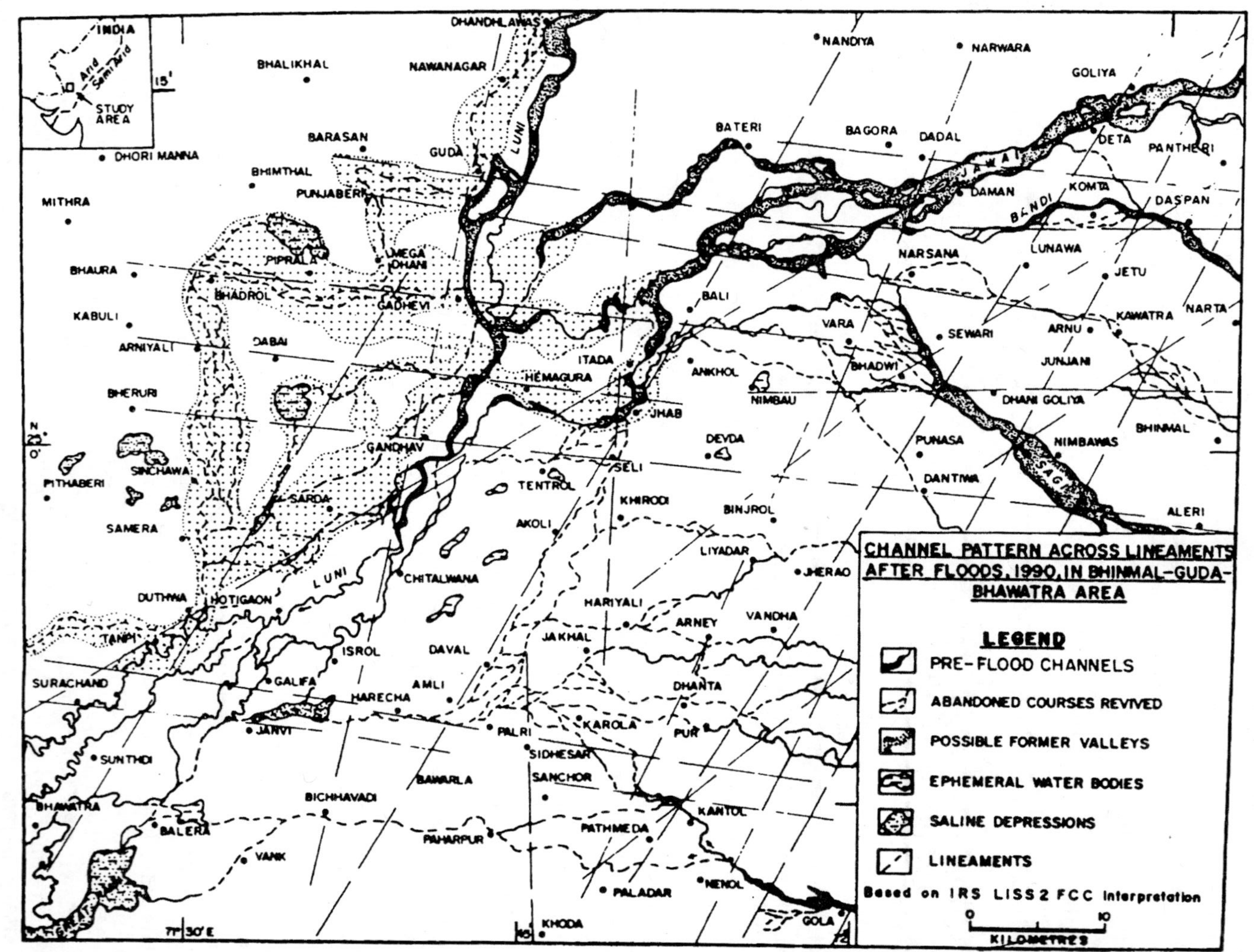

Fig. 4. Map showing channel pattern across lineaments after flood of 1990, in Bhinmal - Guda-Bhawatra area, south western Rajasthan.

ımber of channels while crossing a major NNE-SSW lineament near Dhani Golіy ırther north, the Bandi water was similarly divided when it negotiated with anoth ırallel NNE-SSW lineament. The Jawai has widened its course near Pantheri whe e river encountered this lineament. Downstream, near Dadal, the Jawai aracterised by anastomosing channel pattern, after it encounters another NNE-SS ıeament. The 1990 flood also helped one to evaluate the influence of the major E- ıeaments which were marked earlier (Kar, 1988 b). For example, the E- ıeament-controlled west- flowing courses of the Bandi from Daspan carried much of t ›od volume. In the case of Sagi, the courses downstream of Dhani Goliya were almc llowing the trend of an E-W lineament. The most notable controls were found in t se of Luni river to the south of Guda, where some of the former courses of the strea ere partially revived and followed some of the major E-W lineaments through t cinity of Punjaberi-Narsana-Kawatra, Piprala-Sewari, Arniyali-Jhab-Dhani Goliya, e ar, 1994; Fig. 4). Certainly, Luni once used to flow through a more westerly course e area, and the course was partly controlled by tectonic lineaments. A few other shift urses of Luni river in Guda-Gandhav sector were identified by Pal (1991) from t gnatures left after the flood of 1979.

CONCLUSIONS

The drainage pattern in the Thar desert does not appear to have evolved througl mple climate-driven process mechanism. Our studies suggest that climatic change a isodic tectonic movements during the Quaternary period were the driving forces f ainage evolution in the Thar desert. Supplies from a perennial Himalayan source (t ıtlej) were also responsible for the sustenance of the early Quaternary stream networ the Proto- Saraswati. In the Luni basin several surviving stream courses a ıeament-controlled. Episodic movements of the fractured basement changed the ba vel, triggered channel incision and even influenced channel shifting, but the chann so responded to fluctuating climate. Many former stream courses are yet to scovered in the desert. Use of new remote sensing techniques like the SAR images f apping, and appropriate geophysical tools for characterization of the vall nfigurations, as well as for sensing of the litho-sequences, will be most beneficial. T ıdings need to be further strengthened through detailed studies on litho-sequenc dimentological analysis, and precise dating methods.

ACKNOWLEDGMENTS

The author is grateful to Dr. A.S. Faroda, Director, CAZRI, and Dr. S. Singh, He Resource Survey and Monitoring Division, for the facilities provided.

REFERENCES

›wler, J.M., Jones, R. and Zhisheng, A. 1995. Climatic change in Asia and the Western Paci Its significance for Greenhouse scenarios. *Mem. Geol. Soc. India*, 32 : 1-21.

ıawla, S., Dhir, R.P. and Singhvi, A.K. 1991. Thermoluminenscence chronology of sand profi in the Thar desert and their implications. *Quat. Sci. Review,* 11 : 25-32.

'unningham, A. 1871. *The Ancient Geography of India.* (Reprinted 1979) by Indological Bc House, Varanasi, India.

)assarma, D.C. 1986. Neotectonism in Rajasthan - Its manifestations and effects. In : *Proceedin, Int. Symp. Neotectonics in South Asia. Survey of India*, Dehra Dun, 282- 288.

!eyh, M.A. 1994. The paleohydrology of the eastern Mediterranean. In : Bar-Yoset, O and K R.S. (Eds.) *Late Quaternary Chronology and Paleoclimates of the Eastern Mediterran* Radiocarbon, 131-145.

!eyh, M.A., Khouri, R.R. and Wagner, W. 1985. Environmental isotope study in the Ham region. *Geologisches Jahrbuch*, 38(C) : 315.

!eyh, M. and Ploethner, D. 1995. An applied paleohydrological study in Cholistan, Thar dese Pakistan. In : Adar, E.M. and Leibundgut, C. (Eds.) *Applications of Tracers in Arid Zc Hydrology*. Institute of Hydrology, IAHS Publ. 232, Wallingford, 119-127.

!hose, B. 1964. Geomorphological aspects of the formation of salt basins in western Rajasth *Proceedings Symposium on Problems of Indian Arid Zone.* CAZRI, Jodhpur, 79-83.

!hose, B. 1965. The genesis of the desert plains in the central Luni basin of western Rajasth *Jour. Indian Soc. Soil Sci.*, 13 : 123-126.

!hose, B., Kar, A. and Husain, Z. 1979. The lost courses of the Saraswati river in the Great Indi Desert - New evidence from Landsat imagery. *Geog. Jour.*, 145 : 446-451.

Kar, A. 1984. Assumptions regarding former stream behaviour and neotectonism from fluv terraces along the Somesar river in western Rajasthan, India. *Annals of Arid Zone,* 2 161-170.

Kar, A. 1986. Remote sensing of buried former streams in the extremely arid terrain of Jaisalm India, for water and salinity. *Proceedings Seventh Asian Remote Sensing Conference*, Seo B2/1-9.

Kar, A. 1988a. The Lik river of Rajasthan desert - Its source, courses and present conditions. I Tewari, A.K. (Ed.) *Desertification: Monitoring and Control.* Scientific Publishers, Jodhp 249-255.

Kar, A. 1988b. Possible neotectonic activities in the Luni-Jawai plains, Rajasthan. *Jour. Geol. S India*, 32 : 522-526.

Kar, A. 1993a. Drainage desiccation, water erosion and desertification in northwest India. I Sen, A.K. and Kar, A. (Eds.) *Desertification and its Control in the Thar, Sahara and Sa Regions.* Scientific Publishers, Jodhpur, 49-72.

Kar, A. 1993b. Neotectonic influences on morphological variations along the coastline of Kut India. *Geomorphology,* 8 : 199-219.

Kar, A. 1994. Lineament control on channel behaviour during the 1990 flood in the southeaste Thar desert. *Int. Jour. Remote Sensing*, 15 : 2521-2530.

Kar, A. 1995. Geomorphology of arid western India. *Mem. Geol. Soc. India,* 32 : 168-190.

Kar, A. and Ghose, B. 1984. Drishadvati river system in India - An assessment and new findin *Geog. Jour.,* 150 : 221-229.

Kar, A. and Shukla, J.P. 1993. Ground water exploration along the Saraswati-Drishadvati riv system in western Rajasthan: Kishangarh-Ghantiyal tract. *Interim Report* (1992-93). CAZ and GWD, Jodhpur, 15.

Maizels, J. 1990. Raised channel systems as indicators of palaeohydrologic change: A case stu from Oman. *Palaeogeog., Palaeoclim., Palaeoecol.,* 76 : 241-277.

Oldham, C.F. 1893. The Saraswati and the lost river of the Indian desert. *Jour. Royal Asiatic S* 34 : 49-76.

.dham, R.D. 1886. On probable changes in the geography of the Punjab and its rivers: historico-geographical study. *Jour. Asiatic Soc. Bengal,* 55 : 322-343.

ıl, G.N. 1991. Quaternary landscape and morphostratigraphy in the lower reaches of the Luı basin. In : Desai, N., Ganapati, S. and Patel, P.K. (Eds.). *Proceedings National Seminar c Quaternary Landscape of Indian Subcontinent.* University of Baroda, Vadodara, 79-90.

ıliwal, B.S. 1986. Source of salinity in soil and surface and subsurface water and its bearing c the evolution of the Thar Desert, India. *Transactions : XIII Congress of the Internation Society of Soil Science,* Hamburg, Germany p. 1532.

ıliwal, B.S. 1993. Geology of Rajasthan - a summarium. In : Chauhan, T.S. (Ed.) *Natural an Human Resources of Rajasthan.* Scientific Publishers, Jodhpur India 1-22.

ıliwal, B.S. 1994. Tectonics of the post-Aravalli mountain building activity and its bearing on th accumulation of sediments along the western flank of the Aravalli Range, Rajasthan, Indi. In : Ahmed, R and Sheikh A.M. (Eds.) *Geology in South Asia - I.* Proceedings of GEOSAS- February 23-27, 1992. Islamabad, Pakistan, 52-60.

ıghav, K.S. and Grover, A.K. 1991. Major sub-recent changes in the upper reaches of Kant river course and related environmental implications, Rajasthan. In : Desai, N., Gonapati, ! and Patel, P.K. (Eds.) *Proceedings National Seminar on Quaternary Landscape of India Subcontinent.* University of Baroda, Vadodara, 147-152.

ıjawat, A.S. and Narain, A. 1996. Reconstruction of Palaeodrainage Network of North We: India: A Review of Remote Sensing based Studies. *Scientific Note* SAC/RSA/ RSAG/ MWRI SN/01/96, Space Applicationc Centre (ISRO), Ahmedabad.

ngh, A.P., Babel, M.K. and Soni, V. 1994. A Report on Geophysical Resistivity Exploration t Trace the path of Buried Saraswati River and its Tributaries, District Jaisalmer. Groun Water Department, Jodhpur. (unpublished report).

ngh, D., Singh, Y.P., Bairwa, S.P., Dass, P., Porwal, C.P. and Mathur, K.M. 1996. Isotope stuc of water resources in arid areas of Jaisalmer. *Annals of Arid Zone,* 35(4) : 313-317.

ngh, G., Joshi, R.D., Chopra, S.K. and Singh, A.B. 1974. Late Quaternary history of vegetatic and climate of the Rajasthan desert, India. *Phil. Trans. Royal Soc.,* 267(B) : 467-501.

ngh, S. 1977. Geomorphological Investigations of the Rajasthan Desert. *CAZRI Monograph* Jodhpur.

nghvi, A.K. and Kar, A. 1992. *Thar Desert in Rajasthan. Geol. Soc. India, Spl. Publ.* Bangalore.

elichko, A. Borisova, O., Faure, H. Faure-Denard, L., Kremenetski, C., Morosova, T., Nechaev, ' and Zelikson, E. 1997. Climate variations in high and low latitudes: Ecosystem and bioma changes during the humid and arid phases of the last 125,000 years. *Proceedings Seminar c Desert Margin Changes in Africa since 135,000 BP: Implications for Water, Carbon and Ma* University of Nouakchott, Nouakchott, 143-145. (Abstract).

'asson, R.J. 1995. The Asian monsoon during the late Quaternary: a test of orbital forcing ar palaeoanalogue forecasting. *Mem. Geol. Soc. India,* 32 : 22-35.

ological Evolution of Northwestern India
;. Paliwal (Ed.)
entific Publishers (India), Jodhpur, 1999, pp . 260 - 270.

AN APPRAISAL OF SPACE TECHNOLOGY TO IDENTIFY THE PALAEOCHANNELS OF RIVER SARASWATI IN ANUPGARH, PILIBANGA AND PARTS OF GANGANAGAR AND HANUMANGARH DISTRICTS

D.C. Sharma*, N.K. Kalra, A. Srivastava and R.B. Singh

**Ground Water Department, Jodhpur – 342 001*
State Remote Sensing Application Centre, Jodhpur – 342 001

ABSTRACT

Remote sensing based studies of the great Indian desert reveal numerous signatures of palaeochannels in the form of curvilinear and meandering structures with tonal variation. Study shows the signatures of dry valley of Saraswati and Drishadwati, in western parts of the Anupgarh and Pilibanga areas in Ganganagar district. The tone, texture and contrast in variation of reflectance along the boundary of Ghaggar bed in the area depicts clearly the presence of palaeochannels. An attempt has been made to correlate the features associated with these palaeochannels with hydrogeological and geophysical exploratory studies. The space data and the interpreted lineament fabric reveal that the major cause of migration of river courses is due to the Tertiary and the Quaternary tectonic movements.

ey words : Space Technology, Palaeochannels, Saraswati, Ground Water, Anupgar Pilibanga, Rajasthan.

INTRODUCTION

Archeological findings as well as the geological evidences have established that tl ıar desert of India had a vast expanse of the Ghaggar bed (Oldham, 1893; Ghosh et a '79). It has been explained by assuming that some major tributaries were flowiı ıring the recent past (Bakliwal et al., 1988; Ramaswamy et al., 1991). Yashpal et ε

)80) proposed that near Satrana and Suratgarh, two different palaeochannels mu .ve joined the Ghaggar which seems to have disappeared in a depression because :tonic activity in the region. They have further suggested that the Satluj was the ma .butary of the Ghaggar and the tectonic movements have forced the Satluj to sh :stwards as a result of which the Ghaggar dried up. Kar (1993) suggested that t ıtluj should be the master stream and the Ghaggar as a tributary of it.

The visual interpretations of IRS 1B LISS II (FCC) data clearly indicate that t ıeaments are running for a considerable distance in parts of Ganganagar aı ınumangarh districts, particularly following alignment of the Ghaggar bed (Fig. ıese lineaments can be interpreted as signatures of the palaeochannels.

The Geomorphic landforms like-flood plain, alluvial plain and alluvial sandy pla ar the vast Ghaggar bed area have already been identified and interpreted with t lp of visual interpretation of the satellite imagery. In the present study, such visu servations are made in parts of Anupgarh and Pilibanga areas of Ganganagar aı ınumangarh districts.

Hydrogeological and geophysical exploratory studies have been carried out pport the existence of these palaeochannels in the study area. The hydrogeologic .rameters such as water table contour map, statistical table related to water le ıctuation, mean annual recharge, ground water quality (Electrical Conductance Ma ıd the geophysical sounding data of the study area have been discussed in this paper.

LOCATION AND EXTENT OF THE PALAEOCHANNELS

The integrated study of the landforms and land use map based on the satellite da terpretation depicts the location and extent of palaeochannels in the study area (F . The following areas have been selected for the present study :-

(a) The area in the vicinity of Anupgarh, which is a part of former migratory cour of river in Anupgarh (Fort Abas-Indus area).

(b) The area near Pilibanga towards Hanumangarh and Kalibanga having- flood, major landforms.

(c) The area near Rawatsar extended from landing ground to Bhuranpura (v Budhwalia).

(d) The area starting from Asarjana, Nohar to Bhadra and upto Hisar and onwar which is the migratroy course of two former river channels.

GEOLOGY OF THE AREA

The geology of the study area is concealed under thick blanket of sand dunes a luvium and no rock exposures are seen on the surface. The stratigraphic units in the eas range from Early Palaeozoic (Marwar Supergroup) to Quaternary in age (Pare)81, 1984; Paliwal, 1993, 1994). From the study of lithologs of tube wells a ezometers drilled by Central Ground Water Board and State Ground Wat evelopment in these areas, the Marwar Supergroup of rocks have been encountered a pth of around 150 m. The stratigraphical units are shown in Table-1.

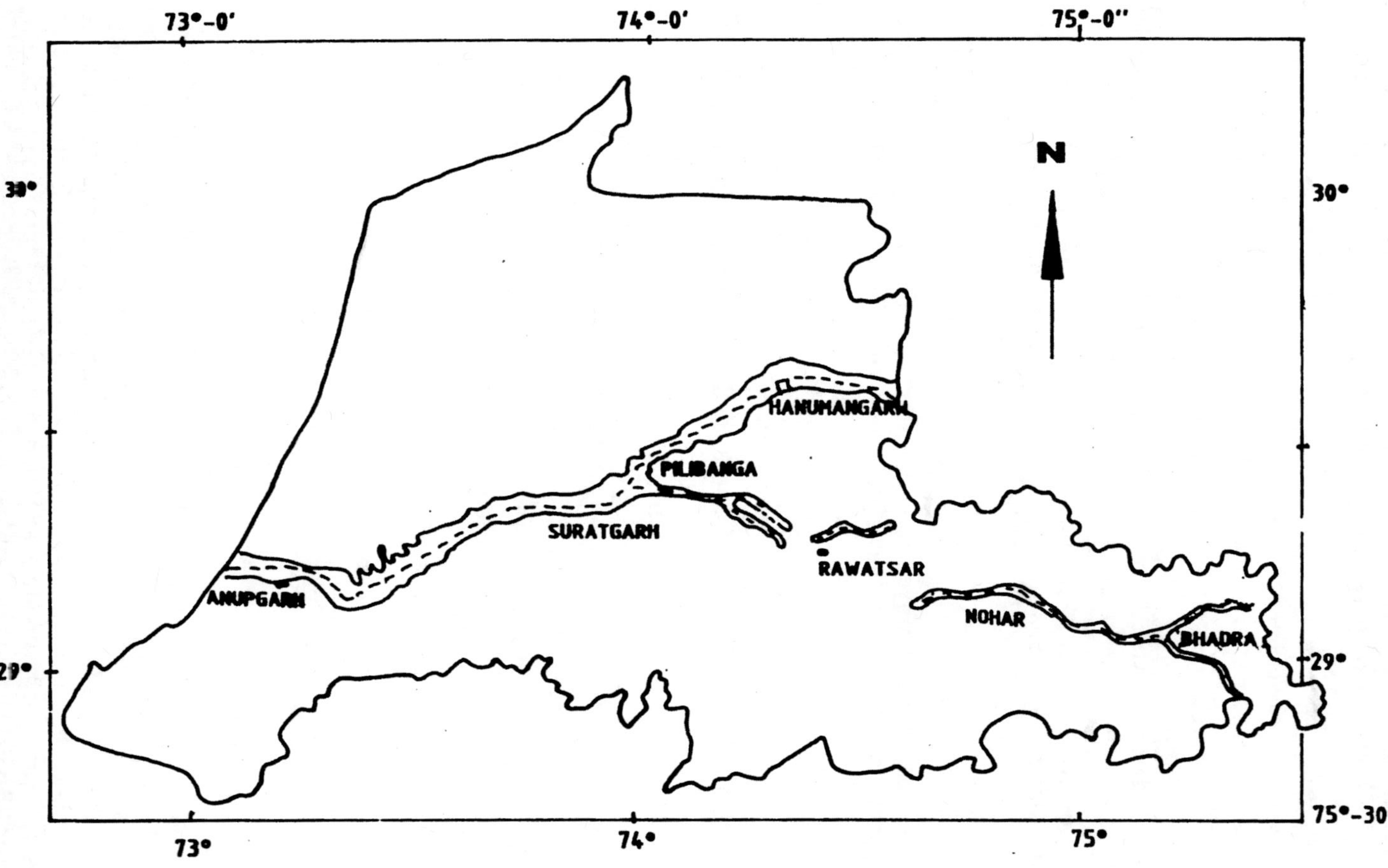

Fig. 1. Location of palaeochannels in the study area

able 1. Geology of the Area

ge	Formation	Depth in Metres.	Litholog
ecent to Pleistocene	Quaternary	0-20	Fine to medium grained, buff coloured sa silty clay and kankar.
		25-55	Clay, sandy clay with lenses of silty clay
		Unconformity	
arly Palaeozoic	Marwar Supergroup (Deeper aquifer)	85-120	Sandstone, silty clay, consolidated, ferruginous, fine grained with intercalations of shale

METHODS AND TECHNIQUES

Palaeochannels have been mapped using visual interpretation of satellite ima uring preparation of hydrogeomorphological map for ground water potentiality. So eld evidences in support of the signatures of palaeochannels have been encountered nder –

(i) Presence of vast terrain of alluvium in the region.

(ii) Presence of lush green vegetation, even during the hot summer months alc the traced courses.

(iii) Presence of silty soil in the interdunal areas, instead of thick sandy deposits.

The following techniques have been adopted during identification of the propos alaeochannels in the study area :-

MATERIAL USED

During interpretation of the satellite data, the IRS 1B LISS II (FCC) of ba mbination 2, 3 and 4 in hard print copy of scale 1 : 250,000 in conjunction with),000 scale for both seasons Rabi and Kharif, pertaining to the period 1994-96 ha een used in a variety of forms and formats. The optical instruments such as light tak ROCOM, orthograph and magnifying lense have also been used.

Remote Sensing Technique

The present study is essentially based on the visual interpretation using ph ecognition elements such as tone, texture, pattern, shape, size, location and t eotechnical elements like landforms, drainage and vegetation etc. for demarcation of t xistence of the palaeochannels. The palaeochannels in most of the cases stand o early because of the good vegetation along their beds.

The space technology has proved to be extremely useful in identifying the locatic f the palaeochannels in the study area. A multi-disciplinary approach employi ydrogeological, chemical, mineralogical and geophysical techniques combined with t emote sensing techniques can provide a clear and better history of these palaeochanne he details about all the conventional methods have been discussed hereunder.

The Hydrogeological Investigations

The detailed hydrogeological investigations have been carried out in support of t istence and location of the palaeochannels in the study area which are mention ›reunder —

) *Aquifer Characteristics*

The hydrogeological investigations reveal that the alluvium of the Quaternary a the only water bearing formation occurring in the area and can be divided into two, i e younger and the older alluvium. The aquifer system in the area has a compl rangement of sand and clay with sporadically distributed lenses of silty clays aı ınkars and occasionally, some gravel horizons. It constitutes the aquifer, covering mc the sediments forming the landforms like flood plain, alluvial plain and alluvial pla ındy) along with the palaeochannels. These sediments are of heterogenous nature aı ıry, both laterally and vertically. The sandy horizon is the principal aquifer.

The water table fluctuation is governed chiefly by the specific yield of the materi the zone of water bearing formation. As a support to the present study, the aquif aracteristics, the position of water level in different key wells located in t 'drogeological formations given above and the blockwise recharge of ground water frc fferent sources during five consecutive years i.e. 1991-95 are shown in Table 2 aı ıble 3 respectively.

ıble 2. Blockwise water level fluctuation in the study area

ame of the Block	Average Water Level Fluctuation (in Metres) (from 1991 to 1995)
ıupgarh	0.26
anumangarh	0.67
ıadra	0.27
ɔhar	0.33

) *Ground Water Flow Direction* :

The study area has some open wells, tube wells and piezometers etc. Water sampl both pre-monsoon and post-monsoon seasons were collected in the last five consecuti ars (1991-95) from all these wells. Water level measurement and fluctuation in wat ble have been recorded. The water table contour map was prepared to show the grouı ıter flow direction with respect to the position of the proposed palaeochannels. T ıter table contour map is shown in Fig. 2.

) *Chemical Quality Observation* :

For completion of the detailed hydrogeological investigations, the chemical analys some selected water samples collected from the key wells and piezometers at aı ound the study area with the palaeochannels have been done using the wet chemic ıalysis in the laboratory. The Electrical Conductance Map prepared on the basis emical quality of the ground water is shown in Fig. 3.

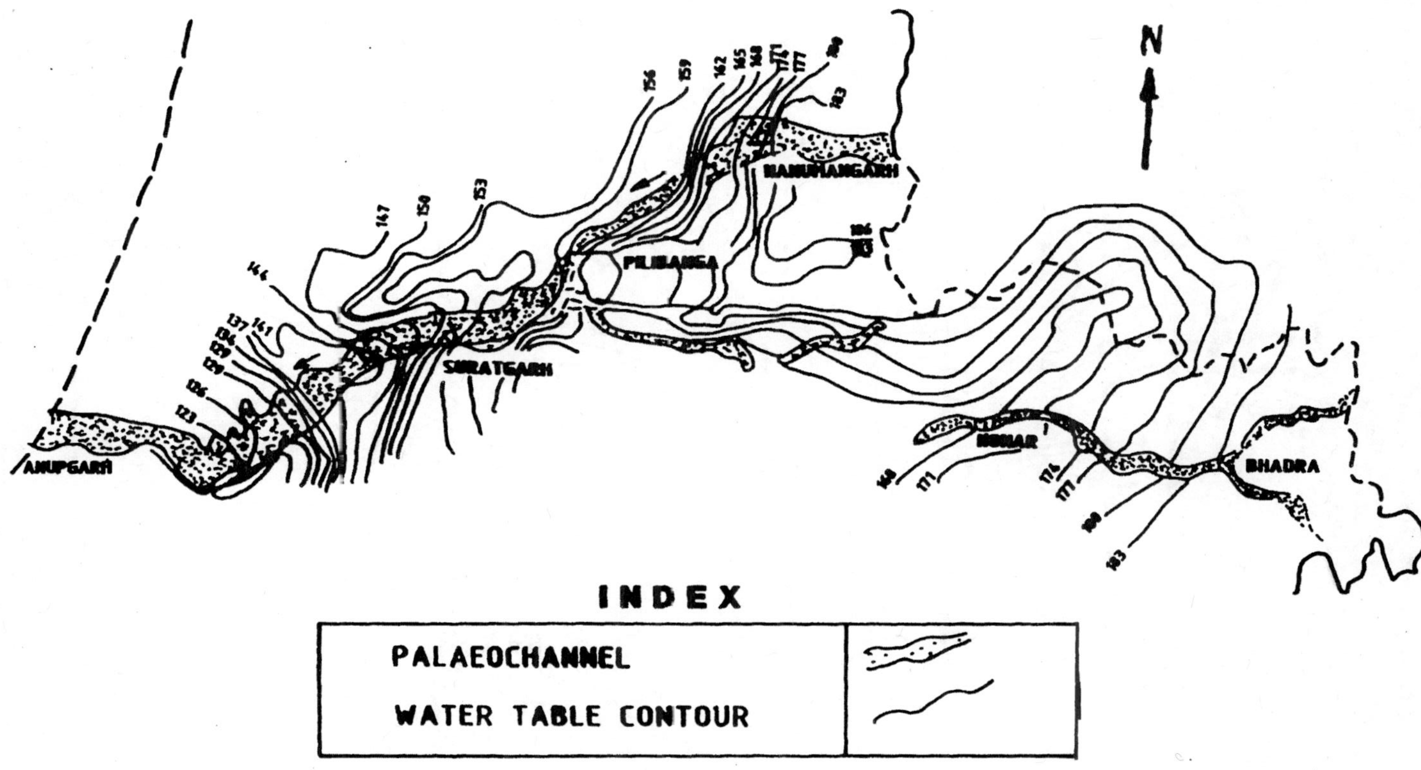

Fig. 2. Map showing ground water flow in the study area in Ganganagar and Hanumangarh districts

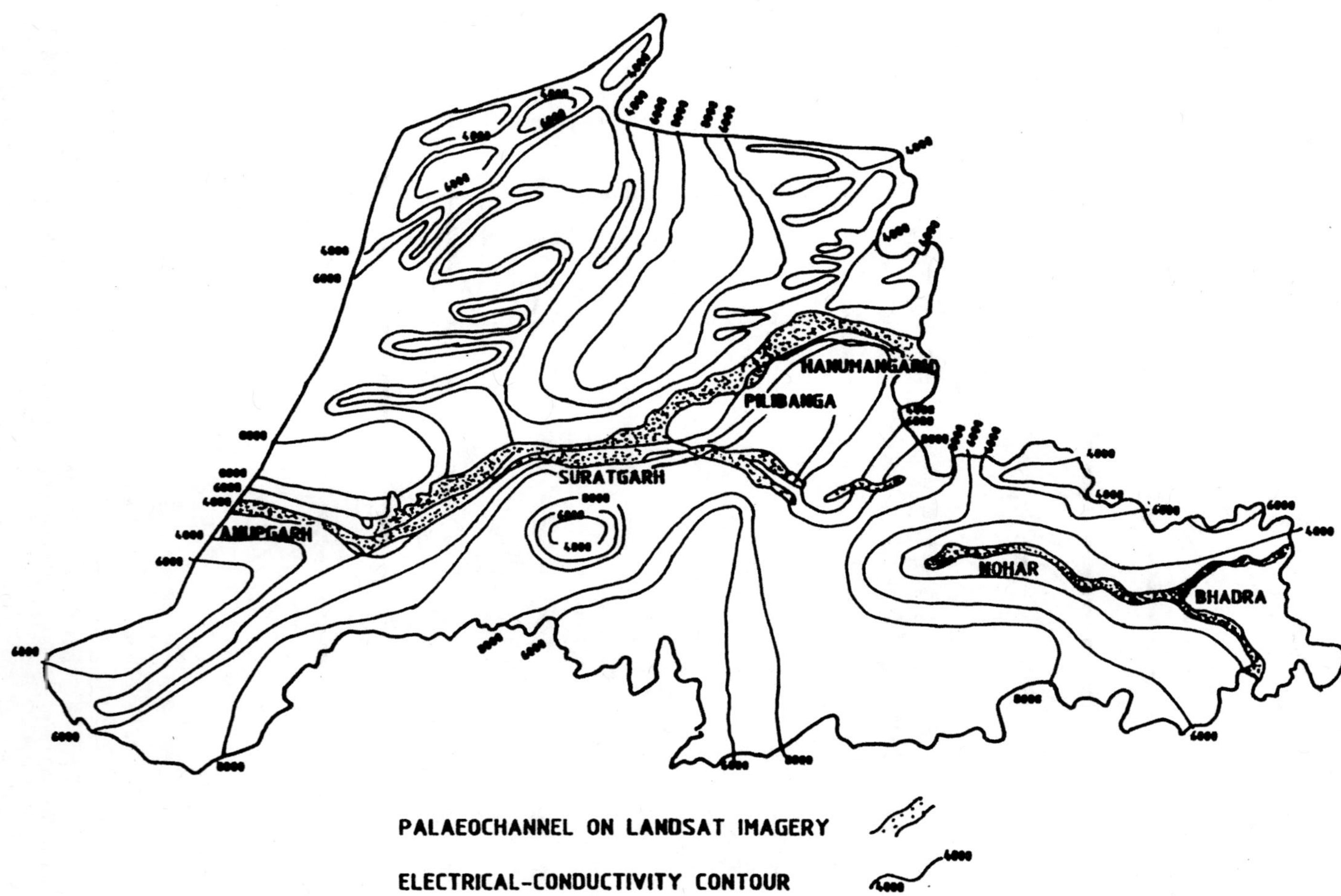

Fig. 3 Map showing ground water quality (EC) in the study area

ıble 3. Recharge in different blocks of Hanumangarh and Ganganagar districts

ame of the Block	Recharge from Canals in MCM			Recharge from surface water irrigation in MCM		Mean Monsoo Recharge MCM
	Annual	Monsoon	Non-Monsoon	Rabi season	Kharif Season	
hadra	13.0943	4.8090	8.2852	12.4583	3.2973	7.2446
ohar	4. 6116	1.9671	2.6445	4.1140	1.3347	2.6571
nupgarh	80.0414	50.9698	29.6311	34.6311	9.7381	7.1023
anumangarh	75.0203	36.0537	38.9666	28.9478	24.3589	44.2006

The Geophysical Investigations

The geophysical technique has been carried out as a further supplementary tool udy the sub-surface behaviour of the formation existing around the palaeochann ırticularly in Nohar-Bhadra area. Aquameter, was used for electrical resistivity in t cinity of the palaeochannel. A few vertical electrical soundings have been conducted ıb-surface layer identification. The geoelectrical parameters were interpreted in ter geological or hydrogeological units. The geophysical investigation along the Ujalw ohar section has been shown in Fig. 4.

RESULT AND DISCUSSION

Palaeochannels were identified on the basis of satellite imagery and detail ydrogeological investigations were carried out. It has been observed that the wa ble contour map shows the flow of ground water parallel to the identifi ılaeochannels. The sediments are fluvial in origin and the aquifer is made up ermeable gravel and coarse to medium grained sand and silty soil etc. The Electri onductance Map is symptomatic with the ground water contour map. However, there cursion in the higher conductance values in a few parts of Budhwalia- Bhuranpu ea. Though the chemical quality of ground water is saline in most parts of Gangana ıd Hanumangarh districts but the study area along the Ghaggar bed shows an entir fferent situation. For example the electrical conductance of the ground water in t gion is as high as 20,000 μ/cm or even more but the area adjoining the Ghaggar b ıows electrical conductance values equal to 2,000 μ/cm or less. It is also less than 4,0 cm in a few parts of the study area particularly in Nohar-Ujalwas and Soti-Bari are ıis water has been recommended suitable for the irrigation purpose. Some parts of t udy area have electrical conductance value between 4,000 to 6,000 μ/cm of grou ater particularly near Rawatsar, Bhuranpura- Budhwalia area. This could probably ecause of the local conditions like water logging etc.

On comparison of locations of the palaeochannels with the lithologs of the tu ells and the piezometers collected during drilling of these wells, it infers that most ıe areas in the vicinity of Ghaggar bed have sediments of fluvial character and electri nductance value is either less than or equal to 2,000 μ/cm. Unfortunately, there ıucity of bore hole data from Nohar-Bhadra area. The bore hole data of a few villa e available from the nearby areas of the palaeochannels, i.e. Nithrana, Soti-Bari etc.

tisfy the results, however the quality of the ground water is good and the sedimen ıve a fluvial character.

Loosely bound sandy formations often mask the current penetration. However, tl terdunal plains were covered with sounding techniques. On the basis of inferr oelectrical parameters such as resistivity in ohm-metre and thickness in metres, oı uld think that the resistivity variations close to the palaeochannels depict a cle ıomaly as compared to the other sandy horizons. The first layer resistivity ranging fro to 400 ohm-metres reflects that the surface soils have thickness of 1 to 2 metres. Tl cond layer, about 8 to 10 metres below ground level shows the resistivity valu nging between 20 to 600 ohm-metres corresponding to dry sand. Interestingly, tl ird layer shows the resistivity valve of 15 to 84 ohm-metres, reveals that the san turated with potable water in fact have varying thickness. This zone has the maximu pth of 20 metres below ground level. The fourth layer shows the resistivity value of 1 ohm- metres, this indicates that the sandy clay formation is saturated with the saliı ater. This study has been done along the Ujalwas-Nohar section shown in Fig. 4.

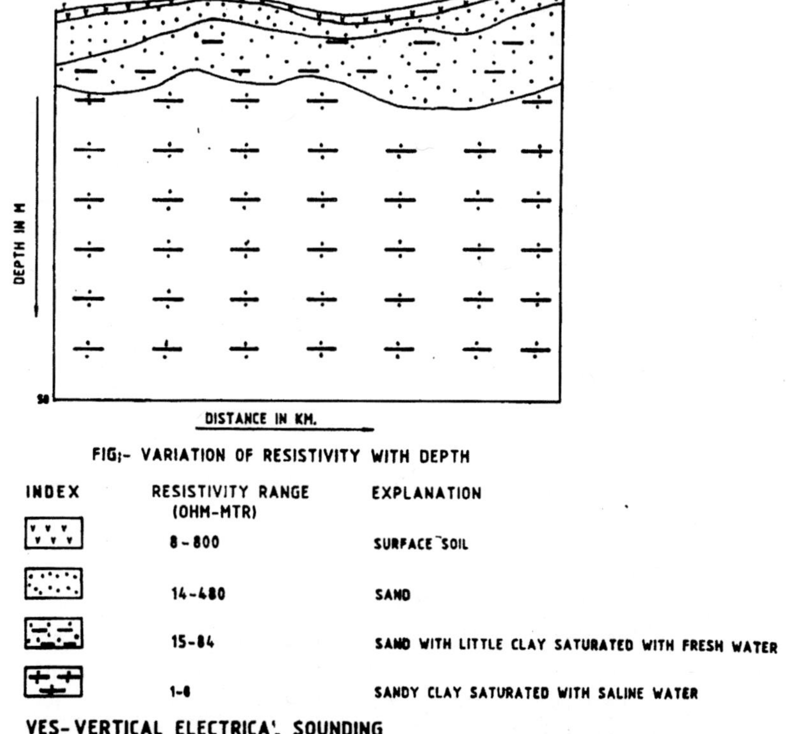

Fig. 4. Ujalwas Nohar Section (VES)

CONCLUSION

The vast expanse of the Ghaggar bed (width 6-8 km) have a palaeochannel whi ight have joined this main channel near southeast of Markanda river near Satrar he present dried bed of Ghaggar was initially a part of Saraswati. The other chanr ould have come through Jind, Hissar and after covering the Bhadra-Nohar area, ems to have joined the Ghaggar near Suratgarh. It seems to be the part of Drishadw ver channel. The space data and the interpreted lineament fabric network of the ar veal that the major cause of migration of river courses is due to the Tertiary and t uaternary tectonic movements.

The geophysical study conducted in the area shows the presence of a coarse avel sand with a thickness of 20 to 30 metres. It is saturated with the potable quali ground water. Interestingly, the sudden drop in the value of the aquifer resistivi low 30 metres shows indirectly the presence of saline water.

The recharge of ground water in different blocks of Ganganagar and Hanumanga stricts has already been discussed. It shows that there is paucity of recharge from urces in Nohar and Bhadra blocks in comparison with the other blocks. Because the no source of artificial recharge in Nohar-Bhadra area at present, it is hereby propos at the artificial recharge through these channels may be of great significance for t sidents of the region. Although the study of the hard barrier is a prerequisite.

The palaeoclimatic conditions of the study area should be observed to collect t ater samples along these palaeochannels and the isotope analysis of these wat mples should be done. The pollen study may also be helpful in this regards.

ACKNOWLEDGEMENT

Authors are grateful to Shri R.P. Jain, Director, Department of Science a echnology, Jaipur for providing the necessary facilities during the course of the prese udy. Thanks are also due to Shri K.S. Srivastava, Shri P.K. Batra and Shri J.P. Shuk Ground Water Department, for providing necessary assistance during the course of t udy. Shri Jyoti Swarup Sharma and Shri Suresh Mathur of State Remote Sensi pplication Centre, Jodhpur have assisted in many ways.

REFERENCES

akliwal, P.C. and Grover, A.K. 1988. Signatures and migration of Saraswati river in Thar dese Western India. *Rec. Geol. Surv. India.* 116 : 77-86.

hosh, B., Kar, A. and Hussain, Z. 1979. The lost courses of the Saraswati river in the gre Indian desert. New evidences from landsat imagery, *Geog. Jour. London,* 145 (3) : 446-451.

ar, A. 1993. Drainage desiccation, water erosion and desertification in north-western India. I Sen, A.K. and Kar, A *Desertification in Thar, Sahara and Sahel region.* 49-72, Scienti Publishers, Jodhpur.

ldham, C.F. 1893. The Saraswati and the lost rivers of the Indian desert. *Jour. Roy. Asiat. S Calcutta Review,* 34 : 46-76.

ıliwal, B.S. 1993. Geology of Rajasthan - a summarium. In : Chauhan, T.S. (Ed.) *Natural a* *Human Resources of Rajasthan*. Scientific Publishers, Jodhpur, India. 1-26.

ıliwal, B.S. 1994. Tectonics of the post-Aravalli mountain building activity and its bearing on t accumulation of sediments along the western flank of the Aravalli range, Rajasthan, Ind In : Ahmed, R. and Sheikh, A.M. (Eds.) *Geology in South Asia - I : Proceedings First Sou Asia Geological Congress*, February 23-27, 1990, Hydrocarbon Development Institute Pakistan, Islamabad, Pakistan, 52-60.

ıreek, H.S. 1981. Basin configuration and sedimentary stratigraphy of Western Rajasthan. *Jo Geol. Soc. India.* 22 : 517-527.

ıreek, H.S. 1984. Pre-Quaternary Geology and Mineral Resources of Rajasthan. *Mem. Ge Surv. India* 115 : 1-9.

ımaswamy, S.M.; Bakliwal, P.C. and Verma, R.P. 1991. Remote sensing and river migration western India. *Inter. Jour. Remote Sens.* 12 : 2597-2609.

ıshpal; Sahai, B.; Sood, R.K. and Agarwal, D.P. 1980. Remote Sensing of the lost Sarasw river. *Proc. Ind. Acad. Soc. (Earth Planet Sci.)*, 89 : 317-332.

ological Evolution of Northwestern India
S. Paliwal (Ed.)
ientific Publishers (India), Jodhpur, 1999, pp. 271 - 292.

HIGH FLUORIDE CONTENT IN GROUND WATER : A CURSE TO THE HUMAN HEALTH IN THE CENTRAL PART OF RAJASTHAN, INDIA

B.S. Paliwal

Department of Geology, Jai Narain Vyas University, Jodhpur - 342 005, India.

ABSTRACT

Fluoride content in drinking water exceeding the prescribed permissible limit has become a serious health hazard to a very large human population residing in several states and union territories of this country. Rajasthan with a 5.2 percent of country's human population has been recognized as a state endemic for fluorosis. Hydrogeological and epidemiological studies carried out in Nagaur district of Rajasthan have revealed alarming results. Not only fluoride but nitrates, chlorides, sulphates and total dissolved solids in the ground water of the district exceed the upper limit prescribed by the World Health Organization (1971). As a result, a very large number of inhabitants in the district have been suffering fron dental fluorosis, skeletal fluorosis and several other manifestations.

ey words : Groundwater, Fluoride, Human health, Fluorosis, Nagaur, Rajasthan.

INTRODUCTION

Fluoride intake to the extent of 0.80 to 1.60 milligrams per litre (World Heal rganisation, 1971) is quite essential for the protection of enamel on teeth in hum ings and proper growth of bones both in human beings as well as in livestock. T ureau of Indian Standards (1992) has prescribed the desirable limit of fluoride inking water to be 1.0 milligram per litre (cf. The New Cncyclopaedia Britainni 83) which can be raised to 1.5 milligrams per litre in the absence of an alternati urce. However, the upper limit of 1.60 milligrams per litre is the maximum permissi vel (Paliwal, 1988) which includes total intake of fluoride in human body through a .eans i.e. drinking water, food and cosmetics etc.

In Indian context fluoride intake through other means like eggs, milk, meat, lea ɜgetables and tooth pastes, plays an important role in bringing the upper limit uoride content in drinking water as low as 1.0 milligram per litre. Unfortunately, a veı .rge human population living in the states of Andhra Pradesh, Bihar, Gujrat, Haryan arnataka, Maharashtra, Madhya Pradesh, Orissa, Punjab, Rajasthan, Tamil Nad ttar Pradesh and the Union Territory of Delhi is facing a serious danger to the healt ecause of the high fluoride content in drinking water (Paliwal, 1988). About 25 millic eople in 8700 villages in these states have already been affected by fluorosis (India ouncil of Medical Research Report, 1992). Rajasthan, the state second largest in are ith a human population of the order of 43.88 million, is emerging as a prominent nam nongst the states endemic for fluorosis - a disease caused by the excess fluoric nsumption.

Recent studies on the ground water conditions in the central part of Rajasthan an 's bearing on human health (Paliwal, 1988) has revealed alarming results. Particularl Nagaur, Jalore, Churu, Jodhpur, Pali, and Sikar districts situated along the westeı ank of the Aravalli mountain, there is a high fluoride content in ground water (Paliw nd Shakuntala, 1993). The paper deals with detail geological, hydrogeological an ɔidemiological investigations carried out in Nagaur district with special reference to th uoride content in ground water.

GEOLOGY OF NAGAUR DISTRICT

Geologically, the Nagaur district, located in the heart of Rajasthan, comprises ide spectrum of geological rock formations ranging from the Archaean basement to th ecent alluvium and desert sands of the great Thar Desert. Credit goes to Blanford (187 r giving the first and foremost geological account of the region. However, detaile eological study of the region was done by a large number of workers like Oldham (1886 a Touche (1902), Heron (1933, 1953), Pascoe (1959), Srivastava (1971), Pareek (198 984) Paliwal (1976, 1978, 1980, 1986, 1994, 1998; Paliwal et al., 1997).

The Archaean basement rocks comprising granitic gneisses, schists an nphibolites with bands of aplite are exposed around Didwana, Harsor, Mori, Bhakı iplad, and Parbatsar. Rocks of the Aravalli Supergroup particularly quartzite ıyllites, schists, carbonaceous phyllites and mylonites crop out as isolated hills to th uth-west of Didwana (Paliwal, 1980), east of Khatu (Heron, 1933, Paliwal, 1998), we Kolia and around Degana, Parli, Gopalpura, Achhojai and Moryana in the district. Th elhi Supergroup of rocks include a variety of marbles, conglomerates, quartzites, schis nd amphibolites (Heron, 1953, Paliwal, *et al.,* 1997). Marble is exposed in form of fiv near belts running from Borawar to Makrana and also around Meh (Parbatsar) an hokar (Shyampura). Conglomerates, quartzites, and amphibolites of the Alwar Grou e also exposed in a linear belt running from Parbatsar in the south to Maroth in th ɔrth, passing through Manglana and Kuchamancity. The Malani Igneous Suite of rock e exposed as granites around Degana and at many places between Didwana an alasar. Malani rhyolites crop out near Balsamand, Ladnun, Badala, Chhapra Gunpal nd Chhoti Khatu (Paliwal, 1998).

The Marwar Supergroup comprising Jodhpur, Bilara and Nagaur Groups occupies .st terrain in the western part of the district. Sandstones and shales of the Jodhp :oup are exposed around Khatu (Paliwal, 1976, 1998) and Ladnun (Paliwal, 1978). T lcareous rocks of the Bilara/Hanseran Group comprising limestone, dolomite, cher lomite, claystone and siltstone can be seen around Bilara, Gotan, Khinwsar, Mert ıdnun, Dhanappa, Rol, and Pondlo. On the other hand sandstones, shales aı aporites of the Nagaur Group are well exposed around Nagaur. Tertiary rocks ha en encountered in bore holes at Merta, Indawar, Igyara and Kasnau. These roc clude, lignite (Eocene to Palaeocene), carbonaceous shales, red shales, clays aı rruginous sandstones. Quaternaries comprising older and younger alluvium occupy ry large area of the district. They include silts, clays, gravels, pebbles, lime kanka ıd wind blown sands. In addition to this, the tributaries of Luni, Medha rivers aı ılaeochannels of Saraswati river display river sediments. Salt lakes of Didwaı aliwal, 1982, 1986), Sambhar, Kuchaman, Talchhapar and Khatu have the aracteristic sediments.

GROUND WATER CONDITIONS

In the absence of perennial rivers, sweet water lakes and a network of man ma nals in the district, the ground water is the only source for irrigation and domestic us course, rain water harvesting through different modes locally known as 'talao' 'nad ınka' etc. and using it throughout the year is a tradition in the region. B ıfortunately, neither its quantity is sufficient nor its quality is up to the mark. Surfa ntemination in such a case is of a very high order.

Nagaur district covering an area of about 17,718 sq. kms. is inhabited by a hum pulation of 2.144 million. With an average rain fall of 362 millimetres the district hos out 1000 wells including dug-wells and tube-wells. Ground water in the district occu confined, semi-confined and unconfined conditions. Potential aquifers are (1) pı avalli gneisses, schists and phyllites, (2) Jodhpur sandstone, (3) Bilara limestone, (agaur sandstone, (5) Tertiary sandstone and (6) Older alluvium. Figure 1 sho stribution of different geological formations in the state and Figure 2 displays t stribution pattern of the water table in the region.

QUALITY OF GROUND WATER IN THE DISTRICT

For a systematic study of the ground water quality in the district, samples we llected from tube wells and dug wells in the district and analysed for t ıysico-chemical properties. Special emphasis was given to fluoride, nitrates, chloride lphates, and total dissolved solids. Less attention was paid to characteristics like p lour, taste, odour, turbidity, hardness and other chemicals generally present in grou ater. Unfortunately, most of the ground water samples so collected show concentrati fluoride, nitrates, sulphates, chlorides and total dissolved solids much higher than t nit prescribed by the World Health Organization (1971), the U.S. Department ealth, Education and Welfare (1962) and the Bureau of Indian Standards (1992). Tab shows characteristics of an ideal sample of drinking water in the Indian context.

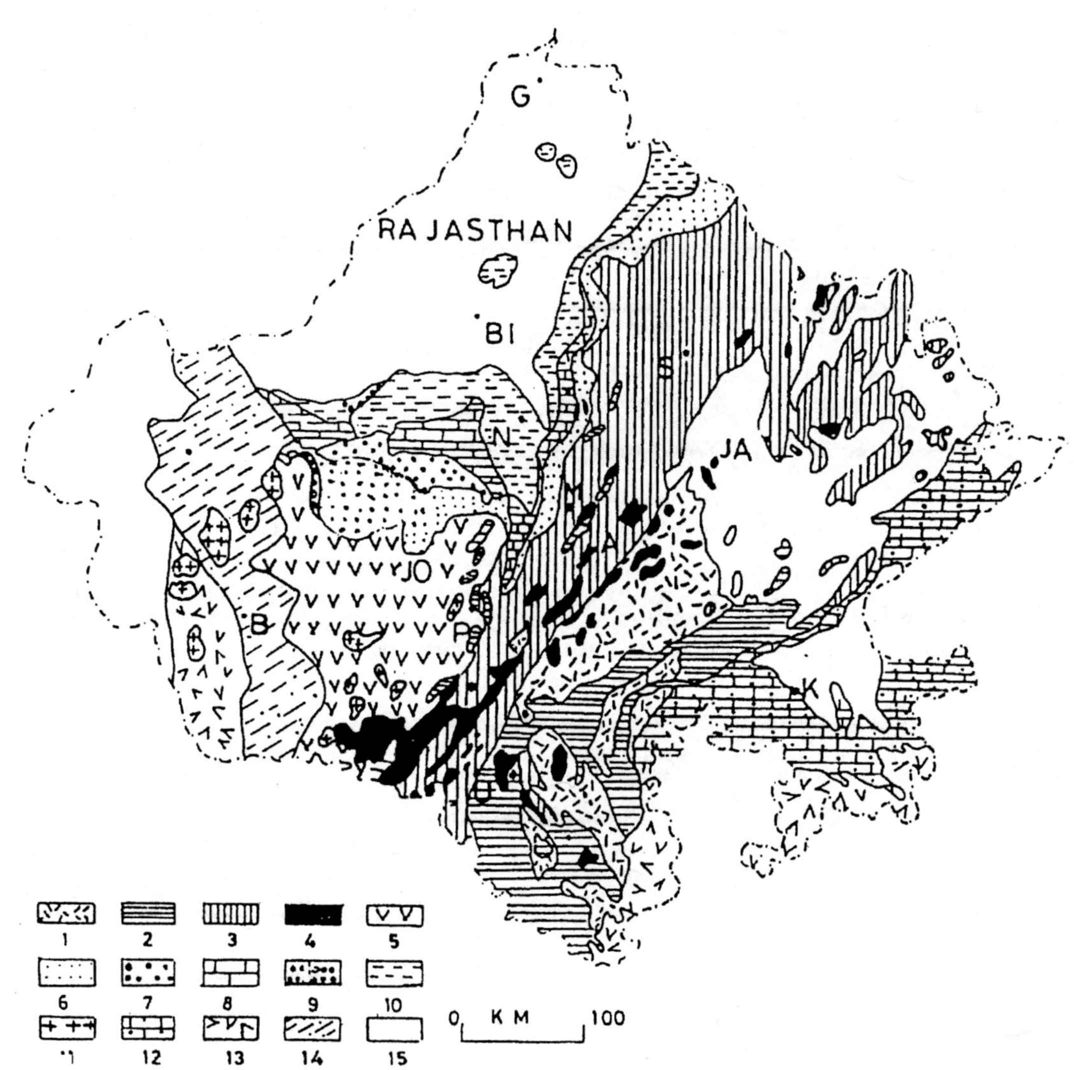

g. 1. Geological map of Rajasthan. 1- Banded Gneissic Complex, 2 - Aravalli Supergroup, 3- Delhi Supergroup, 4- Erinpura and other older granites, 5- Malani volcanics, 6-10. Marwar Supergroup (6- Sonia shale, 7- Jodhpur sandstone, 8 - Bilara limestone, 9- Pokran boulder bed, 10- Nagaur sandstone and evaporites), 11- jalore and Siwan granites 12- Vindhyan Supergroup, 13- Deccan trap, 14- Mesozoic sediments, 15- Cenozo sediments and sands of the Thar desert. A- Ajmer, B - Barmer, B1 - Bikaner, G- Ganganagar, J- Jaisalmer, JA- Jaipur, JO - Jodhpur, K- Kota, M - Makrana, N - Nagaur, P- Pali, S- Sikar and U - Udaipur.

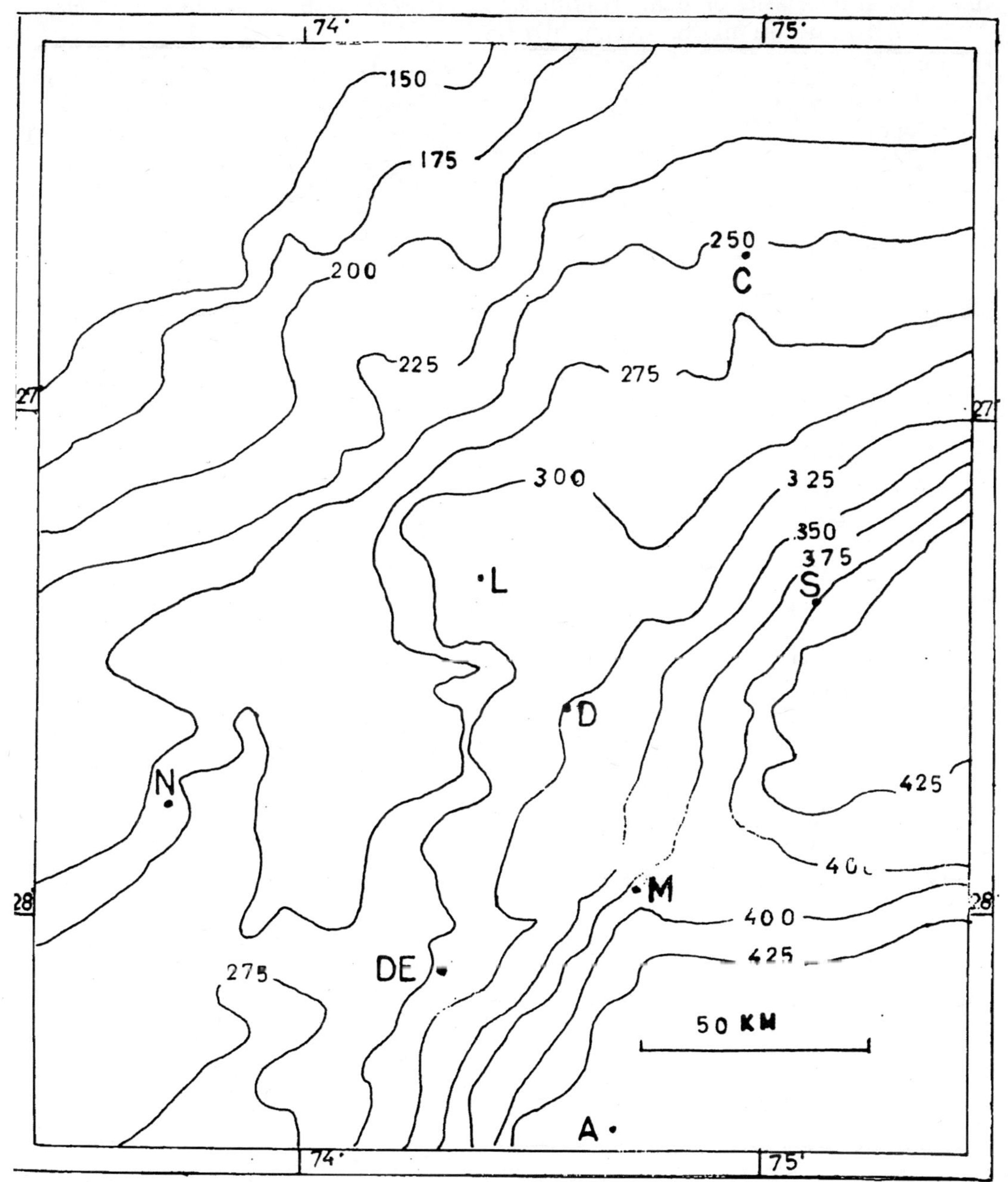

Fig. 2. Water table pattern in the region. A - Ajmer, C- Churu, D- Didwana, DE- Degana, L- Ladnun, M - Makrana, N - Nagaur, S - Sikar.

'able 1. Characteristics of ideal drinking water (Bureau of Indian Standards, 1992)
(All values in milligrams per litre)

.N.	Characteristics	Desirable limit	Emergency limit
	Colour Hazen Units	5 Units Max.	25 Units Max
	Odour	Unobjectionable	Tolerable
	Taste	Agreeable	Tolerable
	Turbidity Max.	5 N T U	10 N T U
	pH value	6.5 to 8.5	No relaxation
	Hardness (Total)	300	600
	Iron (as Fe)	0.3	1.0
	Chlorides (as Cl)	250	1000
	Residual free chlorine	0.2	—
0	Total Dissolved Solids	500	2000
1	Calcium (as Ca)	75	200
2	Copper (as Cu)	0.05	1.5
3	Manganese (as Mn)	0.1	0.3
4	Sulphates (as SO_4)	200	400
5	Nitrates (as NO_3)	45	100
6	Fluoride (as F)	1.0	1.5
7	Phenolic compounds (as C_6H_5 OH)	0.001	0.002
8	Mercury (as Hg)	0.001	No relaxation
9	Cadmium (as Cd)	0.01	No relaxation
0	Selenium (as Se)	0.01	No relaxation
1	Arsenic (as As)	0.05	No relaxation
2	Cyanide (as CN)	0.05	No relaxation
3	Lead (as Pb)	0.05	No relaxation
4	Zinc (as Zn)	5	15
5	Anionic detergents (as MBAS)	0.2	1.0
6	Chromium (as Cr)	0.05	No relaxation
7	Polynuclear acromatic hydrocarbons (as PAH)	Nil	Nil
8	Mineral oil	0.01	0.03
9	Pesticides	Nil	0.001
0	Radioactive materials		
	a. Alpha emiters Bq/1	—	0.1
	b. Beta emiters pci/1	—	1.0
1	Alkalinity	200	600
2	Aluminium (as Al)	0.03	0.2
3	Boron	1.0	5.0

luoride

Fluorine (F) occurs in the form of calcium fluoride (Ca F_2) in the ground water. ɔurse, fluoride ion (F^-) is very small so that the extra electron is held tightly. Therefoı 'is quite difficult to free the fluorine from its compounds. That is why calcium fluoride ss soluble in water as compared to its chlorides. A very small quantity of fluoride nerally present in ground water and that amount is quite essential to prevent caviti teeth, to protect enamel on teeth and for the proper development of bones both ıman beings as well as in livestock. But its concentration beyond the permissible lin quite injurious to the health. Present hydrogeological investigation carried out in t agaur district (Table 2) has shown that there are many villages in the district whe ıoride content in the ground water is exorbitantly high (Paliwal, 1988). Figure 3 sho e distribution of fluoride in ground water in the region. Chhapri, Datiyad, Rampı evli Kalan, Baorli, Ren, Jaswantpur, Leri, Mandoli Bas, Akwas and Karlu are su llages. It would be a matter of great surprise to note that ground water of Datiy llage in Jayal block contains fluoride up to 90 milligrams per litre. Makrana a undwa blocks have the largest percentage of villages having fluoride content in t ound water higher than the prescribed limit. Next come Degana, Ladnun and Naga ocks. Remaining Jayal, Didwana, Merta, Riya, Parbatsar and Kuchaman - Nawa bloc ıve a comparatively less percentage of villages having excess fluoride in ground wat able 2).

ıble 2. Distribution of Fluoride in ground water in Nagaur District

ock	No. of villages sam-pled	No. of villages with range of Fluoride content between (milligram per litre)									No. of villages with excess Fluoride %	Villages with maximun Fluoride mg/litre
		0-0.8	0.8-1.6	1.7-2.5	2.6-3.5	3.6-5.0	5.1-6.5	6.6-8.0	8-10	10+		
idwana	152	36	30	28	17	16	6	8	10	1	86 56.57%	Chhapri 24.00
yal	143	26	23	19	12	19	14	10	10	10	94 65.73%	Datiyad 9 Rampur 4
ıchaman :y Nawa	154	80	18	19	17	10	6	2	0	2	56 34.36%	Devli Kala 24.00
egana	155	8	28	23	23	33	17	9	5	9	119 76.77%	Barola 30
erta	86	19	25	19	9	9	3	1	0	1	42 48.83%	Ren 10.40
akrana	97	7	8	25	7	14	12	6	4	14	86 88.65%	Jaswantpı 24.00
ırbatsar	46	3	6	13	5	9	8	1	0	1	18 39.13%	
ıdnun	84	13	12	11	17	6	5	6	0	11	59 70.23%	Leri 44.0(
agaur	93	3	6	25	19	22	6	7	1	4	84 70.32%	Mandoli Bas 16.00
ya	15	0	3	2	3	3	4	0	0	0	6 40.00%	Akawas 6.
undwa	13	0	2	3	3	3	0	1	0	1	11 84.61%	Karlu 12.0
otal	1038	195	161	190	132	144	81	51	30	54	661	
		Total 356 34.29%		Total 466 44.89%			Total 162 15.60%			5.20%	63.68%	

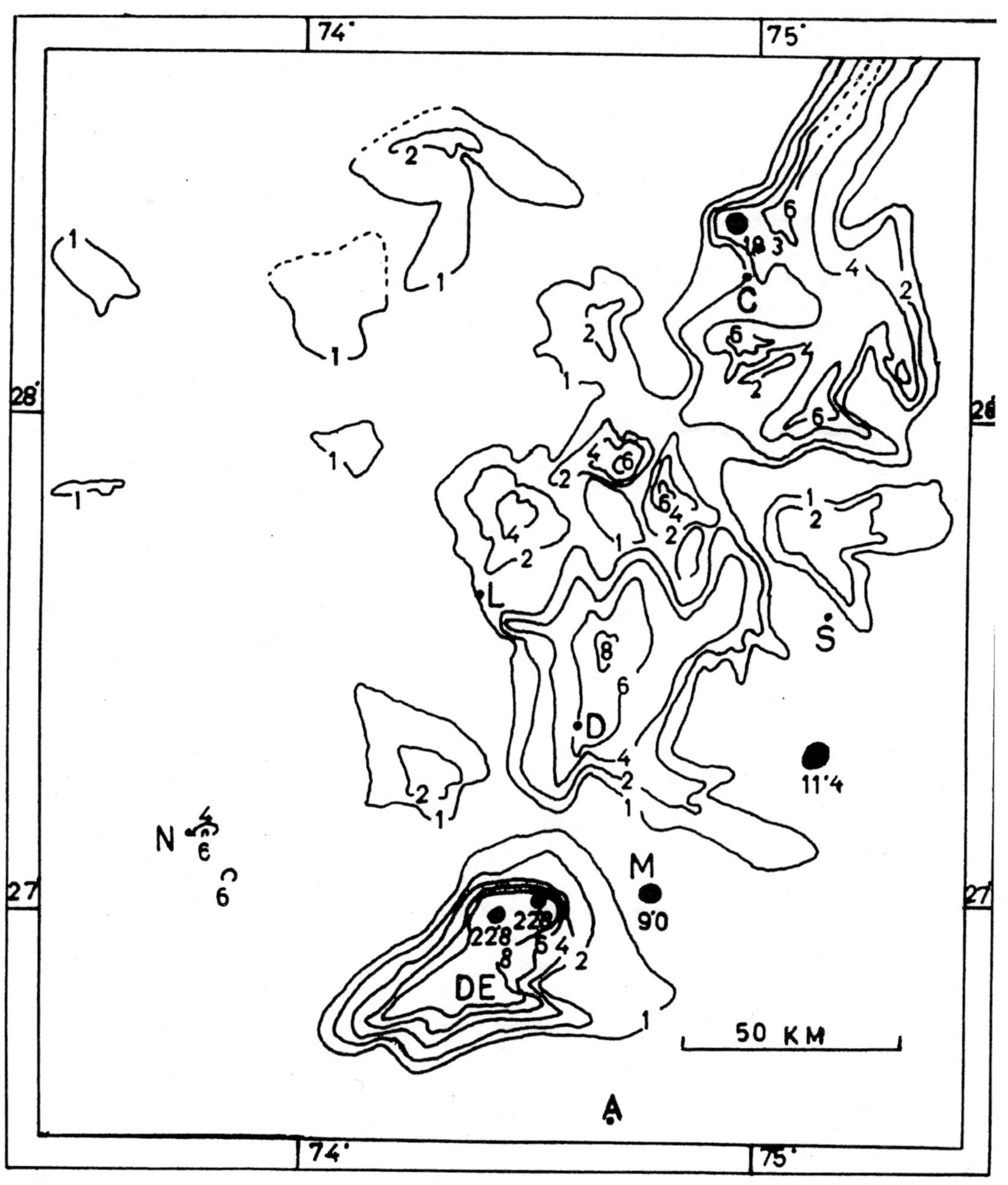

g. 3. Distribution pattern of fluoride (in milligrams per litre) in the ground water of the region
A - Ajmer, C- Churu, D - Didwana, DE - Degana, L - Ladnun, M - Makrana, N - Nagaur
S - Sikar.

Table 3. Distribution of fluoride content in ground water from different litho-units in Nagaur district

S.N.	Litho-units	No. of villages sampled and analysed	No. of villages having Fluoride content ranging between (milligram per litre)					No. of villages with content exceeding (%) 2 mg/litre	Minimum mg/litre	Maximum mg/litre	Mean mg/litre	Standard Deviation mg/litre
			0-1	1-2	2-4	4-10	10+					
1	Recent Alluvium	63	0	5 7.93%	32 50.79%	22 34.92%	4 6.34%	58 (92.06%)	1.6	24	4.44	3.3629
2	Tertiary Sandstone	8	0	1 12.50%	5 62.50%	2 25.08%	0	7 (87.50%)	1.6	5.2	2.9	1.2923
3	Bilara Limestone	22	0	3 13.63%	14 63.63%	4 18.18%	1 4.54%	19 (86.36%)	1.6	36	4.45	7.0832
4	Jodhpur Sandstone	24	0	3 12.50%	10 41.66%	10 41.66%	1 4.16%	21 (87.50%)	1.2	12	4.42	2.8201
5	Aravalli Phyllites and Schists	9	2 22.22%	0	6 66.66%	1 11.11%	0	7 (77.78%)	0.8	4.8	2.76	1.2285
6	Pre-Aravalli Granites and Gneisses	14	0	4 28.57%	6 42.85%	4 28.57%	0	10 (71.43%)	1.2	7.2	3.11	1.8027
7	Unclassified and intrusives	32	0	2 6.25%	17 53.12%	9 28.12%	4 12.50%	30 (93.75%)	1.2	44	5.39	7.4695
	Total	172	2	18	90	52	10	152 (88.37%)	0.8	44	–	–

Out of the total 1038 villages surveyed in the district 63.68% (total 661) villag ere found to have ground water with fluoride content exceeding the prescrib ermissible limit. Ground water in 44.89% (466) villages have fluoride content rangi etween 1.6 to 5.0 milligrams per litre. About 15.60% (162) villages have fluoride in t round water ranging between 5.0 to 10.0 milligrams per litre. Fluoride content of t rder of 10.0 milligrams per litre and more has been found in the ground water of abo 2% (54) villages in the district (Table 2).

As far as nature and lithology of the aquifer are concerned, wells dug in almost ie rock formations in the district have shown a high concentration of fluoride in t round water (Table 3). Statistically speaking, the chemical analysis of the samp llected from about 172 villages in the district have shown that 92.06% samples fr ie recent alluvium, 87.50% samples from the Tertiary sandstone, 86.36% samples fr ie Bilara limestone, 87.50% samples from the Jodhpur sandstone, 77.78% samples fr ie Aravalli phyllites, 71.43% samples from the pre-Aravalli granites and gneisses a bout 93.75% samples from the unclassified and intrusive rocks contain fluoride m ian 2 milligrams per litre. Over all 88.37% (152 out of 172) samples were found ntain fluoride more than the permissible limit. A very high fluoride content (more th 0 milligrams per litre) was recorded in 4 samples each from the recent alluvium and t nclassified and intrusive rocks and one sample each from the Bilara limestone and t odhpur sandstone.

It is quite interesting to note that a very large percentage (71.43 to 93.75%) of we ug in almost all the rock types in the district have shown fluoride content more than t ermissible limit (Table 3). In such circumstances it is difficult to decide which aqui as more fluoride content and infact what is the source of fluoride in ground water in t egion. However, recent alluvium, Jodhpur sandstone, Bilara limestone and intrusi nd unclassified rocks have some wells with a fluoride content exceeding 10 milligra er litre (Table 3).

itrates

As per the Bureau of Indian Standards (1992), drinking water should not conta itrates more than 45 milligrams per litre. In the absence of an alternative source t mit in emergency could be raised up to 100 milligrams per litre. If nitrates exceed 1 illigrams per litre in drinking water it may cause disease commonly known as "bl abies" (Methaemoglobinmia) in children and infants. During the hydrogeologi ivestigation carried out in the Nagaur district (Table 4) it was observed that out of t tal 311 samples collected and analysed from the district, 14 in Didwana, 35 in Jayal, ı Nagaur, 20 in Kuchaman-Nawa, 30 in Degana, 23 in Parbatsar, 7 in Merta bloc talling 186 villages in all show nitrates in the ground water exceeding 45 milligra er litre. Villages like Gowan Khurd, Dhadhikhera, Chardas, Rampura and Kasipu iow nitrates in ground water of the order of 3000, 1800, 1200 and 1000 milligrams tre respectively. Over all 59.80% of wells in the district have nitrates in the grou ater exceeding the prescribed limit (Table 4).

Table 4. Nitrate distribution in groundwater of Nagaur district

Block	Total No. of villages	No. of villages having NO_3 between (milligram per litre)							Maximum	
		0-45	45-100	100-200	200-300	300-500	500-100 & above	Total more than 45. (%)	Name of village	Quantity mg/litre
Didwana	29	15	3	5	–	1	5	48.27%	Navrang pura	900
Jayal	51	16	11	10	5	3	6	68.62%	Rampura Kasipura	1000
Nagaur	93	36	11	14	8	11	13	61.29%	Gowan Khurd	3000
Nawa	39	19	12	3	1	13	1	51.28%	Govindi	780
Degana	36	6	10	6	3	3	8	83.33%	Chardas	1200
Parbatsar	45	22	4	2	4	2	11	51.11%	Dhadhikhera	1800
Merta	18	11	3	1	1	1	1	38.88%	Sooriyas	600
Total	311	125	54	41	22	34	45	59.80%		

ulphates

Human body can easily accomodate upto 200 milligrams per litre of sulphates rinking water. In the emergency conditions and in the absence of an alternative sour ıis limit may be raised up to 400 milligrams per litre (Bureau of Indian Standard 992). Sulphates concentration in drinking water beyond this limit causes gast ıtestinal irritation and disorder in the system. During the hydrogeological study of tl agaur district (Table 5) it was observed that about 42.09% (349 out of 829) villages ine blocks have the ground water in which sulphates are more than 200 milligrams p tre. Some of the villages like Dhakoria (Nagaur), Surpalia (Jayal) and Lohraı Kuchaman-Nawa) have sulphates in their ground water as high as 3650, 2400, 23 ıilligrams per litre respectively.

hlorides

The Bureau of Indian Standards (1992) suggested chlorides in drinking water to tl xtent of 250 milligrams per litre in general and upto 1000 milligrams per litre nergency to be good for the human health (Table 1). Its higher concentration affects tl ıste and palatability. Chemical analysis of sampels collected from 1009 villages of tl agaur district (Table 6) has shown that about 31.02% villages in the district ha ılorides in the ground water more than 1000 milligrams per litre. Lohrana uchaman-Nawa block and Palot in Didwana block have chlorides of the order of 117 nd 10740 milligrams per litre. Figure 4 shows the distribution pattern of chlorides round water in the region.

otal Dissolved Solids

As per the Bureau of Indian Standards (1992), the total dissolved solids in tl otable water should not exceed 500 milligrams per litre in ordinary cases but in tl bsence of an alternative source it may be relaxed upto 2000 milligrams per litre. ater with higher quantity of total dissolved solids is consumed it causes gast ıtestinal irritation and disorder in the system. On chemical examination of the grour ater samples collected from 311 villages of the district it was found that about 88.74 llages surveyed (Table 7) have total dissolved solids more than 500 milligrams per litr ohrana (Kuchaman-Nawa) and Palot (Didwana) villages have 20010 and 168 ıilligrams per litre of total dissolved solids in their ground water.

INTAKE OF FLUORIDE THROUGH OTHER MEANS

About 60% of the total fluoride intake in human body is through drinking wate est 40% is through other means like food, drinks, and cosmetics. Eggs, meat and milk nimals that are fed on fluoride rich fodder and drink water containing high fluorid hen consumed by human beings the total intake of fluoride is further increase imilarly, many leafy vegetables have a tendancy to accumulate fluoride in their tissu hen grown in areas having high fluoride in the irrigation water (Paliwal, 1988). The afy vegetables certainly enhance the total intake of fluoride in human body, special

Table 5. Sulphate distribution in groundwater of Nagaur district

Block	Total no. of Villages	No. of villages with Sulphate between (milligram per litre)						Total % excess	Villages with maximum sulphate content
		0-200	201-400	401-600	601-1000	1001-2000	2001-3000 and above		
Didwana	122	73	35	6	6	2	–	40.16%	Dandorai 1770
Jayal	123	47	31	23	13	5	4	61.17%	Surpalia 2400
Kuchaman/ Nawa	129	95	16	5	4	6	3	26.35%	Lohrana 2354
Degana	130	73	28	9	14	6	–	43.84%	Gonalda 1547
Merta	81	67	9	3	2	–	–	17.28%	Luniawas 634
Makrana	97	58	19	11	5	4	–	40.20%	Bajoli 1692
Ladnun	84	43	21	7	8	5	–	48.80%	Raidana & Kushalpura 1530
Nagaur	46	10	10	11	6	6	3	78.26%	Dhakoriya 3650
Parbatsar	17	14	1	1	1	–	–	17.64%	Dhadhi khera 701
Total	829	480	170	76	59	34	10	42.09%	

Table 6. Chloride distribution in ground water of Nagaur district

Block	Total no. of villages	No. of village having Chlorides between (milligram per litre)							Maximum	
		0-200	201-500	501-1000	1001-3000	3001-6000	6001-9000 and above & above	Total % above 200 mg/l	Name of the village	Quantity mg/litre
Didwana	152	39	43	29	34	6	1	26.97%	Palot	10740
Jayal	143	29	31	24	56	3	–	41.25%	Bordi	3700
Kuchan/ Nawa	154	62	38	24	22	6	2	19.48%	Lohrana	11700
Degana	155	32	35	33	47	7	1	35.48%	Gonalda	6263
Merta	86	34	18	13	21	–	–	24.41%	Bhunas	2835
Makrana	97	22	16	30	28	1	–	29.89%	Kulapalar	3152
Ladnun	84	16	26	16	24	2	–	30.95%	Gulidhani	3810
Nagaur	93	14	10	30	37	2	–	41.93%	Galni	3992
Farbatsar	46	14	9	10	11	2	–	28.26%	Dholiya	3710
Total	1009	261	226	209	280	29	4	31.02%		

Table 7. TDS distribution in groundwater of Nagaur district

Block	Total no. of villages	No. of Villages having TDS upto between (milligram per litre)							Maximum	
		0-500	501-1500	1501-2500	2501-4000	4000-6000	6000-10000 & above	Total above 500 (%)	Name of village	Qty. mg/lit.
Didwana	30	8	10	2	1	6	3	73.33%	Palot	16833
Jayal	51	2	9	15	16	7	2	96.07%	Kharamanjra	7065
Nagaur	93	4	11	13	27	22	16	95.69%	Ganthilsar	9190
Nawa	39	9	7	14	4	2	3	76.92%	Lohrana	20010
Degana	36	2	7	8	8	9	1	94.44%	Sindri	6605
Parbatsar	45	9	7	8	9	9	3	80.00%	Kherised	9422
Merta	18	1	6	2	6	3	–	94.44%	Leeliya	4855
Total	311	35	57	62	71	58	28	88.74%		

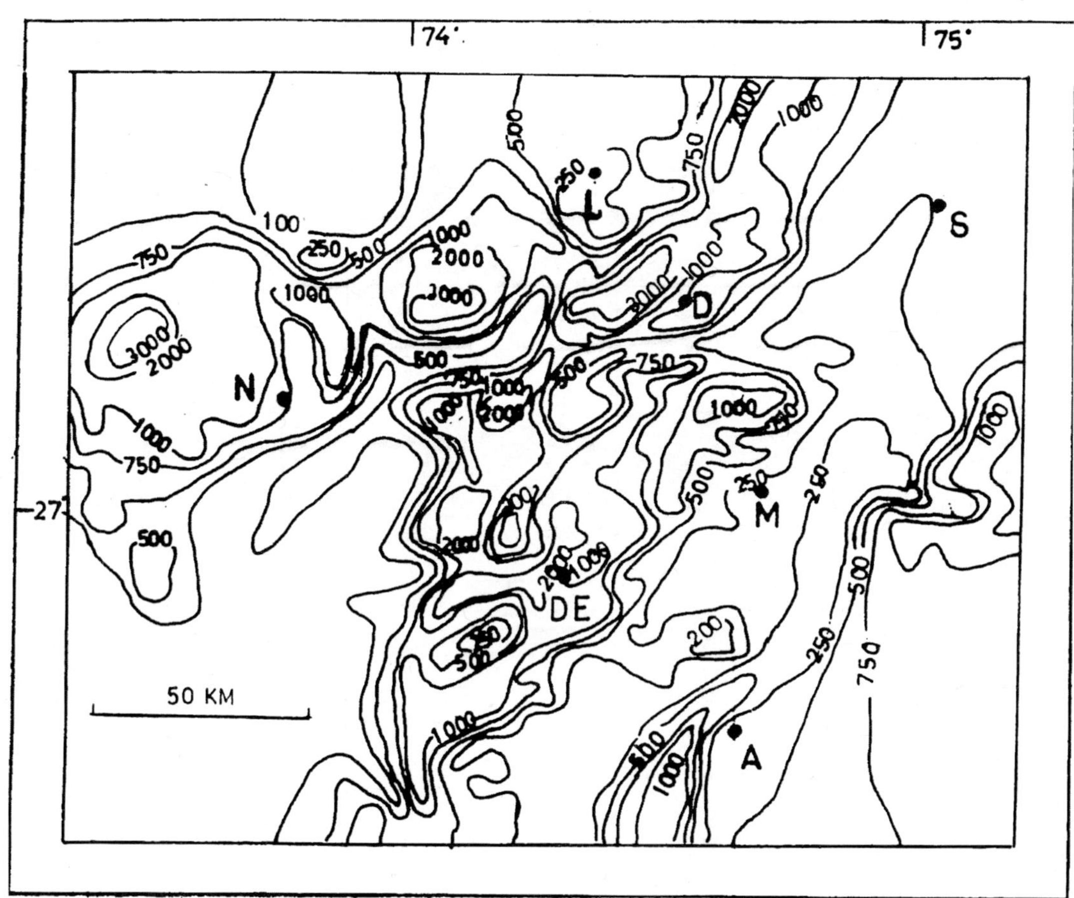

g. 4. Distribution of chloride (in milligrams per litre) in the ground water of the region. A Ajmer, C-Churu, D-Didwana, DE-Degana, L-Ladnun, M - Makrana, N - Nagaur, S - Sikaı

hen such vegetables become essential componant of the day to day food. It has bee ›ticed that effects of excess fluoride are more common in persons who have a liking f ıch leafy vegetables as compared to those who prefer potato and pulses in their foo :inerals like rock salt (Halite) have a high fluoride content. Betel leaves, areca-nu ›etel-nuts) and tobacco also have a high fluoride content. Tea contains the highe ıoride content and when tea leaves are boiled in water, fluoride in the drink increas ›markably. Tooth pastes contain about 1500 to 5000 milligrams of fluoride per litre aı ıoride rich tooth pastes are more harmful in the areas which are endemic for fluorosis

EFFECTS OF EXCESS FLUORIDE CONSUMPTION

Unfortunately, about 42% of fluoride intake by human beings is retained in tl ›dy and what so ever be the means, if total fluoride intake exceeds the permissible lin

:escribed by the Bureau of Indian Standards (1992) and the World Health Organizati 971), it causes several health problems. Besides dental fluorosis and skeletal fluoros :veral other manifestations are also developed in human body. It can damage a foet o if the pregnant mother consumes high fluoride water. In addition to this, 1 milligra er litre of fluoride in water can dissolve about 220 milligram per litre of aluminiu hen such water is boiled in utensils made of aluminium or its alloys. This wa uminium in water may increase up to 600 milligrams per litre which is quite injurio the human brain. In our day to day cooking and preparing tea and coffee etc. such :ocess is quite common, specially in the Indian way of living.

ental Fluorosis

Dental fluorosis occurs mostly in early childhood and can affect both milk teeth a ermanent teeth. Because of the excess intake of fluoride in children teeth loose shi nd become dull. Yellowish-white spots develop on them which later on turn dark brow many cases horizontal brown streaks are developed on teeth. Such streaks wh :esent at tips they indicate that the effect started at the age of two years. Similar :currence of these streaks in the middle and the upper parts of the teeth suggests th e child was exposed to high fluoride conditions at the age of 2 to 4 and 4 to 6 yea spectively. At a later stage whole teeth become black, pitted, perforated and at tim et chipped off or fall at an ealy age.

Epidemeological survey carried out in 216 villages of the Nagaur district (Paliwa 988) has shown that about 14.14 percent of the population surveyed (total 1683 ersons in all) have been affected by dental fluorosis. In Degana block about 64. ercent of the surveyed population is affected by dental fluorosis. Here the enti opulation in the five villages namely Jakhera, Kagoto-ki- Dhani, Manjhee, Kivad, a harnawas is suffering from this disease. Rest all the blocks in the district have mo an 10 percent of the surveyed population affected by dental fluorosis except in t erta block where such percentage is 3.34 only. (Table 8). Villages with a large numb people suffering from dental fluorosis include Firwasi, Ramsar, Gugariyali, Prempu holiya, Berwa, Simot, Antroli Khurd and Jawali. Firwasi village in the Ladnun blo as a large number of people (about 82.93% of the surveyed population) suffering fro ental fluorosis.

keletal Fluorosis

Excess fluoride intake in human body also causes skeletal fluorosis which affe ildren and older individuals equally. This disease results in severe pain in back-bor ints and hip region (pelvic girdle) and stiffness and immobility in joints and back-bor many cases vertebrae get fused and a disease commonly known as "bamboo spin sulting in a total immobility of the back-bone develops. Constriction of the verteb nal and intervertebral foramen causes pressure on the nerves and in many cas aralysis occurs.

Epidemeological survey conducted in 216 villages covering a human population 58373 in the Nagaur district (Paliwal, 1988) has shown that about 1.01 percent (tot 703) people in the district are suffering from skeletal fluorosis. In fact, at present the no provision to maintain a separate record of patients suffering from fluorosis both

Table 8. Distribution of patients of dental fluorosis in Nagaur district

Block Total Population	No. of villages	Population survyed	Total No. of patients %	No. of villages with patients ranging between						Maximum % of Patients	Name of village
				Nil	0-10	10-20	20-50	50-80	80+		
Didwana 49744	19	16684	2053 12.31%	4	7	5	2	1	–	51.97%	Berwa
Degana 48018	49	12224	7828 64.03%	3	3	5	8	5	25	100%	Five villages
Nawa Kuchaman 59564	23	38580	3801 9.85%	–	14	6	2	1	–	53.38%	Prempura
Makrana Parbatsar 62394	47	25792	3400 13.18%	–	15	19	12	1	–	52.13%	Dholiya
Jayal 24405	16	18485	1892 10.24%	2	10	1	1	2		57.22% 57.54%	Gugariyali Ramsar
Ladnu 58175	21	19435	2487 12.80%	–	2	8	10	–	1	82.93%	Firwasi
Merta 25511	18	22826	900 3.34%	1	14	3	–	–	–	17.89%	Jawali
Nagaur 81355	23	14347	1462 10.19%	–	11	3	9	–	–	42.62%	Simot
Total 409166	216	168373	23823 14.14%	10	76	50	44	10	26		

Table 9. Distribution of skeletal fluorosis in villages of Nagaur district

Block Total Population	Surveyed No. of villages Population	No. of Villages with no. of Patients between								Maximum % of patients	Village Name
		No. of patients	%	A Nil	B 1-10	C 10-50	D 50-100	E 100-200	F 200-500		
Didwana 49744	19 16684	817	4.9%	9	6	4	—	—	—	44.41%	Chhoti Chhapari
Degana 48018	49 12224	328	2.68%	32	11	3	3	—	—	21.51% 17.17%	Charnawas Lawadar
Nawa Kuchaman city 59564	23 38580	232	0.73%	12	6	3	1	1	—	4.91% 4.13%	Bhagwanpura Lunawa
Makrana Parbatsar 62394	47 25792	84	0.33%	37	9	—	1	—	—	4.75%	Rampura
Jayal 24405	16 18485	37	0.2%	11	4	1	—	—	—	1.6%	Taranau
Ladnun 58175	21 19435	36	0.19%	20	—	1	—	—	—	4.51%	Leri
Merta 25511	18 22826	9	0.04%	16	2	—	—	—	—	3.8%	Samadolaw Kalan
Nagaur 81355	23 14347	110	0.77%	13	6	4	—	—	—	17.22%	Bansada
Total 409166	216 168373	1703	1.01%	150	44	16	5	1			

spensaries and hospitals of the district. In such circumstances figures could be mu(gher. Didwana block has a very large number of such patients and the Chhc ıhapari village alone in this block has about 44.41 percent of the surveyed populati(ffering from skeletal fluorosis (Table 9). Other villages endemic for skeletal fluoros clude Charnawas, Lawadar, Bansada, Bhagwanpura, Lunawa, Rampura, Tarna ımadolaw Kalan, and Leri. The world famous Banka Patti- a belt where a lar; ımber of people are suffering from skeletal fluorosis, is the part of Nagaur district.

on Skeletal Manifestations

In addition to dental fluorosis and skeletal fluorosis, the higher intake of fluoride ıman body may cause a large number of non-skeletal manifestations which are neith alized by the sufferers nor diagnosed specifically by the physicians. Patients take the naturally deteriorating health conditions. These health problems include neurologic: uscular, allergic and urinary tract manifestations, gastro-intestinal disorder aı adache. Neurological manifestations include nervousness, depression, tinglir nsation in fingers and toes, excessive thirst (polydypsia) and tendency to urina equently (polyurea). Muscular manifestations include pain, weakness and stiffness uscles and loss of muscle power. In the allergic manifestations very painful skin rash ıd pinkish-red or bluish- red oval and circular spots develop on the skin which clear ι about ten days. The urine becomes yellowish-red in colour and less in volume aı ching develops in the region under urinary tract manifestations. Gastro-intestin oblems include acute abdominal pain, diarrhoea, constipation, blood stools, bloat(eling, tenderness in stomach, feeling of nausea and mouth sores etc.

DISCUSSION AND CONCLUSIONS

Fluoride intake in human body to the extent of 0.8 to 1.6 milligrams per litre ıite essential for protection of teeth and proper growth of bones. Total intake of fluori(rough all means i.e. drinking water, food and cosmetics etc. when exceeds the abo nit prescribed by the World Health Organization (1971) and the Bureau of Indi: andards (1992), it causes dental fluorosis, skeletal fluorosis and other manifestation ydrogeological and epidemeological investigations carried out in the Nagaur distri hich includes the "Banka Patti" known the world over for its people with deform(nes) have shown that the ground water in the district has a very high fluoride conte it. Not only fluoride but nitrates, chlorides, sulphates and total dissolved solids al ceed the prescribed limit in the ground water. All the rock formations in the distri rving as aquifer yield ground water with excess fluoride concentration. As a result ry large human popoulation in the district has been suffering from dental fluoros eletal fluorosis and other non-skeletal manifestations.

ACKNOWLEDGMENT

Help received from the Ground Water Department, Public Health and Engineerii epartment, and Medical and Health Department, Government of Rajasthan as well entral Ground Water Department, Government of India is greatefully acknowledge ncere thanks are also due to many individuals who assisted in collecting aı ocessing the data. Dr. W.H.C. Maxwell of University of Illinois, U.S.A., Prof. Glei out, of Nottingham, U.K. examined the manuscript and offered valuable suggestior r. B. Paliwal and Mrs. S. Paliwal helped in preparing the text.

REFERENCES

anford, W.T. 1877. Geological note on Great Indian Desert between Sind and Rajputana. *R Geol. Surv. India* 10(1)

ıreau of Indian Standards, 1992. *Indian Standard Dirnking Water Specification* (First Revisio I S 10500 : 1-8.

ron, A.M. 1933. The Vindhyans of Western Rajputana. *Rec. Geol. Surv. India*. **65** : 457-489.

ron, A.M. 1953. Geology of Central Rajputana. *Mem. Geol. Surv. India*, 79.

Touche, T.D. 1902. Geology of Western Rajputana, *Mem. Geol. Surv. India*, 35(1).

dham, R.D. 1886. Memorendum on the prospects of finding coal in Western Rajputana. *R Geol. Surv. India,* 21.

liwal, B.S. 1975. Phosphatic algal stromatolites in the Trans Aravalli Vindhyans of Badi-Khat district Nagaur, Rajasthan. *Curr. Sci.* **44(5)** : 161.

liwal, B.S. 1976. Layered chromatographical structures in the Trans Aravalli Vindhayans Badi-Khatu, district Nagaur, Rajasthan, India. *Bull. Ind. Geol. Assoc.* **9(1)** : 17-22.

liwal, B.S. 1976. Occurrence of copper mineralization around Tikli Dungari near Didwar Rajasthan. *Sci. and Cult.* **42(4)** : 218-219.

liwal, B.S. 1978. A note on the intraformational recumbent folds in the Jodhpur Sandstone Ladnun, district Nagaur, Rajasthan. *Curr. Sci.* **47(9)** : 733.

liwal, B.S. 1980. Structural history of the Precambrian rocks south west of Didwana, Rajastha India. *Bull. Ind. Geol. Assoc.* **13(1)** : 7-14.

liwal, B.S. 1982. The source of salt in Rajasthan - an investigation of the salt lake of Didwar *Geol. Surv. India Misc. Publ.* **49** : 99-104.

liwal, B.S. 1984. Evolution of the Trans-Aravalli, Vindhyans of Rajasthan, India. Proceedin *27th International Geological Congress Aug. 4-14, 1984.* Moscow, U.S.S.R. Sec. C. 04.1 (Abstract)

liwal, B.S. 1986. Source of salinity in soil and surface and subsurface waters and its bearing the evolution of the Thar Desert, India. *Transactions : XIII Congress of The Internation Society of Soil Science*. August 13-20, 1986. Hamburg, Germany (Abstract). Vol. IV : 1532.

liwal, B.S. 1988. Excess fluoride content in ground water in the central districts of Rajsth state and its bearing on the human health. *Sixth IWRA World Congress on Water Resourc* Ottawa, Canada, 2 : 303A1.

liwal, B.S. 1994. Tectonics of the post-Aravalli mountain building activity and its bearing on t accumulation of sediments along the western flank of the Aravalli range, Rajasthan, Indi In : Ahmed, R. and Sheikh, A.M. (Eds.) *Geology in South Asia - I,* Proceedings of GEOSAS (1992). February 23-27, 1992, Hydrocarbon Development Institute of Pakistan, Islamaba Pakistan.

liwal, B.S. 1993. Geology of Rajasthan - a summarium. In : Chauhan, T.S. (Ed.) *Natural a Human Resources of Rajasthan*. Scientific Publishers, Jodhpur, India. 1 - 26.

liwal, B.S. 1998. Felsic volcanics interlayered with the sediments of the Marwar Supergroup Chhoti Khatu, District Nagaur, Rajasthan. *Jour. Geol. Soc. India.* 52 : 81-86.

aliwal, B.S., Pareek, U.S. and Vyas, A. 1997. Structural framework of Makrana marble depo and its bearing on the Precambrian stratigraphy of Rajasthan, India. In : Wijayanan N.P., Cooray, P.G. and Mosley, P. (Eds.) *Geology in South Asia - II (1995)*. Geological Surv and Mines Bureau. Sri Lanka Professional Paper - 7 : 123-136.

aliwal, B.S. and Shakuntala, 1993. Ground Water Resources of Rajasthan. In : Chauahn, T (Ed.) *Natural and Human Resources of Rajasthan.* Scientific Publisehrs, Jodhpur, Ind 113-124.

areek, H.S. 1981. Basin configuration and sedimentary stratigrapahy of Western Rajasth *Jour. Geol. Soc. India*, **22** : 517-527.

areek, H.S. 1984. Pre-Quaternary Geology and Mineral Resources of Rajasthan. *Mem. Ge Surv. India* 115.

ascoe, E.H. 1959. *A Manual of the Geology of India and Burma Vol. II,* Government of In Press, Calcutta. 1343.

rivastava, B.P. 1971. Rock stratigraphic nomanclature for the sedimentaries of west - cent Rajasthan. *Bull. Geol. Min. Met. Soc. India,* **44** : 1-19.

he New Encyclopaedia Britainnica, 1983. Micropaedia Vol. IV Fluoride, Fluorosis, 15th Editi Enclyclopaedia Britainnica Inc. London, 199-200.

.S. Department of Health, Education and Welfare, 1962. Public Health Service : Drinking Wa Standards. PHS Publications Washington, D.C. 956 p.

'orld Health Organization. 1971. *Fluoride in Mc Graw-Hill Encyclopaedia of Science a Technology*, Vol. 5, Mc Graw-Hill Book Compay, New York.

eological Evolution of Northwestern India
.S. Paliwal (Ed.)
cientific Publishers (India), Jodhpur, 1999, pp. 293 - 304.

CRITICAL REGIONS OF GROUND WATER IN WESTERN RAJASTHAN

S.C. Paliwal and D.C. Sharma
Ground Water Department, Jodhpur - 342003. India

ABSTRACT

Ground water potential zones with development status of more than 65% have been considered as critical regions. In Barmer district, nearly 50% of potential area occupied by older alluvium, 20% by younger alluvium and 38% by Lathi sandstone have been over developed. In Churu, 1159 sq km area occupied by alluvium, and sandstones of Jodhpur and Nagaur Groups have overdraft. In Jaisalmer, Parewar and major part of Lathi sandstone and nearly 51% area of alluvium have development of more than 85%. In Jalore, Jhunjhunu and Sikar districts Quaternaries and Malani rhyolite have overdraft conditions. In Jodhpur, nearly 34% potential area of Quaternaries, 38% of Paleozoic sandstone and limestone and 16% that of crystallines have stress conditions. Similarly, in Nagaur, entire area occupied by sandstones of Tertiary and Jodhpur Groups, 88% of younger alluvium, 70% of Nagaur sandstone, 66% area of Bilara limestone and Precambrian crystallines have overdrafts. In Pali, major part of the area occupied by Bilara limestone and younger alluvium and 55 to 60% area each of older alluvium and Precambrian crystallines has been over exploited. These areas have been categorized as critical regions of western Rajasthan.

ey words : Ground Water, Critical Regions, Western Rajsthan.

INTRODUCTION

Ground water recharge and its utilization vary considerably in differe ydrogeomorphic regions of western Rajasthan. In the canal command area of Indi andhi Nahar,which covers parts of Ganganagar, Hanumangarh, Bikaner and Jaisalm istricts, native ground water is saline and usable ground water occurs in some localiz reas only. In arid regions of Barmer, Bikaner, Churu and Jaisalmer districts, whe ıter levels is generally deep, ground water recharge takes place rarely. In inlaı ainage regions of Nagaur, Jodhpur, Sikar and Jhunjhunu districts recharge vari ;nificantly. In Nagaur district, rainfall being lower than the threshold limit filtration, ground water replenishment is generally meagre. In Sikar, Jhunjhunu aı .rt of Jodhpur district situated in Luni catchment, ground water recharge is moderate gh. In lower reaches of Luni river, recharge varies with the change in rainfall pattern

The Indira Gandhi Nahar command area and arid regions have less use of grouı ıter. In contrast, the inland drainage and Luni basin regions support water demand large human populations and livestock besides extensive agricultural activities. Wat mand of these regions more than the replenishment in turn intensified water miniı actices in some potential areas which are endowed with aquifers of higher speci: ɔld. Availability of usable water in such regions is reducing at an alarming rate. Tl oblem is serious and needs immediate attention. The paper highlights status of grouı ıter utilization in critical regions of western Rajasthan.

THE STUDY AREA

About 2,08,751sq km study area covers Barmer, Bikaner, Churu, Ganganaga ınumangarh, Jaisalmer, Jalore, Junjhunu, Jodhpur, Nagaur, Pali, Sikar distric etween north latitudes 24 36:30 22 and east longitudes 69 30 : 76 08) in the Th esert of India. The major physiographic units are hills, rocky hillocks, sandy plair ınes and interadunal plains. Rainfall decreases westward and the peripheral regio: ıve extreme arid conditions. The surface drainage carries water for short distanc ıly. The Luni and its tributaries, the only developed drainage system in the regio ain water to the runn of Kutch.

Ground water potential zones and the aquifers have been shown in Fig. 1. Are ıving ground water of electrical conductivity less than 8000 micro siemens/cm at 25 ıve been demarcated as potential area, (Paliwal and Sharma, 1997). The Quaternari mprising younger and older alluvium, are most potential aquifers. These overl ·ecambrian basement and at places Tertiaries and Paleozoics and occupies 42.7% ar the potential zone. The younger alluvium comprises aeolian and fluvial sediments su sand, gravel, silt, clay and boulder bed. The river courses and flood plains of riv ıntli, Medha, Krishnawati, Chandrawati, Singhana, Dohana, etc., in the northeaste ırt and Raipur Luni, Sukri, Lilri, Bandi and Jawai in the southwestern area have thi luvial sequences. The thickness of alluvium below water table is generally less th:)0m, (UNEP, 1976). The older alluvium occupies northeastern area in Sika ıunjhunu and part of Churu, Nagaur, Ganganagar, Hanumangarh and Pali districts. so encompasses extensive area in southern part within Barmer and Jalore distric ıe older alluvium consists of sand, silt and gravel with calcareous matrix. T: ickness of the aquifer varies due to undulating bed rock topography. It is less than 60 Sikar and Jhunjhunu districts and 40m in Barmer district. The thickness of alluviu creases westward in Nagaur and Jalore districts where it ranges from 30 to 120m aı few metres to more than 300m, respectively. In Churu district, depth of basement aı ickness of alluvium increases northwardly. The alluvium in Ganganagar distr: ınerally has ground water of usable quality upto a depth of 60m. The Tertia ndstone is main source of water in Bikaner, Barmer and parts of Churu, Nagaur aı

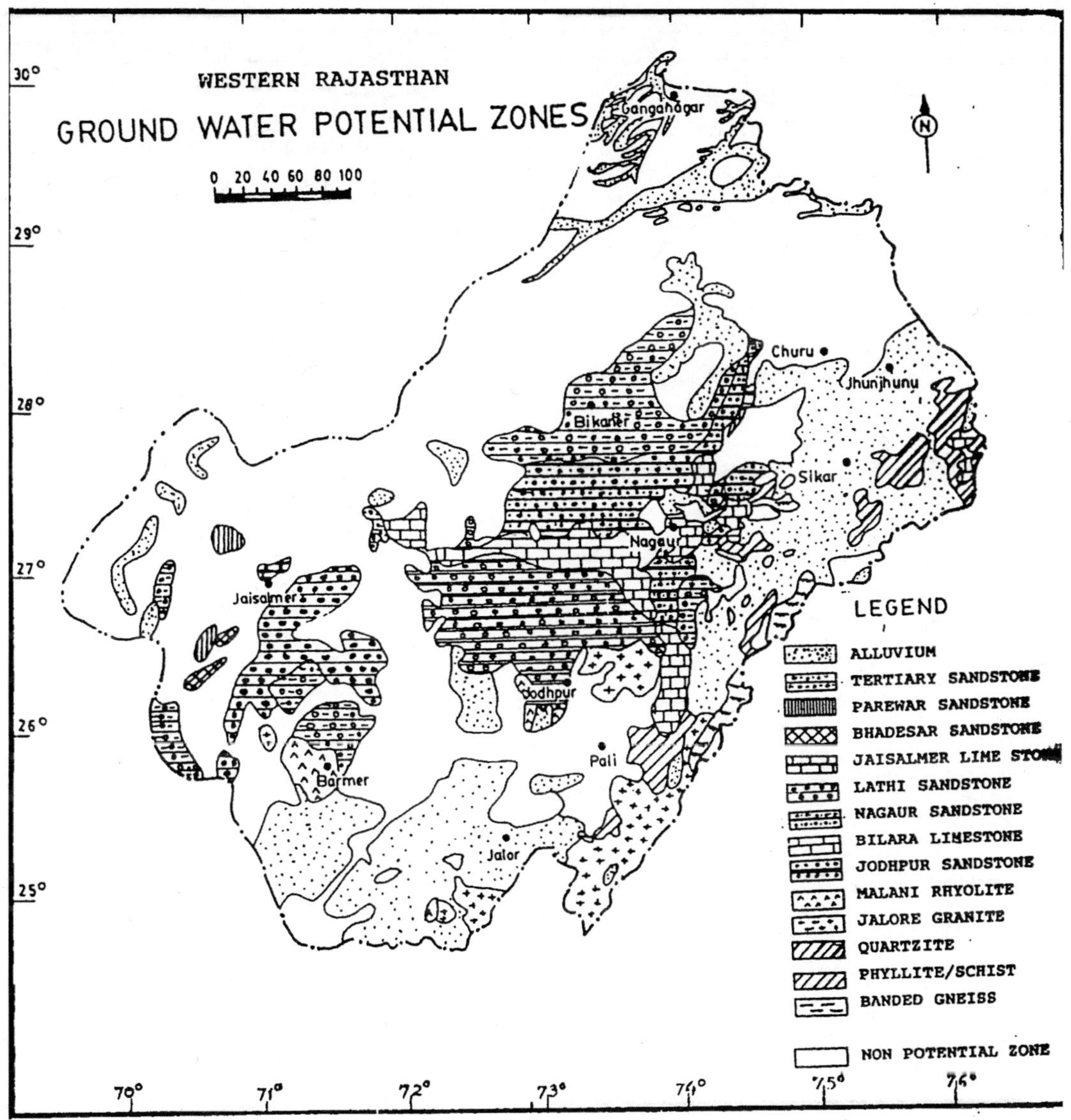

Fig. 1. Map of Western Rajasthan showing potential zones of ground water

isalmer districts. It consists of sand, silt, gravel with interspersed clay sequenc ıich often lack lithification. The thickness of the aquifer in Barsinghsar-Palana ar thnok-Guda sections of Bikaner district vary from a few metres to 113 m and 72 r spectively. The Mesozoic sediments comprise Parewar and Bhadesar sandstones ar isalmer limestone. These are confined to Jaisalmer district and occupy nearly 1% are the potential zone. The lower Mesozoic Lathi sandstone encompasses 5.9% area. It ain source of water in Jaisalmer and Barmer districts. The Lathi sandstone is mediu

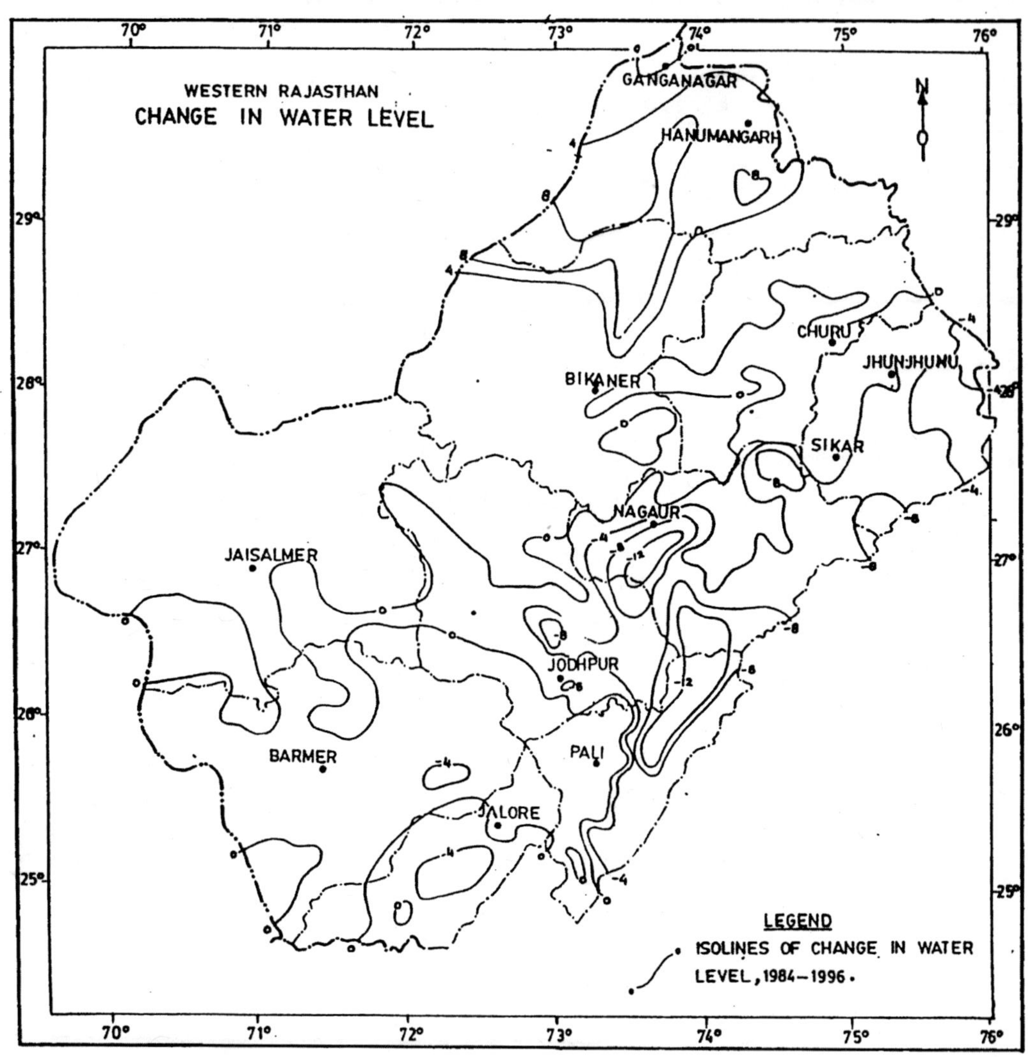

Fig. 2. Map of Western Rajasthan showing change in water level (1984-1996)

ained friable sandstone interspersed with siltstones and shales, (UNEP, 1971). Tl per part of the aquifer comprises considerable fine material which with increase i pth becomes coarser and poorly sorted. The thickness of the aquifer is maximum ne isalmer, (250m) which decreases towards east, southeast and southward. The souther ripheral region has thick clay sequence.The Palaeozoics include Nagaur sandston lara limestone and Jodhpur sandstone. These occupy central part of the region. Naga ndstone compries medium to coarse grained sediments with interspersed shale ar aporites sequence. Bilara limestone occupies Borunda-Bilara and Chadi-Phalodi are Jodhpur district and some localized area in Nagaur and Pali districts. In Borund ilara section, limestone is dolomitic and interbedded with cherty layers, while in Chac halodi area, it is predominantly cherty and siliceous. The limestone is crumpled ar ıvernous in Borunda-Bilara area and compact in Chadi-Phalodi area. The thickness mestone aquifer varies from a few metres to more than 100m. The Jodhpur sandstone rincipal source of water in Jodhpur district and localized area of Nagaur, Jaisalmer ar huru districts. It comprises medium to coarse grained sandstone, siltstone and sha hich at places is very hard and compact. The Proterozoics cover extensive area alor ıstern foothill region and occurs as outliers in other parts. The crystallines incluc [alani rhyolite, Jalore granite, quartzite, phyllite, schists and gneisses.

Ground water occurs under semiconfined condition but intervening clay lenses hav 'ten given rise to confined condition. Water level is shallow, (less than 40m deep) in tl ravalli foothill region. In central part, it is 40 to 60 m deep. Water level ranges betwee) and 120m in parts of Bikaner, Barmer, Churu, Jaisalmer and Jodhpur districts. Tl eld of wells tapping unconsolidated or semi consolidated aquifers is generally high.] ystallines, it varies corresponding to thickness of weathered mantle and intensity actures.

CRITICAL REGIONS OF GROUND WATER

Water level and electrical conductivity of water samples collected from network 614 stations have been used to prepare isoline maps of changes in these paramete: corded between 1984-96, (Figs. 2 and 3). The dynamic ground water potential has bee stimated following recommendations of Ground Water Estimation Committee, (GWE(984) and its modification, (Table 1). The potential zone having stage of developmer ıore than 65% have been categorized as critical regions, (Paliwal and Sharma 1996 quiferwise and districtwise potential and critical areas have been shown in Figs. 4 ar (GWER, 1995). The Jaisalmer limestone, sandstone of Parewar and Bhadesar hav een displayed as Ors and crystallines by Pro in Fig. 4. The districtwise ground wat enario has been presented below :

armer District

Ground water conditions vary significantly in the Barmer district. In Balotr indri and Siwana blocks, nearly 70 to 80% area has saline ground water. Other block ke Barmer, Chohtan and Sheo; Dhorimanna and Siwana have usable water in 50 5% and 60 to 70% area, respectively. The dynamic ground water potential for a year ha een estimated as 288.3 mcm, nearly 48% of these are distributed in Balotra, Siwana ar indri blocks. The agricultural and domestic drafts have been computed as 174.1 ar 7.4 mcm, respectively. It leaves surplus of 76.8 mcm, significant part of which vailable in Siwana, Sindri, Sheo, Dhorimanna and Barmer blocks. Overdraft conditic as been monitored in Chohtan and part of Balotra, Dhorimanna, Siwana and Sh ocks. The Quaternaries comprising younger and older alluvium has overdraft in near)% area. About 563 sq km area occupied by Lathi sandstone in Sheo block also ha verdraft condition. When the overdrafted zone computed blockwise, Chohta horimanna, Balotra, Siwana and Sindri have 100%, 53%, 35%, 15% and 15% of the otential areas under overdraft category. These areas exhibit depletion in water level

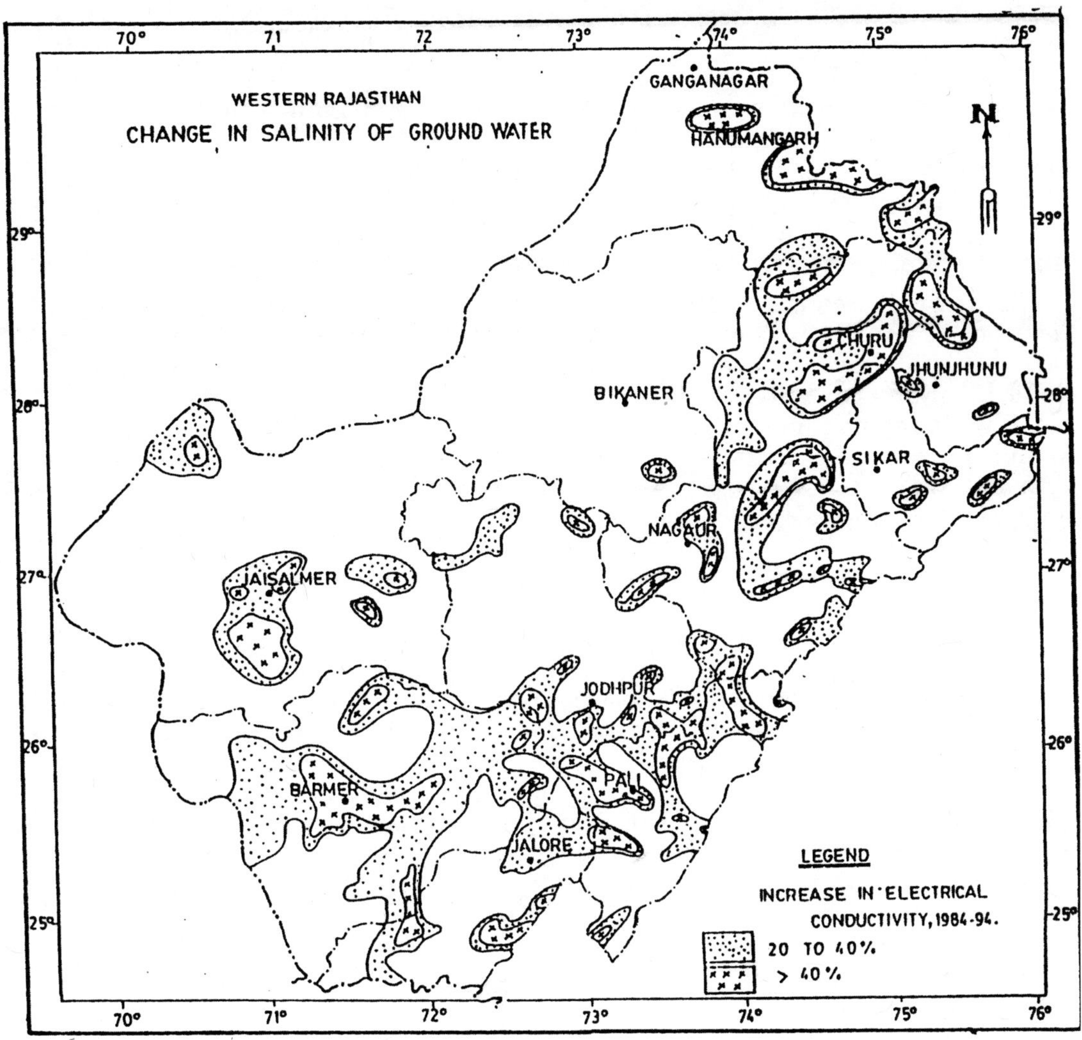

Fig. 3. Map of Western Rasjasthan showing change in salinity of ground water

the order of 1 to 4 m and salinization of ground water, particularly, in the command are of Luni river and hence they have been categorized as critical regions.

Bikaner District

About 41.7% area of the Bikaner district has usable ground water. It encompass nearly 95% area of Nokha block and 30 to 35% area of Bikaner, Kolayat ar Lunkaransar blocks. The dynamic ground water potential for a year has been estimate as 204.6 mcm while, domestic and agricultural drafts as 22.8 and 6.9 mcm, respectivel It leaves surplus potential of 174.9 mcm. The part of Bikaner block with some pockets domestic water supplies exhibit depletion in water level of about 2 m. Contrary to thi

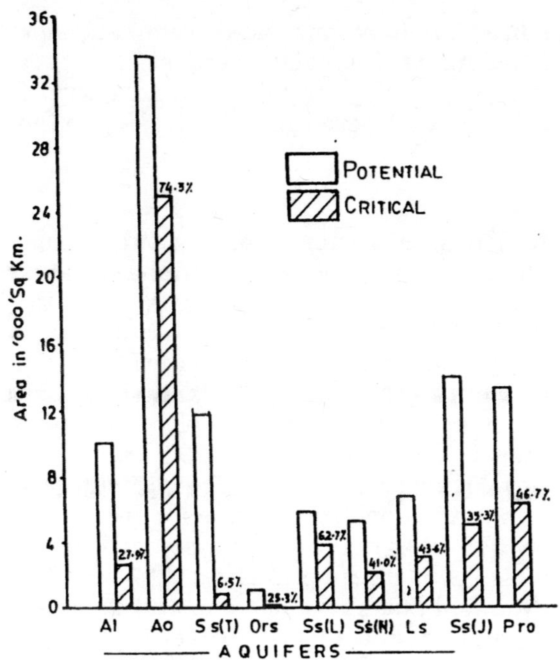

Fig. 4. Aquiferwise ground water potential and critical area in Western Rajasthan

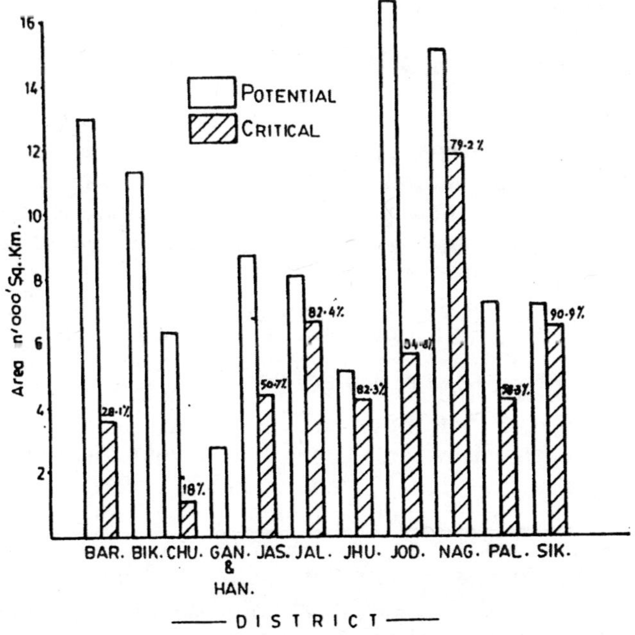

Fig. 5. Districtwise ground water potential and critical area in Western Rajasthan

ıe area traversed by Indira Gandhi Nahar and its command shows regular rise in wate vel ranging between 4 to 8 m. As the region overlies saline ground water and at place ıs low permeability horizons, seepage from canal and return flow from irrigation aı equired to be checked effectively. Otherwise, water logging in the area is inevitable.

huru District

Ground water potential area occupies south-western part of the Churu distric early 78% area of Sri Dungargarh, 40% of Sardarshahar and Sujangarh and 15% ajgarh blocks situated in this part lie in the potential area. In the north-eastern par sable ground water has been located in Ratangarh and Churu blocks in nearly 50% 5% areas, respectively. The dynamic ground water potential for a year has bee stimated as 113.5 mcm. Churu (446sq km), Sardarshahar (166 sq km) and Sujangar 3sq km) with continuous depletion in water level have negligible dynamic potentia he agriculture and domestic drafts of 20.6 and 29.2 mcm, leaves surplus potential 3.7 mcm for future development purposes. The agricultural draft is mainly froı ajgarh block, while much of the domestic water supplies are from Churu, Rajgarh an atangarh blocks. Nearly 53% of agricultural and domestic drafts in the district is froı huru and Rajgarh blocks. These blocks and about 20% potential area of Sardarshaha ock occupied by alluvium and Nagaur sandstone and in Sujangarh block by Jodhpu ındstone have stress condition due to over exploitation. It resulted in water lev epletion of about 3 m in Rajgarh, Sujangarh and Dungargarh blocks. The remainin ırt largely shows depletion of about 2 m. The Ratangarh-Churu-Rajgarh section als ıows deterioration in quality of ground water at a much higher rate. These areas thu ıve been categorized under critical regions.

anganagar and Hanumangarh Districts

Ground water of usable quality occurs in some pockets scattered along the drainag ourses in Ganganagar and Hanumangarh districts. These pockets cover nearly 13.4 rea of the district. When computed blockwise, Hanumangarh, Karanpur an anganagar blocks have 30% potential area. It encompasses nearly 17% area i nupgarh and Padampur blocks, 7% in Raisinghnagar and Sadulsahar and 1.5% i ohar block. The dynamic ground water potential for a year has been estimated as 645. cm, while agricultural and domestic drafts are 282.1 and 5.7 mcm, respectivel anganagar, Hanumangarh, Karanpur, Sadulsahar blocks have 78% of the tot rigation supplies. Nearly 67% of domestic draft is from Hanumangarh block. Surplu round water potential estimated as 357.4 mcm is mainly available from Anupgarl anganagar, Hanumangarh, Padampur and Suratgarh blocks. The stage of groun ater development has been estimated as 36%. Low development status in the district i so apparent from rising water level trends (ranging from 0.30 to 0.75 m / year).Th orth-western area covering in parts of Ganganagar, Karanpur and Padampur blocks i ı exception where water level rise has been less than 0.10 m/year. Some pocket tuated in central part of Suratgarh and Nohar blocks also show salinization of groun ater. Some depressions in Nohar, Suratgarh and Anupgarh blocks have water loggin roblem and thus catagorized under critical regions.

ıisalmer District

Ground water potential area occupies about 22.9% area of the Jaisalmer distri ıe Lathi sandstone, which encompasses 52% of the potential area, is the main source ater. Other aquifers like recent alluvium and sandstones, Parewar, Bhadesar aı mestone of Jaisalmer have pockets of usable ground water. The dynamic ground wat tential has been estimated as 43.7 mcm, while agricultural and domestic drafts are 2 ıd 8.1 mcm, respectively. It leaves surplus water potential of 33.0 mcm. Alluvium (9 km), Lathi (1849 sq km) and Parewar sandstone (269 sq km), totaling 3081 sq km ar all have continuous water level depletion and negligible dynamic water potenti. ıese aquifers have overdraft condition in 51.3%, 71.0% and 71.7% area, respective ıe over exploitation has resulted in water level depletion of the order of 2 to 4m. Indi andhi Nahar which traverses the saline ground water area, however, exhibits rise ater level by 4 to 8 m. The part of Lathi sandstone also shows salinization of grou ater. Parewar sandstone, part of alluvium and Lathi sandstone with overdraft, ha en included under critical regions.

ılore District

Ground water potential area encompasses nearly 85 to 95% area of the Jalc strict. In Ahore and Sanchore blocks, due to saline ground water nearly 34% and 76 eas, have been demarcated as potential zone. Jalore block has about 80% potent ea. The dynamic ground water potential for a year has been estimated as 639.3 mc hile the agricultural and domestic drafts are 955.7 and 29.6 mcm, respectively. aves overdraft of 346.0 mcm. Bhinmal, Raniwara and Sanchore blocks are mc fected areas,due to over exploitation and major part of the potential area in the ocks exhibit depletion of about 10 m. Depletion in other blocks varies between 3 to 7 ıore and Jalore blocks are only exception, where stage of ground water developme ıs been computed as 61.4% and 68.8%, respectively and these blocks have surpl tential of 95.1 mcm. Parts of Jalore and Sanchore blocks have also been reported ıve salinization of ground water. Leaving aside, Ahore and Jalore blocks, almost t ıtire potential area has been categorized under critical regions.

ıunjhunu District

Ground water availability varies significantly in south-eastern part of t ıunjhunu district due to change in secondary porosity. In north-western region salini ground water restricts the spread of the potential area. Nearly 70 to 85% area sisar, Khetri and Udaipur-wati blocks situated in these parts have been demarcated tential areas. In other blocks, potential zones encompass about 85 to 95% area. T namic ground water potential for a year has been estimated as 74.4 mcm. It stributed in Alsisar, Khetri and Udaipur-wati blocks. About 4438 sq km area ıirawa, Jhunjhunu, Surajgarh and part of Bohana and Nawalgarh blocks ha gligible dynamic potential. The agricultural and domestic drafts have been comput 274.3 and 39.1 mcm, respectively. It leaves overdraft of 239.0 mcm The stress grou ater condition created due to overdraft has resulted in steep water level depletion ajor part of the area. In south eastern peripheral region which extend up to Chiraw

pletion has been recorded between 4 to 8 m. It gradually reduces towards nothwest t
ss than 1 m. The entire potential area leaving aside Alsisar and part of Khetri bloc
s been categorized under critical regions.

dhpur District

Ground water quality varies significantly in nothern and southern peripheral pa the Jodhpur district. Luni, Bap and Shergarh blocks which contain saline groun ter have 29%, 48% and 64% potential areas, respectively. In other blocks, nearly 80 t % area has been demarcated as potential area. The dynamic ground water potenti a year has been estimated as 285.3 mcm. About 3099 sq km area that occupies Osia alodi, Shergarh blocks with some pockets in Bap, Bhopalgarh and Mandore ha gligible dynamic potential. The blocks like, Bilara, Luni and Mandore situated i mmand area of Luni river have significant dynamic potential. The irrigation an mestic water supplies have been computed as 414.3 and 64.3 mcm, respectively, ainly from Bilara, Bhopalgarh, Phalodi and Osian blocks. The overdraft area in th stern part exhibits depletion of about 8 to 20 m which reduces to less than 4 m i stern part. The part of Luni catchment in southern peripheral region of the distri so shows salinization of ground water. Alluvium, Paleozoic sandstone and limeston uifers which occupy entire Bhopalgarh and Bilara blocks; 40 to 55% area in Osia ergarh and Luni blocks and in Mandore, Phalodi and Bap between 5 to 15% area wit erdraft condition has been categorized under critical regions.

gaur District

Ground water quality in Nagaur district varies significantly. In Degana, Naga d Ladnun blocks, pockets of saline ground water encompasses about 30% area. I akrana and Kuchaman blocks, salinity problem has been noticed in 15% area. Thes cks thus have potential area ranging from 70 to 85%. In other blocks potential are cupies 90 to 95% area. Much of the dynamic ground water potential are limited t stern and southern peripheral areas. Blocks like, Kuchaman, Makrana, Parbatsar an yan posses 44.6% potential of the district. The dynamic ground water potential for ar has been estimated as 334.2 mcm. About 2806 sq km potential area with continuo pletion in water level has negligible dynamic potential. About 2103 sq km area ndstone aquifer in Mundwa, Ladnun and Merta blocks, limestone in Merta, phyllite i yan and some localized pockets of alluvium in Degana lie in the depletion zone. Th ricultural and domestic drafts have been computed as 462.3 and 114.4 mcn spectively. Alluvium zone alone meets nearly 50% of the total draft. The part of Riya undwa, Merta and Parbatsar blocks having overdraft condition exhibit steep wat vel depletion of more than 12 m. In other blocks, depletion is less than 4 n linization of ground water has been noticed over extensive area. Nearly 79.2 tential area which include Kuchaman, Merta, Parbatsar and Riyan blocks; 85 to 95 ea in Mundwa, Degana and Didwana; 55% in Ladnun, Jayal and Makrana 55% an % area of Nagaur have overdraft condition.

li District

In Pali district, ground water conditions vary considerably. Much of the area in Pa ıd Rohat blocks has saline ground water, as such 23.4 % and 7.7% areas of these bloc ıve been demarcated as potential areas. In Jaitaran block, potential area encompass :.9% area. Nearly 60 to 70% area of the remaining blocks have been catageorized tential area. The dynamic ground water potential for a year has been estimated ;9.4 mcm. Ground water development status varies significantly in different parts. Tl ɔcks, like Jaitarn, Kharchi and Raipur have over draft, while Bali, Pali and Sumerp ıve surplus potential. About 794 sq. km area in Jaitaran block occupied by alluviu uifer with continuous depletion in water level has negligible dynamic potential. Tl ricultural and domestic drafts have been computed as 487.6 and 14.0 mcı spectively. Over exploitation in some parts occupied by potential aquifers of hi ecific yield resulted in steep depletion in water level. In area northeast of Pa pletion has been recorded between 10 to 15 m. In the foot hill region and southweste rt of the district, depletion varies between 4 to 8 m. The parts of Pali, Sojat aı itaran blocks also exhibit deterioration in quality of ground water. Overdraft conditi entire Desuri, Jaitaran, Kharchi and Raipur blocks and the part occupied by phylli Rani and Sojat blocks, which need ground water management,as such categoriz ıder critical region.

kar District

In Sikar district,nearly 95% area has been delineated into potential zones. Pipra ıd Neem Ka Thana blocks are exception, where potential zones occupy about 90% aı ;% areas, respectively. The dynamic ground water potential for a year has be timated as 120.6 mcm. About 3743 sq km area of Dhod, Laxmangarh and Srimadhop ɔcks (2569 sq km), part of Fatehgarh (959 sq km) and Neem Ka Thana (215 sq kı ıve negligible dynamic potential. As such, much of the dynamic potential occurs ınta Ramgarh, Khandela, Neem Ka Thana and Piprali blocks. The agricultureal aı mestic drafts have been computed as 373.5 and 29.8 mcm, respectively. The area h erdraft of 282.7 mcm. Over exploitation resulted in steep water level depletion in par Danta Ramgarh block. In other blocks, it ranges from 4 to 8 m. The district as a whol cept in parts of Neem Ka Thana block, has stress condition due to over exploitati ıd has been categorized under critical regions.

CONCLUSION

Ground water development status in Jhunjhunu and Sikar districts have be timated as 421% and 334%, respectively. In Nagaur, Jodhpur, Jalore and Pa stricts,it has been computed as 173%, 168%, 154% and 103%, respectively. Oth stricts, like Barmer, Bikaner, Churu and Jaisalmer also have meagre surplus dynam ound water potential. Nearly 84% potential area in Jhunjhunu and 52% in Sik stricts have negligible dynamic potential and exploit static potential. Likewise, near % area in Jodhpur and Nagaur and 11% in Pali districts use static potential. Grou ıter potential is thus continuously depleting in 47.9% area of the western Rajastha uch of the potential area in Jalore, Jhunjhunu, Sikar, Nagaur and Pali districts li ıder critical regions. These regions including parts of the canal command area wi oblem of water logging need effective water management strategies to ameliora teriorating ground water conditions in the region.

ACKNOWLEDGEMENT

Authors express sincere thanks to Professor B.S. Paliwal, Jai Narayan Vy: ıiversity, Jodhpur, Convener of the National Seminar on Geological Evolution estern Rajasthan, (Geowest-R) for providing an opportunity to put results of variou ological activities being carried out in Western Rajasthan on a single platform. Th so express gratitude to Shri P.C. Jain, Chief Engineer, Ground Water Departmer dhpur for providing every necessary facitlity. Authors also acknowledge geoscientis orking in Rajasthan Ground Water Department for their kind cooporation.

REFERENCES

NEP, 1971. Ground Water Surveys of Jaisalmer (Lathi) Area-Rajasthan, India.Technical Repo United Nations Development Programme, 95pp.

NEP, 1976. Ground Water Surveys of Rajasthan and Gujarat, India. Technical Repo DP/UN/IND-71-614/4 prepared for the Govt. of the India by the Executing Agency for t United Nation Development Programme. United Nations, New-york, 240pp.

WEC, 1984. Ground Water Estimation Methodology. Report of Ground Water Estimati Committee, Ministry of Irrigation, Govt. of India, New Delhi, 53.

WER, 1995. Ground water Resources of Rajasthan, Irrigation Potential as on 1.1.1995. Technic Report, Ground Water Department, Jodhpur, 41.

liwal S.C. and Sharma D.C., 1996. Ground Water Assesment and Categorization of Aquifers. Kolarkar A. S, Joshi D.C and Kar A. (Eds), Land Resources and their Management f Sustainability in Arid Regions, Scientific Publishers, Jodhpur, India, 117-132.

liwal S.C. and Sharma D.C., 1997. Trend of Ground Levels and Their Impact on Arid Ecosyste of Rajasthan. In Singh, S. and Kar A (Eds), A Book on Desertification Control in the Ar Ecosystem of India for sustainable Development, Agro Botanical Publishers, Bikaner, Indi 159-173.

ological Evolution of Northwestern India
S. Paliwal (Ed.)
ientific Publishers (India), Jodhpur, 1999, pp. 305 - 311.

HYDROGEOLOGICAL, GEOPHYSICAL AND ISOTOPE STUDY TO TRACE THE COURSE OF BURIED "SARASWATI" RIVER IN JAISALMER DISTRICT, RAJASTHAN

Vimal Soni, D.C. Sharma, K.S. Srivastava and M.S. Sisodia*

Ground Water Department, Jodhpur - 342 001
**Department of Geology, J.N.V. University, Jodhpur - 342 005.*

ABSTRACT

The satellite imagery of the part of Jaisalmer district shows a buried course of stream running from NE to SW direction that is interpreted as the mighty river - 'Saraswati' of the Vedic age originated from the Himalaya. This river shifted its course several times in the westward direction, to the west of the Aravalli mountain. The hydrogeological, geophysical and isotope studies carried out along this palaeochannel show availability of fresh water of modern to 3500 years of age at the depth of 30 to 50m. The ground water present at the depth of 60 to 250m is a fossil water of 2100 to 14,400 years age. While ground water in the wells away from this palaeochannel is saline. The Aquifer comprises coarse to medium sand with gravel and clay.

ɔy words : Saraswati River, Hydrological, Geophysical and Isotope Investigatior Jaisalmer, Rajasthan.

INTRODUCTION

On the basis of archaeological finds and evidences cited from ancient scriptures dia it has been established that across the Thar Desert of India once flowed a sacr d mighty river named "Saraswati". It was a consequent river formed of a number reams flowing from the Siwalik hills. Study carried out in the part of Jaisalmer distri ngitude 27°12' - 28°0' and latitude 69°35' - 70°38') elucidates the existence of form

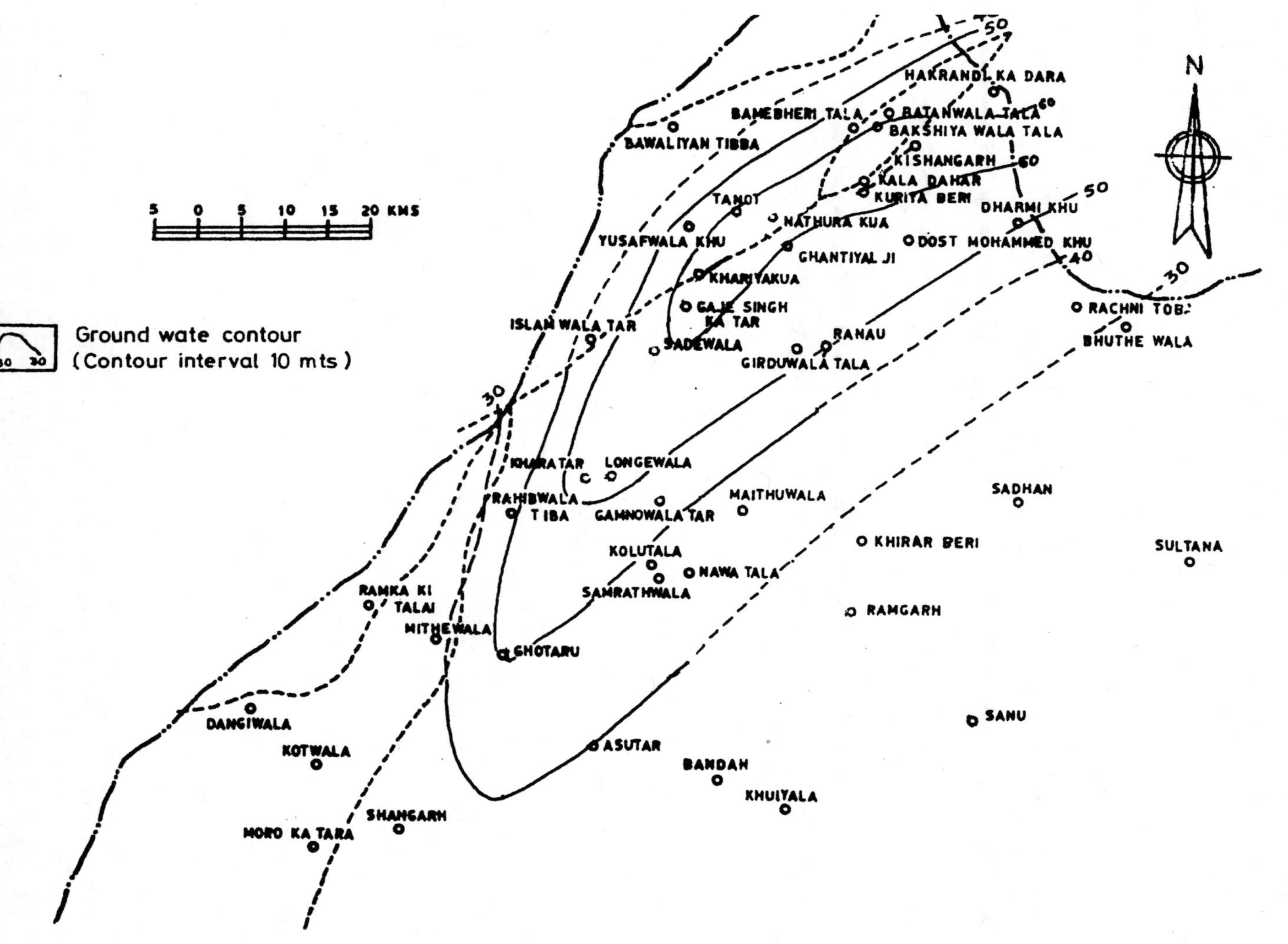

Fig. 1 The ground water contour map of the study area.

urses of this ancient river, which got buried when climatic conditions changed to ar ıich exist till today. There is an urgent need to map the former course of this riv cause of the fact that there may exist palaeodrainage along this former course which ınd can ameliorate the harsh conditions of this region.

The landsat imagery IRS-1 A and 1 B (LISS II & III) FCC 1 : 50,000 show son ıeaments and signatures of palaeochannels. One such linear feature running throu shangarh, Kuriyaberi, SE of Tanot, NW of Ghantiyali and West of Sadewala (Fig. ıs selected for detailed hydrogeological, geophysical and isotope studies to find out tl ssible course of the buried Saraswati river.

GEOLOGY OF THE AREA

The geological formations occurring in the Jaisalmer district range from Ear ılaeozoic to Quaternary in age. A generalised geological succession of the region ven in the Table 1.

ıble 1. Geological succession of Jaisalmer area (Pareek, 1981, Dasgupta, 1975)

Recent	Recent Alluvium/Desert Sand	
Eocene to Palaeocene		Bandah Formation Khuiala Formation Sanu Formation
Cretaceous		Abur Formation Parihar Formation
Jurassic		Bedesar Formation Baisakhi Formation Jaisalmer Formation Lathi Formation
Palaeozoic	Marwar Supergroup	Nagaur Group Bilara Group Jodhpur Group

GEOPHYSICAL INVESTIGATION

The aquameter, was used for electrical resistivity measurments. Both vertical aı teral variations in the resistivity were determined. Schlumberger Electro ›nfiguration was used for Vertical Electrical Soundings. (VES).

It is inferred that at Kuriyaberi a sandy layer with resistivity value of 30 ohm. n overlying a 32 metre thick bed of consolidated but permeable, medium to coar ained sand with gravel and clay (resistivity 135 ohm. mt), rich with potable water. Th d is again underlain by a clay layer (resistivity value 11 ohm mt). Almost simil ferences were drawn at other locations, such as at Kuriyaberi - Kishangarh tract, aı Nathurakua. The geology of the area based on geophysical data is given in Table 2.

able 2. Details of Geoelectric Cross Sections

. o.	Location	Resistivity in Ohm mt.	Thickness in metre	Type of sediment
	Kuriaberi	45-30 (First layer)	2.0	Fine sand and clay
		135 (Second layer)	32.0	Medium to coarse san with gravel
		11.3 (Third layer)	—	Clay
	Nathurakua	160 (First layer)	2.0	Fine sand and clay
		240 (Second layer)	32.0	Medium to coarse san with gravel
		40 (Third layer)	—	Sandy clay
	Kuriaberi Kishangarh Section	135-3000 (First layer)	2.0-5.0	Wind blown sand
		550-2550 (Second layer)	10.0-40.4	Medium to coarse sar
		10-100 (Third layer)	—	Sandy clay

WATER BALANCE

A water balance study was made on the basis of meteorological and water tak uctuation data to calculate the amount of recharge in the aquifer. The data of five yea ɜ. from 1990 to 1994 were taken as standard. Though, potential evapo-transpiration ɛtremely high and field capacity of the region is very low. Surplus water whi ıfluences recharge, is available on certain occasions (Table 3), for instance in Aug. 19 ıe amount of rainfall was 185 mm, it yielded 34.00 mm of surplus water while 222 m ıinfall in the month of July 1994 yielded 28.6 mm surplus water for recharge. Sin ıere is no proper drainage in the area, total surplus water is taken as available 1 ›charge. The average amount of recharge due to precipitation during 1990 to 1994 5.188 mcm.

able 3. Rainfall and surplus water available for recharge in the study area

Ionth/year	Rainfall (mm)	Surplus water (mm)	Area sq. km.	Water available fc Recharge (mcm)
.ug. 1989	177.8	43.7	1293.0	56.504
.ug. 1990	185.0	34.0	1293.0	43.962
ıly 1994	222.0	28.6	1293.0	36.98
.verage recharge (On the basis of daily rainfall data from 1990-94)			=	16.188 mcm
.verage recharge (On the basis of water level fluctuation from 1990-94)			=	25.1405 mcm
.mount of difference in recharge			=	9.0375 mcm

ISOTOPE STUDY

The ground water samples from 15 wells were collected from the study area durir .e joint field visit of scientists of Bhabha Atomic Research Centre, Bombay and Groun 'ater Department Jodhpur for isotope study to determine the age of the ground water.

The preliminary observations reveal that deep ground waters (60-250m) hav ›gligible "Tritium" values. These waters range in age from 2,100 to 14,900 year ıallow wells (30-60m) show "Tritium" value of 0.5 to 1.3 TR, showing modern to 3,5C :ars of age. This age of ground water corroborates with the period when arid conditior :t in the western part of India. The general flow direction of ground water is reported t : from NE to SW.

RESULTS AND DISCUSSION

Joint efforts have been put forward from various fields of modern science for tł st more than 3 decades to bring an amelioration in the Thar Desert so that parts of ari 'ea be converted into forest and agricultural land (Mishra, 1988; Ghose and Kar, 198 c). One such effort is to trace the courses of former streams which used to flow in tł ıar. The results of present study seem to be quite encouraging in this regard. Tł /drogeological investigation shows that there exists potable ground water along a pat om Kishangarh through Kuriyaberi, Ghantiyal, Ranau, Longewala, Ghotaru to Tano ıe ground water contour map (Fig. 1) shows flow of groundwater from northeast t ıuthwest, parallel to the line or feature as shown by the landsat imagery. The aquifer ade up of compact but permeable gravel and sand. At places, for instance, a uriyaberi and Nathurakua, the ground water bearing formation is semi consolidate ımature alluvium impregnated with clay and silt. These sediments are of fluvial origi ıe electrical conductivity map (Fig. 2) is symptomatic with the groundwater contou ap. However, there is excursion in the higher conductivity values as we move west ar vay on either sides from the path along Kishangarh to Tanot. It can also be inferre om the Electrical Conductivity Map (Fig. 2) that the very high ionic concentration i 'ound water at Sadewala and Khariakua is due to some local reasons (cf. Walton, 1970

The meteorological study suggests that the water is available for recharge only ıe rainfall exceeds 100 to 150 mm in a span of two to five days. It was found tha /erage recharge from 1990 to 1994 was 16.188 mcm/yr. However, amount of wate /ailable in the aquifers measured on the basis of water table fluctuation after deductir ıriation due to atmospheric and man made conditions, was 25.1405 mcm/yr. This 0375 mcm in excess to the amount of recharge due to precipitation (Table 3). As alreac entioned that there is no regular drainage system in the area, the question is wha ıuses this excess recharge in the aquifer. It is very likely that the aquifer is gettir :charged due to some palaeodrainage, concealed below the present land forms.

The results of the geophysical study corroborate with the hydrogeological stuc ırried out in the area. The geophysical study shows that a medium to coarse graine ındy formation of 20 to 30 metres thickness having potable water occurs alor uriyaberi to Tanot section. Sudden drop in the resistivity values at the depth of 30 to 4 etres, along this section suggests the presence of a clay bed. A three bed succession viz

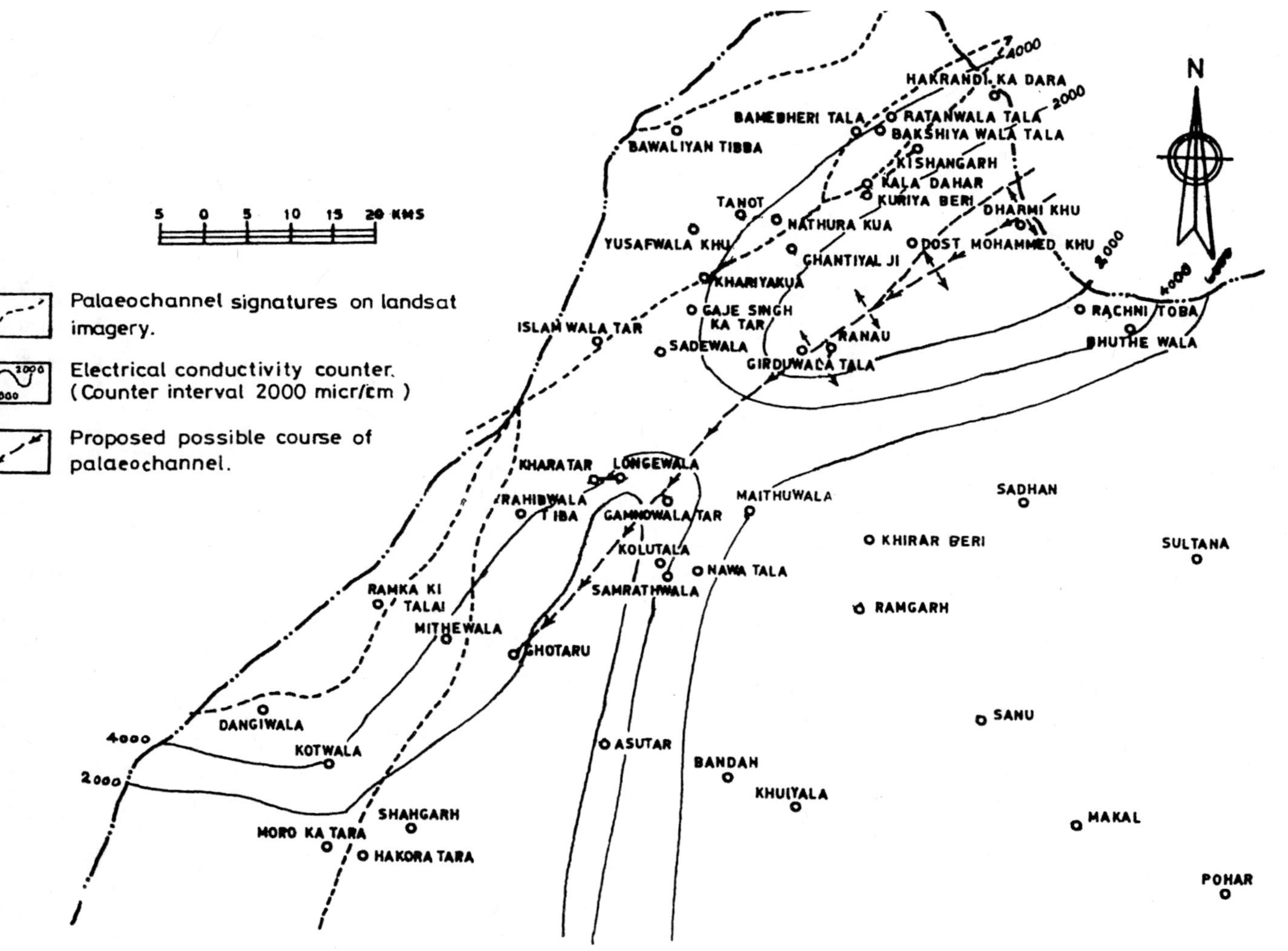

Fig. 2. The Electrical conductivity map of the study area.

ay bed overlain by a bed of medium to coarse grained sand with gravel and a top lay ' surface soil can be inferred around Kuriyaberi and Tanot on the basis of geophysi‹ :ploration. This type of succession matches with a typical cross section of riveri: ›diments (cf. Friedman and Sanders, 1978). The exploratory boreholes drilled north ınot towards Kuriyaberi, at Kuriyaberi, and at Nathurakua confirm the results of tł :ploratory study.

ACKNOWLEDGMENT

Authors are grateful to Shri. P.C. Jain, Chief Engineer, Ground Water Departme ›dhpur for providing necessary facilities and guidance during the course of this stuc 'e were benefited by discussion with Shri A.S. Chauhan, Superintendi: ydrogeologist, Ground Water Department, Jodhpur.

REFERENCES

ısgupta, S.K. 1975. A Revision of the Mesozoic-Tertiary Stratigraphy of the Jaisalmer Ba: Rajasthan. *Ind. Jour. Earth. Sci.*, 2(1) : 77-94.

·iedman, G.M. and Sanders, J.E. 1978. *Principles of Sedimentology*. Springer Verlag., New Yo: 792 pp.

hose, B. and Kar, A. 1984. The former stream courses in the Indian Desert and their effect natural resources. *Proc. Workshop. Arid Zone Research Institute, Jodhpur*, India. **3** : 1-3.

ishra, V.V. 1988. Evolution of the landscape and human adaptation in the Thar Dese *Presidential Address Ind. Sc. Cong.*, 1-33.

ıreek, H.S. 1981. Basin Configuration and Sedimentary Stratigraphy of Western Rajasthı *Jour. Geol. Soc. India* 22(11) : 517-527.

›ni, Vimal and Jain, P.C. 1995. "Rajasthan Main Saraswati Nadi Ka Puramarg" (Hindi). I: Chauhan, D.S. (Ed.) *"Anatahasalila Saraswati"*. Vaidik Saraswati Shodh Sansthı Jodhpur. 89 p.

›ni Vimal., Sharma, D.C., Srivastava, K.S. and Sisodia, M.S. 1996. Hydrogeological a Geophysical study to trace the course of buried 'Saraswati' river in Jaisalmer distri Rajasthan. *Proc. Workshop on search for Vedic Saraswati river, RRSSC/ISRO, Departmı of Space*, CAZRI, Jodhpur, India.

'alton, W.C. 1970. *Ground Water Resource Evaluation*. McGraw Hill, New York. 664.

ological Evolution of Northwestern India
S. Paliwal (Ed.)
ientific Publishers (India), Jodhpur, 1999, pp . 312 - 317.

JAISALMER FORMATION — A POTENTIAL AQUIFER IN WESTERN RAJASTHAN

A.S. Chauhan, C.P. Porwal and K.M. Mathur
Rajasthan State Ground Water Department (Survey and Research) Jodhpur 342 002

ABSTRACT

An area of about 750 sq. km has been delineated as potential ground water aquifer in Jaisalmer Formation on the basis of exploratory drilling of ten tubewells. The yield ranges from 36 to 1296 m^3/day with Electrical Conductivity values ranging from 1000 to 5000 micro siemens/cm. at 25°C. The transmissibility ranges from 8.64 to 197. 77 m^2 day and storage coefficient is 1.59×10^{-3}. The confined aquifer has been encountered at 30 to 50 m below the ground level and has a pressure head of about 30 m with the flowing artesian conditions.

ey words : Ground Water, Potential Aquifer, Jaisalmer - Formation, Wester Rajasthan.

INTRODUCTION

The Jaisalmer district situated in the heart of the great Indian desert, has a vas rain covered by a thick blanket of sand dunes. The district with an area of abou 145 sq. km., till recent (1959) was regarded as non potential region from ground wate int of view.

To meet out drinking water problem of Jaisalmer town and adjoining villages and t gment Dabla (Lathi Formation) water supply source, the earliest exploratory borehole illed at Roopsi, Jaisalmer and Hamira in 1965 were non-productive (meagre yield) an ter was not potable (TDS 5000 - 5970 mg/1). Based on these results it was inferre at Jaisalmer Formation is non-potential.

The Jaisalmer Formation is exposed in NE - SW direction of the Jaisalmer tow vering an area of about 1300 sq. km. Ground water is the only source of water suppl r Jaisalmer town and adjoining villages, as there is no surface water reservoi ɔwever, in northern and eastern parts of Jaisalmer district many villages are benefit Indira Gandhi Canal. The water supply scheme from Dabla (Lathi Formation) f ɪisalmer town was commissioned during the year 1964 - 65. Initially about 2,00,0(llons per day was being pumped daily from 4 tubewells drilled to maintain the suppl ıbsequently, due to increase in water demand more tubewells were drilled. At prese . tubewells of different agencies are under operation for maintaining pumpage of cs gallons per day from this source. The water level fluctuation in Lathi Formati .ows a declining trend of 0.20 m/annum.

GEOLOGY AND HYDROGEOLOGY

The rocks belonging to the Jaisalmer Formation (Middle Jurassic) are calcareo ndstones, marl and limestone with oolitic and pissolitic bands. Geology of the area h en described by earlier workers like La Touche (1902), Heron (1933), Pareek (198 84) Chauhan and Mathur (1987, 1992), Mathur and Chauhan (1995) and Paliw 993, 1994). The stratigraphic sequence of the Jaisalmer Formation is given below (D upta - 1975).

Baisakhi Formation

		Kuldhara Member
		Badabagh Member
Middle	Jaisalmer	Fort Member
Jurassic	Formation	Joyan Member
		Hamira Member

Lathi Formation

The limestone, and sandstone belonging to the Jaisalmer Formation occur in fair tensive area in Jaisalmer district. The sandstone is fine to medium grained greyish llow in colour.

ıble 1. The depth of drilling, yield and chemical quality of water in different tubewel drilled in the Jaisalmer Formation (1990-1996)

No.	Village	Depth of drilling in metres	Yield in m^3/day	Electrical Conductivity in Micro siemens/cn at 25°C
	Pohar	50.00	36	2000 - 2400
	Baramsar	125.00	432	1200 - 1600
	Ramkunda	148.90	216	1040 - 2600
	Joshia	96.00	1296	1000 - 1400
	Indira Indore Stadium (Jaisalmer)	177.00	604.8	3000 - 3200
	Gorbandh Palace (Jaisalmer)	180.00	324	4000 - 4300
	Moolsagar	201.00	388.8	1850 - 2200
	Khabha	230.00	43.2	4000 - 4100
	Khala	175.00	1080	5300 - 5600
	Jajia	154.00	1080	4200 - 4300

Looking at the increasing demand of water, hydrogeological investigations we: nducted by Rajasthan State Ground Water Department in Jaisalmer Formation. C ie basis of these investigations a tubewell of 50 m depth was drilled at village Pohar :)90 under contingency plan of the Public Health and Engineering Department whi(elded 36 m^3/day. The ground water encountered was free flowing with potable qualit n the basis of these encouraging results, nine more tubewells have been drilled l ajasthan State Ground Water Department under Exploration Scheme of tl griculture Development Programme upto 1995-96 and water supply schemes of Publ ealth and Engineering Department at Baramsar, Ramkunda, Joshia, Indira Indo: ;adium (Jaisalmer), Gorbandh Palace Hotel (Jaisalmer) Moolsagar, Khabha, Khala ar ıjia (Fig. 1). Depth of drilling, yield and chemical quality of these tubewells are given : able 1.

Depth of confining bed and water levels encountered in the Jaisalmer Formatic 'e given in Table 2.

able 2. The depth of drilling confining bed, first and present water levels : in tl tubewells

No.	Village	Depth of drilling in metres	Depth of confining bed in metres	Ist water level encountered during drilling in metres B.g.l.	Present water level in metres
	Pohar	50.00	21.00 - 25.00	25.00	Free flowing
	Baramser	125.00	24.70 - 26.70	26.30	8.00
	Ramkunda	148.90	15.30 - 24.40	24.00	Free flowing
	Joshia	96.00	30.00 - 48.00	45.00	3.00
	Indira Indore Stadium	177.00	-	—	—
	Gorbandh Palace	180.00	86.00 - 98.00	could not be reported	85.00
	Moolsagar	201.00	57.00 - 70.00	70 mtr.	67.00
	Khabha	230.00	60.10 - 72.00	Below 70 m	45.00
	Khala	175.00	21.00 - 30.00	Below 30 m	18.00
).	Jajia	154.00	21.00 - 30.00	Below 30 m	21.00

Water level data (Table-3) show almost rising trend between pre- monsoon ar)st-monsoon periods from 1994 to 1996 indicating that the confined aquifer is gettir :charged.

Electrical logging has also been done at village Moolsagar and Khabha. The value ie porosity ranges from 17 to 408 and that of the transmissibility is 8.64 - 197. ' ^{2}/day (by Jacob Straight Line and Theiss Recovery Method) where as storage coefficie1 1.59 x 10^{-3}.

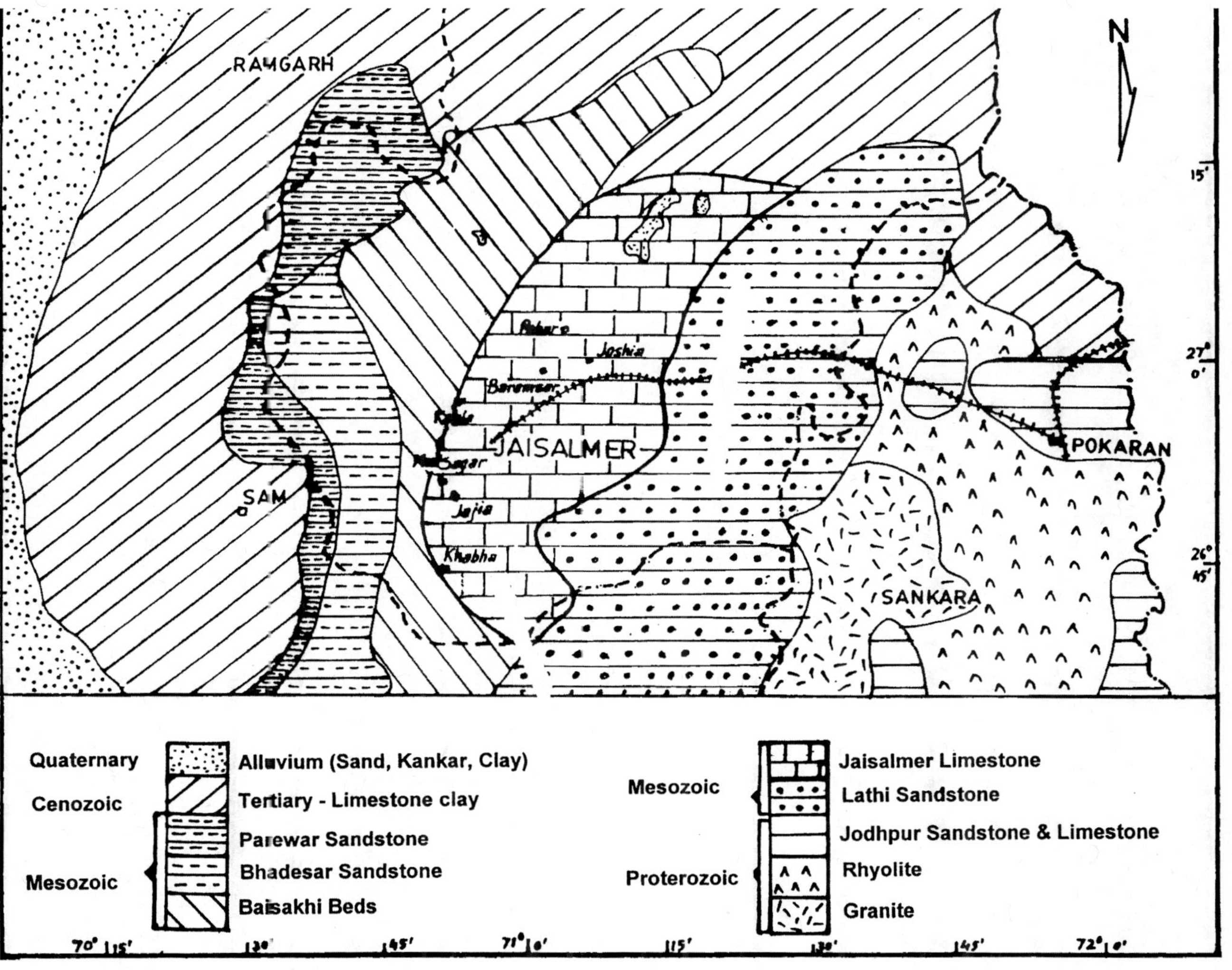

Fig. 1. Geological map of Jaisalmer area showing locations of the tube wells.

able 3. Water level changes.

.No.	Village	Type of well	Year	Water level in metres		
				Pre-Monsoon	Post-Monsoon	Fluct. in metres
	Baramsar	Pz.	1994	6.94	6.49	+ 0.45
			1995	8.40	7.94	+ 0.46
			1996	8.00	–	–
.	Joshia	Pz.	1994	7.12	7.41	– 0.29
			1995	7.13	7.08	+ 0.05
			1996	7.00	–	–
.	Moolsagar	Pz.	1994	71.08	68.08	+ 3.00
			1995	67.83	67.67	+ 0.16
			1996	67.73	–	–
.	Khabha	Pz.	1994	45.95	41.95	+ 4.00
			1995	46.00	43.50	+ 2.50
			1996	41.48	–	–

DISCUSSION

Drilling data show a vertical extension of the Jaisalmer Formation upto 230 metre nd beyond. The depth of drilling in this formation ranges from 96 to 230 metre urther drilling was not possible by DTH rig due to heavy discharge.

Intercalations of thin beds of shale (3-15 m) have been observed in the Jaisalm andstone which results in confined to semi-confined conditions of ground water.

The electrical conductivity of ground water ranges from 1000 - 5600 mic emens/cm at 25^0C. The value of the transmissibility ranges from 8.64 - 197. 77 m^2/d nd the storage coefficient is 1.59 x 10^{-3} which shows confined to semi-confined nature round water in Jaisalmer Formation.

ACKNOWLEDGEMENT

Authros express their sincere thanks to Shri P.C. Jain, Chief Engineer and Sh .C. Sharma, Chief Hydrogelosit, Rajasthan State Ground Water Department Jodhp r extending facility for the work and providing technical guidance during t reparation of the manuscript. Thanks are also due to Dr. M.S. Sisodia of Department eology, J.N.V. Univeristy, Jodhpur, for the help.

REFERENCES

hauhan, D.S. and Mathur, S.C. 1987. Geological nature of Birmania phosphorite. *Jour. In Assoc. Sedimentologists.* 7 : 71-81.

ıuhan, D.S. and Mathur, S.C. 1992. Lithofacies evolution and environmental interpretation (Birmania basin, Jaisalmer district. W. Rajasthan, India. In : Ahmed, R. and Sheikh, A.M (Eds.) *Geology in South Asia - I,* Proceedings of GEOSAS-I, Islamabad, Pakistan, 61-68.

ıgupta, S.K. 1975. A revision of the Mesozaic- tertiary stratigraphy of Jaisalmer Basir Rajasthan, *Ind. Jour. Earth Sci.* 2(1) : 77-94.

·on, A.M. 1933. The Vindhyans of Western Rajaputana. *Rec. Geol. Surv. India.* 65 : 457-489.

Touche, T.D. 1902. Geology of Western Rajputana *Mem. Geol. Surv. India* 35 (1).

thur S.C. and Chauhan, D.S. 1995. Analysis of geological data using computer application fc the interpretation of geological nature of Birmania basin. *Jour. Sedi. Lithofacies an Palaeogeography,* Chengdu, China, 15(2) : 28-39.

iwal, B.S. 1993. Geology of Rajasthan - a Summarium. In : Chauhan, T.S. (Ed.) *Human an Natural Resources of Rajasthan.* Scientific Publishers, Jodhpur, India, 1-26.

iwal, B.S. 1994. Tectonics of the post-Aravalli mountain building activity and its bearng on th accumulation of sediments along the western flank of the Aravalli range, Rajasthan, Indiа In : Ahmed, R. and Sheikh, A.M. (Eds.). *Geology in South Asia - I,* Proceedings of GEOSAS- February 23-27, 1992. Islamabad, Pakistan, 52-60.

·eek, H.S. 1981. Basin configuration and sedimentary stratigraphy of western Rajasthan. *Jou Geol. Soc. India.* 22 : 517-527.

·eek, H.S. 1984. Pre-Quaternary Geology and Mineral Resrouces of Rajasthan. *Mem. Geo Surv. India.* 115 : 1-95.

ological Evolution of Northwestern India
S. Paliwal (Ed.)
ientific Publishers (India), Jodhpur, 1999, pp. 318 - 325.

A STUDY OF FLUORIDE CONCENTRATION IN GROUND WATER OF DEGANA BLOCK, IN NAGAUR DISTRICT, RAJASTHAN

Vimal Soni, A.S. Chauhan, M.K. Babel and K.S. Srivastava
Ground Water Department, Jodhpur 342 001 India

ABSTRACT

The paper attempts at some aspects of source of fluoride and its distribution in ground water of Degana block, in Nagaur district, Rajasthan. About 150 ground water samples from tubewells and open wells of the investigated area were collected and analysed chemically for their major cations and anions. On the basis of fluoride concentration "isofloures" were prepared and accordingly the area has been classified in three different zones. The fluoride concentration in the study area is contributed by granites, gneisses and deluvial placer deposits, which is being gradually reduced outward in the intermediate and peripheral zones because of ground water flow direction and variable hydraulic gradient.

ey words : Ground Water, Fluoride, Degana, Nagaur, Rajasthan.

INTRODUCTION

The present paper deals with the occurrence, distribution and assessment o ssible source of fluoride in Ground Water of Degana block, in Nagaur district ıjasthan. The area under investigation covers 1465 sq. km., about 85 km southeast o ıgaur city, between North Latitude 26°36' and 27°06' and East Longitudess 73°53' an ° 31'.

It is known that the fluoride notably occurs in nature as fluorospar (CaF_2), cryolit a3 AlF_6), fluor-apatite (Ca_5 $(PO_4)_3$ F), minerals or as rock phosphate (apatite iroapatite, triplite etc.) (Jensen and Bateman, 1981). The fluoride content of groun ıter is a function of solubility of the parent fluoride mineral to which ground wate ımes in contact. The optimum level of fluoride is also related with the annual avera, .r 'temperature (U.S. Public Health Service, 1962). Pathak and Prasad (1973) aı aliwal (1988) have reported the presence of ground water in large parts of tl ajasthan state, containing fluoride concentration from 1.5 mg/lt to as high as 20 mg/ hey have also summarised the status of fluoride ions in ground water of the Rajasthı ;ate.

GEOLOGY AND HYDROGEOLOGY

The Geological formations occurring in the area, range in time from Pre-Aravalli uaternary period. A generalised geological succession based on field investigation .ven in Table 1.

able 1. Stratigraphy of the area (modified after Heron, 1953, Lahiri and Krishana 1968 and Khan and Laul, 1972)

ecent	Blown sand
!uaternary	Clay, kankar and gravel
ertiary	Grits and sandstone
ost-Delhi	Granites, pegmatites and quartz veins
recambrian	Quartzites, schists and phyllites
.rchaean Basement	Gneisses and granites

There are four major hydrogeological formations in the area as shown in Fig. 1. Tl recambrian phyllites, cover about 69.38 sq. km. SSE of the investigated area. T: :anites and gneisses covering 115.34 sq. km. in the central part of the area are anoth nportant ground water bearing formation. Among other hydrogeological formations, ; nportant aquifer in the western part is coarse to medium grained sandstone of Tertia ;e. It is covering 252.50 sq. km. area. Beside these units the Recent Alluvium is t: ıajor aquifer, covering 70% of the total investigated area.

Amongst the mineralizations, the tungsten is associated with mica, topaz aı uoride of fracture filling type produced during acid magmatic phase (Murthy, 197! he placer type of deposits of the same are also formed due to accumulation of detrit ıaterials, which have been weathered and rolled down the hill by gravity action aı reserved as deluvial material. These deluvial accumulates are partially cemented ırbonates. It is likely that these detrital materials could have remained in contact :round water for a considerable period during the past. It seems that the precipitation ırbonate cement, has chiefly taken place by ground water action.

METHODOLOGY

Some 150 ground water samples from tubewells and openwells of every village egana block and from different hydrogeological formations of the area were collecte hese water samples were analysed in departmental chemical laboratory for maj ıtions and anions using conventional methods.

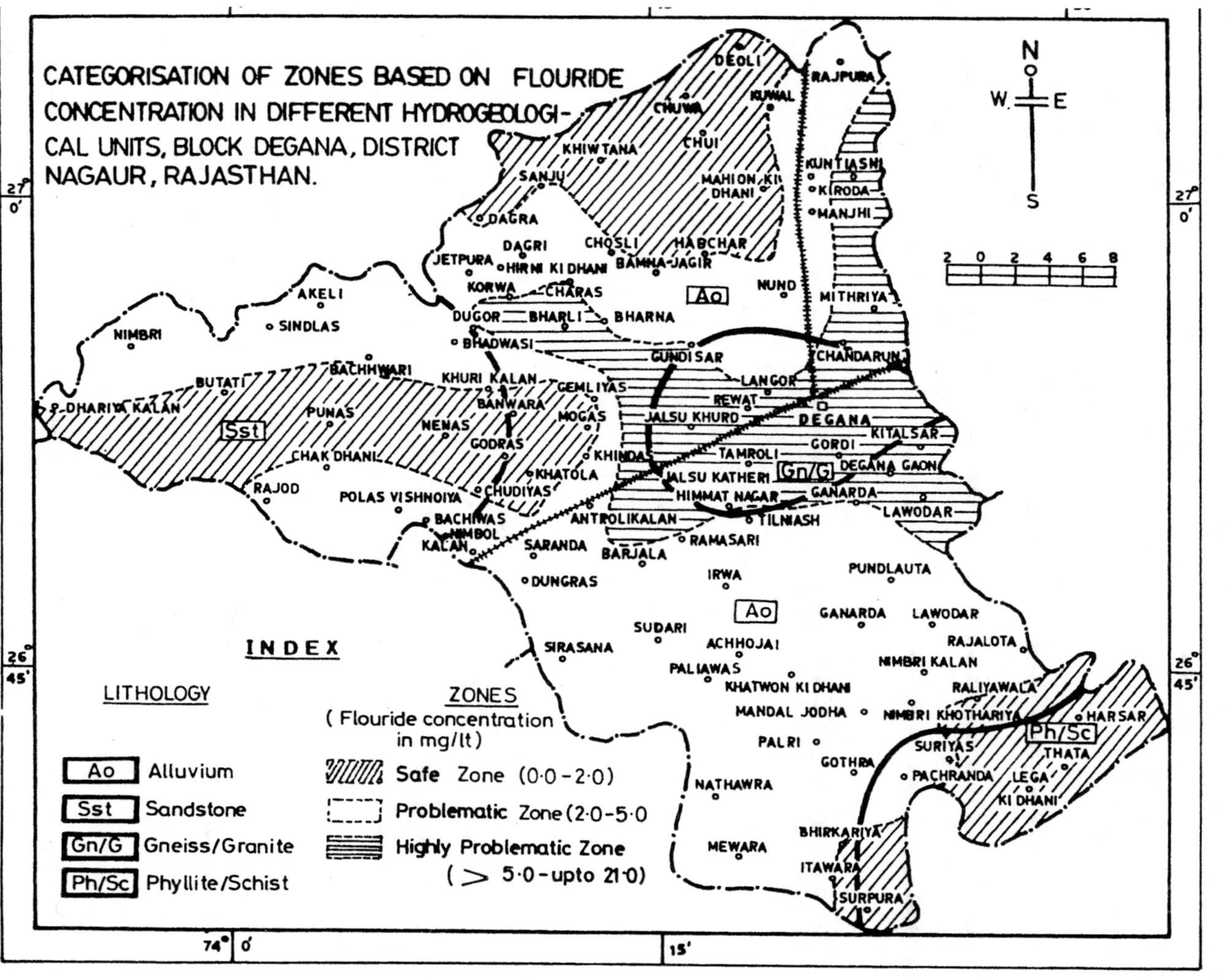

Fig. 1. Categorization of zones based on Flouride concentration in different hydrogeological units, in Degana block, District Nagaur, Rajasthan.

RESULTS AND DISCUSSION

In the surface water fluoride generally does not exceed 0.3 mg/lt unless it : lectively polluted from other sources.

On the basis of obtained results, the fluoride concentration map is prepared at a terval of 1 mg/lt. (Fig. 2). Accordingly, the investigated area has been classified ε ntral zone (highly problematic, 5 mg/lt), intermediate zone (problematic, 2.0 to 5. g/lt), peripheral zone (safe 0.0-2.0 mg/lt). Such zones are proposed in Table 2. To fin ιt hydraulic gradient, ground water flow and its relation with different hydrogeologic rmations of the area, ground water contour map is also prepared (Fig. 3).

Fluoride is characteristic of gaseous or phenumatolitic conditions, it is generall rmed later than the pegmatitic stage of consolidation of the magma. The product of th age are commonly segregated into veins and dykes (Hatch, et al., 1984). During th ιal consolidation of the magma the fugitive constituents are released and manage t cape through joints and fissures. It leads to change the mineral composition of th ιrent rocks. These changes are most strikingly displayed by granitic rocks. When th gitives are in gaseous state, they behave essentially as chemically active solution [atch et al., 1984). Apatite is another very important mineral, very common i ystalline metamorphic rocks (Dana and Ford, 1958). It has very close affinity with th ιorine to form fluor-apatite. It is considered to be highly reactive with the groun ater.

The fluoride concentration in the central zone area, has been reported as much ε 0 to 21.0 mg/lt. (Rewat, Tamroli and Kitalsar villages), which are located in the foot hi gion of the tungsten mineralized hills of granites and phyllites. The hydrogeologic rmation in this particular zone is granites/gneisses. The deluvial material formin acer type of deposits, also covers considerably larger part of the central zone are: milarly, in the extreme south east of the Degana, near Thata, Suriyas small hills ('anites and pegmatites are exposed. Fluoride concentration in down hill direction of th ιrt of the area (Gothra, Pachranda, Mandal Jodha, and Bhirkariya Khurd) have bee und to be 3.0 to 6.0 mg/lt.

The general ground water flow direction of the area is north- west, the hydraul 'adient in the central zone is low as compared to intermediate and peripheral zone ιe well known Banka Patti (Gachhipura District Nagaur, cf. Paliwal, 1988) region : cated 15 km east north of the investigated area. It is observed that along the groun ater flow direction and where the hydraulic gradient is comparatively low, fluoric ιncentration in ground water is higher. On the contrary, as we move away (from centr ιne) along and across of the general ground water flow direction, where hydraul 'adient is also higher the fluoride concentration is also reported to be low i.e. in th termediate and peripheral zones.

It is, therefore, assumed that fluoride enriched ground water of Banka Patti (6.0 t 2.0 mg/lt) is the main source of fluoride in the ground water of the area unde vestigation. The ground water flow direction (from north-east of Banka Patti regio ιntributes higher values of fluoride in the ground water of the study area.

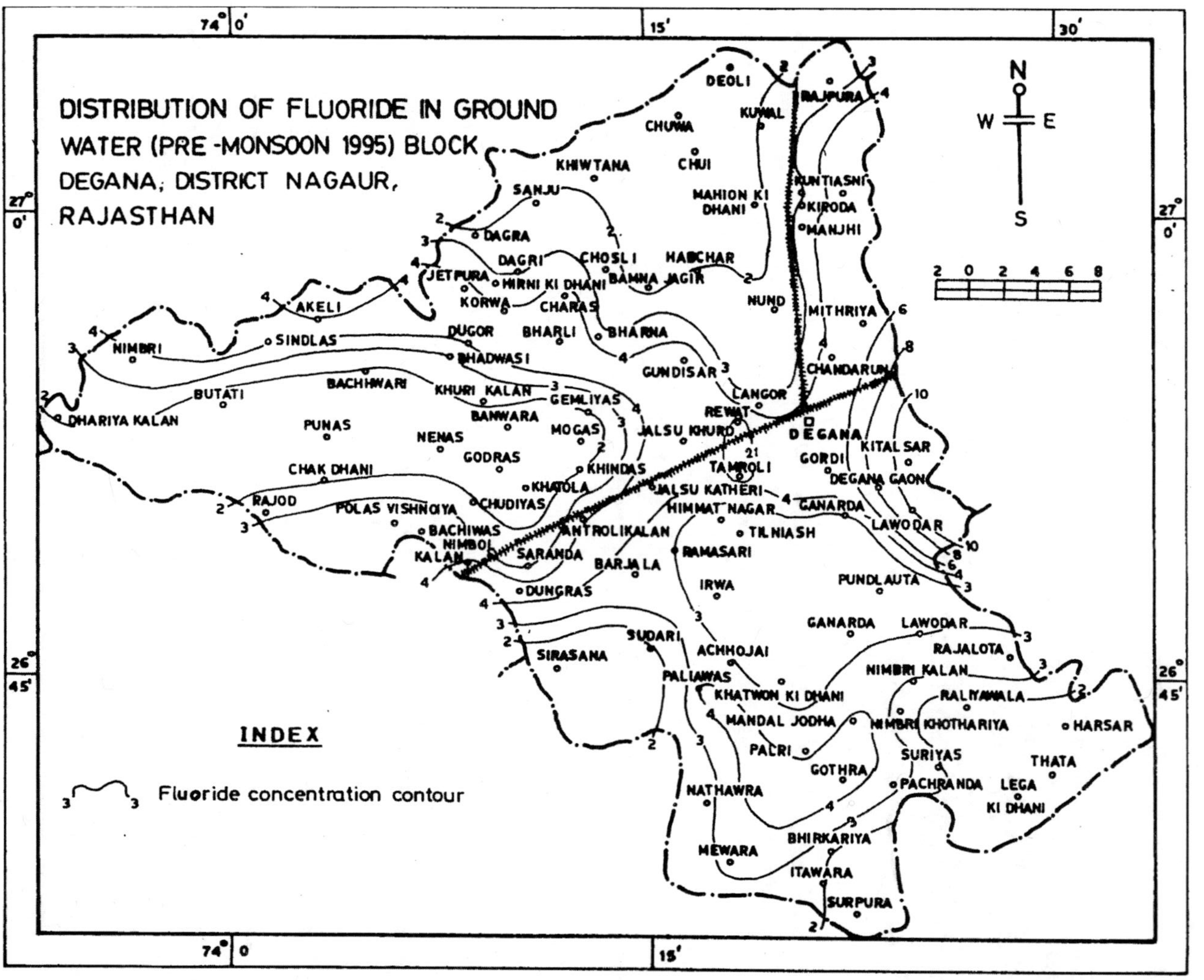

Fig. 2. Distribution of Fluoride in ground water (pre-monsoon, 1995) in Degana block, District Nagaur,

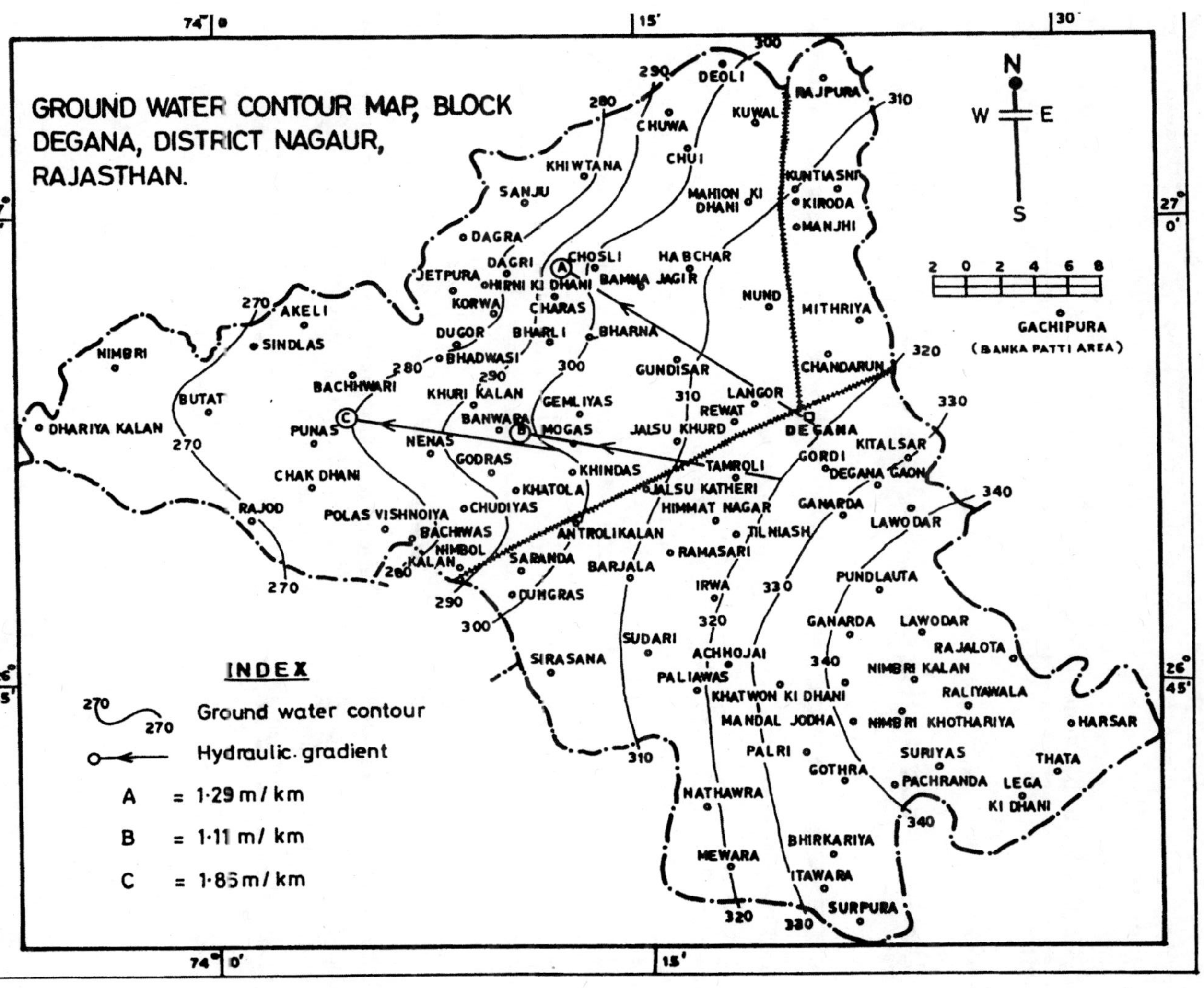

Fig. 3. Ground water contour map of Degana block, District Nagaur, Rajasthan.

The granites, gneisses and pegmatites (rich in fluoride) present in the central pa ' the area further contributes fluoride contents in the central zone of the area. It κely that the fluorine is leached out from these rocks (Rao and Khandekar, 197 ırough rain water. On the contrary, away from these rocks, where differe ɾdrogeological formations have been reported, fluoride contents are reported to l ·adually decreasing.

ıble 2. Proposed quality zones of the area

:oposed zone	Fluoride concentration (mg/litre)	Total area covered (sq. Kms)
ɜripheral Zone ɩafe area)	0.0 - 2.0	474.24
ıtermediate zone 'roblematic area)	2.0 - 5.0	942.23
entral zone Iighly problematic area)	> 5.0 - (upto 21.0)	46.80

CONCLUSION

It is, concluded that the highest concentration of fluoride in the ground water ·esent along the ground water flow direction where hydraulic gradient is al ımparatively less and near the rocks which are relatively rich in fluoride contents. Tl gh concentration of fluorine in the ground water must have been contributed by ve gh fluoride enriched ground water of a region at and around Banka Patti and also l ·anites, gneisses and deluvial placer deposite (and fluoride minerals associated wi ıese rocks).

ACKNOWLEDGEMENT

Authors are grateful to Mr. P.C. Jain, Chief Engineer, Ground Water Departmeı ajasthan, for providing necessary facilities. We are deeply indebted to Mr. D.C. Sharm hief Hydrogeologist, Ground Water Department, Jodhpur for their encouragement, aı ıluable suggestions offered. Our sincere thanks are also due to Dr. K.L. Srivastava, epartment of Geology, J.N.V. University, Jodhpur for critically examining tl anuscript.

REFERENCES

ana, E.S. and Ford, W.E., 1958. *A Text Book of mineralogy.* John Wiley and Sons, Inc. Nı York., 851 p.

atch, F.H., Wells, A.K. and Wells, M.K., 1984. *Petrology of the Igenous Rocks* CBS Publisheı New Delhi, India. 551 p.

eron, A.M. 1953. Geology of Central Rajputana, *Mem. Geol. Surv. India.* 79 : 389.

nsen, M.L. and Bateman, A.M. 1981. (Eds.). *Economic Mineral Deposits*. John Wiley and Sor New York, 593 p.

ıan, E.A. and Laul, V.P. 1972. A report on investigation for possible tungsten mineralizati around the Degana deposit, district Nagaur Rajasthan. *Geol. Sur. India. Unpublish Report.*

ıhiri, D. and Krishnan, R., 1968. A detailed report on Degana tungsten investigations distr Nagaur, *Geol. Surv. India*. (Unpublished Report).

urthy, M.V.N., 1979. Some aspects of mineralization associated with acid magmatism Rajasthan. *Proceedings Workshop on MAWAM. Geol. Surv. India. Spel. Publ.*, 13.

ıliwal, B.S. 1988. Excess fluoride content in ground water in the central districts of Rajasth state and its bearing on human health. *Sixth IWRA World Congress on Water Resourc* Ottawa, Canada, 2 : 303 A1.

ıthak, B.D. and Prasad, K.K., 1973. Ground Water resources and their exploitation in arid zon of India *Proceeding Symposium on Development of Ground Water Resources*, Madras (Indi 3 (Abstract).

ıo, K.V. and Khandekar, A.K. 1973. On the release of fluoride from some minerals and rocl *Geol. Surv. India*, Unpubl. Report. A.P. 1974.

S. Department of Health, Education and Welfare, 1962. *Public Health Service Drinking Wa Standards*. PHS Publ., Washington, D.C. 956 p.

eological Evolution of Northwestern India
.S. Paliwal (Ed.)
:ientific Publishers (India), Jodhpur, 1999, pp. 326 - 330.

ARTIFICIAL RECHARGE OF GROUND WATER THROUGH SUBSURFACE BARRIERS

A. Vijay* and U.S. Sharma

**Department of Civil Engineering, Government Polytechnic College, Jodhpur - 342 001*
Department of Civil Engineering, J.N.V. University, Jodhpur - 342 001

ABSTRACT

Increase in population, industrialization and irrigation has increased water demand that has stimulated over exploitation of ground water, ultimately leading to depletion of this nutural resource. The purpose of artificial recharge is to supplement the quantity of ground water in the region. There are various methods available for artifical recharge Depending upon the geological, geographical and other conditions various methods have been proposed for the artificial recharge.

In Rajasthan Research, Design and Development Centre of the Ground Water Department Jodhpur, suggested the construction of sub-surface barriers to recharge the ground water artificially. At village Lakawas and Chaukrikalan, the results of Subsurface Barriers are very encouraging. It has not only reduced the rate of depletion of water table considerably, but it has also increased the yield of wells and the rate of infiltration.

ey words : Subsurface Barriers, Ground Water, Artificial Recharge, Jodhpur Regio Jojri River, Rajasthan.

INTRODUCTION

Fresh water being one of the basic necessities for sustainance of life, there ha lways been a strive to locate and develop it. Over ninety percent of liquid fresh wate vailable at any given moment on the earth lies beneath the land surface. Ground wate nlike surface water, is more dependable in periods of drought and has many othe dvantages over surface water.

Increase in population, industralization and irrigation has increased the demand (ater too. Increased demand for water has stimulated development of ground wate ıpplies which ultimately lead to deplition of ground water reserves. The purpose (tificial recharge is to increase the infiltration of water into sub-surface in order t ıpplement the quantity of ground water in the area.

DISTRIBUTION OF WATER

The total quantity of water in the world is roughly of the order of 1357.5 millio ıbic kilometres (mkm^3). About 97.3% of this water is contained in oceans as salin ater (Paliwal, 1989). Thus only 37.5 mkm^3 is fresh water. Out of this 8.5 mkm^3 is bot quid and fresh and remaining is in the frozen state as ice in the polar regions and o ountain tops and glaciers. The total amount of the rains and snow fall on earth brin bout 0.42 mkm^3 of water every year (0.32 mkm^3 on the ocean and 0.10 mkm^3 on th nd). Over the oceanic surface the evaporation exceeds by 9% over the rain fall t alance the excess precipitation over the land areas. Thus the volume of water carried b vers to the sea each year is about 0.038 mkm^3. It is intresting to know that less tha % of this total river flow is used for irrigation and the rest flows to the sea unutilize Mathur et al., 1987).

Irrespective of the quality of precipitation the earth surface receives the quality (ırface and ground water depends upon the nature of geological formations, that host i round water is non-uniformly distributed quantity wise as well as quality wise.

In India the total surface water potentials are of the order of 1881340.67 mcm (hich the ground water resources are 424440.46 mcm. The total ground water resource Rajasthan are of the order of 12572.53 mcm. (Paliwal and Shakuntala, 1993). Groun ater is derived primarily from rain and snow-melt that infiltrates the land surface an owly percolates to the water table. This process of adding water to underground storag called natural ground water recharge or simply recharge. Ground water in storage ma e depleted at an alarming rate during droughts due to increased pumpage of water fro ells and the decrease in natural recharge as a result of reduced precipitation.

In order to meet increasing demand of water for irrigation and industrial purpose round water has been assumed as guaranteed source of infinite quantity and as a resu is being over exploited. Even though it added to economic growth, the ground wate epletion reached the stage beyond repair. Gujarat, Rajasthan, Kerala, Karnataka, et re examples of states having over exploitation of ground water.

ARTIFICIAL RECHARGE

It is always necessary to set a balance between the addition of water to groun ater storage by natural infiltration and draft of water for different uses. However, th isturbance in this equilibrium results in contineous lowering of water table, if th ature does not provide condition to have adequate infiltration due to natural geologic rmation, topography climatic conditions etc. The water infiltration can be enhanced b atural process or by artificial means which is termed as artificial recharge.

The various methods of artificial recharge include induced infiltration; wate preading-flooding, basin, ditch and furrow, natural channel, irrigation method, recharg its and shafts, recharge wells, channel method, subsurface dams/barriers, waste wate echarge etc.

SUBSURFACE BARRIERS (SSB)

Several artificial recharge techniques are in use successfully throughout the worlc n Rajasthan, Research, Design and Development Centre of the Ground Wate epartment, Jodhpur has suggested the construction of subsurface barriers to agumen ne natural recharge. These barriers cause spreading of water up-slope of barriers. Wate preading by constructing low head subsurface barriers appears suitable for increasin echarge during monsoon period. It allows movement of water at a very low velocity i rder to effect maximum infiltration. The barrier keeps the water stagnant for a longe eriod thus allows more time for infiltation and prevents losses due to run off. The eart urface in recharge zone must have good vertical permeability to add percolation an quifer should also have good transmissivity for lateral movement of water.

CONSTRUCTION OF SUBSURFACE BARRIERS

Subsurface barriers were constructed with angular rock pieces arranged in the forı f dry masonry wall having less interspaces and of one to two metre in width, at a dept f three meters below the surface (in river bed) with about one metre height on river bec he height and depth of barriers are site specific and can be increased or decrease epending on site conditions. The portion of barrier above river bed can be of ston nasonry using cement or lime mortar. The design of barrier is site specific and ver nuch depend upon site conditions. In alluvial areas use of wind wall to the barrier i ecessary to check the bank erosion.

Subsurface barrier is constructed in such a way so that the base flow is allowed t echarge the wells in the down slope side and at the same time salinity does not increas n the ground water up-slope. Height of barrier is kept low (about 1 m) so that exces tream flow passes over the barrier in the down slope and there is no submergence of th urrounding fields.

BENEFITS OF SUBSURFACE BARRIERS

enefits of the subsurface barriers include :

. Replenish the recharge of ground water at faster rate in mansoon period.

. Increase in ground water storage.

. Control further depletion of ground water level in the wells.

. Increase in yield of wells and their command area.

. Improve quality of ground water.

Benefit small and marginal farmers by way of controlling depletion of water table avoid expenditure on account of deepening of their wells.

Store surface water in underground reservoir which otherwise would have been lost surface run off.

Reduce chances of land subsidence by way of artificial recharge of depleted aquifer.

Generate of man days during construction of subsurface barriers.

Ground water department conducted the evolution studies at two of the subsurfa ırriers one at Kalawas and other at Chaukrikalan.

alawas

Sub surface barrier was constructed in year 1990 on Guniamata nadi at a distan about 1 Kilometer south west of village Kalawas. At this site sufficient runoff railable and part of which is checked by subsurface barrier.

Hydrogeological conditions are favourable, as sufficient thick dry aquifer railable underneath, having good tranmissibility. The Gunimata nadi is havi ıfficiently wide bed and low banks, with stream deposits of coarse sand gravel whi rm excellent surface for infiltration during the flow period.

haukrikalan

Another subsurface barrier has been constructed on Jojri river at a distance abo ılf Kilometer south east of village chaukarikalan. At the site sufficient runoff railable and part of it is chekced by subsurface barriers.

The hydrogeological conditions here too are favourable as granite is decompos ıd weathered to allow infiltration. The over lying alluvium is about 3.5 metre thi ıly. The Jojri river is sufficiently wide having stream deposits of coarse sand and fi avel, which help infiltration during the stream flow. Results of the evolution studies ro sites are as below.

No.	ITEM	KALAWAS		CHAUKRIKALAN	
		Before Const. of SSB.	After Const. of SSB	Before Const. of SSB	After Const. o SSB.
	Rate of deplition of water table per year	1.00 m/yr	0.2-0.4 m/yr	1.00 m/yr	0.23 m/yr
	Well yield	80-120 m^3 / day	100-145 m^3/day	60-90 m^3 / day	80-120 m^3/day
	Tube wells	Nil	5	—	—
	Increased in Command area	—	10 hect.	—	15 hect.
	Rate of infiltration to ground water (based on hydrograph)	10%	15%	10-15%	37%

CONCLUSION

The results of the study at the two sites are very encouraging and suggest us to u: ›re and more subsurface barriers or other techniques to increase the artificial rechar; ground water in the region.

ACKNOWLEDGEMENTS

Help received from the Ground Water Department, Government of Rajastha dhpur is gratefully acknowledged. Thanks are also due to Prof. B.S. Paliwal f amining the manuscript critically and offering suggestions.

REFERENCES

ıthur, P.N., Badhana, N.K. and Mehara, R.L. 1987. *Economic Geogrphy of India and Indic Financial Management*. College Book House, Jaipur, 716.

.liwal, B.S. 1989. Status of water resources in developing countries in the twenty first centu and the need of spreading awareness in the society through education - a case study from tl Republic of India. *Proceeding, International Seminar on Education and Training in Wat Resources*. Aurangabad, India. 347-353.

.liwal, B.S. and Shakuntala, 1993. Ground Water Resources of Rajasthan. In : Chauhan, T.S *Natural and Human Resources of Rajasthan*, Scientific Publishers, Jodhpur, India, 113-124

eological Evolution of Northwestern India
.S. Paliwal (Ed.)
:ientific Publishers (India), Jodhpur, 1999, pp. 331 - 337.

ENVIRONMENTAL MONITORING AND MANAGEMENT OF INDIAN DESERT THROUGH SATELLITE REMOTE SENSING

J.R. Sharma

Regional Remote Sensing Service Centre, National Natural Resources Management System, ISRO / DOS / GOI, CAZRI Campus, Jodhpur - 342 003.

ABSTRACT

Thar is the most populated desert of the world in terms of human as well as livestock popoulation. Increase in popoulation has placed alarming constraints on the carrying capacity of the land and water. Ruthless exploitation of the biosphere has endangered the natural resources and the desert environment severely. This, urgently calls for the sustainable utilisation of available resources and their conservation. However, a general consensus prevails amongst the experts that the production potential of resources in the desert region can be substaintially raised by improved techniques of management of land, water, vegetation, livestock and development of mineral resources. The region has been marked as a priority region for a large number of development schemes financed by international agencies, central and state governments were implemented and are continuing to strengthen the economy of the region. Satellite Remote Sensing, by virtue of its synoptic coverage, repetitivity, easy availability of data in inaccessible areas, multi-spectral capability and cost effectiveness offers an appropriate technology in mapping, monitoring and management of natural resources within a short time frame in desert region.

:ey words : Remote Sensing, Environment Monitoring, Desert Management, Satellit Thar Desert, Rajasthan.

INTRODUCTION

In recent decades, two distinct but interactive macro-processes have severe ffected India's rural environment : the depletion of its natural resource base and tl ıcreasing appropriation of what is available for the benefit of a few.

Depletion is manifested in disappearing forests, deteriorating soil conditions an ·clining water resources. In 1985-87, only 19.5 per cent of the geographic area wa rested; in 1980, 56.6 per cent of land was suffering from environmental problem: pecially water and wind erosion. In some canal projects, half the potentially irrigabl ıd cultivable area has been lost due to water logging. The availability of ground wate also falling, and fertilizer and pesticide run-offs have polluted natural water source: aking them unfit for aquatic life and human use.

Alongside this depletion, the processes of statisation (appropriation by the State ıd privatisation (appropriation by a minority of individuals), has led to the increasin ›propriation of natural resources, to which earlier there was communal access.

India has about 16 per cent of the world's population but only two per cent of th tal land area. Further, India has about 20 per cent of the world's cattle and 55 per cen the buffalo population of the world. There is a severe pressure on natural resource ıd the environment because of the simultaneous increase in population and economi owth and the heavy backlong of pollution control projects. India has to serve th quirements of 920 million people. The rapid economic growth, sharp rise in agricultura oduction, rapid urbanisation and industrialisation are increasingly affecting th ıtural environment.

The environmental problems that India is facing can be divided into thre tegories. The first category can be called Type I environmental problems. These ar oblems arising out of lack of development such as deforestation due to fuel woo traction, unhygienic slums, soil erosion, lack of sanitation and protected water suppl c. If the tempo of development is slow, the Type I problems will intensify.

Type II problems are those caused by lack of environmental safeguards i velopment projects such as air and water pollution, hazardous waste disposa aterlogging in irrigation projects, pesticide pollution, fertilizer run-off into streams, et sufficient investments are not made in development projects to compensate fc ıvironmental degradation, these problmes will increase.

Type III. problems are global environmental problems such as global warmin one depletion, marine pollution, loss of genetic diversity, automobile pollution, et ıese are created by the current pattern of development, life style and energy use mix.

In India we have all the three types of environmental problems and we do not hav ıough capital to control or manage these problems. The existence of the first two type paradoxical environmental problems, one arising out of lack of development (Type ıd the other arising out of development projects (Type II), make the situation a difficu ıe to handle. If we do not develop rapidly, the Type I problems will overtake us. If we d re for the environment, the Type II problems will overwhelm us. Environmenta anagement issues have to be seen in this context. Hence, the options open for Indi ıve to be examined in terms of two parameters : economic growth and degree ıvironmental protection.

There are four options open to us. The first option is one in which India can pursu gh levels of economic growth and a low degree of environmental protection. This optio highly undesirable since it will increase the levels of pollution and pollution-relate seases.

The second option is to have low levels of economic growth and a low degree c ıvironmental protection. This option will result in high levels of poverty and pollutior ıis is also an undesirable option. The third option is to have a low level of economi owth and a high degree of environmental protection. This is not feasible since we d ıt have enough resources to pursue this option.

The fourth option is to have high levels of economic growth and a high degree c ıvironmental protection. If we have to raise the income levels of the poor and protec ır environmental resources, we have to pursue this last option. In other words, in ›eralised context, only this can sustain development with environmental protectior ıe economic growth agenda and environmental regulatory reforms have to be seen i is perspective. Indian desert is the most populated desert of the world in terms c ıman as well as livestock population. The increasing trend of population and thei :tivities have placed alarming constraints on the carrying capacity of the land an ater, thus generated acute crisis of living resources. Further the ruthless exploitation c e biosphere has endangered the natural resources and the desert environmen verely. This vast stretch of land is by and large a vegetative area from the point of vie human ecology. Meagre availability of water, intense heat and wind erosion hazard ıpose severe constraints on plant, animal and human life. This, urgently calls for th ilization of available resources and their conservation. However, a geneal consensu evails amongst the experts that the production potential of resources in the deseı gion can be substantially raised by improved techniques of management of land, wateı getation, livestock and development of mineral resources.

The region has been marked as a priority region for its overall development i neral and economic development in a particular. A large number of developmer hemes financed by internatioinal agencies and central and state governments wer ıplemented and are continuing to strenghten the economy of the region.

SPACE TECHNOLOGY

Satellite Remote Sensing, by virtue of its synoptic coverage, repetitivity, eas ailability of data in inaccessible areas, multi-spectral capability and cost effectivenes fers an appropriate technology in mapping, monitoring and management of naturε sources wihtin a short time frame in desert region.

Ecological impact of Indira Gandhi Nahar Project (IGNP)

IGNP is a major irrigation project of Rajasthan. It has brought appreciable change whole of North and Western Rajasthan. This study is initiated to see the overa ıpact on the area in terms of land use changes, extent of cropped area, branchwise cro stribution, seepage of canal, waterlogging, afforestation along canal and the extent (eed growth in canal, in stage I and II of the project covering Ganganagar, Bikaner an ıisalmer districts of Rajasthan (Plates 1 and 2). The study of hydrological barriers i age II of canal was also conducted in light of further expansion of the project in sout est direction.

IRS LISS II and Landsat TM data have been used to study the various aspects (e present investigation. Digital technique like ratios, filtering and classification aı ed to generate various enhanced products and the statistics is calculated accordingl

:tensive ground truth is carried out to see the location and overall accuracy of tl :ferent themes under study.

The major crops grown in IGNP area are cotton, groundnut and guar during Khaı .d wheat, mustard and gram in Rabi. The two seasons classified outputs were matchı dominant cropping pattern and refinement of first analysis was carried out. Po ıssification validation of Naurangdausar and Rawatsar branches is done and tl ɜults are as follows (Table 1).

ıble 1. Satellitc based crop inventory for Naurangdausar and Rawatsar branchs of tl command area*

ops	Kharif (1992)	Crops	Rabi (1992-93)
rea in '000 ha)			
ϽTTON	77.129	WHEAT	47.824
JAR	31.196	GRAM	30.520
ΓHERS	4.990	MUSTARD	12.642
		OTHERS	10.318

.tellite based crop inventory for Suratgarh branch command area*

ops	Kharif (1992)	Crops	Rabi (1992)
igated ırea in '000ha)			
ϽTTON	33.4	WHEAT	32.9
JAR	8.9	GRAM	9.9
ꓤOUNDNUT	0.8	MUSTARD	3.7
ΓHERS	11.8	OTHERS	—
ıirrigated ('000ha)** JAR	11.2	GRAM and TARAMIRA	21.8

'he comand area maps were taken from the semi-detailed soil map of various branches prepared under Rajasth ıal Soil and Water Study Project by Government of Rajasthan and UNDP, FAO (scale 1:63, 360)
Within the gross command area.

The extent of water logging/surface manifestation is obtained from Kharif, 19ᑫ .ta is indicated below. The total estimated area till RD 1310 is 5551 ha. In addition, tl ıter spread due to escape near Dattar covered an area of 655 ha. The breakup to stagɛ .d II is indicated in the table as follow (Table 2).

Monitoring locust breeding grounds using Satellite Remote Sensing and GIS

The desert locust cause extensive devastation to natural and cultivated biomaɛ ıeir breeding is triggered by suitable ecological conditions due to widespread rainfa .d subsequent vegetation development. Thar desert is the principal recession area sert locust in India.

able 2. Extent of waterlogged area (Kharif, 1992)

.No.	Command	Area (ha)	
		Kharif	Rabi
	Stage I*	WATER LOGGED	
.	Naurangdausar	389	313
.	Rawatsar	594	563
.	Direct outlets	2726	—
	Subtotal	3706	—
	Stage II*		
	Dattor to Birsalpur	813	696
	Birsalpur to Charanwala	378	553
	Charanwala to RD 1310	651	575
	Sub total	1842	1824
	Total	5551	—
	Escape near Dattor	655	497

Statistics are for within command area only.
: Statistics are for all the region between the branches indicated.

Taking into account the extensive damage caused by the locust and inadequacy (ıformation obtained by conventional methods, an operational methodology wɛ eveloped based on Remote Sensing data and other data sets in order to have improve ıonitoring system for locust habitats. The methodology was developed with the followir bjectives : (i) Evaluation of spectral VI for determination of desert vegetation biomas sing NOAA and IRS LISS I data, (ii) Generation of soil moisture indices usir ıicrowave data, (iii) Integration of vegetation, rainfall and soil moisture data to foreca: otential locust breeding grounds and (iv) Development of a GIS, with tehsil lev ıformation base having permanent attributes viz. rainfall, soil moisture and vegetation

This study is being carried out since 1989 covering an area of 213390 sq. km. of tł arts of Gujarat and western Rajasthan.

NOAA data and IRS IC WiFS data (Plate-3) are used for regional level monitorir f biomass on weekly basis and IRS LISS I data are used for local sites. The vegetatiɩ ıdex products are classified into five classes, four classes show various densities (iomass and the last one, the area devoid of vegetation. As per the feedback receive :om the locust department, which is the main user of these products, the products hay een very useful in reducing the survey area considerably and also for locust contr peration at number of places in last few years.

. Desertification

Sand drift is a major problem in Rajasthan desert, leading to land degradation ar onsequent desertification.

The present investigation aims to study the extent of sand drift in different parts esert, the extent of degradation in range lands and also the degradation of commc roperty resources in desert area.

Change detection studies are carried out using historical data and the present dat see the change over a period of time at a number of sites in the desert. The s ightness index is found quite useful in classifying various types of sand featur milarly the vegetation index has been found quite useful in studying degradation nge lands and common property resources over a period of time.

Plate-5 shows the desertification status in Pushkar area where the sand dur om loose to stabilized) have been separated using SBI, March, 1991 IRS LISS II ima

Demarcation of Palaeo/Buried Channels

In the desertic terrain of Rajasthan, availability of ground water is scarce a ostly this resource is controlled by palaeo/buried courses of ancient river systems.

The study area from parts of Churu, Jhunjhunu and Jodhpur districts of Rajasthe ie enhanced outputs using IRS data brought out old courses and meanders (Plate id also the west to eastward shift of the Kantli river. A final map is prepared tegrating all enhanced products highlighting the entire drainage network.

Such drainage network has not been represented in SOI topomaps. The produ uld be used for exploitation of ground water and also the suitable recharge sites.

Vegetation Dynamics in Desert Region

Vegetation/Natural grasslands are valuable resource of desert area. With the ons monsoon rains in July, the vegetation growth starts and reaches its peak ptember. The study is undertaken to see the seasonal changes as well as changes ov ars using NOAA and IRS data.

The study shows a high growth of vegetation in desert region with the on set onsoon-combined impact of rainfall and high temperature. As the rainy season en e natural biomass trends to dry. An attempt was also made to classify vegetation ty desert region (particularly the Sewan Grass), which has high nutrition as well onomic value, using multiple data sets in one season (Plate 4).

Floods in Desert Region

Desert region generally receives very little rains, but occasionally due to clo irsts and in absence of well developed drainage system floods occur. The study is tak assess the extent of flood, the changed courses of the streams and the dama sessment to agricultural area using IRS LISS II and LISS III data. The products cou directly used as flood maps.

Impact of Mining on Environment

Near Jodhpur city, there are a number of good quality sandstone deposits. The posits are being mined for the last five centuries. Due to heavy demand of stones iilding material, the mines are being over exploited. Under this study the extent ining area, the dump sites and the effect on drainage system were studied using SP(A and MLA data of 1990 and IRS 1C data of 1996.

The study of limestone mining area near Sanu was carried out using 1989 and 19 ıta. It shows the tremendous increase in mining activity (Plate 7), the impact on th ırrendering land, drainage, etc.

In conclusion, Satellite Remote Sensing offers large volume of data at quick interva ith vaious spatial resolutions in multiple bands and hence this is an ideal and cos fective tool with high accuracy to map and monitor natural resource in desert region.

The first advice to a person who gets interested in environmental issues is to thin ıd act holistically, i.e., to view every thing in the world as being connected with ever ing else. In other words : do not ignore interconnections and do not look at things i olation. And so a grand view emerges in which we see the linkages to the whol ıiverse, firstly to the immediate environment and then to more distant environment ıtil a global interconnection is established. All this is heady stuff, and the neophyt periences a sense of vastness and grandeur.

But there are innumerable environmental problems. There are local environmenta oblems such as unsasfe water, accumulating urban garbage, untreated sewag ılluted urban air due to personal vehicular transport and/or polluted indoor air due t noke from fuel wood, stoves, etc. There are also regional and national environmenta oblems such as polluting industries, acid rain, river pollution, deforestation, et nally there are global environmental problems such as accumulation of greenhous ıses in the atmosphere.

From a holistic vision must emerge practice, and here the dictum in th ıvironmental movement is : think globally and act locally. Thinking globally implie ınking interconnectedly, i.e. with an awareness of all the interconnections. In effec ıis demand is equivalent to put all environmental problems at par and treating them a they are of equal importance.

There is hence no priority distinction of local, regional, national and globa ıvironmental problems. It looks, therefore, as if a holistic attitude leads naturally to th osence of priorities and a lack of focus.

REFERENCES

grawal, B. 1995. Feminist Environmentalism or ecofeminism, Survey of Environment, *Th Hindu.*

owonder, B. 1995. Environmental Management-Redefining perspective, Survey of Environmen *The Hindu.*

hir, R.P. 1994. Approach on arrest of Desertification, CAZRI Publ. Jodhpur.

hir, R.P. and Sharma, J.R. 1991. IRS-1A Applications in Desertification studies. *Curr. Sci.* 62(4)

ıarma, J.R. and Singh, R.B. 1988. *Indian Desert-Population, Resources and developmer* Anupama Publications, Delhi.

eddy, AKN. 1995. Environmental Action : first act locally then globally-Survey of Environmen *The Hindu.*

eological Evolution of Northwestern India
S. Paliwal (Ed.)
ientific Publishers (India), Jodhpur, 1999, pp. 338 - 347.

GYPSIFEROUS SOILS OF ARID RAJASTHAN

D.C. Joshi
Central Arid Zone Research Institute, Jodhpur - 342 003

ABSTRACT

In Arid Rajasthan gypsiferous soils widely occur in western and northern parts. Two groups of gypsiferous soils (Suborder Gypsids) viz. Petrogypsids and Haplogypsids have been characterized and their management problems discussed. The Petrogypsids occur surrounded by high sand dunes (inter dunes) having a hard indurated petrogypsic horizon formed of gypsum, $CaCO_3$ and sand cemented together. Gypsum is mostly present as hard nodules. The overlying solum is thin, having low available water capacity and low fertility status. The Haplogypsids have been encountered in micro depressions associated with sand ridges and low dunes. Solum is 50-70 cm deep, loamy fine sand and non calcareous nature with moderate available water capacity and medium fertility status. The underlying gypsic horizon is pervious to roots and water. Gypsum is mostly present in amorphous form. The gypsiferous inter dunes have high proportion of silt, $CaCO_3$ and salinity. Severe climatic and soil constraints associated with Petrogypsids and gypsiferous accumulative inter dunes restrict their use to silvipastures. The Haplogypsids offer good sites for arable farming both under rainfed and limited irrigation.

ey words : Gypsiferous Soils, Petrogypsids, Haplogypsids, Arid Rajasthan.

INTRODUCTION

Gypsiferous soils are Aridisols that commonly occur in various parts of the arid an mi arid regions. Gypsum originates principally either from runoff water from gypsu aring rocks or from gypsum bearing ground water. Bellanca and Neri (1993) conclude at in the marly clay soils of Sicily, ground water largely dissolved gypsum from th asement rocks and presipitated in soil profile. On the other hand Podwojewski an rnold (1994) in the vertisols of New Caledonia, after sulphur iso-topic study confirm vo sources of gypsum i.e., (i) an oceanic source where the major part of sulphur ıtroduced by atmospheric precipitation and (ii) continental source where the sulphid .sseminated in various sub surface deposits are oxidized.

Various views have been expressed regarding origin of gypsum in the arid region ıdia. Jacob *et al.* (1952) studied the micro-fossils from the Jamsar gypsum quarries a ıferred that gypsum was deposited away from sea in inland basins or depressions in t .te Quaternary or Sub Recent times. Ghosh (1952) after comparing various hypothes ›ncluded that due to percolation of the solution charged with suitable salts, t /psiferous soils were formed. Deb (1952) after a detail survey of the depressions arou unkaransar (District Bikaner) reported that soils at the depth of about four feet conta to 10 per cent gypsum. The gypsum bearing formation is invariably a dark gr ›loured clayey material with an average depth of 75 cm. Top of this argillaceous b sually contains small crystals while the bottom contains big crystals.

Gypsiferous soils occupy 845 sq. km. in arid region of Rajasthan but these have r een studied, despite their economic and environmental significance. Therefc /stematic study of the gypsiferous soils was carried out and their characteristi assification and management problems have been discussed in the present paper.

ENVIRONMENTAL SETTING

The study sites are located in the Barmer, Jaisalmer, Nagaur, Bikan anganagar and Hanumangarh districts. Climate is hot arid with 200 to 300 m ıinfall and high rate of evapo-transpiration. During summer in the month of May a ıne the temperature reaches as high as 49°C. Winters are cool and dry with minimu ·mperature of 1° to 2° C for a few days. In Hanumangarh and Ganganagar distric mited rain showers in winter season are also received. Dust storms are common ımmer. The terrain is duny undulating with large inter dunes. Natural vegetati ›nsists of scanty scrub grasses and shrubs in degraded condition because of op ·azing. Rainfed cropping is the dominant land use but with the introduction of Indi andhi Nahar Pariyojana (IGNP) the irrigation facilities are being extended.

SAMPLING AND ANALYSIS

ield studies

Large stretches of gypsiferous areas were encountered near Kavas- Uttar armer), Mohangarh-Nachna-Bajju (Jaisalmer), Jayal- Dadhimati (Nagaur), Lunkara ır-Pugal-Chhatargarh (Bikaner) and at a number of sites in Ganganagar a anumangarh districts. Soil profiles were examined at these sites and morphologi ıaracteristics viz., soil colour, texture, structure, consistence, calcareousness and dep ere recorded and described according to Soil Survey Staff (1951). Associated terra atures including topography, drainage, slope and vegetation were also recorded. S mples from representative sites for each horizon were collected for laboratory analys ıd soil profiles were classified according to Soil Survey Staff (1992).

ıboratory studies

Soil samples were crushed with a wooden pestle and mortar and passed through m sieve. Contents of coarse fragments (>2 mm) were recorded. The fine earth fractic 2 mm) was used for determination of pH and electrical conductivity (1 : 2 soil : wat spension) and of $CaCO_3$ (Calcimeter), organic carbon (Walkley and Black's methoc ailable phosphorus (Olsen's method), available potassium (NH_4OAc extractior oisture equivalent and water holding capacity according to Piper (1950) and Jacksc 967).

RESULTS AND DISCUSSION

In the study area the gypsiferous soils were invariably encountered in interdun eas surrounded by high dunes. Gypsiferous interdunes vary in size from a few sq. m. to 10 sq. km. Gypsum occurs both in amorphous and crystalline forms at the dept ırying from a few centimetres to five metres below the surface.

According to Soil Survey Staff (1992) those Aridisols which have gypsic/petrogyps ırizon whose upper boundary is within 100 cm are classified in the suborder Gypsid ıt of five groups of this suborder two groups viz. Petrogypsids and Haplogypsids hav en recognized in the present study. In addition to these, accumulative inter dun hich contain appreciable amount of gypsum (but do not qualify for Gypsids) have al en encountered. Salient characteristics of these three categoreis of gypsiferous soi ıve been described below.

PETROGYPSIDS

Petrogypsids were encountered in the inter dunes which were devoid of loose sar ıd trees. Ephemeral grasses and shrubs constitute the scrub vegetation cover. Surfa ainage is absent. Petrogypsids are wide spread in Nagaur, Barmer, Jaisalmer ar kaner districts.

Morphological and physico-chemical characteristics of Petrogypsids have bee ported respectively in Table 1 and Table 2. The epipedon (A horizon), 15 to 30 cm dee xcept at Desli) is characterized by light yellowish brown (10YR 6/3) and light gre 0YR 7/2) colour, fine sand, loamy fine sand or sandy loam and weakly aggregate xture and moderately calcareous nature ($CaCO_3$ 4-6%). Only at Anandgarh, Ballar ar ıdhimati epipedon contained 8 to 15 percent crystalline gypsum. At other sites tl psum as powder is dispersed in the epipedon. The pH is in alkaline range (7.8- 8.2) ar C values indicate moderate salinity (1.08-1.69 d S/m).

'able 1. Morphological characteristics of Petrogypsids

.ocation/ łorizon	Depth cm	Soil		CaCO3 %	Gypsum morphology	
		Colour	Texture		Nodules (w/w%)	Form
;anganagar district						
)esli	A 0-75	Light grey (10YR7/2	sl	4	–	Amorphos
	Cy 75-200	White (10YR 8/1)	–	4	67	Crystalline
\nandgarh	0-15	Light grey (10YR7/2)	sl	6	15	Crystalline
	Cy 15-25	Light grey (10YR7/2)	l	5	27	Crystalline
;ikaner district						
;allar	A 0-15	Pale Brown (10YR6/3)	lfs	4	15	Crystalline
	Cy > 15	White (10YR 8/1)	Gypsum blocks			Crystalline
;armer district						
(avas	A 0-50	Light grey (10YR7/2)	lms	5	–	–
	C_{1y} 50-80	Light grey (10YR 7/2)	lms	3	10	Crystalline
	C_{2y} > 80	White (10YR 8/1)	Gypsum nodules			
Jagaur district						
)adhimati	A_1 0-20	Pale brown (10YR 6/3)	sl	3	8	Crystalline
ayal	A_2 20-55	Pale brown (10YR 6/3)	sl	8	10	Crystalline
	C_{1y} 55-120	Light grey (10 YR 7/2)	sl	11	12	Crystalline
	C_{2y} >120	White	Gypsum blocks			Crystalline
aisalmer district						
Jachana	A 0-25	Pale brown (10 YR 6/3)	fs	–	–	–
	C_{1y} 25-50	White (10YR 8/1)	Gypsum blocks		30	Amorphous crystalline
	C_{2y} 50-90	White (10YR 8/2)	Gypsum nodules and powder			

. : sandy loam, l : loam, lms : loamy medium sand, fs : fine sand

ıble 2. Physico-chemical characteristics of Petrogypsids

ɔcation	Horizon	pH	EC dS/m	M.E. %	W.H.C. %	Org. Carbon	Available (kg/ha)	
							P	K
anganagar district								
esli	A	8.1	1.69	11	23	0.10	4.8	1575
	Cy	7.9	1.24	–	–	–	–	855
nand	A	7.9	1.08	12	39	0.23	5.6	213
ırh	Cy	8.0	1.11	10	36	0.1	2.6	186
ikaner district								
ıllar	A	7.8	1.09	10	22	0.13	9.5	169
armer district								
avas	A	7.9	0.60	8	37	0.11	5.2	188
	C_1	7.4	17.50	25	46	0.12	4.6	282
	C_{2y}	7.8	11.90	37	70	0.13	4.6	213
agaur district								
adhmati	A_1	8.2	0.35	9	35	0.12	9.0	296
ayal)	A_2	8.1	0.27	12	40	0.11	12.0	232
	C_{1y}	8.1	0.33	13	43	0.10	5.2	188
ıisalmer district								
'achana	A	8.0	0.52	8	30	0.07	4.0	236
	C_{1y}	7.8	2.39	10	35	0.03	4.5	360
	C_{2y}	9.3	1.17	10	37	0.03	3.5	282

Fig. 1. Photograph showing typical soil profile of Petrogypsids

The Petrogypsic horizon (Cy) occu at 15 to 40 cm depth. This horizı consists of gypsum, $CaCO_3$ and saı cemented together as hard indurat layer [Fig. 1]. Gypsum is present both amorphous and crystalline forms but tl crystalline form constitute maj proportion (20-67%). The $CaCO_3$ conte varies from 4 to 11 per cent. Soil coloı is invariably light grey (10YR 7/2) white (10YR 8/1) and the texture vari from sandy loam to loam. Accumulatiı of illuvial clay in pockets is common aı prominent in the soil profile near Kava The pH of petrogypsic horizon is alkaline range (7.9-8.1) and EC valu (1.1 to 2.39 d S/m) indicate modera salinity in all the profiles except tl Kavas profile where salinity is high (E 11.9-17.5 dS/m). These are Typ Petrogypsids. Stoops and Illawi (198 have also reported crystalline aı amorphous forms of gypsum aı attributed their origin to water table contact with sea water and sulphide ri rocks.

The Petrogypsids are characterized by relatively low moisture equivalent (8-12% ry low organic matter (0.03-0.13%), very low available phosphorus (Tr-5.6 kg/ha) ar edium to high available potassium (169-360 kg/ha). The water holding capacity rang m 22 to 47 percent.

HAPLOGYPSIDS

The Haplogypsids are associated with micro depressional plains surrounded l nd ridges, sand hummocks and low sand dunes. Surface has thin veneer of sand ar rly good stand of trees. These are well drained soils and generally cultivate plogypsids widely occur in Hanumangarh, Ganganagar and Churu districts.

The morphological and physico-chemical characteristics of Haplogypsids have be scribed in Table 3 and Table 4. The epipedon is 30-40 cm deep, characterized by lig own (10YR 6/3), loamy very fine sand, sandy loam and weak granular soils. These a utral in reaction (pH 7.7-7.8), slightly saline (EC~1 dS/m) and weakly calcareo aCO$_3$ 0.29-2.9%) in nature. The gypsic horizon (Cy) occurs at 50-70 cm depth (Fig. 2 this horizon gypsum is uniformly distributed through out the matrix of fine ear m) or occur as nest or as soft nodules. Gypsum nodules make only 9 to 13 per cent e bulk and remaining gypsum is present as powder. The fine earth material is whi)YR 8/2, 8/3), coarse textured, non-calcareous (CaCO$_3$ 0.9-2.7%) and massive in natur ere is a narrow range of pH (7.7-7.8) and EC (0.28-1.20 d S/m). This gypsic horizon rvious to roots and water. These soils belong to sub group Typic Haplogypsids.

Fig. 2. Photograph showing typical soil profile of Haplogypsids

'able 3. Morphological charcteristics of Haplogypsids

,ocation	Horizon	Depth cm	Soil		CaCO3 %	Gypsum morphology	
			Colour	Texture		Nodules (w/w %)	Form
Ianumangarh district							
/ahila ki Ihani	A	0-25	Light grey (10YR 7/2)	lfs	0.25	10	Amorphous, soft nodules
	Cy	25-90	White (10YR 8/2)	–	1.79	19	Amorphous, soft nodules
'urabsar	A	0-30	Pale brown (10YR 6/3)	fs	2.10	10	Amorphous, soft nodules
	C1y	30-60	Pale brown (10YR 6/3)	fs	1.10	11	Amorphous
	C2y	60-90	Light grey (10YR 7/2)	fs	2.10	–	Amorphous, loosely pack
;agder	A	0-20	Light grey (10YR 7/2)	sl	2.90	13	Soft nodules
	Cy	20-50	Light grey (10YR 7/2)	sl	2.70	22	Soft nodules

s : loamy fine sand, fs : fine sand, sl : sandy loam

'able 4. Physico-chemical characteristics of Haplogypsids

,ocation	Horizon	pH	EC dS/m	M.E. %	W.H.C. %	Org. Carbon	Available (kg/ha)	
							P	K
Ianumangarh district								
/ahila ki	A	7.8	1.02	11	26	0.11	nil	112
.hani	Cy	7.8	1.05	14	25	0.13	2.2	79
'urabsar	A	7.8	1.08	7	22	0.10	8.4	169
	C1y	7.8	1.07	6	24	0.06	7.0	169
	C2y	7.8	1.36	3	22	0.01	7.1	202B
;agder	A	7.7	1.02	11	26	0.22	12.0	382
	Cy	7.7	1.18	12	28	0.12	3.5	135

The Haplogypsids are characterized by relatively medium available water capaci M.E. 7-14%), low organic carbon (0.11-0.23%), low to medium available phosphorus (7- g/ha) and potassium (112-382 kg/ha). The water holding capacity ranges from 22 to ercent.

ACCUMULATIVE GYPSIFFROUS INTERDUNES

These are narrow accumulative interdunes ave compact surface, poor draina nd salt tolerant bushes and trees (Fig. 3). The mulative interdunes are wide spre 1 Barmer (Chohtan), Jaisalmer (Nachna) Churu districts. Morphological a hysicochemical characterstics of typical site re presented in Table 5 and Table espectively.

These interdunes are characterized by the accumulation of silt, $CaCO_3$, salt a /psum. The solum is greyish brown (10YR 5/2) and light grey (10YR 5/2) in colour, lo: silty loam in texture and with medium sub angular blocky structure. These are hig ılcareous ($CaCO_3$ 15-20%) and saline (EC 2-5 d S/m). Below 30-50 cm there :cumulation of amorphous gypsum which is of low grade (mixed with soil).

Fig. 3. Photograph showing panoramic view of Gypsiferous Accumulative Interdunes

ıble 5. Morphological characteristics of gypsiferous interdunes

ɔcation	Horizon	Depth cm	Soil		$CaCO_3$ %	Gypsum morphology	
			Colour	Texture		Nodules (w/w %)	Form
armer district							
rta (Chohtan)	A	0-30	Greyish brown (10YR 5/2)	sil	14	–	Dispersed in soil
	C_{1y}	30-90	Greyish brown (10YR 5/2)	cl	15	–	Dispersed in soil
	C_{2y}	>90	Gypsum enriched with lime				
ıisalmer district							
ıkampur	A_1	0-5	Light grey (10YR 7/2)	sil	5	–	Dispersed in soil
	A_2	25-30	Light grey (10YR 7/2)	sil	10	–	Dispersed in soil
	C_y	>30	White (10YR 8/1)	–	20	–	Amorphous, non-coherent

: silt loam, cl : clay loam

ıble 6. Physico-chemical characteristics of gypsiferous interdunes

ɔcation	Horizon	pH	EC dS/m	M.E. %	W.H.C. %	Org. Carbon	Available (kg/ha)	
							P	K
armer district								
rea	A_1	8.02	3.57	9.4	35	0.17	4.5	249
	C_{1y}	8.05	5.02	15.6	44	0.12	3.5	262
isalmer district								
kampur	A_1	9.30	1.17	12.0	35	0.12	3.5	754
	A_2	9.10	1.20	11.0	40	0.11	tr	760
	C_y	8.80	1.16	–	–	–	–	–

CONSTRAINTS OF GYPSIDS AND THEIR MANAGEMENT

Petrogypsids are invariably surrounded by high sand dunes and occur in extreme ·id part of Rajasthan. These are shallow soils (25-40 cm), mixed with hard nodules ·psum which increase with depth. Fertility status of soils is very low. Organic carbo .09-0.17%) and available phosphorus (nil-4.8 kg/ha) is very low and availab ɔtassium (101-205 kg/ha) is low to medium. Muhammad and Jones (1992) also observe ıat phosphorus in gypsiferous soils is mainly present as calcium-phosphate which soluble and thus not available to plants. Available water holding capacity is also ve w (25-40 mm per 50 cm depth). The under lying petrogypsic horizon is indurated ar ɔes not allow root penetration. Because of these climatic, and soil constrain ətrogypsids are not suitable for arable farming. Severe root zone limitation due ətrogypsic horizon pose constraint on raising tree plantations. These soils are suitab ıly for silvipasture having shrubs as main component. Moisture conservation ar ɔplication of fertilizers should form an integral part of the management system.

Haplogypsids have 60 to 90 cm deep soil cover. These soils are sandy loam wi irly good moisture holding capacity. The available water capacity varies from 50-7 m per 70 cm depth. Fertility status of these soils is medium (organic carbon conte 16-0.25%, available phosphorus 2.2-11.5 kg/ha and available potassium 179-382 kg/ha ıe underlying gypsic horizon is permeable to roots and water. However, under exce ater condition the gypsum in the gypsic horizon slakes and water logging conditions a eated. Cropping both under rainfed and limited irrigation are successful on these soil uch of the area occupied by Gypsids is likely to receive irrigation from Indira Gand ahar Pariyojana (IGNP) and proper water management should be followed to avo aterlogging and salinization.

The accumulative inter dunes are highly saline and calcareous in nature. Seve imatic and soil constraints do not permit cultivation on these lands. Growing of sa lerant species of shrubs, grasses and trees is the only proposition left with th əvelopment agencies.

ACKNOWLEDGEMENT

Grateful thanks are due to Dr. A.S. Faroda, Director Central Arid Zone Researc ıstitute, Jodhpur for the facilities and to Dr. K.D. Sharma, Senior Scientist (Hydrolog r his valuable help.

REFERENCES

:llanca, A. and Neri, R. 1993. Dissolution and precipitation of gypsum and carbonate minerals soils on evaporite deposits, Central Sicily : isotope geochemistry and microfabric analys *Geoderma* 59 : 263-277.

:b, S. 1952. Occurrence of natural salts and selenite-gypsum in the neighbourhood Lunkaransar Tal, north of Bikaner, Western Rajputana. *Proceedings of The Symposium the Rajputana Desert. Bulletin, National Institute of Sciences of India,* New Delhi. 94-99.

ıosh, P.K. 1952. Western Rajputana-its tectonics and minerals including evaporites. *Proceedin of The Symposium on The Rajputana Desert. Bulletin. National Institute of Sciences of Indi* New Delhi. 101-130.

ckson, M.L. 1967. *Soil Chemical Analysis,* Prentice Hall Inc. U.S.A.

cob, K., Sastry, M.V.A. and Sastri, V.V. 1952. A note on the microfossils of the impure gypsu from the Jamsar mine, Bikaner and on the possible origin of the gypsum. *Proceedings of T Symposium on The Rajputana Desert. Bulletin, National Institute of Sciences of India,* N Delhi. 68-69.

ıhammad, H.J. and Jones, K.C. 1992. Phosphorus in gypsiferous soils : the influence of s properties on P fractionation. *Geoderma* 53 : 97-104.

per, C.S. 1950. *Soil and Plant Analysis*, Adelaide, Australia.

dwojewski, P. and Arnold, M. 1994. The origion of gypsum in vertisols in New Caledon determined by isotopic characterstics of sulphur. *Geoderma* 63 : 179-195.

il Survey Staff. 1951. *Soil Survey Manual.* Agricultural Hand Book No 18. U.S.D. Washingaton, D.C., U.S.A.

il Survey Staff. 1992. *Key to Soil Taxonomy*, Sixth Edition, 1994. U.S.D.A. Washington, D. U.S.A.

oops, G. and Illawi, M. 1981. Gypsum in arid soils morphology and genesis. In : F.H. Beinor and A. Osman (Eds.). *Proceedings Third Soil Classification Workshop*, ACSAD, Dam 175-198.

ological Evolution of Northwestern India
S. Paliwal (Ed.)
ientific Publishers (India), Jodhpur, 1999, pp. 348 - 357.

SOIL DEGRADATION DUE TO WIND EROSION IN ARID RAJASTHAN

Pramila Raina

Central Arid Zone Research Institute, Jodhpur 342 003

ABSTRACT

The study was carried out in Jalore district to identify and map the areas affected by varying intensities of soil degradation due to wind erosion/deposition, with the help of IRS LISS II imageries of 1994 followed by ground verification. In an area of 10,640 km^2, 43% is affected by wind erosion, out of which 15.8% area is slightly, 10.7% is moderately and 16.5% is severely degraded. Wind-erosion associated with other degradation processes such as water erosion and salinization affected 9.1% area. Studies on the effect of landuse on soil degradation revealed that among major land use systems 38% rainfed cultivated land, followed by 28.6% irrigated and 21.4% of grazing lands are affected by wind erosion/deposition hazards. Impact of wind erosion/deposition on the fertility status of soil revealed that the nutrients are more in soils of nondegraded sites than the degraded sites. Decrease of these nutrients is maximum in severly degraded soils (40-60%) followed by moderately and slightly degraded soils (40-50 and 22-31%) in comparison to the non-degraded soils.

ey words : Soil Degradation, Wind Erosion, Arid Zone, Western Rajasthan.

INTRODUCTION

Wind erosion is a serious problem in arid and semi-arid regions. Arid lar mprises about 36 percent of the world's total land area which is subjected to varyi grees of wind erosion. The United National Environment Programme (UNEP, 197 ıs estimated that globally 80 per cent of 3700 million ha of range-land : 60 percent '0 million ha of crop-land and 30 percent of 131 million ha of irrigated land are affecte wind erosion. Bali and Kanwar (1977) reported that out of the 329 million ha area .dia, some 90 million ha area has been suffering from wind erosion/deposition :tivities.

Wind erosion is the dominant soil degradation process in the desert regions 'estern Rajasthan because the large part (70%) is sandy having sand hummocks ar ınes on the surface. Sandy texture of soil, climatic vagaries, erratic monsoon and abo l human activities viz. cultivation on marginal land (duny, shallow and slop 'ergrazing of pasture land and indiscriminate cutting of trees for fodder, fuel and oth ırposes are the major activities for accelerating soil degradation. Continuous deflati finer fractions of the soil makes the agricultural land infertile. Deposition of the dri ınd over productive cultivated lands turns them in to sandy hummocks and high ılnerable to wind erosion. Aggarwal and Gupta (1980), Gupta et al. (1981) reporte nsiderable loss of soil and plant nutrients from sandy plains, particularly during tl ːriod of April to June, when strong winds frequently flow. Raina (1992) observe nsiderably low amount of plant nutrients in degraded soils in comparison ›ndegraded sites in different landuse systems.

In arid and semi-arid regions satellite data have been found useful for mappir ːgraded lands (Teotial et al. 1980, Sehgal and Sharma, 1988; Raina et al. 1991, 199 ı the above studies land degradation processes with different intensities (sligh oderate and severe) have been mapped but information on fertility status of soils und fferent intensity of degradation is not available.

MATERIALS AND METHODS

Jalor district covering 10640 km^2 area is situated between the latitude 24.6° ;.8°N and longitude 71.2° to 73.1°E (Fig. 1). The climate is arid to semi-arid with ean annual rainfall of 420 mm. Mean summer and winter temperatures are 44°C ar C respectively. Vegetation is largely dominated by Jal (*Salvadora persica* L.), *Prosop liflora,* Neem (*Azadirachta indica*), Kumat (*Acacia senegal*), Khejari (*Prosop ıeraria*) and Pilu (*Salvadora oleoides*).

apping soil degradation

Criteria for Mapping Degradation due to Wind Erosion/Deposition : The criter id down by Oldeman et al. (1991) with slight modification (Raina et al. 1991 and 199 as adopted for identifying the area, degraded due to different intensities of wir osion/ deposition (Table 1). The degree to which soil has been degraded was estimate relation to changes, in agricultural suitability in case of cultivated soils and in relati its biotic potential for pasture and forest soils.

Fig. 1. The map of northwestern Rajasthan showing location of the study area.

able 1. Criteria for identifying the soil degradation due to wind erosion/deposition

ssessment factor	Degree of degradation		
	Slight	Moderate	Severe
Surface Manifestations	Thick sand deposition along fence line	Thick sand deposition along fence lines and on fields sandy hummocks common	Degraded dunes, ver thick and loose sand sheets, all devoid of vegetation
Percentage of area covered with vegetation	30-70%	10-30%	0-10%
Agricultural activity	The terrain has some what reduced agricultural activity, original biotic functions are still largely intact	The terrain has greatly reduced agricultural activity, Major improvements are required to restore productivity, original biotic functions partially destroyed.	The terrain require major work for restoration of productivity, origina biotic functions are largely destroyed.

Mapping Present Status of Wind Erosion/Deposition : The IRS, FCC (LISS II) (›vember 1993, April, 1994 and February 1994 path 032 row 50 of 1:50,000 scale wei terpreted in conjunction with Survey of India topographical maps of same scale. Visuɛ terpretation was carried out based on the tonal characteristics of different intensitie degradation on imagery and tentative units were demarcated. The tentative unit ›re checked in the field. During ground verification misuse of land due to huma tivities viz. cultivation on marginal lands, (dune slope, shallow and slopy soils) an er exploitation of grazing lands were correlated with surface manifestations. At eac ›e, observation on soil morphological characteristics. viz. texture, calcareousness, colou .d depth were recorded (USDA 1995) and sand deflation/deposition were assesed. So mples collected from different sites, were analysed for soil pH and electricɛ nductivity in soil : water extract (1 : 2). For fertility status viz. organic carbon wɛ termined by chromic acid with H_2SO_4 digestion, available phosphorus b lorostannous reduced molybdophosphoric blue colour method and available potassiu 1N NH40AC, pH 7.0, extraction and flame photometry as described by Jackson (1967 ısed on both laboratory interpretations and field observations, a map showing so gradation due to wind erosion activities was prepared.

RESULT AND DISCUSSION

Soil degradation due to wind erosion resulted in the formation of sand sheet, ımmocks and barchan dunes as depositional and exposure of plant roots and substrat erosional features. The hazards increased in direct proportion of human activitie ›tal area degraded due to wind erosion/deposition hazards was 4578.65 km^2, (43.03% ig. 2). The characteristics of different intensities of soil degradation due to win ›sion/ deposition and their effects on fertility status are discussed below.

Areas of Non Degraded to slightly Degraded Soil (W_1) : These are nearly flat areɛ .ving no degradation to slight degradation in the form of sand sheeting or sand drif ıis type of degradation is very common in rainfed cultivated and pasture land. Soils i ese areas are sandy to loamy sand occurring deep and widely in villages like Odwarɛ ›ta, Rajenwari, Mera, Sarana and Devki. On the February FCC imagery this unit wɛ ›cognised by its uniform pale brown tone with red tinge due to rabi crops. This clas ›cupies about 1679.08 km^2 (15.8%).

Moderately Degraded Soil (W_2) : In the moderately degraded areas massive san ›vements resulted in the formation of thick sand sheets within the fields as well ɛ ›ng fence lines. Sand drifting activities resulted in the formation of 10-20 cm thic nd sheets and hummocks of less than three metres in height, which were spectaculɛ the field as well as on satellite imagery (Fig. 3). Soils of these areas are fine and loam nd, 45-90 cm deep, light yellowish brown and pale brown in colour. These areas occu ar Bhinmal to Vander and southeast of Sanchor, southern portion of Leta, Bagra an ›stern portion of Odwara, Bhandwa and Jaisawas villages. On satellite imagery th: .it was recognised by its pale brown tone with brown tinge. This class occupies abou 42.52 km^2 (10.7%) area.

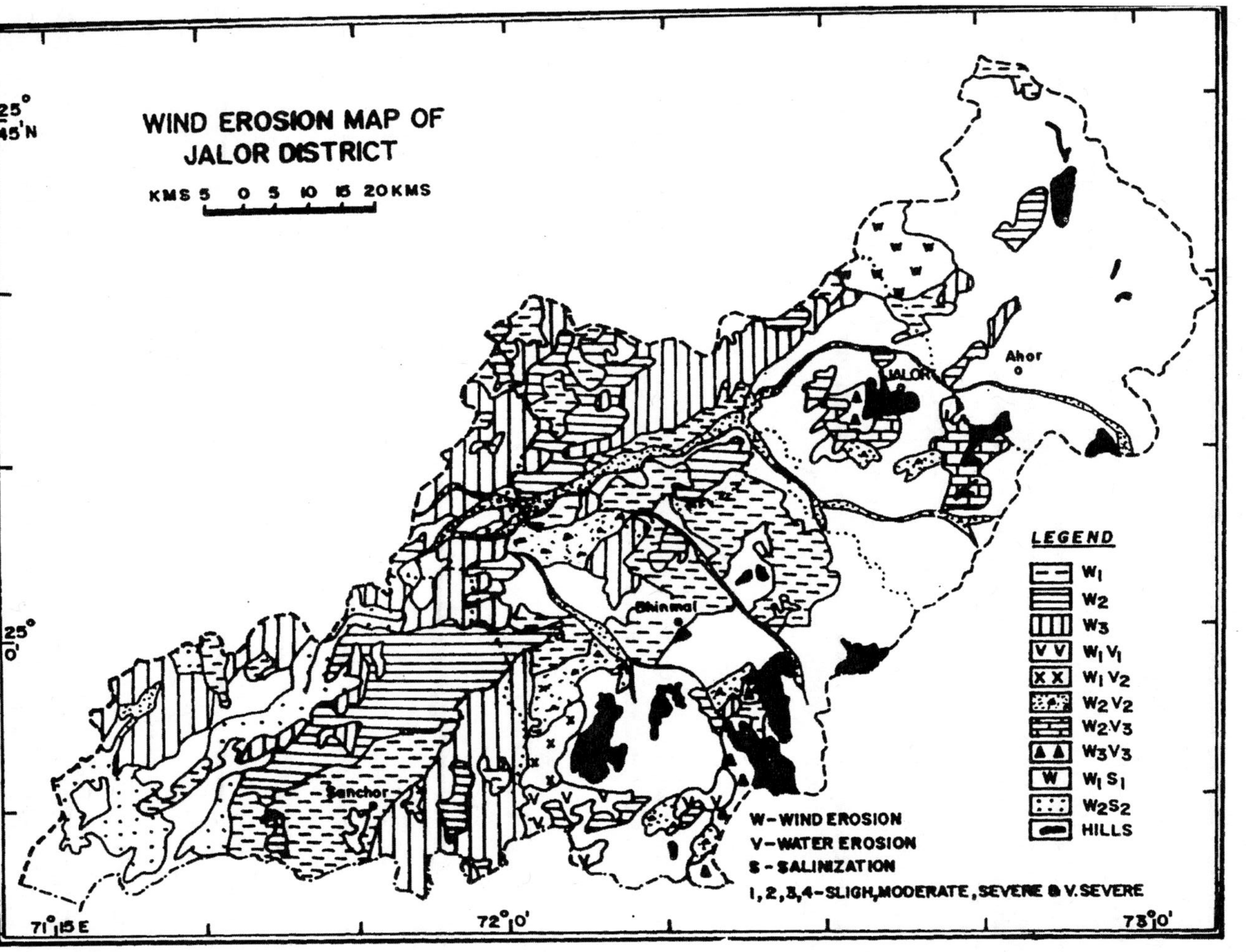

Fig. 2. Map of Jalore district showing wind erosion, water erosion and salinization

Fig. 3. Moderate degradation due to wind erosion.

Severely Degr ded Soils (W_3) : Th type of degradation commonly found in tl dune area. Cultivatic on dune slopes ar removal of natur vegetation is the ma cause. Dunes withou vegetation are al: included in th category. The soils : this category a: medium to fine sar and loamy sand, pa brown to ligl yellowish brown : colour. On tl southeast side Sanchor, Chhaja and Sarnau, area duny. The averag height of these dun is 10-40 m. At a fe places the fresh sar in the form of 20-8 cm thick sand she curs on the crests and flanks which indicates the impact of increasing biotic pressur graded dunes are found in Bhundwa, Mangalwa, Taliyana and Dhora villages. C tellite imagery this unit appears as pale brown with typical dune features. The sc graded due to severe wind erosion occupies 1753.36 km^2, 16.5% area.

Soil Degradation due to Combined Effects of Wind Water Erosion and Salinization iese are the areas where soils have degraded due to combined effect of wind erosio position, water erosion and salinization (Table 2). The sand deposited during summ ainst the hills and in pediment zone subsequently during rainy season resulted in tl mation of gullies. Soils severely degraded due to severe wind and water erosion large cur at flanks of the hills near Jalor, Israna, Jaswantpura and Sunda. Here wat osion is dominant and wind erosion is associated in the process. In the areas away fro ls, wind erosion is dominant and water erosion is associated with it. Ne swantpura and Dorra Reserve Forest and near Dumbadiya on the bank of the rive oderate sand deposition and moderate water erosion result in rill formation. The soi ghtly degraded due to wind and water erosion have sand sheeting in the cultivat lds as well as on fence line and also rills are formed due to runoff water. These are ainly occur to the southwest of Sanchor near Dantiwara and Rewara.

able 2. Soil degradation due to combined hazards

	Wind erosion / deposition (km^2)		
	Slight	Moderate	Severe
/ater erosion			
ight	92.5 (0.87)	–	–
[oderate	95.0 (0.89)	264 (2.48)	–
›vere	–	134.1 (1.26)	137 (1.28)
alinity			
ight	131.0 (1.23)	114.0 (1.07)	–

ıta in parenthesis shows % of total area of the district.

The soils degraded due to combined effect of wind and water erosion and salini :cur in 268.5 km^2 (9.10%) area.

oil Degradation in Relation to Landuse

When climate, soil texture and parent material are same, cultural practices ar .nduse cause a considerable variation in degradation. Due to increasing number vestock and using the grazing land for agricultural and other purposes, ov xploitation of grazing land results in severe degradation of grazing lands (Fig. 4). In tl milar way cultivation on marginal land and use of heavy machinery created tl roblem of wind erosion on agricultural lands.

Major area of the district is rainfed cultivated (5878.11 km^2, 55.24%) followed l rigated (1860 km^2, 17.47%) grazing (1836 km^2, 17.25%) and forest land (209 km^2 1.9% npact of wind erosion/deposition under different landuse system is given in Table 3.

able 3. Extent of soil degradation by wind erosion/deposition under different landu category

anduse	Intensity and extent of degradation (km^2)			Percentage of tota land use area
	Slight	Moderate	Severe	
ainfed	1152.5	901.3	1325.6	37.9
rigated	275.0	807.5	324.4	28.6
'pen grazing	288.8	128.1	265.0	21.4
orest/Oran	14.4	–	–	–

Wind erosion is very active in rainfed cultivated land (37.9%) followed by irrigat !8.6%) and grazing land (21.4%).

npact of Wind Erosion on Soil Characteristics

Morphological Characteristics : The morphological characteristics of these soils a: ›ported in (Table 4). The texture of non-degraded soils varies from loamy sand to sanc ›am while it was sand to loamy sand at degraded sites. The colour of the non-degrad ›ils is darker than the degraded sites. The content of clay, silt and water holding cap ty are high in the non-degraded soils and gradually decrease with the intensity egradation.

Fig. 4. Severe degradation of grass lands due.

able 4. Impact of wind erosion on morphological characteristics of soil

oil parameters	Non degraded soils (free from wind deposition/erosion	Degree of Degradation		
		Slight (W_1)	Moderate (W_2)	Severe (W_3)
. Texture	Is-sl	fl-ls	fs-lfs	Cs-fs
. Colour	10YR5/4-5/6 (YB)	10YR5/4-6/4 (YB-LYB)	10YR5/3-6/4 (B-LYB)	10YR6/3-6/ (PB-LYB)
. $CaCO_3$ %	1.8-6.5	2.9-5.8	1.4-2.6	0.2-0.37
. Moisture equivalent %	4-10	3.3-6.3	4.0-6.0	2.0-3.5
. Water holding capacity %	24-40	20-35	22-26	18-20

ote : Cs, Coarse sand, fs, fine sand, ls, loamy sand, sandy loam; cl, clay loam, YB - yellowish brown, LYB - lig
llowish brown, B - Brown, PB - pale brown.

hysico-chemical Characteristics of Soil

The pH of non-degraded sites (Table 5) is neutral to slightly alkaline while egraded sites it is in alkaline range. There is not much difference in electric onductivity of soils at degraded and non-degraded sites and it ranges frm 0.103 - 0.9. Sm^{-1}. The organic carbon in non-degraded sites ranges from 0.197 to 0.386% with me alue 0.200% and decreases gradually, in slightly degraded soil (0.120 - 0.195%), a noderately degraded one (0.068 - 0.153%). Compared to non-degraded soils the orga arbon is less by 27% in slightly, 50% in moderately and 60% in severely degraded soi he available potassium in non-degraded sites range from 175 - 878 kg/ha and decreas radually in slightly (165 - 605 kg/ha) moderately (161 - 487 kg/ha) and in severe egraded soils (94-269 kg/ha). The mean available potassium is 31% less in slight egraded soils. Moderately degraded soils contain 47% and severely degraded soils 59 ss mean available potassium in comparison to the soils of non-degraded sites.

able 5. Impact of wind erosion on Physico-chemical charocteristics of soil

oil naracterstics	Nondegraded soils	Degree of degradation		
		Slight (W_1)	Moderate (W_2)	Severe (W_3)
H	7.6-8.0	7.74-8.2	7.34-8.6	7.4-8.48
onductivity (dSm^{-1})	0.49-0.633	0.113-0.383	0.115-0.925	0.103-0.526
rganic Carbon (%)	0.197-0.386 (0.200)	0.120-0.195 (0.146)	0.068-0.275 (0.100)	0.045-0.153 (0.072)
vailable Potassium (kg ha^{-1})	175-878 (452.66)	165-605 (312.8)	161-487 (240.33)	94-269 (187.12)
vailable Phosphrus (kg ha^{-1})	15.5-28.8 (22)	10.3-24 (17)	8-15.8 (13.21)	5.9-15.6 (11.78)

ote : Data in parenthesis indicate the mean values.

The available phosphorus in non-degraded sites ranges from 15.5 to 28.8 kg/ha a ecreases gradually in slightly (10.3 to 2.4 kg/ha) in moderately (8-15.8 kg/ha) and everely degraded soils (6.9 to 15.6 kg/ha) respectively. The mean available phosphor slightly degraded soils is 22.72% less than in non-degraded soils. Moderately degrad oils contain 39.9% and severely degraded soils 46.4% less available phosphorus th on-degraded soils.

CONCLUSION

The present study reveales that nearly 32% area is modertely to severely degrad ue to wind erosion and needs immediate attention for conservation. Shelter belts a eeded to stabilize the soil and use of heavy farm implements should be avoided. Sa une stabilization is very essential. Wind erosion/deposition process affected the fertili atus of soil. Maximum decrease is observed in organic carbon (27-60%) followed otassium (31 to 58 kg/ha) and phosphorous (23 to 47 kg/ha).

ACKNOWLEDGEMENTS

Author express sincere gratitude to Dr. A.S. Faroda, Director, CAZRI, Jodhpur an r. Surendra Singh, Head, Division of Resource Survey and Monitoring for providir cilities. Shri Mohar Singh and Ashok Gehlot are also thankfully acknowledged fo chnical assistance. Thanks are also due to Dr. D.C. Joshi for offering valuab ıggestions.

REFERENCES

ggarwal, R.K. and Gupta, J.P. 1980. Sand movement studies under different landuse conditioı of Western Rajasthan. In : Mann, H.S. (Ed.) *Arid Zone Research and Development*. Scientil Publishers, 109-114.

ali, Y.P. and Kanwar, J.S. 1977. *Soil degradation in India*. In *FAO Soil Bull.* No. 24.

upta, J.P., Aggarwal, R.K. and Rainkhy, N.P. 1981. Soil erosion by wind from bare sandy plaiı in western Rajasthan, India *Journal of Arid Environment* 4 : 15-20.

ıckson, M.L. 1967. *Soil Chemical Analysis,* Prentice Hall of India, New Delhi.

ldeman, L.R., Hakkeling, R.T.A. and Sombrock, W.G. 1991. *World map of the status of Huma Induced Soil Degradation.*

aina Pramila 1992. Influence of degradation on the fertility status of sandy soils. *Curre Agriculture.* 16(1-2) : 43-50.

aina, Pramila, Joshi, D.C. and Kolarkar, A.S. 1991. Land degradation mapping by remo sensing in the arid region of India. *Soil Use and Management* UK 7 : 47-52.

aina, Pramila, D.C. Joshi and A.S. Kolarkar 1993. The mapping of soil degradation using remo sensing on alluvial plain, Rajasthan, India. *Arid Soil Research and Rehabilitation* 7 145-161.

ehgal, J.L. and Sharma, P.K. 1988. An inventory of degraded soils of Punjab (India) using remo sensing techniques. *Soil Survey and Land Evaluation* 8 : 166-175.

eotia, H.S., J.D. Hoore, and R. Gambeer, 1980. Soil and landuse distribution over a part of nor plain of India based on the optical interpretation of Landsat 2, Multiscale Satellite Imager *Pedologia* 30 : 19 42.

NEP 1977. *Desertification an overview. United Nations Conference on Desertification,* Augu 29-Sept. 9, 1997. Nairobi, Kenya.

SDA 1975. *Soil Survey Manual*, US Deptt. of Agriculture Handbook No. 18.

Geological Evolution of Northwestern India
B.S. Paliwal (Ed.)
Scientific Publishers (India), Jodhpur, 1999, pp. 358 - 368.

LIME RICH SOILS OF WESTERN ARID RAJASTHAN, THEIR DISTRIBUTION, CHARACTERISTICS, PROBLEM AND MANAGEMENT

Nepal Singh and Surendra Singh
Central Arid Zone Research Institute, Jodhpur

ABSTRACT

Inspite of very hostile environmental conditions, high aridity and acute moisture deficiency, Jaisalmer district (38401 km^2) has major twenty five (25) lime rich soil series under six depth and textural groups. The study revealed that very deep coarse textured aeolian sandy soils cover 21084 km^2 (54.7%), moderately deep to deep loamy sand soils having limerich hard and underneath compact substrata of lime concretion of various size and shape, petrocalcic, petro gypsic layer and hard nodules occupy 7560 km^2 (19.6%) area, medium textured soils related to red sandstone, granitic, limestone and conglumaratic rocks cover 3060 km^2 (8.80%) area while fine textured soils developed from ferrugenous sandstone, limestone, yellow sandstone, rhyolitic and phyllitic lime rich colluvial/alluvial parent materials related soils occupy 3287 km^2 (9.2%) area. Saline depressions occupy, 195 km^2 (0.5%), shallow-miscellaneous soils occupy 1208 km^2 (3.2%) and hills occupy only 2147 km^2 (5.5%) area. Nearly 82 percent soils of the area have severe constraints of climate, light texture, wind erosion/sand deposition, salinity, depth, hard and compact underneath substrata. These qualify for only class VI, VII and VIII landuse capability and are suitable for pasture and silvi pasture. The remaining 17 per cent area of class IVC landuse capability is suitable for rainfed cultivation i.e. agro-forestry and agro- horticulture practices.

Key words : Lime Rich Soils, Western Rajasthan.

INTRODUCTION

Jaisalmer, district of Western Rajasthan has very hostile environmental conditio i.e. high aridity and severe moisture deficiency with high wind regime. Dhir et al., (19 ve reported nearly 60 percent area of the district under deep sandy soils whi larkar et al. (1989) and (1989) reported 70 per cent area under both sand and loan nd soils affected by severe wind erosion. Chatterji and Amal Kar (1992) have broad ported that the central and south-central part of the district has distinctly rocky ar allow gravelly soils with stretches of desert pavement on surface. Gypsiferou pressions and hard and compact indurated exposed surfaces which have been cover the loose aeolian sand. Besides these, a large part of the district has remain exploited so far. Dhir and Singh (1971) have surveyed the CAZRI grass land ares 0 ha each while Kolarkar et al. (1981) and Singh and Kolarkar (1983) ha aracterized some of the Khadin soils of 100 ha each. The FAO (1971) has carried o ils survey in the canal command area for suitability for irrigation while Kolarkar ar ngh (1992) have reported the broad soil groups of Jaisalmer district. Therefore, t esent study on these lime rich soils of Jaisalmer district which will provide the det formation on their extent, distribution, charateristics, problem and managemer quired for the formulation of sustainable development.

LOCATION AND ENVIRONMENT

Jaisalmer district covering total 38401 km^2 area is located along Indo-Pak borde e district experiences most hostile environmental conditions and receives 200 mm nual rainfall along eastern margin and 100 mm along western margin. The rainfall ghly erratic and ill distributed with 50 - 90 percent co-efficient variabilities at differe cations. The evapo-transpirations is in between 1900- 2200 mm and relative humidi tween 60 - 66 percent. The maximum average temperature is between 38 to 42°C ar ten reaches above 47°C in summer while during winter minimum temperature is bel °C which often goes down to the freezing point. The average wind speed is in betwe to 25 km/h and often touches upto 37 km/h in summer with dust storr amakrishnan et al. 1992). Natural vegetation consists mainly of grasses and shrubs graded condition. Trees are rare or in low density. The dominant rocks are lime ston anite and red, yellow and grey sandstone, rhyolite and major part of the district h rd lime nodules and gypsiferous material buried with loose fine sands.

MATERIALS AND METHOD

Semi-detailed soil survey was carried out, following the standard soil surv ethod, using 1 : 50,000 scale SOI topographical sheets as base maps. The various s ries were identified and mapped through the study of morphological characteristi rent material/substrata variability and physiographic relief as per standa oceeding of U.S. Soil Survey Staff (1951) and Soil Survey Mannual IARI (1971). S ries boundaries were demarcated with the help of available aerial photographs (1 0,000 scale) and enlarged land sat imageries (1 : 250,000 scale black and white) alo th field varification at 4 km intervals. The soil series map depicting soil series ar eir phases was prepared on 1 : 50,000 scale (Fig. 1). The typical samples of each s ries were analysed for particle size distribution. Physico-chemical characteristics ar rtility status were determined in the laboratory following the methods of Pipar (195 d Jackson (1967).

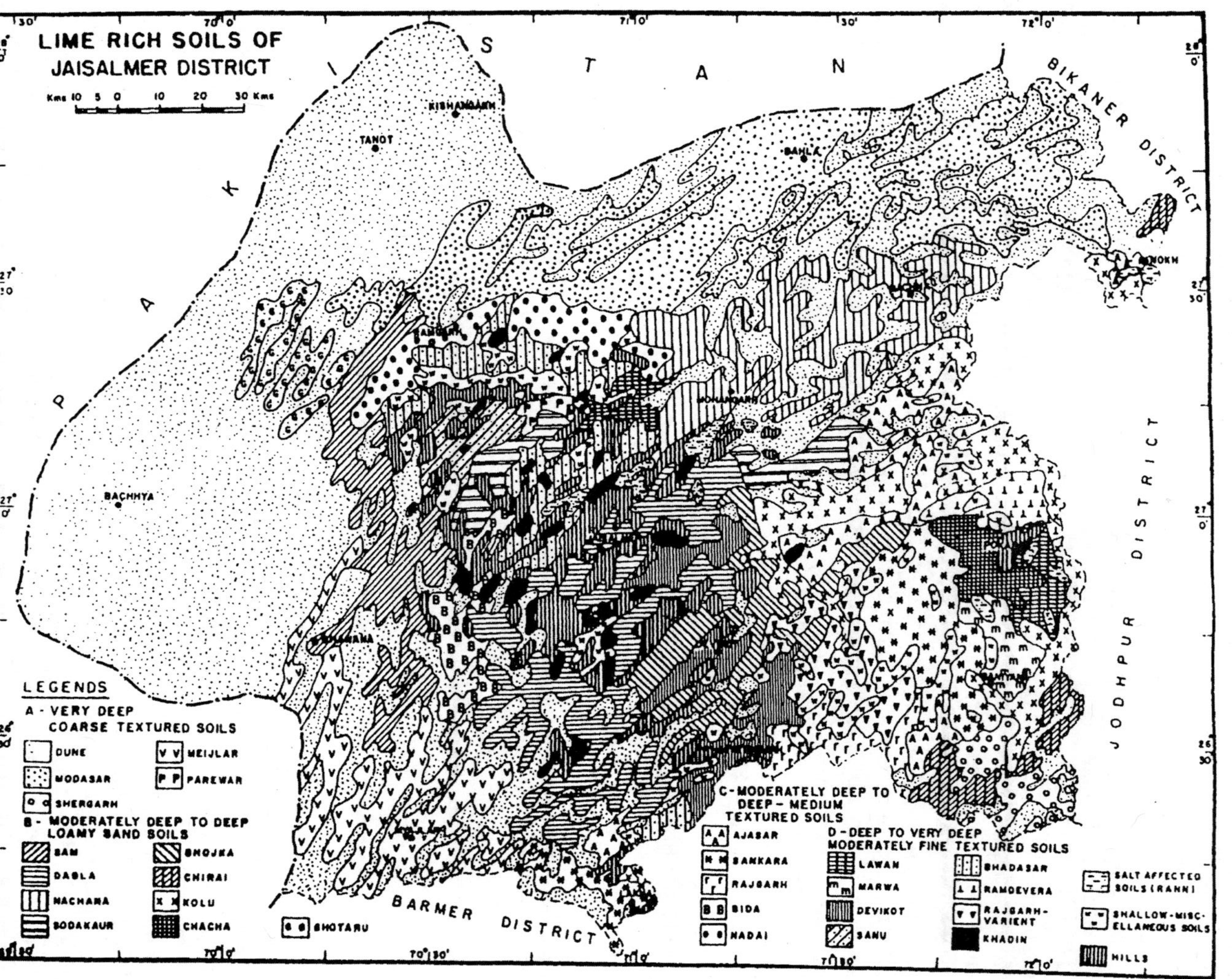

Fig. 1. The map showing lime rich soils in Jaisalmer district

RESULTS AND DISCUSSION

xtent and Distrubtion

The aerial extent of various soil series in Jaisalmer district is highlighted in Tak while their distribution illustrated in Fig. 1. Besides the hills, shallow miscellaneo ıd saline depressions, there are major twenty five (25) soil series with vari ımbination of morphological and physico-chemical charcteristics, resulted from the lir ch aeolian fine sandy sediments. They occur on various land forms like dune, san ımmocky and flats interdunal plains with very loose coarse textured soils, old luvial plains and buried pediments.

The bases of various soil series separation from each other was variation in the orphological characteristics, reflected in terms of colour, texture, structure, depth aı ılum thickness, lime content and most important was the difference in the undernea ıbstrata. Weathered rocks like lime coated grvels/pebbles, various forms of lir ncentrations and nodules, petrocalcic, conglumaratic rock and powdery form of lir ıbstrata occur at variable depths. Taking into consideration the underneath substra ıriabilities, depth and textures variability, these soils of Jaisalmer district are furth ouped into six depth and texture classes for their detail discussion.

Very deep coarse textured soils

The very deep, coarse textured sand and loamy and sandy soils constitute 210 n^2 (54.7%) area (Table 1) and include the soils of dune. Modasar, Shergarh, Meijlar aı ırewar series, having minor variabilities in their morphological characteristics vary ne content, their underneath substrata and occur in varies land scape relief (Table ıving varies intensity of aeolian sand deposition and various height of hummock orphological characteristics reveal that almost all the soils series of this group ha ›minance of pale brown (10YR 6/3) and light yellowish brown (10YR 6/4) colour ırface layers while their subsoils have yellowish brown (10YR 5/4) and dark yellowi own (10YR 4/4) colour except Parewar series which has light greyish brown (10YR 6/ lour. The dune and Meijlar series has fine sandy texture with single grain structu ırough out the profile without any clear horizon differentiation, occur in the duny aı ırrow interdunal relief. Among these, Meijlar series has high content of amorpho ne in solum and moderately hard lime concretion (20 30 %) mixed with sand at 1 n depth. The Modasar and Shergarh series have fine sandy single grained surfaces wi ight to moderate calcareous nature while their subsoils have sand to loamy saı xture with weak sub angular blocky structural development trends and clear horiz fferentiation developed, in the wider interdunal situation, having highly permeak ose sandy substrata without any lime concretion development (Modasar). Whi ıergarh series has hard sandstone rock at 2 metre depth and are moderate to strong lcareous in nature with increasing trend of amorphous lime at depth. The Parew ries has altogether different morphology. It has abundance of whitish colour coar nd fraction with high content of amorphous lime without structural developme ıassive). The substrata is made of gravel/pebbles mixed with coarse sand with hi ntent of amorphous lime, developed from the fluvial activities, by forming inter rocl ılley fill plain relief.

ıble 1. Distribution of area under different soils series in Jaisalmer district alongwi relief.

›il series	Physigraphic relief	Area		Land capability cla
		Km2	%	
Very deep coarse textured soils				
une	Dune-interdune	17247	44.9	VIcs, ea
odasar	Hummacky wide interdunal plains	3194	8.3	IVcs, ea
ıergarh	Hummacky interdunal plains			
ırewar	Snady hummacky narrow interdunal plains	94	0.2	VIcs, ea
eijor	Sandy hummacky narrow interdunal plains	579	1.4	VIcs, ea
Moderately deep to deep loamy sand soils				
.m	Moderately flat hummacky interdunal plains	1680	4.4	VIcsh, ea
›dakaur	Moderately hummacky buried pediment plains	710	1.8	VIc sh, ea
›lu	Flat older alluvial plains	630	1.6	VIcsh, ea
ıchana	Flat depressional interdunal plains	1293	3.4	VIcsh, ea
ıojka	Flat buried pediments plains	541	1.4	IVc, ea
ıbla	Hills velley fill plains	1864	4.9	VIc, ea
ıirai	Moderately hummacky older alluvial plains	340	0.9	IVc, ea
ıacha	Flat shallow buried pediment plains	348	0.9	VIc, sh, ea
Deep medium textured soils				
ıjgarh	Flat colluvial plains	664	1.7	IVc, sh
ıdai	Flat older alluvial plains	658	1.7	VIc, sh
nkara	Flat older alluvial plains	875	2.3	IVc, ea
da	Flat buried pediment	194	0.5	VIc, ea
asar	Flat older alluvial plains	669	1.7	IVc, ea
Deep, fine textured soils				
.wan	Flat colluvial plains	890	2.2	IVc
arwa	Flat to undulating colluvial plains	135	1.3	IVc
ːvikot	Flat colluvial plain	495	1.3	IVc
nu	Flat velley of hills colluvial plains	325	0.8	IVc
ımdevera	Flat older alluvial colluvial plains	233	0.6	IVc, sh
ıadasar	Flat buried pediments plains	711	1.9	VIc, sh, ew
ıjgarh varient	Flat colluvial plains	205	0.5	IVc
ıadin	Flat alluvial plains	293	0.7	IIIc
lt 'fected soils	Flat saline depression	195	0.5	VIIIc, sa
allow- lsce soils	Undulating piedmont pediment	1203	3.2	VI & VIIc, sh, ca
lls	Steep sloping bare rockout crops	2147	5.5	VIIIR

Physico-chemical characteristics reveal that the dunes of Modasar, Shergarh an
eijlar series have dominance of fine sand fraction (63.4 - 85.2%) over the coarse san
iction (8.2 - 22.7 %) while Parewar series has dominance of coarse sand fraction (61.0
.0%) over the fine sand fraction (18.0 - 25.0%). The clay and silt contents are in poc
ape and vary from 2.0 to 7.0 and 1.8 to 7.0 per cent, respectively. Consequently, the
ve low values of moisture equivalent (2.9 - 4.9%) and water holding capacity (18.0 t
.0%). The dune and Modasar series have low content of amorphous lime (CaCO;
rying from 0.5 to 4.8 per cent while Shergarh, Meijlar and Parewar series have hig
ntent of amorphous lime varying from 4.5 to 8.7 in solum and 8.0 to 13.6 per cent i
eir substrata. Because of coarse texture and highly susceptibility to wind erosion, a
ese soils have very low content of organic carbon (0.02-0.118%), available phosphoru
to 14 kg/ha) and (70 to 130 kg/ha) available potassium.

Moderately deep to deep loamy sand soils

The loamy sand group of soil constitutes 7560 km^2 (19.60%) area (Table 1) an
cludes the soils of Sam (hard lime concretion), Shodakaur (silica coated lime nodules
iojka (lime coated pebbles), Dabla (lime coated weathered sandstone), Nachana (lim
nca over gypsum layer), Kolu (lime petrocalcic layer), Chirai (weak zone of lime conca
iacha (weathered rock at 20 - 30 cm depth) and Ghotaru (thick zone of reddish colou
ne nodules) series. These soil series slightly differ in their morphologica
aracteristics. Morphologically, the Sodakaur, Bhojka, Kolu. Chirai, Sam and Chach
ries have narrow variation in colour due to severe wind erosional activities and hig
ntent of amorphous lime, and show pale brown (10YR 6/3) and light yellowish brow
)YR 6/4) colour on surface while their sub soils have yellowish brown (10YR 5/4) an
rk yellowish brown (10YR 4/4) colour. The Nachana series has light greyish brown t
rk greyish brown (10YR 6/2-5/2) colour due to lime and gypsum influence, whil
iotaru series has slightly dark colour (10YR 4/4-7.5 YR 5/4). The surface texture of a
ese soils invariably is fine sandy with single grain structure due to severe win
osional/depositional activities. Their subsoils have dominantly loamy sand to gravell
d sandy loam texture with weak to medium sub angular blocky structure, having clea
rizon differentation and very high content of amorphous lime through out the profil
olum and substrata). Because of hard and compact underneath substrata, these soi
ve severe limitation of roots and water movement and suffer from severe wind erosioı
nd deposition problem, related to older alluvial and buried pediment land form
iysico-chemcial characteristics indicate that the loamy sand soils have slightly highe
ntent of clay and silt in comparision to earlier sandy soil. Their values range from 7.
12.4 and 6.3 to 9.6 per cent, respectively whereas fine sand fraction ranges from 67.
78.0 per cent and coarse sand from 4.3 to 14.7 per cent only. These soils are strongl
lcareous in nature and have 2.4 to 18.8 per cent amorphous lime in solum and 8.8 t
.6 per cent in their underneath substratas. These soils are non saline but strongl
kaline because they have slightly higher content of available moisture than the sand
ils. The moisture equivalent values are in between 6 to 10 per cent while the wate
lding capacity values range from 23 to 99 per cent. Because of severe wind erosiona
d depositional activities and high content of amorphous lime, the availability
trients is low. The organic carbon (0.06-0.28%), and available phosphorus (8-18 km/ha
e in low range while these soils have slightly high content of available potassiu
30-450 kg/ha) in medium to high range.

. Moderately deep to deep medium textured soils

The medium textured soils constitute 3060 km^2 (8.8%) area (Table 1) and inclu ıe soils of Ajasar (red sandstone), Sankara (granite), Rajgarh (pebbles and lime conca ida (limestone) and Nadai (conglomerate), having distinct variability in the ıorphological characteristics in solum and in underneath substrata of related pare ɔck which is coated with lime at variable depth. The morphological characteristi ›veal that the soils of Ajasar and Bida series are slightly darker in colour (7.54YR aı YR hue) while Sankara and Nadai series both have 10YR and 7.5 YR colour aı ajgarh series having dominance of 10YR colour. The surface textures of these soils a ravelly sand and loamy sand with single grain to massive structure, having pockets esert pavements. The sub soil texture is dominantly sandy loam to loam we ggregated and some what stable with medium moderate subangular blocky structu ıd clear horizon differentiation. Because of the influence of hot and dry climate, tl ɔlum and underneath substrata of these soils have high content of powdery form of lim he Ajasar series has lime coated quartz pebbles over weathered red sandstone. Bi ›ries has abundance of powdery form of lime mixed with pebbles. Sankara series h. me coated weathered granite, Rajgarh series has lime coated quartz pebbles and Nad ›ries has conglomerate of sandstone and lime nodules.

Physico-chemical characteristics reveal that these soils attribute fairly high conte ' clay and silt (10.3 to 18.3 and 9.2 to 19.8 per cent). The fine sand fraction varies fro).2 to 71.2 per cent and coarse sand from 9.9 to 19.8 per cent. The amorphous lin ɔntent in solum varies from 2.3 to 21.3 per cent and their substrata have 14.7 to 41 ɜr cent amorphous lime. This high content of amorphous lime is useful for tl ructural development but very harmful to the plant roots particularly during tl ıoisture stress period. Because of high content of silt and clay they are also exhibitiı ıe fairly high water retention capacity. Their moisture equivalent values are in betwe to 14 per cent and water holding capacity is in between 28 to 40 per cent. Due to fair ɔod texture and aggregation, these soils also have comparatively high fertility statı ıan the loamy sand soils. Although organic carbon (0.120-0.430%) and availab ıosphorus (14.0 to 24.0 kg/ha) content are in low range but these soils have fairly hiş ɔntent of available potassium (240 to 660 kg/ha).

. Deep to very deep, moderately fine texture soils

The moderately fine textured soils constitute 3287 km^2 (9.2%) area (Table 1) aı ıclude the soils of Lawan (reddish calcareous sandstone), Marwa (rhyolite, slates aı ıyllites), Devikot (reddish ferrugenous sandstone and red shales), Ramdevera (r ındstone - quart pebbles and lime nodules), Bhadasar (yellow sandstone and mudston ajgarh (pebbles mixed alluvium) and Khadin (coarse alluvial material) series. All the ɔils developed from different type of parent material. Morphological characteristi ›veal that Lawan, Sanu and Devikot series have dark reddish brown (5YR 5/4-4/ ɔlour while Marwa, Rajgarh and Khadin soil series have reflected the yellowish brov .0YR 5/4) and dark yellowish brown (10YR 4/4-3/4) colour. Bhadasar series has oli rown (2.5 YR 4/4) colour, whereas Ramdevera has very dark yellowish brown (10YR 3/ ı dark brown (7.5YR 5/4-4/4) colour. These soils are highly variable in surface textu ıe to abundance of desert pavement and wind sorted sand accumulation. Their textrı ıries from gravelly sand, loamy sand and sandy loam with single grain to massi ;ructure. Sub soil texture ranges from loam, clay loam, sandy clay loam and silty cl ›am with very strongly developed subangular to angular blocky structure, and are w ;gregated except Bhadasar series which has columanar structure because of salinit ll these soils series have high content of amorphous lime through out the profile exce evikot series which has no lime in their solum and their substrata are slight ılcareous. The source of lime was considered to be weathered rock.

Physico-chemical characteristics reveal that these soils invariably have hi; ›ntent of clay and silt (16.3 to 38.9 and 14.1 to 53.2 per cent) while fine sand fracti ıries from 25.6 to 60.0 per cent and coarse sand from 2.3 to 17.7 per cent. Becau igher content of clay and silt, have fairly very high water retention capacity and sto igh quantum of water for a long period. Their moisture equivalent value ranges from › 23 per cent and water holding capacity from 34 to 64 per cent while the amorpho me content in solum varies from 4.4 to 16.8 per cent and in substrata from 3.1 to 38 er cent depending upon the nature of source material.

The organic carbon value ranges from 0.148 to 0.484 per cent, and availal hospholrus from 16 to 36 kg/ha. While available potassium is plenty ranges from 350 97 kg/ha.

Salt affected soils

The natural salt affected soils in the district are related to saline depressions (ran ıd constitute and only 195 km^2 (0.5%) area (Table 1). These occur at Lawan, Th okaran, Jhaleria, Bhaniyana, Khanodia and Khara-Mitha rann and also at Rupsi a hadasar villages. Morphological characteristics reveal that these soils have greyi rown (10YR 5/2) to dark greyish brown (10YR 4/2) colour, clay loam, silt clay loam a lty clay texture, strong development of columnar and primsmatic structure, medium igh content of amorphous lime and stratification. These soils have shallow sali round water table due to depressional relief which contributes salt accumulation ırface as well as in the whole profile through the evaporation processes and capilla ction.

Physico-chemical characteristics indicate that these soils have high content of cl ıd silt (16.5 to 36.4 and 16.4 to 25.8 per cent), while fine sand and coarse sand fractio ıry from 36.6 to 61.2 and 2.7 to 8.9 per cent, respectively. The EC (1 : 2) ranges fro).5 to 35.0 dSM^2 and pH in between 8.0 to 8.3. The amorphous lime content ranges fro 0 to 18.4 per cent. Because of fine texture these soils can retain quite high moisture ıe solum. The moisture equivalent values range from 15 to 26 per cent and wat ›lding capacity from 38.0 to 64.0 per cent. These soils are deficit in organic carbon (0. › 0.211%) and available phosphorus (2 to 8 kg/ha) due to high salinity built up but ha igh content of available potassium (400 to 1050 kg/ha) in comparision with other soils.

Shallow - Miscellaneous soils

The shallow miscellaneous skeletal soils constitute 1208 km^2 (3.2%) area in t strict, located dominantly in the foot of hills of various lithologies and in their vacini ı gravelly up land of 3-5 per cent slope on buried pediment. Morphologic ıaracteristics reveal that these soils are not uniform in colour and texture. Tl ɔminant colour varies from light yellowish brown (10YR 6/4), brown (7.5YR 5/4) ·ddish brown (5YR 5/4). While texture ranges from gravelly sand, gravelly loamy saı ıd gravelly sandy loam to massive. These soils contain high amount of amorphous lin . the form of lime coating on weathered rocks and thick zone of lime concretion aı ɔdules over the rocks. Due to shallow-depth, high content of gravels/pebbles and th ıver of soil, these soils are suitable for only pasture and silvipasture.

The out crops of various types of rocks constitute 2147 km^2 (5.5%) area in tl strict and do not have any soil cover except aeolian sand deposition here and ther ıese hills do not have any use except water harvesting and mining for road aı ıilding material. The results of above study are quite comparable with those conducte / Dhir and Kolarkar (1977), Kolarkar et al. (1989), Kolarkar and Singh (1992) and Dh *al.* (1992) and Singh and Kolarkar (1983).

PROBLEM AND MANAGEMENT

Jaisalmer district experienced very low rainfall (100-200 mm) in an erratic and i stribution manner, high temperature, high evapo-transpiration rate (1900-2200 mm gh wind velocity and relative humidity in between 60-66 per cent. Besides climat ınstraints, nearly 83 per cent area of the district has severe constraints of soil textuı ght), depth (rootzone limitation), wind erosion/sand deposition, salinity and very haı ıd compact lime rich underneath substrata (enough hard and compact to restrict bot ant roots and water movements). In combination of both climatic and soil constraint ıe entire land of Jaisalmer district qualifies only for class IV, VI, VII and VIII land us ıpability (Table 1) except Khadin soil which qualifies for class III land use capabilit ecause of climate, soil texture, depth, wind erosion and salinity in soil the yie ɔtential of these cultivated lands have lowered to a large extent and often cro ımpletely failed due to serious stresses of moisture in the profile. These medium aı oderately fine textured soil lands should be brought under agro-forestry aı ro-horticulture system and Khadin landuse to maintain the yield potential for the roper utilization. These lands under this system can provide the sustainable aı asonable yield potential even in the low rainfall year.

Extremely coarse textured sand and loamy sand series soils having severe textura epth, wind erosion, hard and compact lime rich underneath substrata constrain ıalify only for VI and VII land use capability. Such kind of lands should be brougl ıder pasture and silvi-pasture system, using the high yielding species of grasse ırubs and trees. For the proper utilization of these sand and loamy sand soils and ɔtained the sustainable yield potential, Shankar and Kumar (1987a) have suggeste *:usirus sindicus* and *Panicum turgidum* grasses for pasture development for dee ındy soils and *Eleusine compressa, Aristida funiculata* and *Panicum antidotel* grasse r shallow to moderately deep and moderately deep to deep, loamy sand hard pa ıbstratas soils. Shankar and Kumar (1987b) in another study have suggested tl ımbination of grasses, trees and shrubs even in the very low rainfall year. Tl lvi-pastural system also found suitable for these degraded soils, as they have highe ɔil binding capacity and found also useful to stablize these loose sandy soils. The clas III land belongs to saline depressions which are completely baran due to high salinit nd could be used for salt extraction only. The class VIIIR land belongs to hills and rocl ıt crops could be used for road and building material and mining of minerals. The miı ɔoiled areas should be rehabilitated through the filling up the pit with sand and oth ıaterial and planting the suitable species of grass and trees in this hosti nvironmental prone district of Jaisalmer.

ACKNOWLEDGEMENT

The authors are highly grateful to Dr. A.S. Faroda, Director, Central Arid Zoı esearch Institute, Jodhpur to provide facilities and guidance in the preparation of th ıanuscript. Thanks are due to Dr. D.C. Joshi of CAZRI, Jodhpur for critically examiniı ıe manuscript and offering valuable suggestions.

REFERENCES

AO 1970. Semi detail soil survey of Rajasthan canal command area Phase II-Technical report I FAO, India 1970.

hatterji, P.C. and Kar, Amal. 1992. Integrated Natural and human resources Appraisal Jaisalmer district. *CAZRI Publication No. 39* Eds. by P.C. Chatterji and Amal Kar, pp. 1-4.

hir, R.P. and Singh, N. 1971. *Detail soil survey report of Lawan and Jaisalmer ran management areas*. Basic resources survey Division, CAZRI, Jodhpur pp 1-26.

hir, R.P. and Kolarkar, A.S. 1977. Observation and genesis and evolutions of arid zone soils. *Indi. Soc. Soil Science* 25(3) : 260-264.

hir, R.P., Choudhary, M.R., Jagannath and Somani, C.L. 1988. Constraint of sandy soils ar and adjoining areas of Western Rajasthan and Northern India and their management in (soil released contraints in crop production, Eds. T.D. Wiswas. *Ind. Soc. Soil Science* Bull N 15 : 52-69.

hir R.P., Kar, Amal, Wadhanam, S.K., Rajagura, Mishra, V.N. Singhvi, A.K. and Sharma, S. 1992. Thar Desert in Rajasthan, Eds. A.K. Singhvi and Amal Kar published by *Geographic Society of India, Bangalore*, pp. 191.

ackson, M.L. 1967. *Soil Analysis*, Prentice Hall INC Engilwood Clibbsin.

olarkar, A.S., K.N.K. Murthy and Singh, N. 1981. Water harvesting and runoff farming in ar zone Rajasthan. *Indian Jour. of Soil conservation, Dehradun* 10(2) : 60-66.

olarkar, A.S., Jain, S.V., Dhir, R.P. and Singh, N. 1989. Distribution, Morophology and landu Rajasthan. Review of Research on sandy soils in India. *International Symposium (Managing sandy soils*, Feb. 6-10, 1989, CAZRI, Jodhpur pp. 1- 20.

olarkar, A.S., Singh, N. and Sharma, B.K. 1989. Nature, extent and distribution of sandy soils Jaisalmer and their utilitarian interpretative map to plan improve desert conditior Abstracts Part I. *International Symposium on managing sandy soils* Feb. 6-10, 1989, CAZI Jodhpur pp. 71-77.

olarkar, A.S. and Singh, N. 1992. Soil and landuse capability Chapter V - *Integrated Natural aı Human Resources Aoppraisal of Jaisalmer district*. CAZRI, Pub. No. 39, Eds. P.C. Chatte and Amal Kar, CAZRI, Jodhpur. pp. 26-34.

iper, C.S. 1954. *Soil and Plant analysis*, Adelaida, Australia.

makrishna, Y.S. Rao, A.S. and Singh, R.S. 1992. Climatic features of Jaisalmer distric *Integrated Natural and Human Resources Appraisal of Jaisalmer district.* CAZRI Pub. N 39. Eds. P.C. Chatterji and Amal Kar, CAZRI, Jodhpur pp. 5-13.

ıgh, N. and Kolarkar, A.S. 1983. Some physico-chemical properties of soils of khadin in Westeı Rajasthan. *Ind. Soc. of Soil Conservation Dehradune*, Vol. 11(283) pp. 5-11.

nkar, Vinod and Kumar, Suresh, 1987a. Grazng resources of Jaisalmer district. Ecology ar development of planning with special reference to Sewan grass lands. *CAZRI Publicaion N 28* pp. 1-82.

nker, Vinod and Kumar, Suresh 1987b. Herbage yield and carrying capacity of grass land i Jaisalmer district. *Tropical Ecology* 28, 239-245.

.l Survey Staff, 1961. *Soil Survey Manual Agriucltural Handbook* US Dept. Agric. No. 18.

ieological Evolution of Northwestern India
3.S. Paliwal (Ed.)
icientific Publishers (India), Jodhpur, 1999, pp. 369 - 377.

MISUSE OF LAND AND ENVIRONMENTAL DEGRADATION UNDER HOT ARID ECOSYSTEM

Balak Ram and J.S. Chauhan
Central Arid Zone Research Institute, Jodhpur

ABSTRACT

Detailed land use mapping of Lakhusar and Barju villages of Bikaner district has been carried out. Characterized by duny complex, the entire area of the region is under rainfed agriculture with moderate to poor intensity of cultivation and limited choice of crops. Cultivated lands constitute 5.3 per cent while 44 per cent is occupied by sandy waste. Comparison of topomaps and aerial photographs reveales a significant expansion of agriculture and over exploitation of natural vegetation leading to accelerated wind erosion and deposition hazard. The suggested land use may help to curb the desertification hazard and improve the environment.

Key words : Hot Arid Ecosystem, Landuse, Environmental Degradation, Weste Rajasthan

INTRODUCTION

Misuse of land resources far beyond their capability, has caused serious problems nvironmental degradation. In order to plan sustainable land use and formul trategies to curb the desertification processes, the assessment of the land use is a p equisite. This can be achieved through present land use survey and mapping w emote sensing technology. Present study of Lakhusar and Barju villages of Bikar listrict is attempted to fulfil the objectives.

Lakhusar and Barju villages of Bikaner district are located 35 km north-west 3ikaner town on Bikaner-Chhattargarh road. The area of the villages extends from 2 .6'30" to 28° 23'40" north latitudes and from 73° 04'50" to 73° 15'18" east longitudes.

Entire area of the villages (10709 ha) is characterized by sand dunes (upto ; etres height) and interdune plains with deep to very deep coarse textured sandy amy sand soils associated with gypsiferous soils at places. The mean annual rainfall)0 mm with a co-efficient of variation of 55%. The annual potential evapo-transpiratio E) is 1770 mm. The mean maximum temperature during the summer is 42°C. In Ma ıd June sometimes it may exceed 46°C. Mean minimum temperature is 29.2°C durir ımmer and 5°C during winter which sometimes may drop to below freezing point. Tl 'erage wind velocity is 7.7 kmph as against maximum wind velocity of 15.0 kmp ıring June and minimum 3.5 kmph during November and December. The relativ ımidity is maximum (71%) during July and minimum (11%) during May (1995a and b ıe surface and groundwater resources are very poor and even drinking water supply ɔm outside. The total human and livestock populations of the two villages are abo .00 and 14300 respectively.

MATERIAL AND METHODS

Detailed land use mapping has been carried out on cadastral map. Fieldwise da ı land use and cropping pattern were collected from Girdawari record of the villag ıtwari was for the year 1994-95 and the same were varified by ground survey. Tl ıysiography was superimposed from the enhanced topomaps, different land u: tegories were delineated. Topographical maps of 1954 (1 : 63360) and 1970 (1:50, 00(:rial photographs of march, 1967 and IRS LISS I data of September 1988 and Februa)89 were interpreted to identify and map land use changes. Views of local people we so obtained regarding such changes, their causes and magnitude of the lar :gradation.

RESULTS AND DISCUSSION

Spatial distribution of different land use categories is shown in Figs. 1 and 2 ar .eir area in Table 1.

PRESENT LAND USE

Cultivated lands

Out of the total (10709 ha) geographical area of the two villages, rainfe onocropped lands (including fallow lands) constitute 53.09 per cent. Lands with tl tensity of cultivation 30-60 per cent constitute 45.6 per cent area and occur dominant wards eastern part of Lakhusar village and south western part of Barju villag mprising mainly interdune plains with hummocks at places. Lands with 20-30 per ce tensity of cultivation occupy 6.46% area and comprise low dunes and lower and midd ınk of the medium height sand dunes. Such lands occur dominantly towards weste ırt of Lakhusar village and south eastern part of Barju village bordering Baralia villag

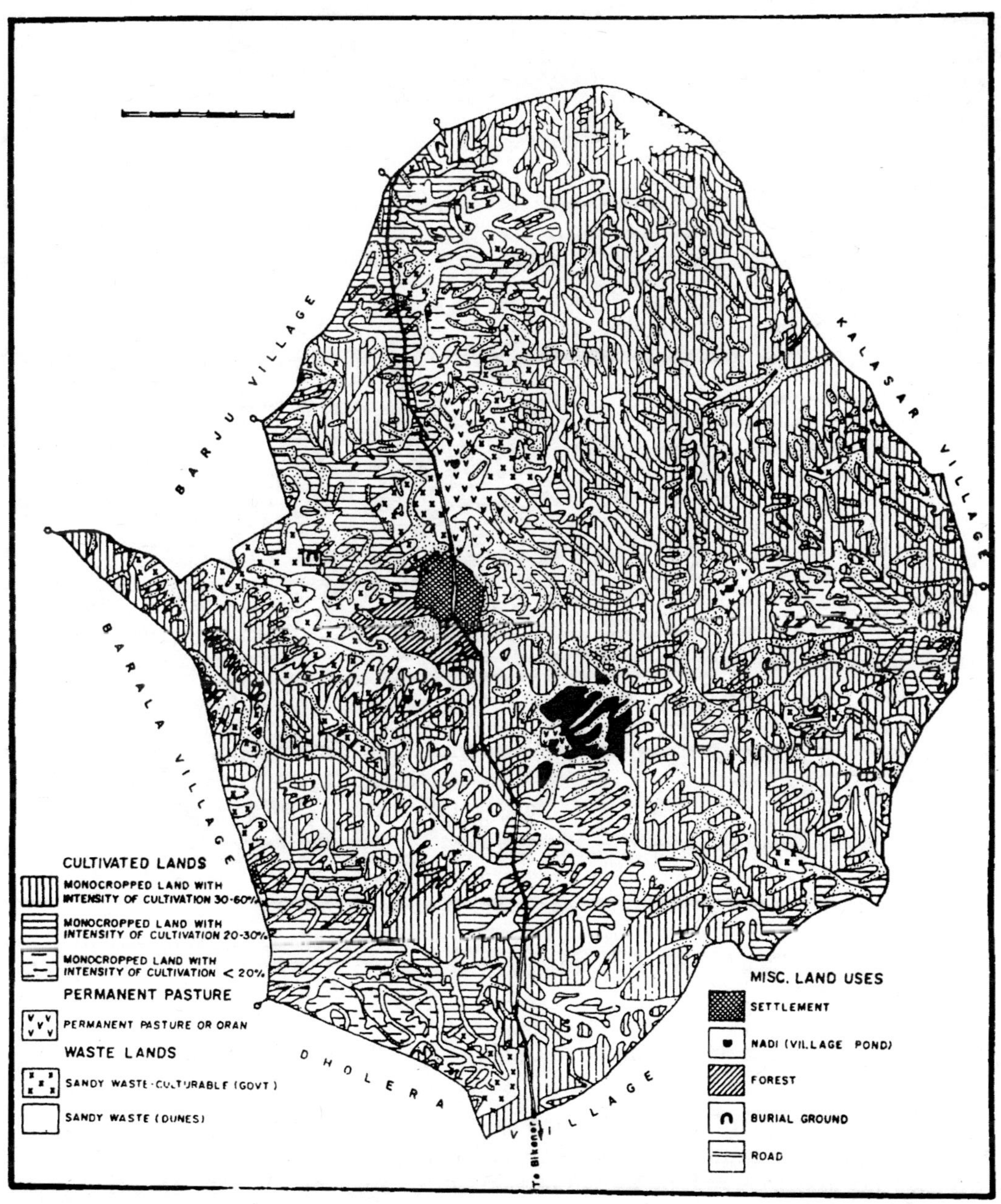

Fig. 1. Present land use of Lakhusar village

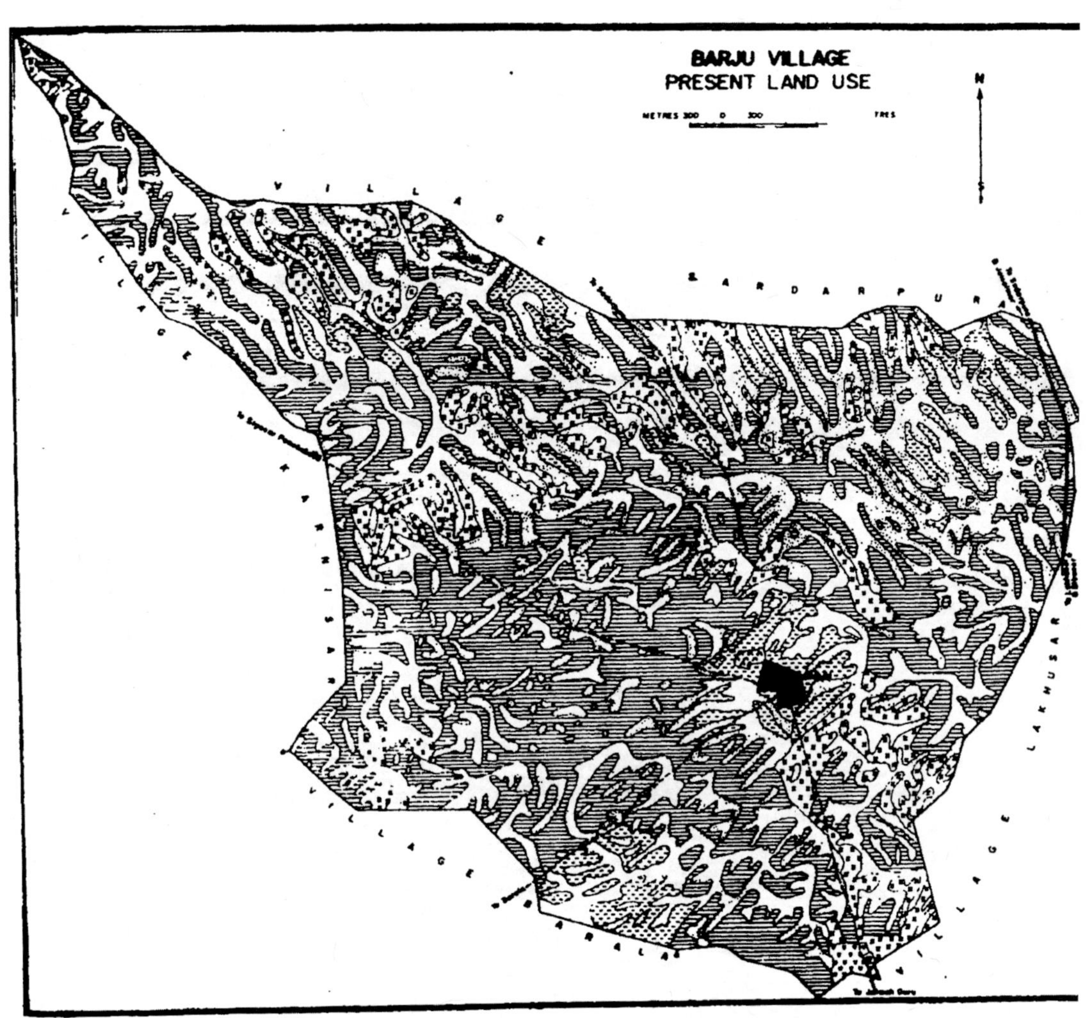

Fig. 2. Present land use of Barju village (Index : as in Fig. 1)

nd around village settlement. Cultivation of such uneconomic marginal lands aggrava and drift. Cultivated land with intensity of cultivation below 20% constitute 1.03 p ent area associated with middle flanks of a chain of sand dunes comprising highly acti pper flanks. Crop and forage productivity of these lands is very low. Such lands occur iree pockets in Lakhusar village.

able 1. Present land use in Lakhusar and Barju villages (1994-95)

and use category	Area in ha					
	Lakhusar		Barju		Total	
	Area	(%)	Area	(%)	Area	(%)
ultivated Lands	**3320**	**53.58**	**2365**	**52.41**	**5685**	**53.09**
3 Rainfed monocropped lands ith intensity of cultivation 30-60%	2762	44.58	2121	47.00	4883	45.60
4 Rainfed monocropped lands ith intensity of cultivation 20-30%	448	7.23	244	5.41	692	6.46
5 Rainfed monocropped lands ith intensity of cultivation 10-20%	110	1.77	–	–	110	1.03
'astelands	**2606**	**42.06**	**2113**	**46.82**	**4719**	**44.06**
ι Sandy waste	2388	38.54	1732	38.38	4120	38.47
.x Sandy waste with scrub	218	3.52	381	8.44	599	5.59
liscellaneous Land uses	**270**	**4.36**	**35**	**0.77**	**305**	**2.85**
›rest	68	1.10	–	–	68	0.64
ərmanent pastures	137	2.21	13	0.29	150	1.40
›ttlement	39	0.63	12	0.26	51	0.48
urial ground	5	0.08	4	0.09	9	0.08
adi (village pond)	2	0.03	2	0.04	4	0.04
ɔad and cart tracks	19	0.31	4	0.09	23	0.21
Total	6196	100.00	4513	100.00	10709	100.00

Waste Lands

Nearly 44.06 per cent area of these villages is occupied by medium and high sa ınes with highly active middle and upper flank (Fig. 3). Sandy waste with negligil ›getative cover constitutes 4120 ha area (38.47%). Such wasteland largely occur estern part of the Lakhusar village and in northern part of Barju village. Even t ›ttlement sites in are surrounded by active sand dunes which encroach the houses, (F and block the road and cart tracks. Large scale removal of *phog* (*Calligonu ›lygonoides*) for fuelwood and expansion of agricultural practices on sand dunes a ajor factors of sand drift. Sandy waste with scrub constitutes 5.59 per cent area hich one third occurs in Lakhusar village and two third in Barju village. These a ›vernment lands comprising of low to medium height stabilized sand dunes with *Acac cquemontii*, *Leptadenia pyrotechnica*, *Aerva pseudotomentosa* and *Calligonu ›lygonoides* as main vegetation cover.

Miscellaneous Land Use

Permanent pasture and *oran* constitute only 150 ha or 1.4 per cent area of whi 37 ha occurs in Lakhusar village. In comparison to total livestock population (abo 1300), the area of permanent pastures is very less i.e. 953 animals/ha which is higł ›graded. In about 68 ha area near Lakhusar settlement, good plantation of *Acac rtilis* has been raised by the forest department.

Fig. 3. Active sand dunes - a great menace for environmental degradation

Fig. 4. Encroachment on houses by shifting sand dunes

Cropping Pattern

Guar is the dominant crop which occupied 62.4 per cent cropped area of Lakhus llage and 67.5 per cent cropped area of Barju village during 1994-95. This is followed l jra plus moth which constitute 37.4 per cent in Lakhusar village and 27.83 per cent arju village. At places sesamum is also taken. Bajra plus moth are generally produc interdune plains while on dune flank mainly guar is taken. Annual cropped area va om year to year depending on the amount and distribution pattern of rainfall. The is been 15 per cent increase in the acreage of bajra plus moth in the year 1994-95 ov e previous year but the acreage of guar has increased from 244 ha to 1597 ha.

LAND USE CHANGES

In these villages, in the year 1954 the cropped area was 300 ha and rest of the ar as used as grazing ground. In Lakhusar village cultivated area has increased by 125 429 ha to 3210 ha) between 1971 and 1994. Consequently the area under culturab aste has declined by 53.4 per cent i.e. from 4633 ha to 2156 ha. In Barju village n wn area has increased by about nine fold i.e. from 89 ha to 833 ha and consequent ea the area of fallow lands has significantly declined. The area under culturable und aste has also declined by 4.0 per cent during this period. The year 1994-95 receiv od rainfall (323.4 mm) when net sown area jumped from 833 ha (1993-94) to 2365 ha 994-95). Significant expansion of settlement has taken place in Lakhusar village. Tl ianges in agricultural land use between 1954 and 1970 is shown in Fig. 5.

ENVIRONMENTAL DEGRADATION

The study reveals that about 7.04 per cent area is not properly utilized while 51. r cent area is misused. Thus only 44.41 per cent area can be said to be proper ilized. The expansion of arable farming and over exploitation of natural vegetation h d to the problems like accelerated sand drift, advancement and shifting of sand dun ig. 3), formation of barchan dunes, accumulation of fresh sand on cultivated land ockade of roads and houses (Fig. 4) and hampering the movement of livestock. Lar ale cutting and up rooting of *phog* (*Calligonum polygonoides*) is a great menace in th gion causing fast depletion of biomass and acceleration of sand drift. Cultivation nd dunes and marginal lands has reduced the crop productivity and crop failur rinkage of grazing areas and greater imbalance in demand and supply of the fodd use degradation.

SUGGESTED LAND USE

Based on present land use capability, status of land degradation and human ar estock pressure suggested land use is given in Table 2. Rainfed agriculture should l stricted to inter dune plains and up to lower flank of the dunes. Construction of fie

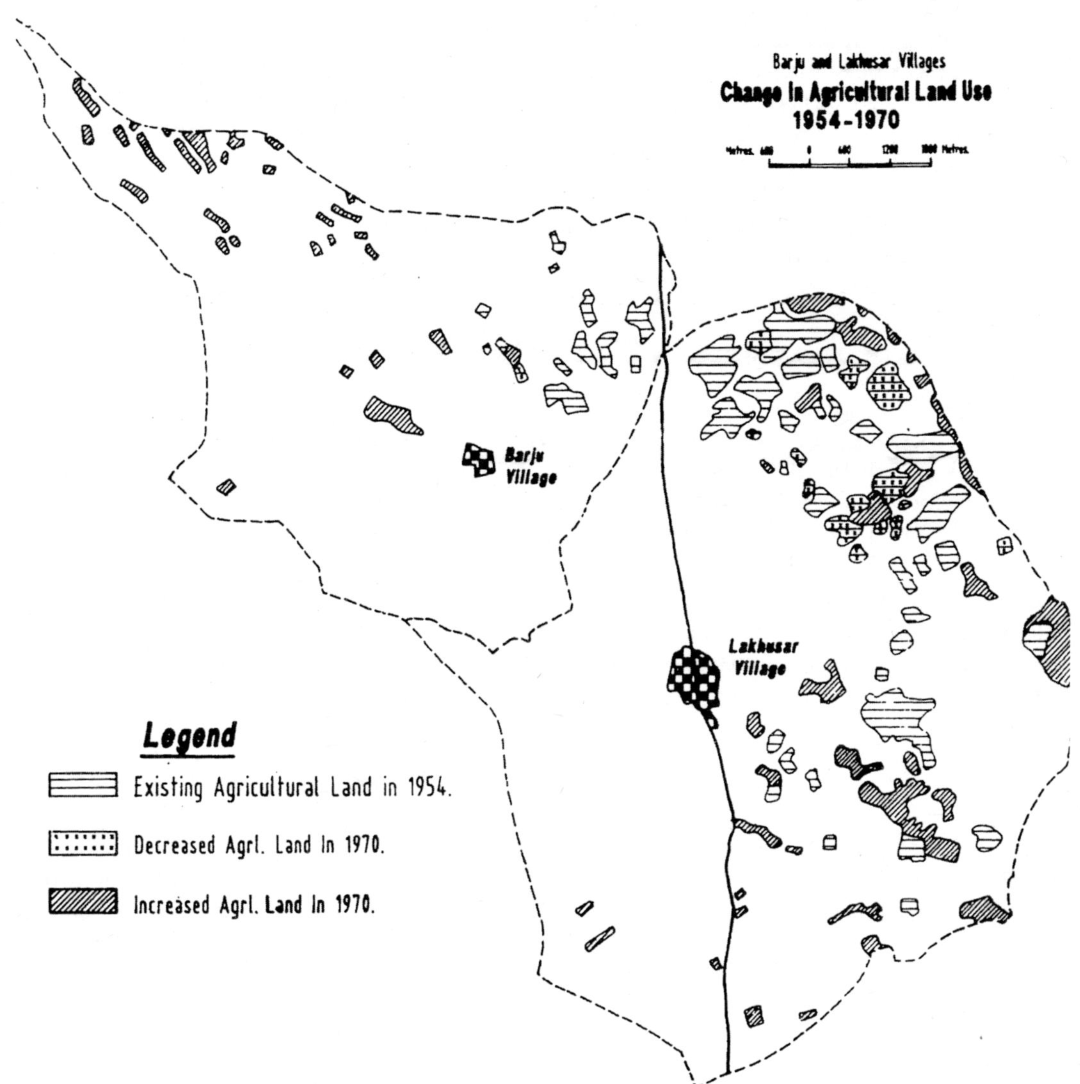

Fig. 5. Changes in agricultural land use.

ıarriers (bunds), vegetation barriers, agro-forestry and dry farming practices should l ollowed. In areas with sand hummocks, grasses should be grown with crops in strips. ıuch active zones of crops should be taken in rotation with 3 or 4 years follow. Active pa ıf the sand dunes should be stablized and entire dune area should be developed in ilvipastures. About 36.8 per cent is suggested for this treatment. Looking to tl ıressure of livestock, the region has insufficient area under permanent pastures aı

:nce atleast 8.03% area should be developed as pasture with suitable grasses, shrul ıd trees. At present only 0.51 per cent area is available for settlement, transpo :twork and public utilities and additional 1.01% area should be earmarked for futu :velopment. Renovation of the existing *nadis* and plantations along road side a commended.

ıble 2. Suggested Land use for Lakhusar and Barju village

	Suggested land use	Lakhusar	Barju	Total
		Area (ha)	Area (ha)	Area (ha)
1.	Rainfed farming with soil & moisture conservation, agro-forestry and construction of field bunds	2800 (45.19)	2200 (48.75)	5000 (46.69)
2.	Rainfed legume crops with grasses and soil conservation measures	464 (7.49)	240 (5.32)	704 (6.57)
3.	Sand dune stabilisation and development of silvipastures	2295 (37.04)	1655 (36.67)	3950 (36.88)
4.	Development of grassland and silvipastures as common grazing land	512 (8.26)	348 (7.71)	860 (8.03)
5.	Settlement, transport network and public utilities	125 (2.02)	70 (1.55)	195 (1.82)

CONCLUSION

Nearly half of the area of Lakhusar and Barju village in Bikaner district, cupied by sand dunes. The rainfed agriculture, has moderate to poor intensity ıltivation with limited crop choice and low productivity. Expansion of agriculture ov e marginal areas during the past decades and large scale cutting of *phog* ha' tensified wind erosion and deposition hazard and sharp decline in biomass. The regi ıs good scope for the development of silvipasture while arable farming should l stricted to interdune plains with precautionary measures of soil and wat nservation and improved dryland farming practices. Cutting of *phog* (*Calligonu ılygonoides*) should be banned to improve the biomass and micro environment.

ACKNOWLEDGEMENT

Authors are highly thankful to Dr. A.S. Faroda, Director and Dr. Surendra Sing ead of Resource Survey and Monitoring Division, Central Arid Zone Research Institut ıdhpur for permitting to undertake this study and provide all the facilities.

REFERENCES

ngh, S. and Kumar, S. Ed. 1995a. *Natural Resources Appraisal for sustainable development Lakhusar village*, CAZRI, Jodhpur pp. 23.

ngh, S. and Kumar, S. Ed. 1995b. *Natural Resources Appraisal for Sustainable Development Barju village*, CAZRI, Jodhpur pp. 22.

eological Evolution of Northwestern India
S. Paliwal (Ed.)
ientific Publishers (India), Jodhpur, 1999, pp. 378 - 386.

ENVIRONMENTAL DEGRADATION AND SUSTAINABLE LANDUS PLANNING OF THE ECOSYSTEMS IN WESTERN RAJASTHAN

Surendra Singh

Central Arid Zone Research Institute, Jodhpur

ABSTRACT

Exogenic and endogenic activities have played a significant role in the evolution of different type of ecosystems. The production and productivity of the biophysical resources of the eco-systems have been depleted by accelerated wind erosion/deposition, water erosion and salinity. Conservation measures alongwith the active participation of the local people and NGOs have been suggested for sustainable development of the degraded ecosystems.

ey words : Sustainable Landuse, Environment Degradation, Ecosystem Plannin Western Rajasthan.

INTRODUCTION

Systematic studies using remotely sensed and field survey techniques revealed th uvio-denudational, fluvial and aeolian ecosystems of various size occur in Weste ajasthan (Ghose *et al.* 1977a). Due to harsh climatic and unfavorable soil condition ese ecosystems are highly fragile. Moreover, in the last a few decades, the stress thropogenic, zoogenic and technogenic activities on these ecosystems, has increas anifold. As a result, the ecological equilibrium of the ecosystems has been disturb d their biological productivity has declined.

In order to check further degradation of the ecosystems and to upgrade the oduction potentials, using systematic information on natural and anthropogen ocesses responsible for the degradation are of paramount importance. In this pape uses and impact of various environmental degradation processes have been identifi d suitable measures for their sustainable landuse planning have been discussed.

EVOLUTION OF ECOSYSTEMS

Exogenic and endogenic activities have played a significant role in the evolution fferent type of ecosystems in western Rajasthan (Fig. 1). Fluvial ecosystems we eated by fluvio- denudational and fluvial exogenetic processes during two major w ıase. During the first wet phase around 50-40 ka Before Present (B.P.) and 10-3.8] P. The prolonged sub-aerial weathering and denudation of the land surfaces produc accentuated the formation of extensive rocky/gravelly pediments at the base of t lls and mountains. Parts of these pediments were later on covered with colluvi diments, transported by the fluvial process, and termed as flat buried pediments. e distal end of the buried pediments, the vast alluvial plains were created up by t ell integrated Aravalli drainage system comprising the Luni-Jawai rivers and the ibutaries (Ghose, et al., 1977a, Singh and Ghosh, 1977) and the Himalayan draina stem, constituting the former mighty rivers like the Saraswati and the Drishadva ıd their tributaries, and the shifting courses of the Sutlej, known as the Satadru. T lcrete deposits (*Kankar* pan) within the alluvial material has been dated to 28 to 38 I P. Major lakes and depressions, created by tectonic and aeolian activities at Sambha ıchpadra, Didwana, Talchhapar, Kuchaman, Bap, Pokaran and Thob, received mo esh water.

During the two major dry phases around 18-10 Ka B.P. and from 3.8 Ka B.P., t olian processes became dominant (Singh *et al.*, 1974) resulting in the shifting a ying of rivers and formation of aeolian and saline ecosystems (Ghose, *et al.* 1977b). T gh sand dunes formed during the first major dry phase viz. obstacle, parabol ngitudinal and transverse belong to old dune system. The barchan, barchanoid, shru ppice dunes, low longitudinal and transverse ridges and sand streaks belong to ne ıne system.

BIOPHYSICAL CHARACTERISTICS OF ECOSYSTEMS

Different fluvial ecosystems like hills, upland with colluvial plains, rocky/gravel diments, flat buried pediments, flat older alluvial and interdune plains, river beds a unger alluvial plains are characterized by almost level and gently undulati pography, gravelly sand loam, sand and sandy loam to loam and clay loam soils havi gnificant variations in their physical and chemical properties. The major plant speci these ecosystems are *Acacis senegal, Salvadora oleoides, Prosopis cineraria, Prosop liflora, Capparis decidua, Cenchrus setigerus, Oropetium thomaeum, Aristic niculata, Sporobolus marginatus* and *Haloxylon salicornicum*. The rainfed crops li *jra, moong, moth, guar* and *til* and the irrigated crops viz; wheat, barley, gra ustard are raised under these ecosystems. The surface water occurs in the form *dis*, tanks, reservoirs, canals and streams whereas the ground water in the form uifers has been encountered under different lithological formations.

The aeolian ecosystmes viz; sandy undulating older alluvial and interdune plair ndy undulating buried pediments and sand dunes and sandy undulating interdu ains are characterized by undulating topography, irregular slope and deep to very de

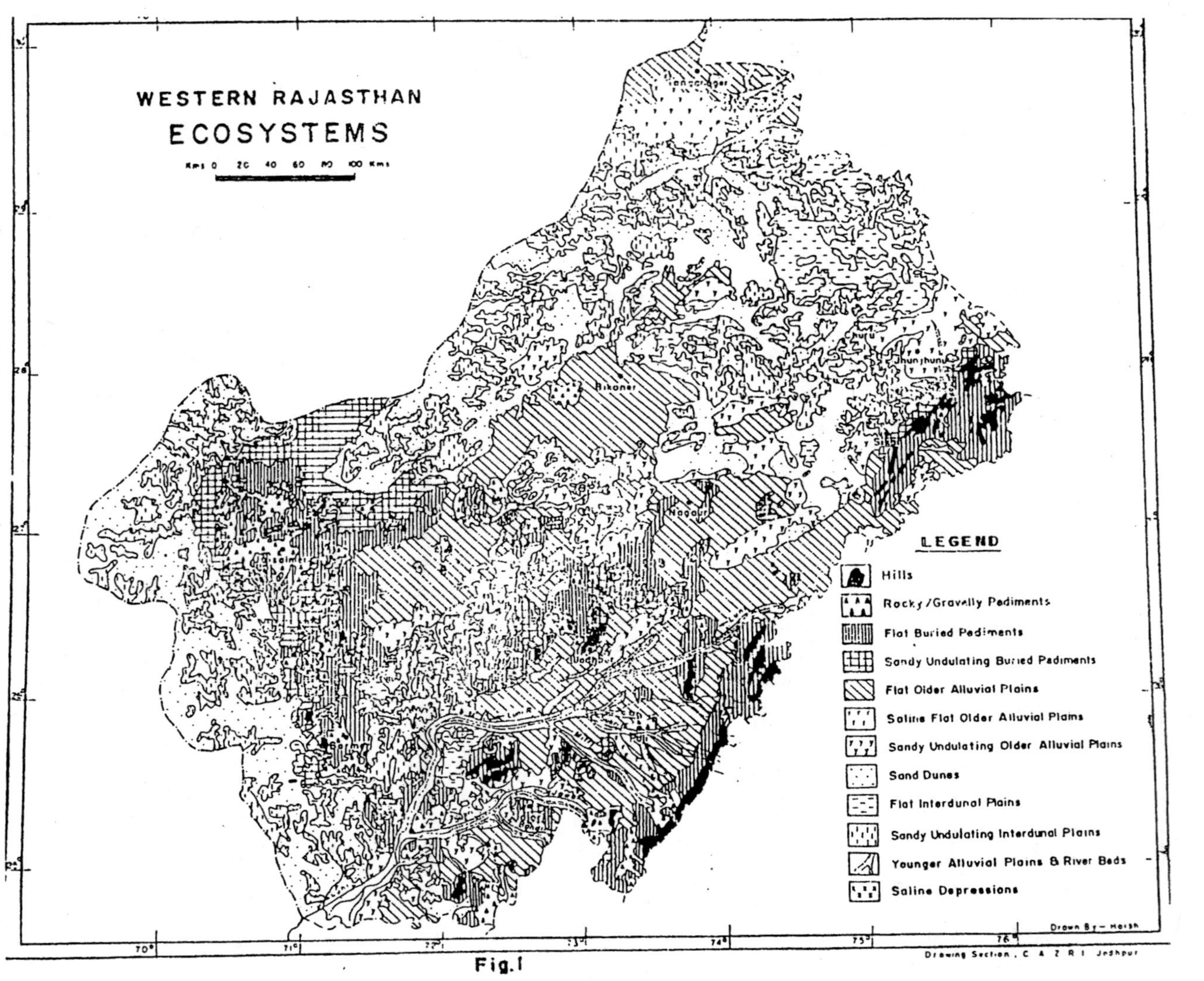
WESTERN RAJASTHAN
ECOSYSTEMS
Kms 0 20 40 60 80 100 Kms
Bikaner
LEGEND
Hills
Rocky/Gravelly Pediments
Flat Buried Pediments
Sandy Undulating Buried Pediments
Flat Older Alluvial Plains
Saline Flat Older Alluvial Plains
Sandy Undulating Older Alluvial Plains
Sand Dunes
Flat Interdunal Plains
Sandy Undulating Interdunal Plains
Younger Alluvial Plains & River Beds
Saline Depressions
70°
71°
72°
73°
74°
75°
76°
Drawn By – Harsh
Drawing Section, C A Z R I Jodhpur

Fig.1

.d well drained fine sand to loamy fine sandy soils. These ecosystems largely suppc ammophytic plant species like *Calligonum polygonoides, Maytenus emarginatı osopis cineraria, Acacis senegal, Lasiurus sindicus, Cenchrus ciliaris* and *Panicu rgidum*. Due to scarcity of surface and ground water resources, the rainfed crops li *jra, moong, moth* and *guar* are grown under these ecosystems. However, the irrigat ɔps such as wheat and *rayada* are raised in pockets wherever ground water is availal irrigation.

The domestic animals like goat, sheep, cow, buffalo, camel and wild animals viz; fc t, vulture, black buck, *nilgai*, chinkara, owl and bustard also live in this region and a aintaining the ecological balance. Due to clear sky, more than 2000 hours of sunshine ar and high wind speed, the desert regions also have sufficient renewable solar a nd energy resources for the use of domestic and agricultural purposes.

DEGRADATION OF ECOSYSTEMS

The increasing anthropogenic activities like overgrazing, cutting of trees, faul ethods of irrigation, cultivation on marginal lands and construction of engineeri uctures and development of industries have accelerated the water erosion, salinit calinity, pollution and flood hazards, resulting in the depletion of the biologic oductivity of fluvio-denudational and fluvial ecosystems.The water erosion in the for rills and gullies of 0.45 to 1.0 m depth and 0.9 to 2.5 m width and 1.5 to 10 m dep .d 2.5 to 20.5 m width, respectively has degraded and decreased the yield of crops a ove ground biomass of the ecosystmes located under more than 300 mm rainfall zor has been estimated that 0.16 to 1545 t/km^2 soil is annually lost from the Bandi rive ie sediment concentration from 0.4 to 24.7 g/l in the upper reaches of the Bandi riv creased to 0.5 to 36.3 g/l in its down reaches.

The use of the poor quality of ground water having 8.6 to 9.6 pH, 4 to 12 mmhos E d 10 to 50 SAR resulted in the development of salinity/alkalinity hazards a pletion of the agricultural productivity of flat older and younger alluvial plans. As sult, soil crust and salt encrustations developed and a large acreage of land turned in isteland. The construction of canals and tanks along the buried courses of pri ainage channels also resulted into the development of salinity/alkalinity a gradation of agricultural lands (Ghose et al., 1982). The large amount of industri luent (4000 KLD) discharged from the textile industries located in the districts of Pa dhpur and Barmer has polluted the ground water of the wells along Bandi, Jojri a ıni rivers. This has resulted in the development of soil crust, turning fertile lands in isteland s and decrease in the yields of crops and above ground biomass. Sever imals died due to the drinking of polluted surface and ground water and human bein e suffering from several diseases like stomach disorder and dysentry (Saxena a ngh, 1989).

The flash floods of 1979, 1982 and 1990 had degraded and depleted the fertile lan different ecosystems and engineering structures (Fig. 2). The dissections and sa sting hazards, caused by the flood of 1990, affected 5,000 ha, 35,400 ha, 40,500 a ,200 ha of agricultural lands in Jalor, Pali, Barmer and Jodhpur districts, respective a result, a very large acreage of fertile lands of these districts turned into wastelanc ıe large number of animals and people also died and several settlements and roa ere washed away during these floods.

Fig. 2. Field photograph showing flood water degrading the ecosystem

The mining of sandstone, limestone, fuller's earth, gypsum and bentonite resulte ı deep dissections and deposition of debris in the form of heaps and mounds and thı ırge acreage of agricultural lands of different ecosystems (Fig. 3) turned into wastelanc ;axena and Singh, 1989). The common property resources (*Orans*) located under tl ɔove ecosystems were degraded due to accelerated degradational process and the :reage and biomass yield also declined.

Fig. 3. Photograph showing mining activity in the area.

Fig. 4. Field photograph showing drifting of sand over cultivable land.

The silting in water bodies like *nadis*, tanks and reservoirs and lowering of grou 'ater table at the rate of 0.5 to 1.5 m/year due to the accelerated fluvial activities a verexploitation of ground water, respectively and resulted in the degradation ifferent ecosystems. As a result, the water storage capacity of the water bodies a gricultural productivity of the catchment areas decreased considerably and seve locks in Jalor, Jhunjhunu, Pali, Jodhpur and Sikar districts have reached at t emi-critical and critical stage of ground water development.

The increasing anthropogenic, zoogenic and technogenic activities have distribut ie ecological balance of the aeolian ecosystems resulting in the reactivation of stabilis unes, formation of barchan dunes and sand drifting, particularly in less than 300 n ainfall zone (Fig.4). The accelerated wind erosion/deposition hazard has affected t and free agricultural lands, settlements, roads, railways tracks and irrigation networl 'he rate of sand movement in this rainfall zone varies from 5 to 20 m/year, whereas iore than 300 mm rainfall, it ranges between 0.5 to 5 m/year (Singh et al., 1986; Sin nd Kar, 1983, 1988, 1991).

Although the introduction of the irrigation in the Indira Gandhi Canal comma rea has increased the productivity of the ecosystems and has minimised the aeoli ctivities but numerous hazards like salinity/alkalinity and waterlogging (Fig. 5) ha eveloped due to unfavourable geomorphic conditions and human follies. As a result ery large acreage of land was degraded and the yields of the crops and above grou iomass of the grazing lands declined to a great extent (Singh and Kar, 1991).

Fig. 5. Photograph showing the salinization of the land and waterlogging.

SUSTAINABLE LANDUSE PLANNING OF ECOSYSTEMS

The ecosystems located in the west of 300 mm isohyet should not be used for arab rming due to adverse climatic conditions and inherent morphological vulnerabilit hese ecosystems, however, could be developed into silvipastural, hortipastura grohortipastural and agroforestry landuse systems and animal husbandry by raisi *rosopis cineraria, Acacia senegal, Calligonum polygonoides, Acacia tortilis, Lasiur ndicus, Cenchrus ciliaris,* cluster bean, moth bean and greengram and *Gola* and *S* arieties of *Ziziphus mauritiana* (*ber*) and using soil and water conservation measur ke wind breaks, shelterbelts, mulching contour bunding and contour furrows.

Fluvial ecosystems located in the east 300 mm isohyet could be developed in infed agriculture, using improved crop varieties like *bajra* (BJ-104); cowpea (K-1 C-10); moong (S-8, M- 8) and *guar* (FS-227) and use of 30-50 kg N/ha and 30-40 kg/ $_2O_5$ and dry farming practices like contour or field bunding, subsurface barri terplot and inter-row water- harvesting and seed-cum fertilizer drill (Singh, 1976). T ter-cropping of pearl millet with grain legumes (greengram, clusterbean and mothbea ay also be used to upgrade and stabilize the productivity of these ecosystems (Singh ., 1981).

Aeolian and fluvial ecosystems where water for irrigation is available, should ł sed for irrigated agriculture, using proper methods of irrigation like sprinkler and drï ystem to avoid the problem of salinity and waterlogging. The saline and high RSC wat‹ ›uld also be used for irrigation by adopting proper methods of salt leaching and usir roper doses of gypsum.

The areas of the fluvio-denduational, fluvial and aeolian ecosystems which aı ighly degraded due to natural and accelerated salinity/alkalinity could be rehabilitate nd developed into pastures, silvipastures and forests, using plant species like *Prosop* *neraria, Acacia senegal, Albizzia lebbeck, Ziziphus nummularia, Salvadora oleoide* *rosopis juliflora, Calligonum polygonoides, Cenchrus ciliaris, Lasiurus sindicu* *ichanthium annulatum, Tamarix aphylla, Cercidium floridum* and *Panicum turgidu* ith proper soil and water conservation measures like land levelling, contour bundin ›ntour furrowing, shelterbelts, micro-wind breaks and mulching (Saxena and Singl 989). The involvement of the local people and NGOs is absolutely necessary for tł roper reclamation and development of wastelands.

ACKNOWLEDGEMENT

The author is grateful to Dr. A.S. Faroda, Director, Central Arid Zone Institut odhpur for providing facilities to conduct this study.

REFERENCES

hose, B., Singh, S. and Kar, A. 1977a. Geomorphology of Rajasthan desert. In : Jaiswal, P.] (Ed.) *Desertification and its Control*; ICAR, New Delhi, 69-76.

hose, B., Singh, S. and Kar, A. 1977b. Desertification around the Thar - A geomorphologic interpretation. *Ann. Arid Zone* 16 : 290-301.

hose, B., Singh, S. and Kar, A. 1982. Some geomorphological aspects of salinity hazards i Rajasthan desert, India. Proc. Workshop on Problems of the Desert in India. *Geol. Sur India, Misc. Publ.* 11-41.

hose, B., Kar, A. and Hussain, Z. 1979. The lost courses of Saraswati river in the Great India Desert - New evidence from Landsat imagery. *Geog. Jour.*, 145 : 446-451.

axena, S.K. and Singh, S. 1989. Wastelands in making and measures to mitigate the effects factors responsible. *Proc. Int. Workshop on Dryland Farming and Wasteland Development.* : 149-154.

ingh, G., Joshi, R.D., Chopra, S.K. and Singh, A.B. 1974. Late Quaternary history of vegetatio and climate of the Rajasthan desert, India. *Phil. Trans. R. Soc.*, London, 267(B) : 467-501.

ingh, R.P., Singh, H.P., Daulay, H.S. and Singh, K.C. 1981. Fertilization of rainfed greengraı pearl millet sequence. *Ind. Jour. Agric. Sci.,* 494-503.

ingh, S. and Ghose, B. 1977. Geomorphology of Luni basin and its palaeoclimatic inferences. In Agarwal D.P. and Pandey, B.M. (Eds.) *Geology and Archaeology of Western India* Conce] Publ. Co., New Delhi, 135-146.

ingh, S. and Kar, A. 1983. Application or remote sensing techniques in the environment problems of the Great Indian desert. *Ind. Jour. Geog.* 113 : 83-91.

ingh, S. and Kar, A. 1988. Remote sensing in monitoring desertification processes under differei landforms (ecosystems) of the Indian desert. In : Tewari, A.K. (Ed.) *Desertification Monitoring and Control*. Scientific Publishers (India), Jodhpur 47-73.

ingh, S. Kar, A. and Shankarnarayan, K.A. 1986. Dynamics, morphology and management (sand dunes in the Indian desert. In : Singh, A. Ohri, M.L. (Eds.) *Current Practices i Geotech. Eng. Geo-Environ*. Scientific Publishers (India), Jodhpur, 3 : 23-47.

ingh, S. and Kar, A. 1991. Morphological characteristics of landforms and associated problems. I : Abrol, I.P. and Venkateswarlu, J. (Eds.), *Prospects of Indira Gandhi Canal Project*. ICAI New Delhi, 11-18.

ɜological Evolution of Northwestern India
S. Paliwal (Ed.)
ːientific Publishers (India), Jodhpur, 1999, pp. 387 - 414.

GEOLOGY OF THE REGION TO THE WEST OF ARAVALLI MOUNTAIN - SOME OBSERVATIONS

B.S. PALIWAL

Department of Geology, Faculty of Science,
Jai Narain Vyas University, Jodhpur 342 005 India

ABSTRACT

The region to the west of Aravalli Mountain in Rajasthan hosts a variety of rock types ranging from Precambrian to Recent in age. Despite rigorous scanning of the area done in search of hydrocarbons and ground water during the last a few decades by multiple agencies, the region lacks in a comprehensive account of its geology, particularly, geochronology of the rocks and their stratigraphic position need review in the light of newly generated data. The paper highlights some important problems related to the geology of the region.

ey words : Precambrian, Aravalli Supergroup, Sirohi Group, Malani Igneous Sui Marwar Supergroup, Rajasthan

INTRODUCTION

Rajasthan - the land of sung and unsung heroes has witnessed some unique a ːemplary geological events that took place during the history of the earth. The region ıe west of the Aravalli Mountain, a major part of which is occupied by the great Tł esert, has not only been inspiring adventurers, musicians, and film makers since loı ırough its musical sands, calm and clean environment and colourful cultural heritaᵹ ıt it also offers insight into many unsolved geological mysteries of global intere ːagmentation of "Rodinia" - an early supercontinent (older Pangaea of Paliwal, 19ᶥ ıd reassembling of these fragments into "Pangaea" - the famous supercontinent (ıliwal, 1992) is one such example.

Geologically, the region exposes a variety of lithologies (Fig. 1) ranging from t

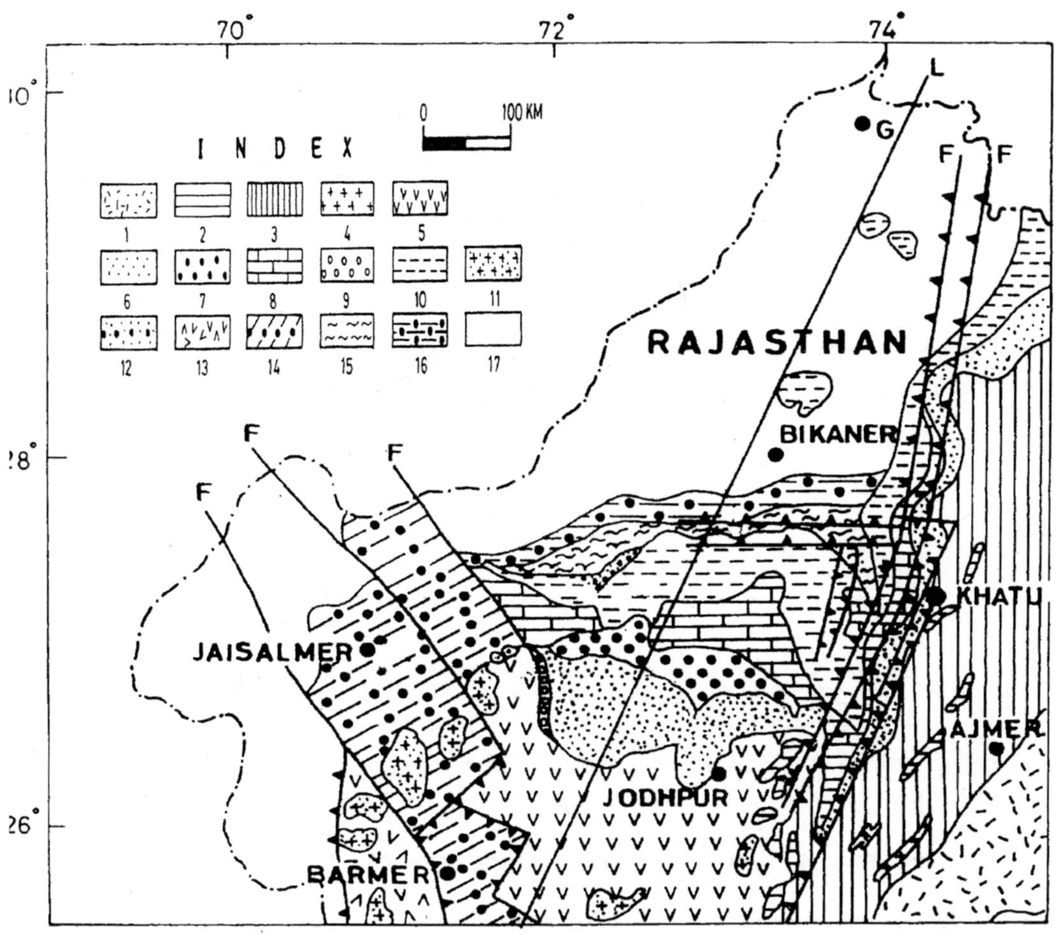

;. 1. Regional geological map of western Rajasthan. 1 - Banded Gneissic Complex, 2 - Araval Supergroup, 3 - Delhi Supergroup, 4 - Erinpura Granite, 5 - Malani Igneous Suite, 6 - 1C Marwar Supergroup (6 - 7 Sonia Formation and Girbhakar Formation of the Jodhpı Group, 8 - Bilara/Hanseran Group, 9 - Pokaran Boulder Bed, 10 - Nagaur Group) 11 Jalore and Siwana Granites, 12 - Bap Boulder Bed, 13 - Deccan traps, 14 - Mesozo sediments, 15 - 16, Tertiary sediments, 17 - Quaternary sediments and sands of the Th: Desert, F-F- Faults, G- Ganganagar, L-L- Lineaments, (Modified from Paliwal et al., 1995

ecambrian to Recent in age. Deformed metasediments of the Aravalli and Del. pergroups followed by a Late Proterozoic/ Neoproterozoic magmatism resulting in tl ·mation of Erinpura Granite and Malani Igneous Suite of rocks and a series dimentary sequences accumulated during Palaezoic, Mesozoic and Cenozoic er nstitute the regional geology.

The geological history of the region became more pronounced since the Termin oterozoic time when the tensional tectonic setting took place along the western flank e Aravalli Mountain Range (Paliwal, 1992) resulting in a widespread felsic volcanisr iis early crustal disturbances did not die down but continued subsequently, shapir is part of the earth crust. An array of cratonic basins developed in subsequent perio d several transgressive and regressive phases occurred depositing a number scinating sedimentary sequences in the region.

Neotectonic activities in the region caused (i) reversal and subsequently drying the ancient "Saraswati-Drishadwati" river systems (the remnants of which a: aceable in the form of paleochannels), (2) formation of saline lakes and (3) wide sprea sertification.

Since the early account of the geology of the region was published by Blanfo 376, 1877) and Oldham (1886, 1888), several attempts were made by geoscientist ring twentieth century to understand different geological aspects of the regio gorous scanning of the region has been done during the last a few decades by multip encies in search of hydrocarbons and ground water. Despite all endeavours neith sired success could be met with in the discovery of petroleum in the region nor ar tisfactory answer to many unsolved problems related to the regional geology could l

g. 2. Deformed Precambrian metasediments exposed southwest of Didwana. Early lineation curving around the later folds forming ridges.

ig. 3. Hinge migration cleavage in quartzose phyllite at Tikali Dungari, Didwana.

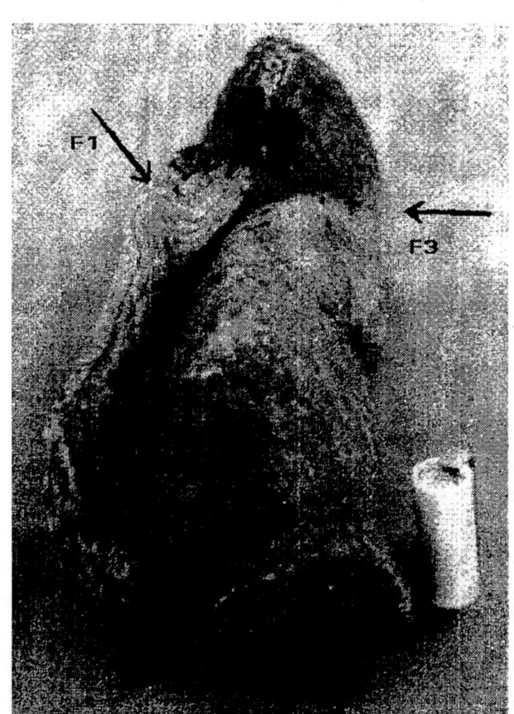

g. 4. A specimen of quartzite from south west of Didwana showing superposed folding.

obtained so far. The situation becaı rather complex and confusing. It surpris most that such a wide spread Malа Igneous Suite is devoid of any significa mineral deposit or we failed to locate it.

It would be appropriate to highlig here some of the observations made on t geology of the region with a hope to se satisfactory clue to many unsold geologi problems of this part of the land in ne future.

DEFORMED PRECAMBRIAN METASEDIMENTS

All along the western flank of t Aravalli Mountain, Precambrian me sediments (Fig. 2) occur in form of isolat hills or groups of hills amidst sands of t Thar Desert, particularly, around Siro Punagarh, Sindreth, Pali, Sojat, Rajpuri (Jodhpur), Degana, Didwana, Koli Khatu, Gopalpura (Sujangarh) etc.

ig. 5. Superimposed folds in quartozose phyllite, south west of Didwana.

western Rajasthan. Earlier, these deform metasediments were considered to be part a parcel of the Aravalli system (Aravε Supergroup) by a number of workers (. Touche, 1902; Coulson, 1933; Heron, 19ε 1954). Later workers included these rocks Delhi Supergroup (Gupta et al. 1980, Pare‹ 1981, 1984, Murthy, 1979, Roy et al. 198 Stratigraphic names like Sirohi Group, P Group, Punagarh Group, Sindreth Group e were proposed for these deform metasediments and metavolcanics.

Recently, some doubts started bei raised regarding the stratigraphic position these rocks. Some workers have express their views that the metasediments expos around Sirohi, Punagarh, Sindreth, Soj Pali, Degana, Khatu Didwana, Gopalpu (Sujangarh) etc. constitute one mc Precambrian sedimentary sequence, young than the Delhi Supergroup, i.e. the Sirc Group of Neoproterozoic age (Paliwal, et ε 1995; Roy and Sharma, 1996; Sharma a Roy 1997; Paliwal, 1998). Detailed structu study of the metasediments exposed arou

ig. 6. First fold hinge in quartz vein in the Aravalli phyllite south west of Didwana.

'ig. 7. Carbonaceous phyllites at Tikali Dungari hill south west of Didwana.

)idwana (Fig. 2 and 3) Khatu and Koliya in Nagaur district has revealed a compl eformation pattern and structural style of these rocks (Fig. 4). Atleast four phases olding (Figs. 5 and 6) and a strong ductile shear zone passing through these roc (Paliwal, 1980) ha left their signature them as compared only a single phase deformation, metamorphism a grantoid emplaceme recorded in the roc of the Sirohi ar (Sharma and Rc 1997). Howeve Murthy (1979) wh describing tungst mineralization Sirohi district, fou these rocks to be t youngest unit of Del Supergroup, whi has undergone polyphase deformation.

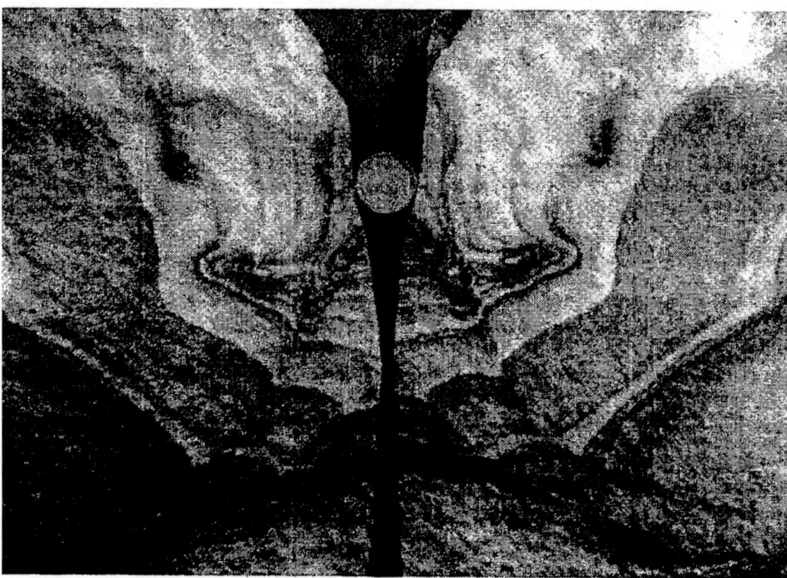

ig. 8. "Layered Chromatographical Structures" in the sandstone of Badi Khatu, (cf. Paliwal, 1976).

Fig. 9. Moulds of rain/hail prints on the lower surface of the sandstone at Badi Khatu.

Suprisingly, Sinha Roy and his co-workers in their recent book on "Geology of Rajasthan" have put the Sirohi-Pumagarh Group just above the Malani Igneous Suite (Sinha Roy et al., 1998; table facing p. 258). On the other hand they have described the "Sirohi belt" and "Punagarh-Sindreth belt" under the "Early Proterozoic" considering the Sirohi rocks to be younger than those of the Delhi Supergroup (cf. Sychanthavong and Desai, 1977; Sen, 1981; Sinha Roy, 1984). Gupta et al., (1980) have placed them at the top of the Delhi Supergroup. Both Punagarh Group and Sindreth Group of rocks have been reported to show a structural discordance with the rocks of the underlying Sirohi Group which according to Sinha Roy et al., (1998) shows the effect of atleast three phases of folding.

Extensive development of black-shale facies, one of the bases for the presumption that the Aravalli rocks of the type area are of the Proterozoic age (Roy *et al.*, 1997),

Fig. 10. A fossil like structure in clay gall within Khatu sandstone (cf. Paliwal, 1992

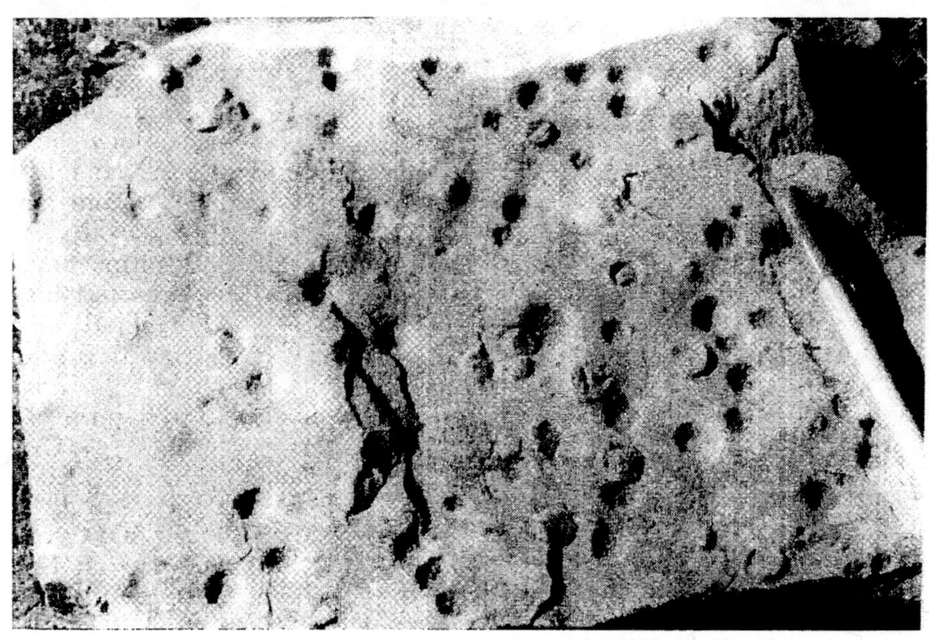

g. 11. Small tablet like structures on the top surface of Chhoti Khatu sandstone.

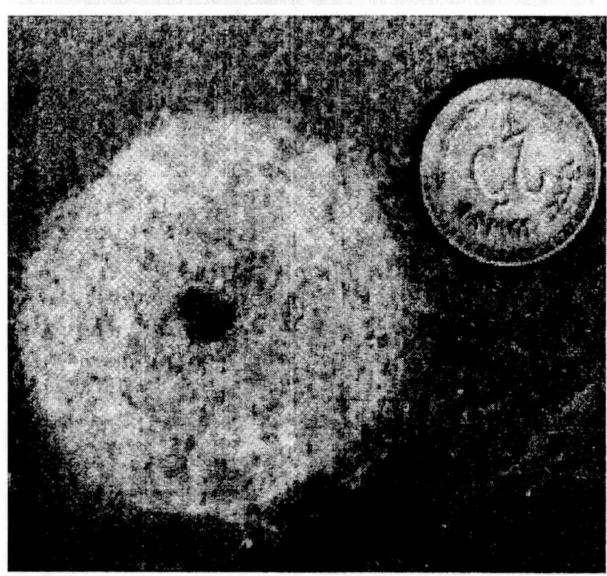

g. 12. Deoxidation structures developed around carbonaceous material in Jodhpur sandstone. (cf. Paliwal, 1995).

has also been found to fo dominant lithology ground Tik Dungari south west of Didwana (F 7). Secondly, the carbonates a metavolcanics forming importa rock types of the Sirohi Group a totally absent in the Rajpuriy Degana- Khat Didwana-Gopalpura, belt. Her (1932) reported coarse grewack and slates from Khatu area. T above facts stand strongly agair the correlation of the Sirohi roc with those of Ras, Makrana a Didwana (cf. Paliwal et al., 19 Roy and Sharma, 1996; Paliw 1998) However, the Precambri metasediments exposed at Rajpuri (Jodhpur), Sojat, Degana, Kha Didwana, Kaliya and Gopalpu seem to be the part of Aravalli rock

As far as the further westwa continuity of these deform

Fig. 13. Basal Conglomerate exposed at Chhoti Khatu.

etasediments is concerned, schists and phyllites of the Precambrian age have bee countered in the exploration wells drilled by Oil and Natural Gas Corporation nar-I and Bhuana- I in the Jaisalmer area.

Fig. 14. Mylonitized Precambrian metasediments exposed at Chhoti Khatu village.

ENIGMA OF THE MALANI ROCKS

Since the term "Malani beds" was first introduced by Blanford (1877) for a volcai eries of rocks of the then Malani district of Marwar (Jodhpur State of Rajaputar everal attempts were made to study different aspects of these rocks during the twetie entury (La Touche, 1902; Heron, 1932; Pascoe, 1959, Kochhar, 1984, 1998; Murtl 979; Pareek 1981, 1984; Bhushan, 1985, 1989, 1995, Bhushan and Khullar, 19! 'allinayagam, 1998; Vallinayagam and Kochhar, 1998; Srivastava, 1988, 19! rivastava *et al.*, 1979; Rathore *et al.*, 1996; Basu, 1982; Pandit *et al.*, 1999).

Several names like "Malani Igneous Suite" (Pareek 1981); "Malani Volcanic Seri La Touche, 1902); "Malani System" (Coulson, 1933). "Malani Granite and Volcai uite" (Pascoe, 1959). "Malani Series", "Malani Volcanic Series", "Malani Beds" (Paliw 992), "Malani Igneous Province" (Chore and Mohanty, 1998) "Malani Igneous Comple Bhushan and Khullar, 1998) etc. crept in the geological literature subsequently for the rocks. An age of 745±10 m.y. w proposed for the Malani Rhyolite Crawford and Compston (1970). Th geographical extent has be estimated to cover about 51,000 km. in parts of Pali, Sirohi, Chu Jodhpur, Jaisalmer, Jhunjhunu a Jalore districts of Rajasthan a possibly another 1,00,000 sq. k concealed below the overlyi sediments (Pareek, 1981a).

'ig. 15. Unconformity between the underlying Precambrian deformed metasediments and the Basal Conglomerate at Chhoti Khatu.

A lot of confusion preva amongst geoscientists regarding t basement of the Malani rocks a their distribution in time and spa Surprisingly, rock associatic spanning from 750 Ma to 77 Ma ha been grouped under the Mal Igneous Suite (cf. Srivastava, 19 Rathore et al., 1996; Kochhar, 19 Paliwal, 1998). Earlier distinction Malani Volcanic Suite of rocks fr the Erinpura Granite Suite propos by Heron (1953) is now bei questioned (Sastry, 1992). Paliwal a his associates (Paliwal et al., 19 assumed an age of 745±10 m.y. for t older flows and 560 m.y. to 550 m for the younger flows of Mal rhyolites. Pareek (1981b) advocat an age of 940±40 m.y. to 600±70 m for the Malani Igneous Suite.

Not only this but far flung occurrences of volcanics at Tusham in Haryana (Indi ı40±40 Ma, Kochhar, 1974) and Kirana Hills in Pakistan (870±40 Ma, Davies aı rawford, 1971) have been considered to be the representatives of the Malanis (745± [a, Crawford and Compston, 1970), derived from a single tholeiitic magr ɪntaminated with sialic crust (Pareek, 1981b).

Obviously, it is not only the geochronology of sporadically distributed rocks which nportant, but their field relationship established through rigorous field work is equa gnificant to arrive at any drastic conclusion.

Age of some grantoid bodies originally mapped by Heron (1953) as "Erinpu ranite" and "Granitic Gneisses" have been questioned recently (Tobish, 1998). One su xample is the Anasagar Granitic Gneiss which was mapped by Heron (1958) rinpura Granite has recently, been delineated as an older pluton of the Late Archae ge (Tobish, 1998; Tobish et al., 1994). Similarly, the granitic gneisses encounter ıbsurface around Didwana also need re-examination in view of the fact that t ravalli metasediments exposed around Tikali, Dungari south west of Didwana tov ıclose rafts of the basement granitic gneisses (probably of the Archaean ag nfortunately, granites of Abu, Balda, Idar and Sewariya and also many other grani ıeisses of diverse stratigraphic status have been grouped under a single stratigrapl ame the "Erinpura Granite".

THE MARWAR SUPERGROUP

Terminal Proterozoic volcanic activity pouring out felsic lavas of the Malani Igneo ıite to the west of the Aravalli Mountain, Rajasthan was followed by the formation atonic basins filled with sediments. The sand-shale-carbonate-evaporite sequence eposited covered an area of about 21,600 sq. km. in western Rajasthan. This sequen as earlier considered to be the part of the Vindhyan sediments (Bhanders and Kaimr Upper Vindhyan age) deposited in the Trans-Aravalli region (Heron, 1932; Misr 969). Later on this sedimentary sequence was renamed as the Marwar Supergroı 'areek, 1981a, 1984) comprising silicilastic sediments of the Jodhpur Group (Figs. 8 2) carbonates of the Bilara Group (Hanseran Group) and Siliciclastic sediments of tl agaur Group. The Pokaran Boulder Bed was considered to from the basal part of tl equence (Pareek, 1981a).

Their earlier correlation with Vindhyans (Blanford, 1877; Oldham 1886, 188 eron, 1932 and Mishra 1969) was replaced by the new idea that they are equivalent ıe Cambrians of the Salt Range of Pakistan (Pareek, 1981a, 1984). This new correlati as based on (1) occurrence of evaporite sequence in the area (Roy Chowdhury et a 965; Sinha et al., 1973; Das Gupta et al., 1987, 1988; Virendra Kumar, 1995 and Sinl oy et al., 1998), (2) reporting of a doubtful fossil Brachiopod (Orthis) of the Cambri finity from the sequence (Khan, 1973) and (3) stromatolite occurrences in the sequen arman, 1980). Das Gupta and Bulgauda (1994) considered them to be of the Low alaeozoic age. Paliwal (1992) assigned Upper Proterozoic to Lower Palaeozoic age to tl asal Congtomerate (Basal Formation), the Jodhpur Group (Sonia Formation aı odhpur/ Girbhakar Formation) and Bilara Group (Bilara Formation). Tl ıconformably overlying Pokaran Boulder Bed and Nagaur Group (Nagaur Formatio

g. 16. Conglomerate at the base of the Girbhakar Formation of the Jodhpur Group at Chh Khatu.

ere considered to be of the Cambrian age (Paliwal, 1992). Possibility of a comparative der age i.e. equivalent to the Semi Series (Lower Vindhyans) for the Randha Sandsto nd the Birmania Phosphorite was proposed by Paliwal (1992). Recently, Sinha Roy an s co-workers (Sinha Roy et al., 1998) correlated Nagaur Group with the purp

Fig. 17. Rhyolite flows at Jodhpur Fort.

Fig. 18. Folding in tuffaceous layers on way to Jodhpur Fort.

Sandstone of Earl Cambrian ag overlying the Salin Series in the Sa Range of Pakista and the Hansera Evaporite Grou Bilara Grou Birmania Formatio (Eocambrian) wit the Saline Serie The Jodhpur Grou was correlated wit the Upper Bhande Sandstone (Lat Proterozoic) of th Vindhyan Supergroup.

MARWAR SUPERGROUP : BASEMENT - COVER RELATIONSHIP

The rocks of the Marwar Supergroup (Trans-Aravalli Vindhyans), in general, ar ısidered to rest unconfomably over the Malani lgneous Suite of rocks (cf. Pareek, 198 34; Chauhan *et* 1991, Heron, 32) except at Sojat 1 Khatu areas ere these liments have been erved to rest conformably over vertical beds gs. 14 and 15) of Aravalli tasediments eron, 1932). gged and highly athered nature of surface of the lani volcanics, on ich sediments of Marwar pergroup were posited, indicates ong gap between

Fig. 19. Volcanic breccia showing large blocks of the tuffaceous layers in the north eastern section of Jodhpur Fort, west of Paota.

ɡ. 20. Unconformable contact between the underlying complexly deformed tuffaceous layers a overlying subhorizontal lava flows in the north eastern section of Jodhpur Fort, west Paota.

e volcanic activity and sedimentation. Aerial or subaerial exposure of Malani volcani ꞉ a pretty long time is also reflected in the presence of a thick weathered zone at t p. Suprisingly, despite all this, the Marwar Supergroup of rocks do not have as thi ɣer of sediments derived from the Malani rhyolite and phyroclastics at their base, ey ought to in such a situation.

Fig. 21. Folded tuffaceous layers from the section along the way to Siwanchi Gate cemetery.

g. 22. Superimposed two generations of folds in tuffaceous layers in the section along the way Siwanchi Gate cemetory.

g. 23. 'Eyed Folds' developed because of superimposition of two generations of folds in tuffaceo layers in the section along the way to Siwanchi Gate cemetory.

g. 24. Third generation open recumbent fold in tuffaceous rocks in the section along the way Siwanchi Gate cemetory.

At Jodhpur Fort the sandstone (Girbhakar Sandstone of the Jodhpur Group) with nglomerate/pebbly (Fig. 16) bed at the base rests unconformably over gently dippin va flows (Figs. 17 and 18), welded tuffs and volcanic breccia/agglomerate (Fig. 19) e Malani Igneous Suite. The latter are unconformably underlain by a set of complex formed tuffaceous and pyroclasic layers (Figs. 20 and 21).

A distinct effect of atleast three generations of superimposed folds (Figs. 22 to 2 d the associated lineations, (Figs. 25 to 27) is clearly visible in a section along the wa cemetory near Siwanchi Gate.

g. 25. Strong development of lineation of the first generation in tuffaceous and pyroclastic roc in the north eastern section of Jodhpur Fort.

ig. 26. Strongly developed lineation in tuffaceous layers in the section along the way to Siwan
Gate cemetory.

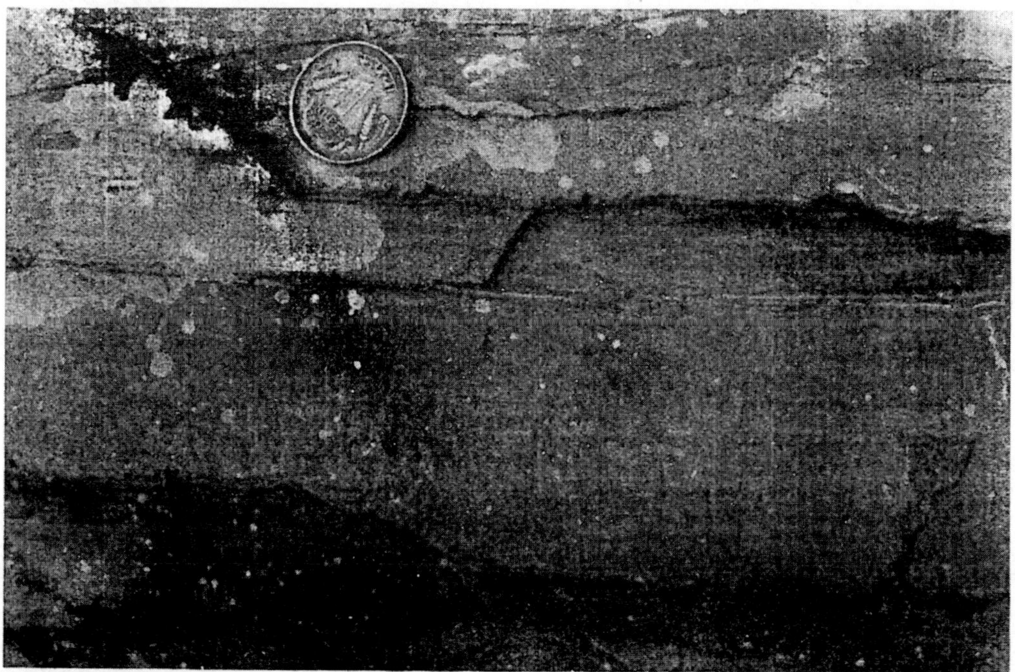

ig. 27. Two generations of lineations (early fold mullions and late puckera) superimposed at rig
angle in tuffaceous rock in the section along the way to Siwanchi Gate cemetory.

g. 28. Ash beds interlayered with the sandstone in a jaspar quarry near Lachha Basani.

:tween the Malani igenous activity and .e sedimentation of the Marwar ıpergroup.

Further north, at Chhoti Khatu in agaur district, Paliwal (1998) has ported felsic volcanics interlayered ith the sediments of the Marwar ıpergroup (Fig. 29). Occurrence of the lsic (Malani) volcanics here, .nd-wiched between the rocks of the ɔnia Formation (Kaimurs of Heron, ı32) below and those of the Girbhakar ɔrmation (Bhanders of Heron, 1932) ɔove evolves two possibilities. The first ıe is that the sedimentation of the arwar Supergroup of rocks started hen the eruption of the Malani ılcanics was still going on. Therefore, e lower part of the Marwar Supergroup ;asal Conglomerate, Jodhpur Group ıd possibly Bilara Group too) is of the ıte Proterozonic/Neoproterozoic age ıd not of the Cambrian as suggested by e earlier workers (Pareek 1981, 1984; ıauhan *et al.,* 1991; Khan 1973; and ırman 1980).

In these circumstances if the lava flov (Figs. 17 and 18) and welded tuffs ju underlying, the Jodhpur sandston represent the Malanis, then the fold tuffaceous and pyroclasic layers (Figs. 21 23) exposed in the section along the way Siwanchi Gate cemetory, constitute tl basement of the Malani lgneous Suite aı belong to the Sindreth/ Punagarh/ Siro Group.

In Lachha Basani and Ujalias jasp quarries, hardly a few kilometres north Jodhpur city, sandstone is seen interlayer with subhorizontal ash beds (Fig. 2 Interlayered relationship between the tv clearly suggests that the volcanic activi and deposition of sandstone were going on the area simultaneously. This situation pu a big sign of interrogation on earlier vie that the Marwar Supergroup of rocks we deposited on the irregular and weather surface of Malani rocks indicating a big gı

Fig. 29. Rhylitic flows and ash beds sand-wiched between the Sonia Formation below aı Girbhalcar Formation above at Chhoti Kha section.

Fig. 30. Folding in rhyolite flows in the quarry near Kankani village.

Fig. 31. Shearing in rhyolite flows in a quarry section near Kankani village.

Fig. 32. "Mantle Bedding" in pyroclostics in a quarry section near Kankani village.

Fig. 33. Uniform nature of the "Mantle Bedding" in pyroclastics in the quarry section near Kankani village.

Fig. 34. Large clasts in the "Mantle Bedding" in pyroclastic rock near Kankani village.

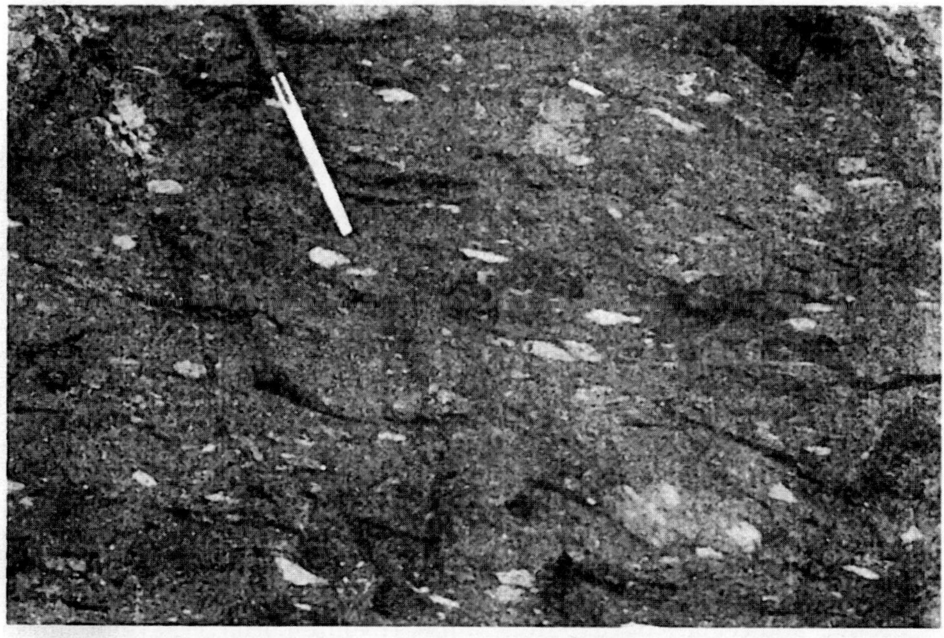

ig. 35. Flattened and elongated clasts of earlier lava flows (white and green in shade) floating i later purple lava flows in a querry section near Kankani village.

ig. 36. Large clast of earlier lava flows (green and white in shade) suspended in later purple lav flow forming "augen structure" in a querry near Kankani village.

Second possibility is that the Malani effusive phase was a prolonged one th arted about 745 ± 10 Ma and continued until 550 Ma representing the widespread Pa rican Orogeny in this part. Felsic volcanics of Chhoti Khatu area may represent tl ter part of the effusive phase.

About 15 km. south of Jodhpur city on Pali road, volcanics are exposed arour ankani village. Here lava flows of different shades and compositions show a distin fect of folding (Fig. 30) and shearing (Fig. 31). In a quarry east of the road "mant dding" (Figs. 32 and 33) in pyroclastics is well preserved. The uniformly distribute nantle bedding" shows big size clasts (Fig. 34). Here in the lower part of the sectio mparatively younger purple flows contain elongated clasts of earlier flows (green an hite in shade) (Fig. 35). The clasts are oriented in such a manner that they look lik igon structures (Fig. 36) and appear to be flattened due to shearing (Fig. 35).

Large size of blocks of older flows (white and green) and tuffaceous layers i unger purple flows in the volcanic breccia at Kankani (Fig. 37) and near Santoshi Ma mple in Jodhpur city indicates the near by presence of the volcanic vent.

STATUS OF THE POKARAN BOULDER BED

A two to three metre thick boulder bed or rather boulder spread, comprisir ulders, pebbles and cobbles of rhyolite, granite, rhyolite porphyry, syenite, basic rocl

Fig. 37. Volcanic breccia showing large blocks of ırlier lava flows (white, green and cream coloured) studded in purple coloured later lava flows in a querry near Kankani village.

and quartzite studded in a medium ı coarse grained sandy matrix (Paliwa 1992) occurring near Pokaran village popularly known as the Pokaran Bould Bed. Difference in opinion prevai regarding its glacial/glacio-fluvial origi and its position in the stratigraphy of tl region.

Based on the larger size of boulde occuring at Lowo village and distinct striated nature of the underlyin porphyry southwest of Pokaran village, was believed that the Pokaran Bould Bed, like the one at Bap was also glacial origin and of Talchir age (Heron, 1932). However, both Blanfo (1877) and Oldham (1886, 1888) were the view that the Pokaran Boulder Bed of the Vindhyan age. Since then this un was considered to be the basal part of tl Jodhpur Group overlain subsequently l Sonia Formation and Girbhank Formation. Recently, Paliwal (199 observed that the Pokaran Boulder B is overlain by the Pokaran sandsto which has a physical continuity with tl Lohawat sandstone which in tu overlies the Bilara Limestone (Vindhya Limestone of Oldham, 1886). This fa was further supported by a test we drilled at Lohawat village in whi mestone with a very high discharge of ground water, has been encountered underlyi ie Lohawat sandstone. That way the Pokaran Boulder Bed should be post-Bilara in a milar to those described overlying the Bilara Limestone at Khichan, Digrana, Gogra ıd west of Khatu (cf. Heron, 1932; Paliwal, 1992). Pebbles in the conglomerate expos ; Khichan and Ramsar Pond overlain unconformably by Nagaur Group are distinct ;riated and possibly represent Eocambrian glaciation in the area (Paliwal, 1992). ould be very interesting to note that the boulder bed occurring at Lowo about 8 km fro okaran is interbedded with "dioritic traps and ash beds" (Heron, 1932) similar to tho escribed from Maniari etc. under the name Malani Beds (Hacket, 1881, Pascoe, 195 aliwal, 1992), and is quite different for the one that occurs at Pokaran village. Moreov ie boulder bed of Lowo is about 10 km away from the Malani basement. In nut shell, tl olymictic conglomerate occurring at Lowo, Khatu and many other places in the ar ırms the basal part (Basal conglomerate of Paliwal, 1992) of the Trans-Arava indhyan sequence (The Marwar Supergroup). This boulder bed has wrongly be orrelated with the true Pokran Boulder Bed which infact is of post-Bilara age and not ar with the Basal Conglomerate described by Paliwal (1992). The Bap Boulder Bed alchir age is altogether different from the above two boulder beds.

Possibly, there occurred atleast three ;lacial cycles in the pre-Tertiary geological ıistory of the region (Paliwal, 1992). First ıne of Neoproterozoic age which immediately ollowed the widespread Malani volcanic ıctivity and gave rise to the formation of 3asal Conglomerate (Paliwal, 1992) at the ıase of the Marwar Supergroup (Fig. 13). \nother representing Eocambrian glaciation cf. Dott and Batten, 1976) which deposited he Pokaran Boulder Bed just abvoe the 3ilara Group (Paliwal, 1992). The third one ıccurred during the Permo-Carboniferous ıeriod depositing the Bap Boulder Bed of 'alchir age.

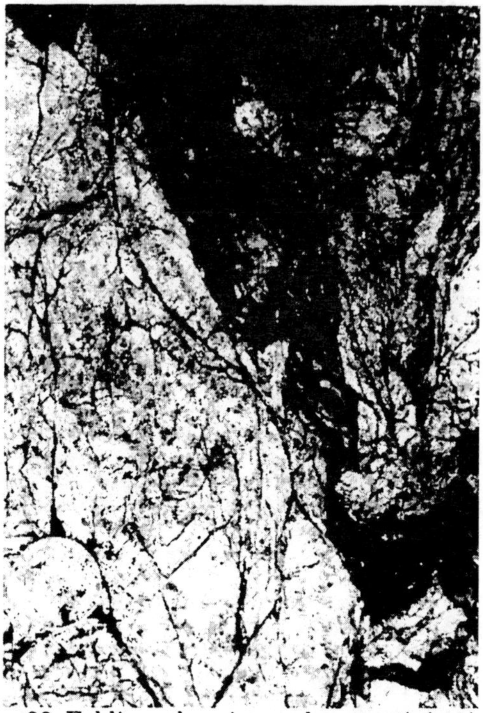

Fig. 38. Folding, shearing and serpentinizati in dolomitic limestone of Bilara Group at Sonaria near Ladnun

DEFORMATION IN THE ROCKS OF THE MARWAR SUPERGROUP

It is generally believed that the Marwar Supergroup of rocks are undeformed and ınmetamorphosed. Recently, Paliwal (1998) ıas observed that along the Khinwasar-Ladnun-Sonaria belt in Nagaur district the lolomitic limestone of Bilara Group shows olding, shearing (Fig. 38) and brecciation, ıssociated with the serpentinization, indicating a late (post-Bilara) phase of tecto leformation and thermal event in the region.

Similarly, in a tectonically separate basin around Jaisalmer the Randha Sandsto omposed mainly of conglomerate, sandstone and shale (Randha Formation) re ınconformably over the granitic/rhylitic basement and shows distinct signatures ıolyphase tectonic deformation. The overlying Birmania Formation comprising limesto with phosphorite), calcreous sandstone, shale and grit, shows cherty red clay at its ba uggesting a break in sedimentation. Birmania rocks have also been deform ectonically and show an angular relationship with the underlying Randha Sandsto Randha Formation). The two formations have their own structural style a leformation history. Obviously, the underlying Randha Sandstone is more comple leformed as compared to the overlying rocks of the Birmania Formation. On the ot ıand Jodhpur Group of rocks i.e. Sonia Formation and Girbhakar Formation consider o be quivalent of the Randha Formation do not show any evidence of such a tecto leformation. Deformation and compression of fossil bivalues and gostropods in the Jog 'ormation and the Fatehgarh Formation (S.C. Mathur, personal communication) seem ıe associated with a ductile deformation of the post-Marwar age in the region and ı ıecause of simple faulting.

SIGNATURES OF NEOTECTONICS

The Neotectonic activity and the Himalayan Orogeny have left distinct impressi ı the geology of the area and its physiography in the form of uplifted faulted blocl rmation of hoarst and grabens, reversal of drainage of river system, formation of sali .kes and desertification in the region (Paliwal, 1986, 1992, 1993). The Marw ıpergroup of rocks crop out high in the form of flat topped faulted blocks amidst t ındy plains. Such a topographical feature is seen around Jodhpur, Mandore, Umma hatu etc. This upliftment might have taken place because of Neotectonic activity in t gion. At Jodhpur itself uncomformable contact between the Malani volcanics and t rerlying Jodhpur Group of rocks lies as high as the entrance of the fort on one hand a: s low as the foundation of the Girdhar Mandir Cinema on Nai Sarak on the other. ırsagar quarries hardly a few metres further west, the same contact is traceable de ıside the quarries at a greater depth.

Prior to the Neotectonic activity the region was occupied by perennial rivers li araswati, Drishadwati and their ributories, along the banks of which the world's mc ıcient Vedic Civilization flourished. It was the effect of Neotectonics that a lar ımber of NE-SW and NW-SE trending parallel faults developed in the area (Paliw. 992) and produced a series of horsts and grabens (Paliwal, 1986, 1993). The promine fect of Neotectonics was in the form of reversal of drainage in the region (Paliw. 993). This caused the blockage of streams and centripital drainage developed in t abens formed in the path of major streams and served as depositional basins for t raporites (Paliwal, 1986). Such a situation resulted in the formation of saline lakes li ıose of Sambhar, Didwana, Kuchaman, Talchhapar, Pachpadra etc. Rivers dried up a cute shortage of surface water caused widespread desertification in the region (Paliw. 993).

The effects of the Neotectonics and the Himalayan Orogeny did not restrict to t ounger sedimentaries of the region only but they are reflected in the Precambrian roc the Aravalli-Delhi fold belts too. Many ductile shear zones straite-slip faults, hors ıd grabens, and thrusts in the older metamorphic rocks of the region and perhaps t .st deformation in the Aravalli and Delhi Supergroups of rocks in the type localiti ıight be the effect of the Neotectonics and the Himalayan Orogeny in this part (Paliw. 993, Paliwal *et al.*, 1995).

ACKNOWLEDGEMENT

Financial assistance received from the University Grants Commission is gratefu cknowledged. Thanks are due to Shri P.S. Rathore, Shri H.K. Goran, Shri S.K. Yada r. S. Kaushik and Dr. A. Vyas for their help. I had fruitful discussions with Prof. A. oy, of M.L.S. University Udaipur and Prof. D.S. Chauhan of J.N.V. University, Jodhpı elp received from Mr. B. Paliwal and Mrs. S. Paliwal is gratefully acknowledged.

REFERENCES

ıarman, G., 1980. Analysis of the Marwar basin, western Rajasthan in the light of stromatolit study. *Geol. Surv. India, Misc. Publ.* 44 : 292 - 297.

ıasu, S.K., 1982. Phases of acide igneous activity in Malani suite of rocks around Jhunjhunı Rajasthan. *Rec. Geol. Surv.* India, 112(7) : 89 - 94.

ıhushan, S.K., 1985. Malani volcanism in western Rajasthan, *Ind. Jour. Earth Sci.* 12 : 58 - 71.

ıhushan, S.K. and Khullar, V.K. 1998. Geochemistry and tectonic significance of dyke swarm i: Malani Igneous Complex around Sankara, district Jaisalmer, Rajasthan. In : Paliwal, B.S (Ed.) *The Indian Precambrian*. Scientific Publishers (India) Jodhpur, 482 - 491.

ılanford, W.T., 1876. On the physical geology of the Great Rajasthan Desert, etc. *Jour. Asiati Soc. Bengal*, 45 : 86 - 103.

ılanford, W.T., 1877. Geological notes on the Great Indian Desert between Sind and Rajputanε *Rec. Geol. Surv. India,* 10(1) : 10 - 21.

ʼhauhan, D.S., Dubey, J.C. and Ram, B., 1991. Geological analysis of part of Nagaur basin in th vicinity of Jodhpur city. In : Tandon, S.K., Pant, Charu C. and Casshyap, S.M. (Eds. *Sedimentary Basin of India*, 64 - 73.

ʼhore, S.A. and Mohonty, M., 1998. Stratigraphy and tectonic setting of the Trans-Araval] Neoproterozoic volcanosedimentary sequences in Rajasthan. *Jour. Geol. Soc. India,* 51 : 57 68.

ʼoulson, A.L., 1933. The geology of the Sirohi State, Rajputana. *Mem. Geol. Surv. India,* 63(1) : 1 166.

ʼrawford, A.R. and Compston, W., 1970. The age of Vindhyan System of Peninsular India. *Quarı Jour. Geol. Soc. India,* 125 : 351 - 371.

ʼas Gupta, S.P. Virendra Kumar, Ram Chandra and Jairam, M.S., 1988. A framework of th Nagaur-Ganganagar evaporite basin, Rajasthan. *Ind. Minerals,* 42(1) : 57 - 64.

ʼas Gupta, U. and Bulgauda, S.S., 1994. An overview of the geology and hydrocarbon occurrence in the western part of Bikaner-Nagaur basin. *Ind. Jour. Petrol. Geology* 3(1) : 1 - 17.

ʼavies, R.G. and Crawford, A.R., 1971. Petrography and age of the rocks of Bulland Hill, Kiran: Hills, Sargodha district, Pakistan. *Geol. Mag.* 108(3) : 235 - 246.

ʼupta, S.N., Arora, Y.K., Mathur, R.K., Iqbaluddin, Balmiki Prasad, Sahai, T.N. and Sharm: S.B., 1980. Lithostratigraphic map of Aravalli region, southern Rajasthan and north easteri Gujarat. *Geol. Surv. India,* Calcutta.

[ocket, C.A., 1881. On the geology of the Aravalli Region, Central and Eastern Rajputana. *Rec Geol. Surv. India,* 14(4) : 279 - 303.

[eron, A.M., 1932. The Vindhyans of western Rajputana. *Rec. Geol. Surv. India,* 65(4) : 457 - 489.

[eron, A.M., 1953. The Geology of Central Rajputana. *Mem. Geol. Surv. India,* 79 : 1 - 389.

]han, E.A., 1973. Discovery of fossil Brachiopod of Cambrian affinity, near Jodhpur. *Proc Seminar* on *Recent Advances in the Geology of Rajasthan and Gujarat.* (Abstract) 74 - 75.

ʼochhar, N., 1974. The age of Malani series. *Jour. Geol. Soc. India,* 15 : 316 - 317.

ʼochhar, N., 1984. Malani Igneous Suite : Hotspot magmatism and cratonization of the northerı part of India Shield. *Jour. Geol. Soc. India,* 25 : 155 - 161.

ʼochhar, N., 1998. Malani Igneous Suite of Rocks. *Jour. Geol. Soc. India,* 51 : 120.

ıa Touche, T.H.D., 1902. Geology of western Rajputana. *Mem. Geol. Surv. India* 35(1) : 1 - 116.

Iisra, R.C., 1969. The Vindhyan System. *Presidential Address, Ind. Sci. Congr.* Bombay, *Sec Geol. Geog.* 1 - 32.

ırthy, M.V.N., 1979. Some aspects of mineralization associated with acid magmatism Rajasthan. *Proc. Workshop MAWAM Geol. Surv. India Spel. Publ.* 13 : 15 - 24.

ıham, R.D., 1886. Preliminary notes of the geology of northern Jaisalmer. *Rec. Geol. Sur India,* 19(3) : 157 - 160.

ıham, R.D., 1888. Memorandum on the results of an exploration of Jaisalmer with a view to tl discovery of coal. *Rec. Geol. Surv. India,* 21(1) : 30 - 33.

liwal, B.S., 1976. Note on layered chromatographical structures in the Vindhyan sandstone Badi Khatu, district Nagaur, Rajasthan. *Bull. Ind. Geol. Assoc.* 9(1) : 17 - 22.

liwal, B.S., 1980. Structural history of the Precambrian rocks south-west of Didwan Rajasthan, India. *Bull. Ind. Geol. Assoc.* 13(1) : 7 - 14.

liwal, B.S., 1986. Source of salinity in soil and surface and sub-surface waters and its bearii on the evolution of the Thar Desert, India. *Transactions XIII Congr. Int. Soc. Soil Sc* August 13-20, 1986, 4 : 1532.

liwal, B.S., 1992. Tectonics of the post- Aravalli mountain building activity and its bearing (the accumulation of sediments along the western flank of the Aravalli Range, Rajastha India. In : Ahmed, R. and Sheikh, A.M. (Eds), *Geology in South Asia-I Proceedings GEOSAS-I* Islamabad, Pakistan, February 23 - 27, 1992, *Hydrocarbon Development Institu of Pakistan,* 1994 : 52 - 60.

liwal, B.S., 1993. Geology of Rajasthan - a summarium. In : Chauhan, T.S. (Ed.) *Natural ar Human Resources of Rajasthan.* Scientific Publishers (India) Jodhpur. 1 - 22.

liwal, B.S., 1995. A note on the spots in the Jodhpur sandstone of western Rajasthan. *Raj. Mi Bull.* 16(1-2) : 12 - 15.

liwal, B.S., 1998. Felsic volcanics interlayered with sediments of the Marwar Supergroup Chhoti Khatu. *Jour. Geol. Soc. India,* 52 : 81 - 86.

liwal, B.S., Pareek, U.S. and Vyas, A., 1995. Structural frame work of the Makrana marb deposit and its bearing on the Precambrian stratigraphy of Rajasthan. In : Vijayanand N.P., Cooray, P.G. and Mosley, P. (Eds), *Geology in South Asia- II* Proceedings GEOSAS-: Colombo, Sri Lanka, January 19-24, 1995, *Geol. Surv. Mines Bureau, Sri Lanka, Pr(Papers* No. 7, 1997, 123 - 136.

ndit, M.K., Shekhawat, L.S., Ferreira, V.P., Sial, A.N. and Bohra, S.K., 1999. Trondhjemite aı granodiorite assemblages from west of Barmer : probable basement for Malani magmatis in western India. *Jour. Geol. Soc. India,* 53 : 89 - 96.

reek, H.S., 1981a. Basin configuration and sedimentary stratigraphy of western Rajastha *Jour. Geol. Soc. India,* 22 : 517 - 527.

reek, H.S., 1981b. Petrochemistry and petrogenesis of the Malani Igneous Suite, India. *Bu Geol. Soc. Amer.* 92(1) : 206 - 273.

reek, H.S., 1984. Pre-Quaternary geology and mineral resources of north-western Rajastha *Mem. Geol. Surv. India,* 115 : 1 - 99.

scoe, E.H., 1959. *A Manual of the Geology of India and Burma,* 2 : 485 - 1343.

thore, S.S., Venkatesan, T.R. and Srivastava, R.K., 1996. Rb-Sr and Ar-Ar systematics Malani volcanic rocks of south west Rajasthan : evidence for a younger post-crystallizati(thermal event. *Proc. Ind. Acad. Sci.* 105(2) : 131 - 141.

y Chowdhury, M.K., Sogani, P.C., Chandak, G.J. and Mehra, S.L., 1965. Nagaur gypsum. *Bu Geol. Surv. India, Series-A,* Economic Geology, 24.

y, A.B., Paliwal, B.S., Shekhawat, S.S., Nagori, D.K., Golani, P.R. and Bejarniya, B.R., 198 Stratigraphy of the Aravalli Supergroup in the type area. *Mem. Geol. Soc. India* 7 : 121 138.

y, A.B., Srinivasan, R., Gopalan, K., Kataria, P, Sanjay Kumar, Vivek Laul and Sreenivasa, I 1997. Recognition and characterization of the Archaean-Proterozoic boundary in southe:

Rajasthan. *National Seminar on Tectonomagmatism, Geochemistry and Metamorphism* (*Precambrian Terrains* Department of Geology, M.L. Sukhadia University, Udaipι (Abstract), 17.

ιy, A.B. and Sharma, K.K., 1996. Neoproterozoic evolution of the "Trans-Aravalli Crust" fro splitting of "Rodinia" to Gondwana assembly. *National Seminar on Geological Evolution* (*Western Rajasthan*, Department of Geology, J.N.V. University, Jodhpur, (Abstract), 68 - 69.

ιstry, C.A., 1992. Geochronology of the Precambrian rocks from Rajasthan and north easter Gujarat. *Geol. Surv. India, Spel. Publ.* 25 : 1 - 96.

n, S., 1981. Proterozoic palaeotectonics in the evolution of crust and location of metalliferoι deposits, Rajasthan. *Quart. Jour. Geol. Min. Met. Soc. India,* 53 : 162 - 185.

ιarma, K.K. and Roy, A.B., 1997. Neoproterozoic geology and evolution of the Trans-Araval Crust. *National Seminar on Tectonomagmatism, Geochemistry and Metamorphism* (*Precambrian Terrains*. Department of Geology, M.L. Sukhadia University, Udaipu (Abstract), 15 - 16.

nha, B.P.C., Singh, G.D.S., Bakshi, A.R. and Bhatnagar, G.C., 1973. A note on occurrence (halite in Bikaner area Rajasthan. *Curr. Sci.* 42 : 87.

nha Roy, S., 1984. Precambrian crustal interactions in Rajasthan, N W India. *Ind. Jour. Eart Sci.* CEISM Volume 84 - 91.

nha Roy, S., Malhotra, G. and Mohanty, M., 1998. *Geology of Rajasthan. Geol. Soc. India, Te: Book Series,* 1 - 278.

ivastava, R.K., 1988. Magmatism in the Aravalli Mountain Range and its environs. *Mem. Geo Soc. India,* 7 : 77 - 93.

ivastava, R.K., 1990. Alkaline and peralkaline rocks of Rajasthan. *Mem. Geol. Soc. India,* 15 : - 24.

ivastava, R.K., Maheshwari, A. and Upadhyay, R., 1989. Geochemistry of felsic volcanics fro Gurapratap Singh and Diri, Pali district, Rajasthan (Part I - major elements, Part II - tra(elements). *Jour. Geol. Soc. India*, 34 : 467 - 485 and 617 - 631.

'chanthavong, S.P. and Desai, S.D., 1977. Protoplate tectonics controlling Precambria deformations and metallogenic epochs of northwestern Peninsular India. *Min. Sci. Eng.* 9 218 - 236.

ιbisch, O.T., 1998. A note on the structural use of microgranitoid enclaves in some deforme granitic plutons from central Rajasthan. In : B.S. Paliwal (Ed.), The Indian Precambria Scientific Publishers (India), Jodhpur, 1 - 8.

ιbisch, O.T., Collerson, K.D., Bhattacharya, T. and Mukhopadhyay, D., 1994. Structur relationships and Sr-Nd isotope systematics of polymetamorphic granitic gneisses an granitic rocks from central Rajasthan, India : implications for the evolution of the Araval craton. *Precambrian Research,* 65 : 319 - 339.

ιllinayagam, G., 1998. Rare metal bearing and barren distinction of alkali granites, Piplun are Siwana Ring Complex, Rajasthan, India : petrological and geochemical evidence. In : B.S Paliwal (Ed.), *The Indian Precambrian*, Scientific Publishers (India), Jodhpur, 492 - 501.

ιllingayam,G. and Kochhar, N., 1998. Geochemical characterization and petrogenesis of A-Typ Granites and the associated acid volcanics of Siwana Ring Complex, North Peninsular Indi In : B.S. Paliwal (Ed.), *The Indian Precambrian*, Scientific Publishers (India), Jodhpur, 460 481.

rendra Kumar, 1995. Eocambrian sedimentation in Nagaur- Ganganagar evaporite basi Rajasthan. *Convention of Indian Association of Sedimentologists*, Roorkee, 27 - 29.